WELCOME to CALCULUS 2^n

Calculus 2nd semester is a continuation of the 1st semester and follows a similar structure of Concepts, Applications and Techniques. The major topics this semester are

> Applications of integrals and methods of integration
> Solutions to simple and separable Differentiable Equations
> The inverse trigonometric functions and calculus with these functions
> The calculus of functions given in Polar Coordinates and as Parametric Equations
> Infinite Sequences and Series

So welcome back for more calculus.

Succeeding at calculus.

Probably more than in your previous mathematics classes you need
* to think about the concepts as well as the techniques,
* to think about the patterns as well as the individual steps,
* to think about the meaning of the concepts and techniques in the context of particular applications,
* to think about how the ideas and techniques apply to functions given by graphs and tables as well as by formulas, and
* to spend enough time (1 to 2 hours each day) doing problems to sort out the concepts and to master the techniques and to get better and more efficient with the algebra skills that are vital to success.

Sometimes all this mental stretching can seem overwhelming, but stick with it (and do lots of problems). It can even become fun.

This text book

A free, color PDF version is available online at
> http://scidiv.bellevuecollege.edu/dh/Calculus_all/Calculus_all.html

This text is licensed under a Creative Commons Attribution-Share Alike 3.0 United States License.

You are **free**:
> **to Share** – to copy, distribute, display and perform the work
> **to Remix** – to make derivative works

Under the following conditions:
> **Attribution:** You must attribute the work in the manner specified by the author
>> (but not in any way that suggests that they endorse you or your work)
>
> **Share Alike:** If you alter, transform or build upon this work, you must distribute the resulting work only under the same, similar or compatible license.

Copyright © 2013 Dale Hoffman   12/02/2013

CONTEMPORARY CALCULUS 2nd Semester: Contents

How to Succeed in Beginning Calculus
    Calculus takes time. Do not get behind. Use the textbook intelligently.
    Working the problems. Work together. Study with a friend.
    Work in small groups.

Chapter 5: Applications of Definite Integrals

5.0   Introduction
5.1   Volumes
5.2   Arc Length
5.3   Work
5.4   Moments and Centers of Mass
5.5   Additional Applications
Odd numbered solutions for Chapter 5

Chapter 6: Introduction to Differential Equations

6.0   Introduction
6.1   The Differential Equation $y' = f(x)$
6.2   Separable Differential Equations
6.3   Exponential Growth, Decay and Cooling
Odd numbered solutions for Chapter 6

Chapter 7: Inverse Trigonometric Functions

7.0   Introduction to Transcendental Functions
7.1   Inverse Functions
7.2   Inverse Trigonometric Functions
7.3   Calculus with the Inverse Trigonometric Functions
Odd numbered solutions for Chapter 7

Chapter 8: Improper Integrals and Integral Techniques

8.0   Introduction
8.1   Improper Integrals
8.2   Integration Review
8.3   Integration by Parts
8.4   Partial Fraction Decomposition
8.5   Trigonometric Substitutions
8.6   Trigonometric Integrals
Odd numbered solutions for Chapter 8

Chapter 9: Polar Coordinates, Parametric Equations & Conic Sections

    9.1   Polar Coordinates
    9.2   Calculus with Polar Coordinates
    9.3   Parametric Equations
    9.4   Calculus with Parametric Equations
    9.4.5 Bezier Curves
    9.5   Conic Sections
    9.6   Properties of the Conic Sections
Odd numbered solutions for Chapter 9

Chapter 10: Infinite Series & Power Series

    10.0   Introduction to Sequences & Series
    10.1   Sequences
    10.2   Infinite Series
    10.3   Geometric and Harmonic Series
    10.3.5 An Interlude and Introduction
    10.4   Positive Term Series: Integral & P-Tests
    10.5   Positive Term Series: Comparison Tests
    10.6   Alternating Sign Series
    10.7   Absolute Convergence and the Ratio Test
    10.8   Power Series
    10.9   Representing Functions with Power Series
    10.10 Taylor and Maclaurin Series
    10.11 Approximation Using Taylor Series
    10.12 Fourier Series
Odd numbered solutions for Chapter 10

Appendix A: Trigonometry Reference Facts

Appendix B: Derivative and Integral Reference Facts

How to Succeed in Beginning Calculus

The following comments are based on over thirty years of watching students succeed and fail in calculus courses at universities, colleges and community colleges and of listening to their comments as they went through their study of calculus. This is the best advice we can give to help you succeed.

Calculus takes time. Almost no one fails calculus because they lack sufficient "mental horsepower". Most people who do not succeed are unwilling (or unable) to devote the necessary time to the course. The "necessary time" depends on how smart you are, what grade you want to earn and on how competitive the calculus course is. Most calculus teachers and successful calculus students agree that 2 (or 3) hours every weeknight and 6 or 7 hours each weekend is a good way to begin if you seriously expect to earn an A or B grade. If you are only willing to devote 5 or 10 hours a week to calculus outside of class, you should consider postponing your study of calculus.

Do NOT get behind. The brisk pace of the calculus course is based on the idea that "if you are in calculus, then you are relatively smart, you have succeeded in previous mathematics courses, and you are willing to work hard to do well." It is terribly hard to **catch up** and **keep up** at the same time. A much safer approach is to work very hard for the first month and then evaluate your situation. If you do get behind, spend a part of your study time catching up, but spend most of it trying to follow and understand what is going on in class.

Go to class, every single class. We hope your calculus teacher makes every idea crystal clear, makes every technique obvious and easy, is enthusiastic about calculus, cares about you as a person, and even makes you laugh sometimes. If not, you still need to attend class. You need to hear the vocabulary of calculus spoken and to see how mathematical ideas are strung together to reach conclusions. You need to see how an expert problem solver approaches problems. You need to hear the announcements about homework and tests. And you need to get to know some of the other students in the class. Unfortunately, when students get a bit behind or confused, they are most likely to miss a class or two (or five). That is absolutely the worst time to miss classes. Come to class anyway. Ask where you can get some outside tutoring or counseling. Ask a classmate to help you for an hour after class. If you must miss a class, ask a classmate what material was covered and skim those sections before the next class. Even if you did not read the material, come back to class as soon as possible.

Work together. Study with a friend. Work in small groups. It is much more fun and is very effective for doing well in calculus. Recent studies, and our personal observations, show that students who **regularly** work together in small groups are less likely to drop the course and are more likely to get A's or B's. You need lots of time to work on the material alone, but study groups of 3–5 students, working together 2 or 3 times a week for a couple hours, seem to help everyone in the group. Study groups offer you a way to get and give help on the material and they can provide an occasional psychological boost ("misery loves company?"). They are a place to use the mathematical language of the course, to trade mathematical tips, and to "cram" for the next day's test. Students in study groups are less likely to miss important points in the course, and they get to know some very nice people, their classmates.

Use the textbook effectively. There are a number of ways of using a mathematics textbook:

i. to gain an overview of the concepts and techniques,
ii. to gain an understanding of the material,
iii. to master the techniques, and
iv. to review the material and see how it connects with the rest of the course.

The first time you read a section, just try to see what problems are being discussed. Skip around, look at the pictures, and read some of the problems and the definitions. If something looks complicated, skip it. If an example looks interesting, read it and try to follow the explanation. This is an exploratory phase. Don't highlight or underline at this stage — you don't know what is important yet and what is just a minor detail.

The next time through the section, proceed in a more organized fashion, reading each introduction, example, explanation, theorem and proof. This is the beginning of the "mastery" stage. If you don't understand the explanation of an example, put a question mark (in pencil) in the margin and go on. Read and try to understand each step in the proofs and ask yourself why that step is valid. If you don't see what justified moving from one step to another in the proof, pencil in question marks in the margin. This second phase will go more slowly than the first, but if you don't understand some details just keep going. Don't get bogged down yet.

Finally, worry about the details. Go quickly over the parts you already understand, but slow down and try to figure out the parts marked with question marks. Try to solve the example problems before you refer to the explanations. If you now understand parts that were giving you trouble, cross out the question marks. If you still don't understand something, put in another question mark and **write down** your question to ask your teacher, tutor, or classmate.

Finally it is time to try the problems at the end of the section. Many of them are similar to examples in the section, but now you need to solve them. Some of the problems are more complicated than the examples, but they still require the same basic techniques. Some of the problems will require that you use concepts and facts from earlier in the course, a combination of old and new concepts and techniques. Working lots of problems is the "secret" of success in calculus.

Working the Problems: Many students read a problem, work it out and check the answer in the back of the book. If their answer is correct, they go on to the next problem. If their answer is wrong, they manipulate (finagle, fudge, massage) their work until their new answer is correct, and then they go on to the next problem. **Do not try the next problem yet!** Before going on, spend a short time, just half a minute, thinking about what you have just done in solving the problem. Ask yourself, "What was the point of this problem?", "What big steps did I have to take to solve this problem?", "What was the **process**?" Do <u>not</u> simply review every single step of the solution process. Instead, look at the <u>outline</u> of the solution, the **process**. If your first answer was wrong, ask yourself, "What about this problem should have suggested the right process the <u>first</u> time?" As much learning and retention can take place in the 30 seconds you spend reviewing the **process** as took place in the 10 minutes you took to solve the problem. A correct answer is important, but a **correct process, carefully used, will get you many correct answers**.

There is one more step which too many students omit. **Go back and quickly look over the section one more time.** Don't worry about the details, just try to understand the overall logic and layout of the section. Ask yourself, "What was I expected to learn in this section?" Typically this last step, a review and overview, goes quickly, but it is very valuable. It can help you see and retain the important ideas and connections.

Good luck. Work hard. Have Fun.
Calculus is beautiful and powerful.

5.0 Applications of Definite Integrals

The last chapter introduced the idea of a definite integral as an "area" and a limit of Riemann sums, showed some of the properties of integrals, showed some ways to calculate values of definite integrals, and started to examine a few of their uses. This chapter focuses on several common applications of definite integrals. An obvious goal of the chapter is to enable you to use integration when you encounter these particular applications later in mathematics or in other fields. A deeper goal is to illustrate the process of going from a problem to an integral, a process much broader than these particular applications. If you understand the process, then you can understand the use of integrals in many other fields and can even develop the integrals needed to solve problems in new areas. A final goal is to give you additional practice evaluating definite integrals.

Each section in this chapter follows the same basic format. First a problem is described and some background information presented. Then the solution to the basic problem is approximated using a Riemann sum. An exact answer comes from taking a limit of the Riemann sum, and we get a definite integral. After looking at several examples of the same basic application, we will examine some variations of it.

5.1 VOLUMES OF SOLIDS

The last chapter emphasized a geometric interpretation of definite integrals as "areas" in two dimensions. This section emphasizes another geometrical use of integration, calculating volumes of solid three–dimensional objects such as those shown in Fig. 1. Our basic approach is to cut the whole solid into thin "slices" whose volumes can be approximated, add the volumes of these "slices" together (a Riemann sum), and finally obtain an exact answer by taking a limit of the sums to get a definite integral.

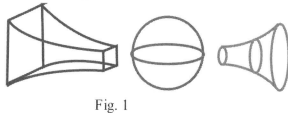

Fig. 1

The Building Blocks: Right Solids

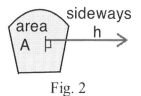

Fig. 2

A **right solid** is a three–dimensional shape swept out by moving a planar region A some distance h along a line perpendicular to the plane of A (Fig. 2). The region A is called a **face** of the solid, and the word "right" is used to indicate that the movement is along a line perpendicular, at a right angle, to the plane of A. Two parallel **cuts** produce one **slice** with two **faces** (Fig. 3): a slice has volume, and a face has area.

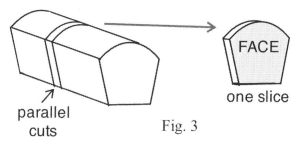

Fig. 3

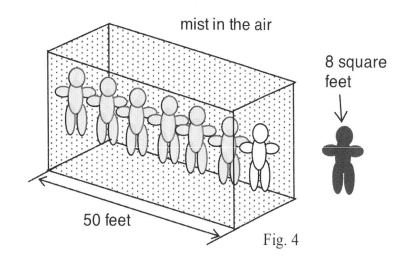

Fig. 4

Example 1: Suppose there is a fine, uniform mist in the air, and every cubic foot of mist contains 0.02 ounces of water droplets. If you run 50 feet in a straight line through this mist, how wet do you get? Assume that the front (or a cross section) of your body has an area of 8 square feet.

Solution: As you run, the front of your body sweeps out a "tunnel" through the mist (Fig. 4). The volume of the tunnel is the area of the front of your body multiplied by the length of the tunnel: volume = (8 ft^2)(50 ft) = 400 ft^3. Since each cubic foot of mist held 0.02 ounces of water which is now on you, you swept out a total of (400 ft^3)·(0.02 oz/ft^3) = 8 ounces of water. If the water was truly suspended and not falling, would it matter how fast you ran?

5.1 Volumes

If A is a rectangle (Fig. 5), then the "right solid" formed by moving A along the line is a 3–dimensional solid box B. The volume of B is

(area of A)·(distance along the line) = (base)·(height)·(width).

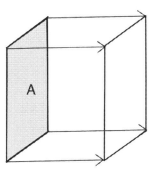

Fig. 5: Solid box

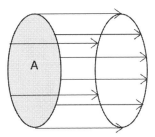

If A is a circle with radius r meters (Fig. 6), then the "right solid" formed by moving A along the line h meters is a right circular cylinder with volume equal to

$$\{\text{area of A}\}\cdot\{\text{distance along the line}\}$$
$$= \{\pi(r\text{ ft})^2\}\cdot\{h\text{ ft}\} = \{\pi r^2\text{ ft}^2\}\cdot\{h\text{ ft}\} = \pi r^2 h\text{ ft}^3.$$

Fig. 6: Solid cylinder

If we cut a right solid perpendicular to its axis (like cutting a loaf of bread), then each face (cross section) has the same two–dimensional shape and area. In general, if a 3–dimensional right solid B is formed by moving a 2–dimensional shape A along a line perpendicular to A, then the volume of B is *defined* to be

volume of B = (area of A)·(distance moved along the line perpendicular to A).

The volume of each right solid in Fig. 7 is (area of the base)·(height).

Example 2: Calculate the volumes of the right solids in Fig. 7.

Solution: (a) This cylinder is formed by moving the circular base (area = $\pi r^2 = 9\pi$ in^2) along a line perpendicular to the base for 4 inches, so the volume is (9π in^2)·(4 in) = 36π in^3.

(b) volume = (base area)·(distance along the line) = (8 m^2)·(3 m) = 24 m^3.

(c) This shape is composed to two easy right solids with volumes $V_1 = (\pi 3^2)\cdot(2) = 18\pi$ cm^3 and $V_2 = (6)(1)\cdot(2) = 12$ cm^3, so the total volume is $(18\pi + 12)$ cm^3 or approximately 68.5 cm^3.

Practice 1: Calculate the volumes of the right solids in Fig. 8.

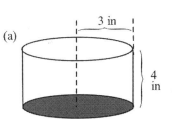
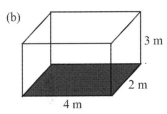
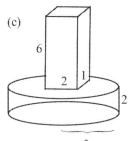

Fig. 7

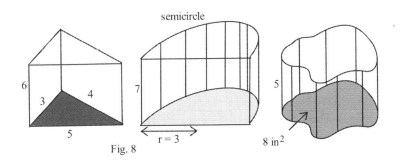

Fig. 8

Volumes of General Solids

A general solid can be cut into slices which are almost right solids. An individual slice may not be exactly a right solid since its cross sections may have different areas. However, if the cuts are close together, then the cross sectional areas will not change much within a single slice. Each slice will be almost a right solid, and its volume will be almost the volume of a right solid.

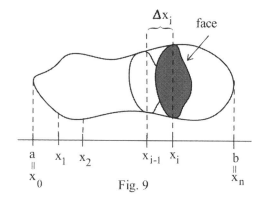

Fig. 9

Suppose an x–axis is positioned below the solid shape (Fig. 9), and let $A(x)$ be the **area of the face** formed when the solid is cut at x perpendicular to the x–axis. If $P = \{ x_0=a, x_1, x_2, \ldots, x_n = b\}$ is a partition of $[a,b]$, and the solid is cut at each x_i, then each slice of the solid is almost a right solid, and the volume of each slice is approximately

(area of a face of the slice)·(thickness of the slice) $\approx A(x_i) \Delta x_i$.

The total volume V of the solid is approximately the sum of the volumes of the slices:

$$V = \sum \{\text{volume of each slice}\} \approx \sum A(x_i)\Delta x_i \quad \text{which is a Riemann sum.}$$

The limit, as the mesh of the partition approaches 0 (taking thinner and thinner slices), of the Riemann sum is the definite integral of $A(x)$:

$$V \approx \sum A(x_i)\Delta x_i \longrightarrow \int_a^b A(x)\, dx .$$

Volume By Slices Formula

If   S is a solid and $A(x)$ is the area of the face formed by a cut

at x and perpendicular to the x–axis,

then   the volume V of the part of S above the interval [a,b] is $V = \int_a^b A(x)\, dx$.

If S is a solid (Fig. 10), and $A(y)$ is the area of a face formed by a cut at y perpendicular to the **y–axis**, then the volume of a slice with thickness Δy_i is approximately $A(y_i) \cdot \Delta y_i$. The volume of the part of S between cuts at c and d on the y–axis is

$$V = \int_c^d A(y)\, dy .$$

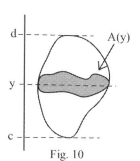

Fig. 10

Example 3:   For the solid in Fig. 11, the face formed by a cut at x is a rectangle with a base of 2 inches and a height of cos(x) inches. (a) Write a formula for the approximate volume of the slice between x_{i-1} and x_i. (b) Write and evaluate an integral for the volume of the solid for x between 0 and $\pi/2$.

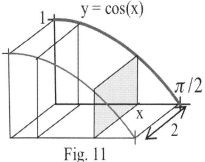

Fig. 11

Solution:

(a) The volume of the slice ≈ (area of the face)·(thickness)
$$= \text{(base)} \cdot \text{(height)} \cdot \text{(thickness)}$$
$$= (2 \text{ in}) \cdot (\cos(x_i) \text{ in}) \cdot (\Delta x_i \text{ in})$$
$$= 2\cos(x_i) \Delta x_i \text{ in}^3.$$

(b) Volume $= \int_a^b A(x)\,dx = \int_0^{\pi/2} 2\cos(x)\,dx = 2\sin(x)\Big|_0^{\pi/2}$
$= 2\sin(\pi/2) - 2\sin(0) = 2 \text{ in}^3.$

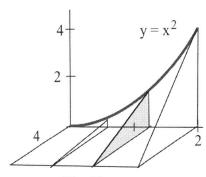

Fig. 12

Practice 2: For the solid in Fig. 12, the face formed by a cut at x is a triangle with a base of 4 inches and a height of x^2 inches.
(a) Write a formula for the approximate volume of the slice between x_{i-1} and x_i. (b) Write and evaluate an integral for the volume of the solid for x between 1 and 2.

Example 4: For the solid in Fig. 13, each face formed by a cut at x is a circle with <u>diameter</u> $\sqrt{x}$.

(a) Write a formula for the approximate volume of the slice between x_{i-1} and x_i.

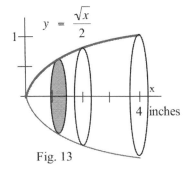

Fig. 13

(b) Write and evaluate an integral for the volume of the solid for x between 1 and 4.

Solution: (a) Each face is a circle with diameter $\sqrt{x_i}$, and the area of the circle is
$$A(x_i) = \pi \cdot (\text{radius})^2 = \pi(1/2 \text{ diameter})^2 = \pi(1/2 \sqrt{x_i})^2 = \pi x_i / 4.$$
The volume of the slice ≈ (area of the face)·(thickness) $= (\pi x_i / 4) \cdot (\Delta x_i)$

(b) Volume $= \int_a^b A(x)\,dx = \int_1^4 \frac{\pi x}{4}\,dx = \frac{\pi}{4} \cdot \frac{x^2}{2}\Big|_1^4 = \frac{\pi}{4}\frac{16}{2} - \frac{\pi}{4}\frac{1}{2} = \frac{15\pi}{8} \approx$ 5.89 in^3.

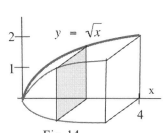

Fig. 14

Practice 3: For the solid in Fig. 14, each face formed by a cut at x is a square with height $\sqrt{x}$.

(a) Write a formula for the approximate volume of the slice between x_{i-1} and x_i.

(b) Write and evaluate an integral for the volume of the solid for x between 1 and 4.

Example 5: Find the volume of the square–based pyramid in Fig. 15.

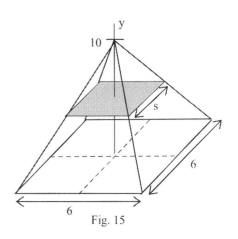

Fig. 15

Solution: Each cut perpendicular to the y–axis yields a square face, but in order to find the area of each square we need a formula for the length of one side s of the square as a function of y, the location of the cut. Using similar triangles (Fig. 16), we know that

$\frac{s}{10-y} = \frac{6}{10}$ so $s = \frac{6}{10}(10-y)$.

The rest of the solution is straightforward.

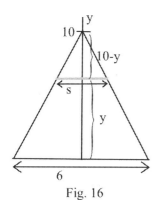

Fig. 16

$A(y) = (\text{side})^2 = \{ \frac{3}{5}(10-y) \}^2 = \frac{9}{25}(100 - 20y + y^2)$ and

$V = \int_0^{10} A(y)\,dy = \int_0^{10} \frac{9}{25}(100 - 20y + y^2)\,dy = \frac{9}{25}(100y - 10y^2 + \frac{y^3}{3}) \Big|_0^{10}$

$= \frac{9}{25}(1000 - 1000 + \frac{1000}{3}) - (0) = \frac{9}{25}\cdot\frac{1000}{3} = 120 \text{ ft}^3$.

Example 6: A solid is built between the graphs of $f(x) = x+1$ and $g(x) = x^2$ by building squares with heights (sides) equal to the vertical distance between the graphs of f and g (Fig. 17). Find the volume of this solid for $0 \le x \le 2$.

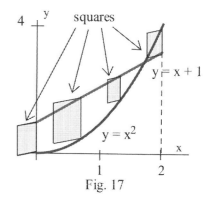

Fig. 17

Solution: The area of a square face is $A(x) = (\text{side})^2$, and the length of a side is either $f(x)-g(x)$ or $g(x)-f(x)$, depending on which function is higher at x. Fortunately, the side is squared in the area formula so it does not matter which is taller, and $A(x) = \{ f(x) - g(x) \}^2$. Then

$V = \int_a^b A(x)\,dx = \int_0^2 \{ f(x) - g(x) \}^2\,dx = \int_0^2 \{ (x+1) - x^2 \}^2\,dx = \int_0^2 \{ (x+1) - x^2 \}^2\,dx$

$= \int_0^2 (1 + 2x - x^2 - 2x^3 + x^4)\,dx = x + x^2 - \frac{x^3}{3} - \frac{x^4}{2} + \frac{x^5}{5} \Big|_0^2 = \frac{26}{15} = 1\frac{11}{15}$.

We saw earlier that areas can have nongeometric interpretations such as distance and total accumulation. Similarly, volumes can have nongeometric interpretations. If x represents an age in years, and $f(x)$ is the number of females in a population with age exactly equal to x, then the "area," $\int_a^b f(x)\,dx$, is the total number of females with ages between a and b (Fig. 18). If the birth rate for females of age x is $r(x)$, with units "births per female per year," (Fig. 19) then the "volume" of the solid in Fig. 20 is $C = \int_a^b r(x) \cdot f(x)\,dx$. C is the number of births during a year to females between the ages a and b, and the units of C will be "births."

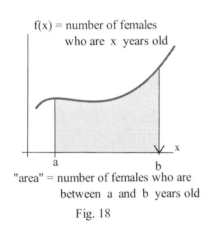

"area" = number of females who are between a and b years old

Fig. 18

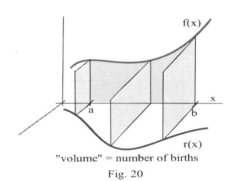

Fig. 19 r(x) = reproductive rate of females at age x

"volume" = number of births
Fig. 20

Volumes of Revolved Regions

When a region is revolved around a line (Fig. 21) a right solid is formed. The face of each slice of the revolved region is a circle so the formula for the area of the face is easy: $A(x)$ = area of a circle = $\pi(\text{radius})^2$ where the radius is often a function of the location x. Finding a formula for the changing radius requires care.

Example 7: For $0 \le x \le 2$, the area between the graph of $f(x) = x^2$ and the horizontal line $y = 1$ is revolved about the horizontal line $y=1$ to form a solid (Fig. 22). Calculate the volume of the solid.

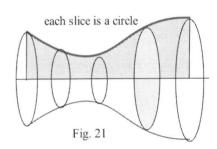

each slice is a circle

Fig. 21

Solution: The radius function is shown in the figure for several values of x. If $0 \le x \le 1$, then $r(x) = 1 - x^2$, and if $1 \le x \le 2$ then $r(x) = x^2 - 1$. Fortunately, however, $A(x) = \pi \{r(x)\}^2$ always uses the square of $r(x)$ and the squares of $1-x^2$ and $x^2 - 1$ are equal.

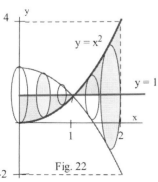

Fig. 22

$A(x) = \pi \{ r(x) \}^2 = \pi \{ x^2 - 1 \}^2 = \pi \{ x^4 - 2x^2 + 1 \}$, and

$V = \int_0^2 \pi \{ x^4 - 2x^2 + 1 \} \, dx = \pi \left(\frac{x^5}{5} - \frac{2}{3} x^3 + x \right) \Big|_0^2 = \frac{46}{15} \pi \approx 9.63$.

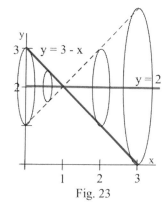

Practice 4: A solid of revolution is formed when the region between $f(x) = 3 - x$ and the horizontal line $y = 2$ is revolved about the line $y=2$ for $0 \le x \le 3$ (Fig. 23). Find the volume of the solid.

Fig. 23

Volumes of Revolved Regions ("Disks")

If the region formed between f, the horizontal line $y = L$, and the interval $[a, b]$ is revolved about the horizontal line $y = L$, (Fig. 24)

then the volume is $V = \int_a^b A(x) \, dx = \int_a^b \pi \cdot (\text{radius})^2 \, dx = \int_a^b \pi \{ f(x) - L \}^2 \, dx$.

This is called the "disk" method because the shape of each thin slice is a circular disk. If the region between f and the x–axis (L=0) is revolved about the x–axis, then the previous formula reduces to

$V = \int_a^b \pi \{ f(x) \}^2 \, dx$.

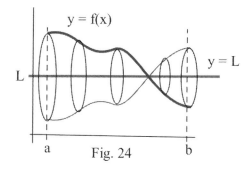

Fig. 24

Example 8: Find the volume generated when the region between one arch of the sine curve $(0 \le x \le \pi)$ and the x–axis is revolved about (a) the x–axis and (b) the line $y=1/2$.

Solution: (a) $V = \int_a^b \pi \cdot (\text{radius})^2 \, dx = \int_0^\pi \pi \{ \sin(x) \}^2 \, dx = \pi \int_0^\pi \sin^2(x) \, dx = \frac{\pi}{2} \int_0^\pi 1 - \cos(2x) \, dx$

$= \frac{\pi}{2} \left\{ x - \frac{\sin(2x)}{2} \right\} \Big|_0^\pi = \frac{\pi}{2} \{ \pi - 0 \} = \frac{\pi^2}{2} \approx 4.93$.

(b) $V = \int_a^b \pi \cdot (\text{radius})^2 \, dx = \int_0^\pi \pi \{ \sin(x) - \frac{1}{2} \}^2 \, dx = \pi \int_0^\pi \{ \sin^2(x) - \sin(x) + \frac{1}{4} \} \, dx$

$= \pi \{ \frac{\pi}{2} - 2 + \frac{\pi}{4} \} \approx 1.12$.

Practice 5: Find the volumes swept out when

(a) the region between $f(x) = x^2$ and the x–axis, for $0 \le x \le 2$, is revolved about the x–axis, and

(b) the region between $f(x) = x^2$ and the line $y=2$, for $0 \le x \le 2$, is revolved about the line $y=2$.

Example 9: Given that $\int_1^5 f(x)\,dx = 4$ and $\int_1^5 \{f(x)\}^2\,dx = 7$. Represent the volumes of the solids (a), (b) and (c) in Fig. 25 as definite integrals and evaluate the integrals.

Solution: (a) $V = \int_1^5 \pi(\text{radius})^2\,dx = \int_1^5 \pi\cdot\{f(x)\}^2\,dx = \pi\int_1^5 f^2(x)\,dx = 7\pi$.

(b) $V = \int_1^5 \pi(\text{radius})^2\,dx = \int_1^5 \pi\cdot\{f(x)-(-1)\}^2\,dx = \pi\int_1^5 \{f^2(x) + 2f(x) + 1\}\,dx$

$= \pi\left\{\int_1^5 f^2(x)\,dx + 2\int_1^5 f(x)\,dx + \int_1^5 1\,dx\right\} = \pi\{7 + 2\cdot 4 + 4\} = 19\pi$.

(c) $V = \int_1^5 \pi(\text{radius})^2\,dx = \int_1^5 \pi\cdot\{f(x)/2\}^2\,dx = \frac{\pi}{4}\int_1^5 f^2(x)\,dx = \frac{7\pi}{4}$.

Practice 6: Set up and evaluate the integral for the volume of (d) in Fig. 25.

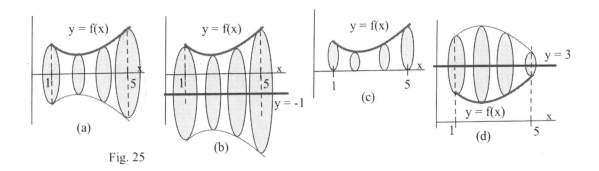

Fig. 25

Solids With Holes

The previous ideas and techniques can also be used to find the volumes of solids with holes in them. If $A(x)$ is the area of the face formed by a cut at x, then it is still true that the volume is $V = \int_a^b A(x)\,dx$. However, if the solid has holes, then some of the faces will also have holes and a formula for $A(x)$ may be more complicated.

Sometimes it is easier to work with two integrals and then subtract: (i) calculate the volume S of the solid without the hole, (ii) calculate the volume H of the hole, and (iii) subtract H from S.

Example 10: Calculate the volume of the solid in Fig. 26.

Solution: The face for a slice at x, has area

$A(x)$ = {area of large circle} – {area of small circle}

$= \pi\{\text{large radius}\}^2 - \pi\{\text{small radius}\}^2$

$= \pi\{x+1\}^2 - \pi\{1/x\}^2 = \pi(x^2 + 2x + 1 - 1/x^2)$.   Then

Volume = $\int_a^b A(x)\,dx = \int_1^2 \pi(x^2 + 2x + 1 - 1/x^2)\,dx$

$= \pi\{\frac{1}{3}x^3 + x^2 + x + 1/x\}\Big|_1^2 \approx 18.33$.

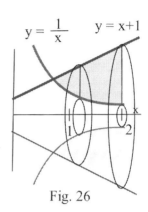

Fig. 26

Alternately, the volume of the solid with the large circular faces is $\int_1^2 \pi(x^2 + 2x + 1)\,dx = \frac{19\pi}{3} \approx 19.90$,

and the volume of the hole is $\int_1^2 \pi(1/x^2)\,dx = \frac{\pi}{2} \approx 1.57$ so the

volume we want is $19.90 - 1.57 = 18.33$.

Practice 7:   Calculate the volume of the solid in Fig. 27.

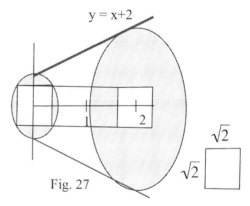

Fig. 27

WRAP UP

At first, all of these volumes may seem overwhelming — there are so many possible solids and formulas and different cases. If you concentrate on the differences, it is very complicated. Instead, focus on the pattern of **cutting, finding areas of faces, volumes of slices, and adding**. With that pattern firmly in mind, you can reason your way to the definite integral. Try to make cuts so the resulting faces have regular shapes (rectangles, triangles, circles) whose areas you can calculate. Try not to let the complexity of the whole solid confuse you. Sketch the shape of **one** face and label its dimensions. If you can find the area of **one** face in the middle of the solid, you can usually find the pattern for all of the faces and then you can easily set up the integral.

PROBLEMS

In problems 1 – 6, use the values given in the tables to calculate the volumes of the solids. (Fig. 28 – 33)

Table 1: (Fig. 28)

box	base	height	thickness
1	5	6	1
2	4	4	2
3	3	3	1

Table 2: (Fig. 29)

box	base	height	thickness
1	5	6	2
2	5	4	1
3	3	3	1
4	2	2	1

Table 3: (Fig. 30)

disk	radius	thickness
1	4	0.5
2	3	1
3	1	2

Table 4: (Fig. 31)

disk	height	thickness
1	8	0.5
2	6	1
3	2	2

Table 5: (Fig. 32)

slice	face area	thickness
1	9	0.2
2	6	0.2
3	2	0.2

Table 6: (Fig. 33)

slice	rock area	min. area	thickness
1	4	1	0.6
2	12	1	0.6
3	20	4	0.6
4	10	3	0.6
5	8	2	0.6

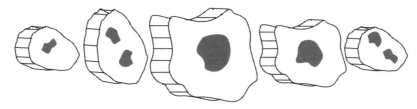

Fig. 28   Fig. 29   Fig. 30   Fig. 31   Fig. 32

Fig. 33

5.1 Volumes   Contemporary Calculus   11

In problems 7 – 12, represent each volume as an integral and evaluate the integral.

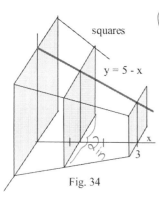

7. Fig. 34. For $0 \le x \le 3$, each face is a rectangle with base 2 inches and height $5-x$ inches.

8. Fig. 35. For $0 \le x \le 3$, each face is a rectangle with base x inches and height x^2 inches.

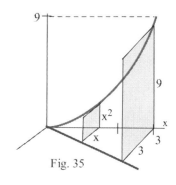

Fig. 34

Fig. 35

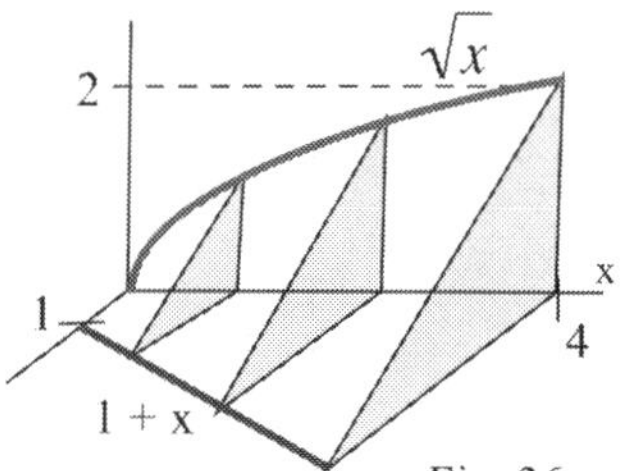

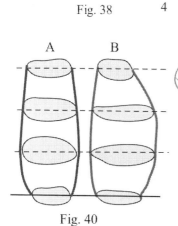

Fig. 36

9. Fig. 36. For $1 \le x \le 4$, each face is a triangle with base $x+1$ meters and height $\sqrt{x}$ meters.

10. Fig. 37. For $0 \le x \le 3$, each face is a circle with height (diameter) $4-x$ meters.

11. Fig. 38. For $2 \le x \le 4$, each face is a circle with height (diameter) $4-x$ meters.

12. Fig. 39. For $0 \le x \le 2$, each face is a square with a side extending from $y=1$ to $y=x+2$.

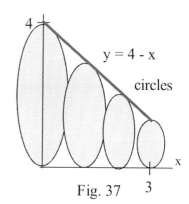

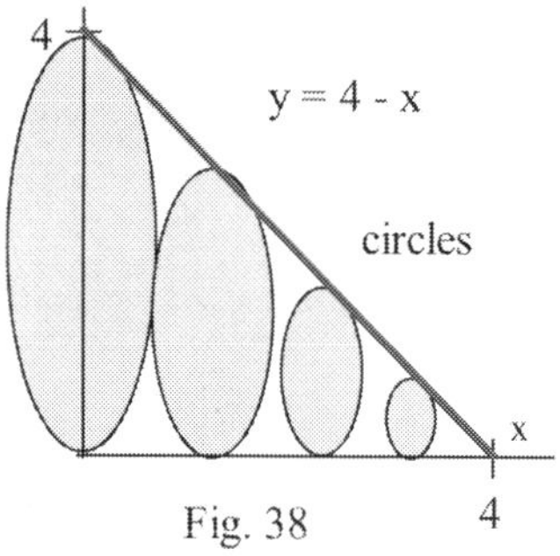

Fig. 37

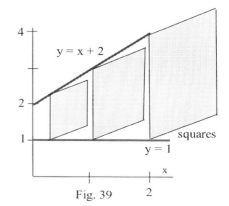

Fig. 38

Fig. 39

13. Suppose A and B are solids (Fig. 40) so that every horizontal cut produces faces of A and B that have equal areas. What can we conclude about the volumes of A and B? Justify your answer.

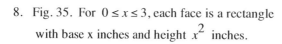

Fig. 40

5.1 Volumes   Contemporary Calculus   12

In problems 14 – 22, represent each volume as an integral and evaluate the integral.

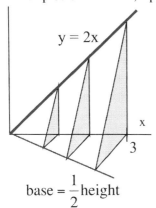

$y = 2x$, base = $\frac{1}{2}$ height
Problem 14

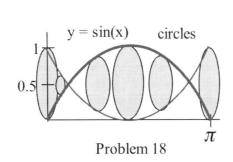

$y = \frac{1}{x}$, base = height
Problem 15

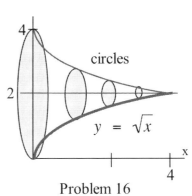

circles, $y = \sqrt{x}$
Problem 16

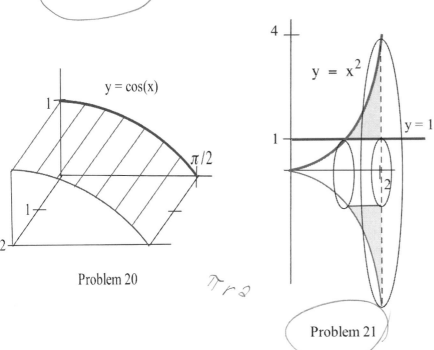

Problem 17 ($y = \sqrt{x}$, circles)

Problem 18 ($y = \sin(x)$, circles)

Problem 19 (height = x^2, base = x)

Problem 20 ($y = \cos(x)$)

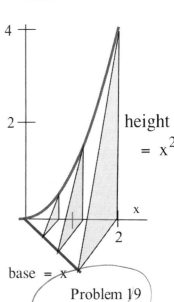

Problem 21 ($y = x^2$, $y = 1$)

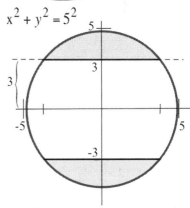

$x^2 + y^2 = 5^2$

A sphere with radius 5 has a hole of radius 3 drilled through its center

Problem 22

23. Calculate the volume of a sphere of radius 2. (A sphere is formed when the region bounded by the x–axis and the top half of the circle $x^2 + y^2 = 2^2$ is revolved about the x–axis.)

24. Determine the volume of a sphere of radius r. (A sphere is swept out when the region bounded by the x–axis and the top half of the circle $x^2 + y^2 = r^2$ is revolved about the x–axis.)

25. Calculate the volume swept out when the top half of the elliptical region bounded by $\dfrac{x^2}{5^2} + \dfrac{y^2}{3^2} = 1$ is revolved around the x–axis (Fig. 41). ($y = +3\sqrt{1 - (x^2/25)}$)

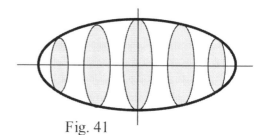

Fig. 41

26. Calculate the volume swept out when the top half of the elliptical region bounded by $\dfrac{x^2}{a^2} + \dfrac{y^2}{b^2} = 1$ is revolved around the x–axis. ($y = +b\sqrt{1 - (x^2/a^2)}$)

27. Determine the volume of the "doughnut" in Fig. 42. (The top half of the circle is given by $f(x) = R + \sqrt{r^2 - x^2}$ and the bottom half is given by $g(x) = R - \sqrt{r^2 - x^2}$. (It is easier to use a single integral for this problem.)

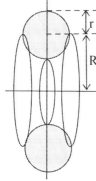

torus ("doughnut")
Fig. 42

28. (a) Find the **area** between $f(x) = 1/x$ and the x–axis for $1 \leq x \leq 10$, $1 \leq x \leq 100$, and $1 \leq x \leq A$. What is the limit of the area for $1 \leq x \leq A$ as $A \to \infty$?

    (b) Find the **volume** swept out when the region in part (a) is revolved about the x–axis for $1 \leq x \leq 10$, $1 \leq x \leq 100$, and $1 \leq x \leq A$. What is the limit of the volumes for $1 \leq x \leq A$ as $A \to \infty$?

29. Personal Calculus:" Describe a **practical** way to determine the volume of your hand and arm up to the elbow.

30. Personal Calculus:" Most people have a body density between .95 and 1.05 times the density of water which is 62.5 pounds per cubic foot. Use your weight to estimate the volume of your body. (If you float in fresh water, your body density is less than 1.)

Volumes of "right cones"

31. Calculate (a) the volume of the right solid in Fig. 43a, (b) the volume of the "right cone" in Fig. 43b, and (c) the ratio of the "right cone" volume to the right solid volume. (square cross sections)

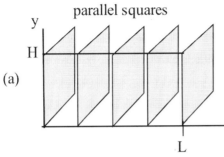
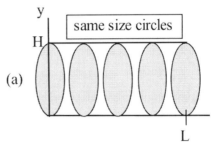
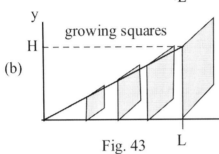
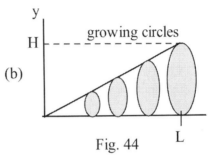

Fig. 43

Fig. 44

32. Calculate (a) the volume of the right solid in Fig. 44a, (b) the volume of the "right cone" in Fig. 44b, and (c) the ratio of the "right cone" volume to the right solid volume. (circular cross sections)

33. The "blob" in Fig. 45 has area B.
    (a) Calculate the volume of the right solid in Fig. 45a.
    (b) If a "right cone" is formed (Fig. 45b), then the cross section area at x is $A(x) = (B/L^2)x^2$. Find the volume of the "right cone".
    (c) Find the ratio of the "right cone" volume to the right solid volume.

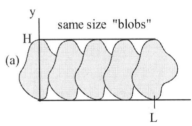
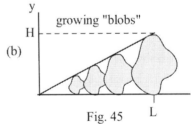

Fig. 45

34. Represent each volume as a definite integral.

A.
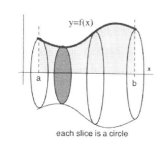

B.
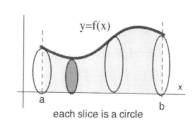

C.
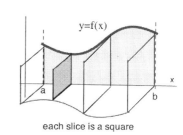

D.
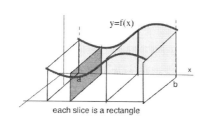

E.
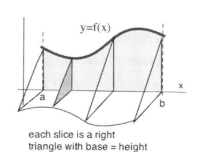

F.
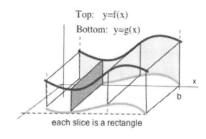

G.
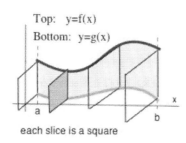

H.
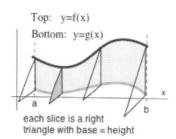

I.
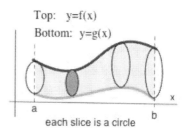

J.
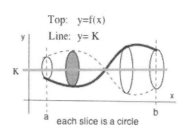

K.
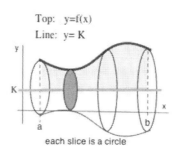

L.
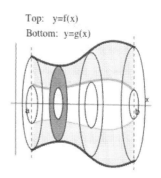

Section 5.1   **PRACTICE Answers**

Practice 1: (a) Triangular base: v = (base area)·(height) = $(\frac{1}{2} \cdot 3 \cdot 4) \cdot (6) = 36$.

(b) Semicircular base: v = (base area)·(height) = $\frac{1}{2} \cdot (\pi \cdot 3^2) \cdot (7) \approx 98.96$.

(c) Strangely shaped base: v = (base area)·(height) = $(8 \text{ in}^2) \cdot (5 \text{ in}) = 40 \text{ in}^3$.

Practice 2: (a) $v_i \approx$ (area of face)(thickness) $\approx (\frac{1}{2} \cdot 4 \cdot x_i^2)(\Delta x_i) = 2x_i^2 \Delta x_i$.

(b) $v = \int_1^2 2x^2 \, dx = \frac{2}{3} x^3 \Big|_1^2 = \frac{16}{3} - \frac{2}{3} = \frac{14}{3}$ **cubic inches**.

Practice 3: (a) $v_i \approx$ (area of face)(thickness) $\approx (\sqrt{x_i})^2 (\Delta x_i) = x_i \Delta x_i$.

(b) $v = \int_1^4 x \, dx = \frac{1}{2} x^2 \Big|_1^4 = 8 - \frac{1}{2} = \mathbf{7.5}$.

Practice 4: $v_i \approx$ (area of face)(thickness) $\approx (\pi r^2)($ thickness $)$

$= (\pi \cdot ((3-x_i) - 2)^2 (\Delta x_i) = \pi(1 - 2x_i + x_i^2) \Delta x_i$.

Then volume $= \int_0^3 \pi(1 - 2x + x^2) \, dx = \pi(x - x^2 + \frac{1}{3} x^3) \Big|_0^3 = 3\pi \approx \mathbf{9.42}$.

Practice 5: (a) $v = \int_a^b \pi($ radius $)^2 \, dx = \int_0^2 \pi(x^2)^2 \, dx = \frac{\pi}{5} x^5 \Big|_0^2 = \frac{32\pi}{5} = \mathbf{20.1}$.

(b) $v = \int_a^b \pi($ radius $)^2 \, dx = \int_0^2 \pi(2 - x^2)^2 \, dx = \int_0^2 \pi(4 - 4x^2 + x^4) \, dx$

$= \pi(4x - \frac{4}{3} x^3 + \frac{1}{5} x^5) \Big|_0^2 = \frac{56\pi}{15} \approx \mathbf{11.73}$.

Practice 6: (d) $v = \int_a^b \pi($ radius $)^2 \, dx = \int_1^5 \pi(3 - f(x))^2 \, dx = \pi \int_1^5 9 - 6f(x) + f^2(x) \, dx$

$= \pi \int_1^5 (9 - 6f(x) + f^2(x)) \, dx = \pi \int_1^5 9 \, dx - 6\pi \int_1^5 f(x) \, dx + \pi \int_1^5 f^2(x) \, dx$

$= \pi(36) - 6\pi(4) + \pi(7) = 19\pi \approx \mathbf{59.69}$.

(The values "7" and "4" are given in Example 9).

Practice 7: The volume we want can be obtained by subtracting the volume of the "box" from the volume of the truncated cone generated by the rotated line segment.

volume of truncated cone $= \int_a^b \pi($ radius $)^2 \, dx = \int_0^2 \pi(x + 2)^2 \, dx$

$= \pi \int_0^2 x^2 + 4x + 4 \, dx = \pi \{ \frac{1}{3} x^3 + 2x^2 + 4x \} \Big|_0^2 = \frac{56}{3} \pi \approx 58.64$.

volume of "box" = (length)(width)(height) = $2 (\sqrt{2})(\sqrt{2}) = 4$.

The volume we want is $\frac{56}{3} \pi - 4 \approx \mathbf{54.64}$.

5.2 LENGTHS OF CURVES & AREAS OF SURFACES OF REVOLUTION

This section introduces two additional geometric applications of integration: finding the length of a curve and finding the area of a surface generated when a curve is revolved about a line. The general strategy is the same as before: partition the problem into small pieces, approximate the solution on each small piece, add the small solutions together in the form of a Riemann sum, and finally, take the limit of the Riemann sum to get a definite integral.

ARC LENGTH: How Long Is A Curve?

In order to understand an object or an animal, we often need to know how it moves about its environment and how far it travels. We need to know the length of the path it moves along. If we know the object's location at successive times, then it is straightforward to calculate the distances between those locations and add them together to get a total distance.

Example 1: In order to study the movement of whales, a scientist attached a small radio transmitter to the fin of a whale and tracked the location of the whale at 1 hour time intervals over a period of several weeks. The data for a 5 hour period is shown in Fig. 1. How far did the whale swim during the first 3 hours?

Solution: In moving from the point (0,0) to the point (0,2), the whale traveled at least 2 miles. Similarly, the whale traveled at least $\sqrt{(1-0)^2 + (3-2)^2} \approx 1.4$ miles during the second hour and at least $\sqrt{(4-1)^2 + (1-3)^2} \approx 3.6$ miles during the third hour. The scientist concluded that the whale swam **at least** $2 + 1.4 + 3.6 = 7$ miles during the 3 hours.

Practice 1: How far did the whale swim during the entire 5 hour period?

The scientist noted that the whale did not swim in a straight line from location to location so its actual swimming distance was more than 7 miles for the first 3 hours. The scientist hoped to get better distance estimates in the future by determining the whale's position over shorter, five–minute time intervals.

Our strategy for finding the length of a curve will be similar to the one the scientist used, and if the locations are given by a formula, then we can calculate the successive locations over very short intervals and get very good approximations of the total path length. In fact, we can get the exact length of the path by evaluating a definite integral.

Suppose C is a curve, and we pick some points (x_i, y_i) along C (Fig. 2) and connect the points with straight line segments. Then the sum of the lengths of the line segments will approximate the length of C. We can think of this as pinning a string to the curve at the selected points, and then measuring the length of the string as an approximation of the length of the curve. Of course, if we only pick a few points as in Fig. 2, then the total length approximation will probably be rather poor, so eventually we want lots of points (x_i, y_i), close together all along C.

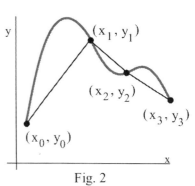

Fig. 2

Suppose the points are labeled so (x_0, y_0) is one endpoint of C and (x_n, y_n) is the other endpoint and that the subscripts increase as we move along C. Then the distance between the successive points (x_{i-1}, y_{i-1}) and (x_i, y_i) is $\sqrt{(\Delta x_i)^2 + (\Delta y_i)^2}$, and the total length of the line segments is simply the sum of the successive lengths. This is an important approximation of the length of C, and all of the integral representations for the length of C come from it.

The length of the curve C is approximately $\sum \sqrt{(\Delta x_i)^2 + (\Delta y_i)^2}$.

Example 2: Use the points $(0,0), (1,1)$, and $(3,9)$ to approximate the length of $y = x^2$ for $0 \leq x \leq 3$.

Solution: The lengths of the two linear pieces in Fig. 3 are
$\sqrt{1^2 + 1^2} = \sqrt{2} \approx 1.41$ and
$\sqrt{2^2 + 8^2} = \sqrt{68} \approx 8.25$ so the length of the curve
is approximately $1.41 + 8.25 = 9.66$.

Practice 2: Get a better approximation of the length of $y = x^2$ for $0 \leq x \leq 3$ by using the points $(0,0), (1,1), (2,4)$, and $(3,9)$. Is your approximation longer or shorter than the actual length?

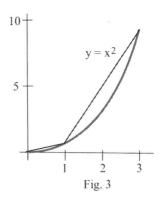

Fig. 3

5.2   Arc Length & Surface Area

The summation does not have the form $\sum f(c_i) \cdot \Delta x_i$ so it is not a Riemann sum. It is, however, algebraically equivalent to several Riemann sums, and each one leads to a definite integral representation for the length of C.

(a) **y = f(x)**: When y is a function of x, we can factor $(\Delta x_i)^2$ from inside the radical and simplify.

$$\text{Length of C} \approx \sum \sqrt{(\Delta x_i)^2 + (\Delta y_i)^2} = \sum \sqrt{(\Delta x_i)^2 \cdot \left\{ 1 + \left(\frac{\Delta y_i}{\Delta x_i}\right)^2 \right\}}$$

$$= \sum \Delta x_i \sqrt{1 + \left(\frac{\Delta y_i}{\Delta x_i}\right)^2} \quad \text{(Riemann sum)} \longrightarrow \boxed{\int_{x=a}^{x=b} \sqrt{1 + \left(\frac{dy}{dx}\right)^2}\, dx}$$

(b) **x = g(y)**: When x is a function of y, we can factor $(\Delta y_i)^2$ from inside the radical and simplify.

$$\text{Length of C} \approx \sum \sqrt{(\Delta x_i)^2 + (\Delta y_i)^2} = \sum \sqrt{(\Delta y_i)^2 \cdot \left\{ \left(\frac{\Delta x_i}{\Delta y_i}\right)^2 + 1 \right\}}$$

$$= \sum \Delta y_i \sqrt{\left(\frac{\Delta x_i}{\Delta y_i}\right)^2 + 1} \quad \text{(Riemann sum)} \longrightarrow \boxed{\int_{y=c}^{y=d} \sqrt{\left(\frac{dx}{dy}\right)^2 + 1}\, dy}$$

(c) **Parametric equations**: When x and y are functions of t, $x = x(t)$ and $y = y(t)$, for $\alpha \le t \le \beta$ we can factor $(\Delta t_i)^2$ from inside the radical and simplify.

$$\text{Length of C} \approx \sum \sqrt{(\Delta x_i)^2 + (\Delta y_i)^2} = \sum \sqrt{(\Delta t_i)^2 \cdot \left\{ \left(\frac{\Delta x_i}{\Delta t_i}\right)^2 + \left(\frac{\Delta y_i}{\Delta t_i}\right)^2 \right\}}$$

$$= \sum \Delta t_i \sqrt{\left(\frac{\Delta x_i}{\Delta t_i}\right)^2 + \left(\frac{\Delta y_i}{\Delta t_i}\right)^2} \quad \text{(Riemann sum)} \longrightarrow \boxed{\int_{t=\alpha}^{t=\beta} \sqrt{\left(\frac{dx}{dt}\right)^2 + \left(\frac{dy}{dt}\right)^2}\, dt}$$

The integrals in (a), (b), and (c) each represent the length of C, and we can use whichever one is more convenient. Unfortunately, for most functions these integrands do not have easy antiderivatives, and most arc length integrals must be approximated using one of our approximate integration methods or a calculator.

Example 3: Represent the length of each curve as a definite integral.

(a) The length of $y = x^2$ between $(1,1)$ and $(4,16)$.

(b) The length of $x = \sqrt{y}$ between $(1,1)$ and $(4,16)$.

(c) The length of the parametric curve $x(t) = \cos(t)$ and $y(t) = \sin(t)$ for $0 \le t \le 2\pi$.

Solution: (a) Length = $\int_{x=1}^{x=4} \sqrt{1+(dy/dx)^2}\, dx = \int_{x=1}^{x=4} \sqrt{1+4x^2}\, dx \approx 15.34$.

(b) Length = $\int_{y=1}^{y=16} \sqrt{(dx/dy)^2+1}\, dy = \int_{y=1}^{y=16} \sqrt{\tfrac{1}{4y}+1}\, dy \approx 15.34$.

The values of the integrals in (a) and (b) were approximated using Simpson's rule with n = 10. It is not an accident that the lengths in (a) and (b) are equal. Why not?

(c) Length = $\int_{t=0}^{t=2\pi} \sqrt{(-\sin(t))^2 + (\cos(t))^2}\, dt = \int_{t=0}^{t=2\pi} \sqrt{\sin^2(t) + \cos^2(t)}\, dt = \int_{t=0}^{t=2\pi} 1\, dt = 2\pi$.

The graph of $(x(t), y(t))$ for $0 \le t \le 2\pi$ is a circle of radius 1 so we know that its length is exactly 2π.

Practice 3: Represent the length of each curve as a definite integral.

(a) The length of one period of $y = \sin(x)$.

(b) The length of the parametric path $x(t) = 1 + 3t$ and $y(t) = 4t$ for $1 \le t \le 3$.

AREAS OF SURFACES OF REVOLUTION

Rotated Line Segments

Just as all of the integral formulas for arc length came from the simple distance formula, all of the integral formulas for the area of a revolved surface come from the formula for revolving a single straight line segment. If a line segment of length L, parallel to a line P, (Fig. 4) is revolved about the line P, then the resulting surface can be unrolled and laid flat. The flattened surface is a rectangle with area $A = 2\pi \cdot r \cdot L$.

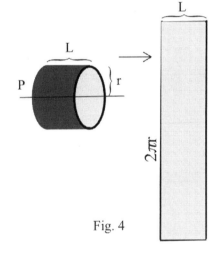

Fig. 4

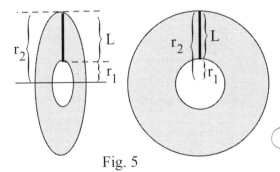

Fig. 5

If a line segment of length L, perpendicular to a line P and not intersecting P, (Fig. 5) is revolved about the line P, then the resulting surface is the region between two concentric circles and its area is

A = (area of large cirle) − (area of small circle) = $\pi(r_2)^2 - \pi(r_1)^2$

$= \pi\{(r_2)^2 - (r_1)^2\} = 2\pi\{\tfrac{r_2+r_1}{2}\}(r_2-r_1) = 2\pi \cdot \left(\dfrac{r_1+r_2}{2}\right) \cdot L$

In general, if a line segment of length L which does not intersect a line P (Fig. 6) is revolved about P, then the resulting surface has area

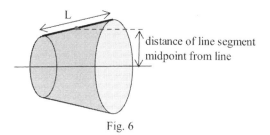
Fig. 6

A = surface area

= (distance traveled by midpoint of the line segment)·(length of the line segment)

= **2π·{ distance of segment midpoint from the line P }· L**

There are several integral formulas for the surface area of a curve rotated about a line, but all of the formulas come rather easily and quickly from this one fundamental formula for a surface area of a rotated line segment.

Example 4: Find the surface area generated when each line segment in Fig. 7 is rotated about the x–axis and the y–axis.

Solution: Line segment B has length L=2 and its midpoint is at (2,1).

When B is rotated about the x–axis, the surface area is

2π·(distance of midpoint from x–axis)·2 = 2π(1)2 = 4π.

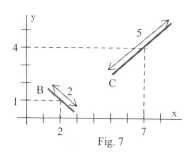
Fig. 7

When B is rotated about the y–axis, the surface area is

2π·(distance of midpoint from y–axis)·2 = 8π.

Line segment C has length 5 and its midpoint is at (7,4). When C is rotated about the x–axis the resulting surface area is

2π·(distance of midpoint from x–axis)·5 = 2π(4)5 = 40π.

When C is rotated about the y–axis, the surface area is

2π·(distance of midpoint from y–axis)·5 = 70π.

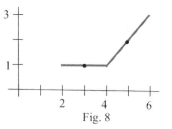
Fig. 8

Practice 4: Find the surface areas generated when the graph in Fig. 8 is rotated about each axis.

Rotated Curves

When a curve is rotated about a line P, we can use our old strategy again (Fig. 9). Select some points (x_i, y_i) along the curve, connect the points with line segments, calculate the surface area of each rotated line segment, and add together the surface areas of the rotated line segments. This final sum can be converted to a Riemann sum, and the limit of the Riemann sum is a definite integral for the surface area of the rotated curve.

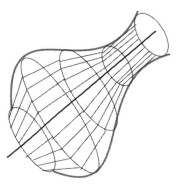

Fig. 9

Suppose we select points (x_i, y_i) along C and number them so (x_0, y_0) is one endpoint of C, (x_n, y_n) is the other endpoint, and the subscripts increase as we move along C. Then each pair of successive points $(x_{i-1}, y_{i-1}), (x_i, y_i)$ are the endpoints of a line segment with

$$\text{length} = \sqrt{(\Delta x_i)^2 + (\Delta y_i)^2} \quad \text{and} \quad \text{midpoint} = \left(\frac{x_{i-1}+x_i}{2}, \frac{y_{i-1}+y_i}{2}\right).$$

The midpoint is $(y_{i-1}+y_i)/2$ units from the x–axis and $(x_{i-1}+x_i)/2$ units from the y–axis.

About the x–axis: If C is rotated about the x–axis, then each segment generates an area equal to

$$2\pi \cdot (\text{distance of midpoint from the x-axis}) \cdot (\text{length of the segment}) = 2\pi \left\{\frac{y_{i-1}+y_i}{2}\right\} \sqrt{(\Delta x_i)^2 + (\Delta y_i)^2}.$$

The sum of these surface areas is

$$\sum 2\pi \left\{\frac{y_{i-1}+y_i}{2}\right\} \sqrt{(\Delta x_i)^2 + (\Delta y_i)^2} = \sum 2\pi \left\{\frac{y_{i-1}+y_i}{2}\right\} \sqrt{(\Delta x_i/\Delta x_i)^2 + (\Delta y_i/\Delta x_i)^2} \, \Delta x_i$$

$$\boxed{\int_a^b 2\pi y \sqrt{1+(dy/dx)^2} \, dx = \text{area of the surface of revolution of curve C}}.$$

About the y–axis: If C is rotated about the y–axis, then each segment generates an area equal to

$$2\pi \cdot (\text{distance of midpoint from the y-axis}) \cdot (\text{length of the segment}) = 2\pi \left\{\frac{x_{i-1}+x_i}{2}\right\} \sqrt{(\Delta x_i)^2 + (\Delta y_i)^2}.$$

The sum of these surface areas is

$$\sum 2\pi \left\{\frac{x_{i-1}+x_i}{2}\right\} \sqrt{(\Delta x_i)^2 + (\Delta y_i)^2} = \sum 2\pi \left\{\frac{x_{i-1}+x_i}{2}\right\} \sqrt{(\Delta x_i/\Delta x_i)^2 + (\Delta y_i/\Delta x_i)^2} \, \Delta x_i$$

$$\boxed{\int_a^b 2\pi x \sqrt{1+(dy/dx)^2} \, dx = \text{area of the surface of revolution of curve C}}.$$

Example 5: Use definite integrals to represent the areas of the surfaces generated when the curve $y = 2 + x^2$, $0 \leq x \leq 3$, is rotated about each axis.

Solution: Surface area about the x–axis is $\int_a^b 2\pi \cdot f(x)\sqrt{1+(dy/dx)^2}\, dx = \int_0^3 2\pi(2+x^2)\sqrt{1+4x^2}\, dx \approx 383.8$.

Surface area about the y–axis is $\int_a^b 2\pi \cdot x \sqrt{1+(dy/dx)^2}\, dx = \int_0^3 2\pi \cdot x \sqrt{1+4x^2}\, dx \approx 117.32$.

Parametric Form For Surface Area of Revolution

If the curve C is described by parametric equations, $x = x(t)$ and $y = y(t)$, then the forms of the surface area integrals are somewhat different, but they still follow from the fundamental surface area formula for the surface area of a line segment rotated about a line P:

Surface area $= 2\pi \cdot \{$ **distance of midpoint from the line P** $\} \cdot$ **L** .

About x–axis: Starting with the previous equation for the area of a segment rotated about the **x–axis**,

$$2\pi \cdot \left\{ \frac{y_{i-1}+y_i}{2} \right\} \sqrt{(\Delta x_i)^2 + (\Delta y_i)^2} \quad,$$

we can factor Δt_i from the radical, sum the pieces and take the limit, as the mesh approaches 0, to get

$$\int_{t=\alpha}^{t=\beta} 2\pi y \sqrt{(dx/dt)^2 + (dy/dt)^2}\, dt = \text{area of the surface of revolution of curve C}.$$

About y–axis: Starting with the previous equation for the area of a segment rotated about the **y–axis**,

$$2\pi \left\{ \frac{x_{i-1}+x_i}{2} \right\} \sqrt{(\Delta x_i)^2 + (\Delta y_i)^2}$$

we can factor Δt_i from the radical, sum the pieces and take the limit, as the mesh approaches 0, to get

$$\int_{t=\alpha}^{t=\beta} 2\pi x \sqrt{(dx/dt)^2 + (dy/dt)^2}\, dt = \text{area of the surface of revolution of curve C}.$$

WRAP UP

One purpose of this section was to obtain a variety of integral formulas for two geometric quantities, the length of a curve and the area of the surface generated when a curve is rotated about a line. The integral formulas are useful, but a more basic and fundamental point was to illustate again how relatively simple approximation formulas can lead us, via Riemann sums, to integral formulas. We will see it again.

PROBLEMS

For arc length

1. A squirrel was spotted at the backyard locations in Table 1 at the given times. The squirrel traveled at least how far during the first 15 minutes?

time (min)	location relative to oak tree	
	north	east (feet)
0	10	7
5	25	27
10	1	45
15	13	33
20	24	40
25	10	23
30	0	14

Table 1: Locations of Squirrel

2. The squirrel in Problem 1 traveled at least how far during the first 30 minutes?

3. Use the partition $\{0, 1, 2\}$ to estimate the length of $y = 2^x$ between the points $(0, 1)$ and $(2, 4)$.

4. Use the partition $\{1, 2, 3, 4\}$ to estimate the length of $y = 1/x$ between the points $(1, 1)$ and $(4, 1/4)$.

The graphs of the functions in problems 5 – 8 are straight lines. Calculate each length (a) using the distance formula between 2 points and (b) by setting up and evaluating the arc length integrals.

5. $y = 1 + 2x$ for $0 \le x \le 2$.

6. $y = 5 - x$ for $1 \le x \le 4$.

7. $x = 2 + t, y = 1 - 2t$ for $0 \le t \le 3$.

8. $x = -1 - 4t, y = 2 + t$ for $1 \le t \le 4$.

9. Calculate the length of $y = \frac{2}{3} x^{3/2}$ for $0 \le x \le 4$.

10. Calculate the length of $y = 4x^{3/2}$ for $1 \le x \le 9$.

Very few functions $y = f(x)$ lead to integrands of the form $\sqrt{1 + (dy/dx)^2}$ which have elementary antiderivatives. In problems 11 – 14, $1 + (dy/dx)^2$ is a perfect square and the resulting arc length integrals can be evaluated using antiderivatives. Do so.

11. $y = \frac{x^3}{3} + \frac{1}{4x}$ for $1 \le x \le 5$.

12. $y = \frac{x^4}{4} + \frac{1}{8x^2}$ for $1 \le x \le 9$.

13. $y = \frac{x^5}{5} + \frac{1}{12x^3}$ for $1 \le x \le 5$.

14. $y = \frac{x^6}{6} + \frac{1}{16x^4}$ for $4 \le x \le 25$.

In problems 15 – 23, (a) represent each length as a definite integral, and (b) evaluate the integral using your calculator's integral command.

15. The length of $y = x^2$ from $(0,0)$ to $(1,1)$.

16. The length of $y = x^3$ from $(0,0)$ to $(1,1)$.

17. The length of $y = \sqrt{x}$ from $(1,1)$ to $(9,3)$.

18. The length of $y = \ln(x)$ from $(1,0)$ to $(e,1)$.

19. The length of $y = \sin(x)$ from $(0,0)$ to $(\pi/4, \sqrt{2}/2)$ and from $(\pi/4, \sqrt{2}/2)$ to $(\pi/2, 1)$.

5.2 Arc Length & Surface Area

20. The length of the ellipse $x(t) = 3\cos(t), y(t) = 4\sin(t)$ for $0 \le t \le 2\pi$.

21. The length of the ellipse $x(t) = 5\cos(t), y(t) = 2\sin(t)$ for $0 \le t \le 2\pi$.

22. A robot was programmed to follow a spiral path and be at location $x(t) = t\cos(t), y(t) = t\sin(t)$ at time t. How far did the robot travel between $t = 0$ and $t = 2\pi$?

23. A robot was programmed to follow a spiral path and be at location $x(t) = t\cos(t), y(t) = t\sin(t)$ at time t. How far did the robot travel between $t = 10$ and $t = 20$?

24. As a tire of radius R rolls, a small stone stuck in the tread will travel a "cycloid" path, $x(t) = R \cdot (t - \sin(t))$, $y(t) = R \cdot (1 - \cos(t))$. As t goes from 0 to 2π, the tire makes one complete revolution and travels forward $2\pi R$ units. How far does the small stone travel?

25. As a tire with a 1 foot radius rolls forward 1 mile, how far does a pebble stuck in the tire tread travel? ($x(t) = 1(t - \sin(t))$ and $y(t) = 1(1 - \cos(t))$)

26. (Calculator) Graph $y = x^n$ for $n = 1, 3, 10$, and 20. As the value of n gets large, what happens to the graph of $y = x^n$? Estimate the value of $\lim_{n \to \infty} \left(\int_{x=0}^{x=1} \sqrt{1 + (n \cdot x^{n-1})^2} \, dx \right)$.

27. (Calculator) Find the point on the curve segment $f(x) = x^2$ for $0 \le x \le 4$ which will divide the segment into two equally long pieces. Find the points which will divide the segment into 3 equally long pieces.

28. Find the pattern for the functions in problems 11 – 14. If $y = \dfrac{x^n}{n} + \dfrac{1}{Ax^B}$, then how are A and B related to n? ($A = 4(n-2)$ and $B = n-2$)

29. Use the formulas for A and B from the previous problem with $n = 3/2$ and write a new function
$y = \dfrac{2}{3} x^{3/2} + \dfrac{1}{Ax^B}$ so that $1 + (dy/dx)^2$ is a perfect square.

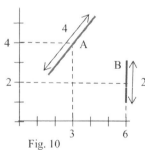

Fig. 10

For surface area of revolution

30. Find the surface area when each line segment in Fig. 10 is rotated about the (a) x–axis and (b) y–axis.

31. Find the surface area when each line segment in Fig. 11 is rotated about the (a) x–axis and (b) y–axis.

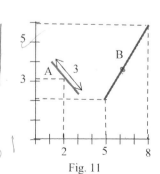

Fig. 11

32. Find the surface area when each line segment in Fig. 10 is rotated about the lines (a) y = 1 and (b) x = –2.

33. Find the surface area when each line segment in Fig. 11 is rotated about the lines (a) y = 1 and (b) x = –2.

34. A line segment of length 2 has its center at the point (2,5) and makes an angle of θ with horizontal. What value of θ will result in the largest surface area when the line segment is rotated about the y–axis? Explain your reasoning.

35. A line segment of length 2 has its one end at the point (2,5) and makes an angle of θ with horizontal. What value of θ will result in the largest surface area when the line segment is rotated about the x–axis? Explain your reasoning.

In problems 36 – 44, (a) represent each surface area as a definite integral, and (b) evaluate the integral using your calculator's integral command.

36. Find the area of the surface when the graph of $y = x^3$ for $0 \le x \le 2$ is rotated about the y–axis.

37. Find the area of the surface when the graph of $y = 2x^3$ for $0 \le x \le 1$ is rotated about the y–axis.

38. Find the area of the surface when the graph of $y = x^2$ for $0 \le x \le 2$ is rotated about the x–axis.

39. Find the area of the surface when the graph of $y = 2x^2$ for $0 \le x \le 1$ is rotated about the x–axis.

40. Find the area of the surface when the graph of $y = \sin(x)$ for $0 \le x \le \pi$ is rotated about the x–axis.

41. Find the area of the surface when the graph of $y = x^3$ for $0 \le x \le 2$ is rotated about the x–axis.

42. Find the area of the surface when the graph of $y = \sin(x)$ for $0 \le x \le \pi/2$ is rotated about the y–axis.

43. Find the area of the surface when the graph of $y = x^2$ for $0 \le x \le 2$ is rotated about the y–axis.

44. Find the area of the surface when the graph of $y = \sqrt{4 - x^2}$ is rotated about the x–axis
    (a) for $0 \le x \le 1$, (b) for $1 \le x \le 2$, and (c) for $2 \le x \le 3$.

45. (a) Show that if a thin hollow sphere is sliced into pieces by equally–spaced parallel cuts (Fig. 12), then each piece has the same weight. (Show that each piece has the same surface area).

    (b) What does the result of part (a) mean for an orange cut into slices with equally–spaced parallel cuts?

    (c) Suppose a hemispherical cake with uniformly thick layer of frosting is sliced with equally–spaced parallel cuts. Does everyone get the same amount of cake? Does everyone get the same amount of frosting?

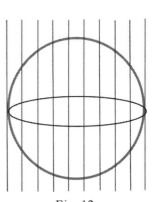

Fig. 12

3–D Arc Length Problems (Optional)

The parametric equation form of arc length extends very nicely to 3 dimensions. If a curve C in 3 dimensions (Fig. 13) is given parametrically by $x = x(t), y = y(t)$, and $z = z(t)$ for $a \le t \le b$, then the distance between the successive points $(x_{i-1}, y_{i-1}, z_{i-1})$ and (x_i, y_i, z_i) is

$$\sqrt{(x_{i-1} - x_i)^2 + (y_{i-1} - y_i)^2 + (z_{i-1} - z_i)^2}$$

$$= \sqrt{(\Delta x_i)^2 + (\Delta y_i)^2 + (\Delta z_i)^2} \; .$$

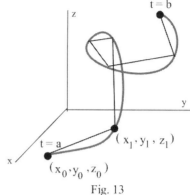

Fig. 13

We can, as before, factor $(\Delta t_i)^2$ from each term under the radical, sum the pieces to get a Riemann sum, and take a limit of the Riemann sum to get a definite integral representing the length of the curve C.

$$\sum \sqrt{(\Delta x_i/\Delta t_i)^2 + (\Delta y_i/\Delta t_i)^2 + (\Delta z_i/\Delta t_i)^2} \; \Delta t_i \longrightarrow \int_{t=a}^{t=b} \sqrt{(dx/dt)^2 + (dy/dt)^2 + (dz/dt)^2} \; dt \; .$$

The length of the curve C is $\int_{t=a}^{t=b} \sqrt{(dx/dt)^2 + (dy/dt)^2 + (dz/dt)^2} \; dt$.

In problems 46 – 50, (a) represent each length as a definite integral, and (b) evaluate the integral using your calculator's integral command.

46. Find the length of the helix $x = \cos(t)$, $y = \sin(t)$, $z = t$ for $0 \le t \le 4\pi$. (Fig. 14)

47. Find the length of the straight line segment $x = t, y = t, z = t$ for $0 \le t \le 1$.

48. Find the length of the curve $x = t, y = t^2, z = t^3$ for $0 \le t \le 1$.

49. Find the length of the "stretched helix" $x = \cos(t)$, $y = \sin(t)$, $z = t^2$ for $0 \le t \le 2\pi$.

50. Find the length of the curve $x = 3\cos(t), y = 2\sin(t), z = \sin(7t)$ for $0 \le t \le 2\pi$.

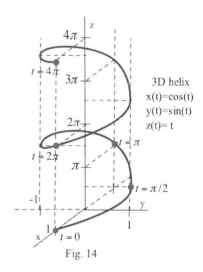

Fig. 14

Section 5.2   PRACTICE Answers

Practice 1: At least $2 + \sqrt{2} + \sqrt{13} + 1 + \sqrt{2} \approx 9.43$ miles.

Practice 2: Using the points $(0,0), (1,1), (2,4),$ and $(3,9)$,

$L \approx \sqrt{2} + \sqrt{10} + \sqrt{26} \approx 9.68 <$ actual length.

Practice 3: (a) Using $\int_{x=a}^{x=b} \sqrt{1 + \left(\frac{dy}{dx}\right)^2} \, dx$ with $y = \sin(x)$, $L = \int_{x=0}^{x=2\pi} \sqrt{1 + \cos^2(x)} \, dx$.

(b) Using $\int_{t=a}^{t=b} \sqrt{\left(\frac{dx}{dt}\right)^2 + \left(\frac{dy}{dt}\right)^2} \, dt$ with $x(t) = 1 + 3t$ and $y(t) = 4t$ for $1 \leq t \leq 3$,

$L = \int_{t=1}^{t=3} \sqrt{3^2 + 4^2} \, dt = \int_{t=1}^{t=3} 5 \, dt = 10$.

Practice 4: surface area of revolved segment $= 2\pi \cdot \{$ distance of segment midpoint from the line $P \} \cdot L$

surface area of horizontal segment revolved about **x–axis** $= 2\pi(1)(2) = 4\pi \approx 12.57$.

surface area of other segment revolved about **x–axis** $= 2\pi(2)(\sqrt{8}) \approx 35.54$.

The total surface area about the x–axis is approximately $12.57 + 35.54 = 48.11$ square units.

surface area of horizontal segment revolved about **y–axis** $= 2\pi(3)(2) = 12\pi \approx 37.70$.

surface area of other segment revolved about **y–axis** $= 2\pi(5)(\sqrt{8}) \approx 88.86$.

The total surface area about the y–axis is approximately $37.70 + 88.86 = 126.56$ square units.

5.3 MORE WORK APPLICATIONS

In Section 4.7 we introduced the problem of calculating the **work** done in lifting an object using a cable which had weight. This section continues that introduction and extends the process to handle situations in which the applied force or the distance or both may be variables. The method we used before is used again here. The first step is to divide the problem into small "slices" so that the force and distance vary only slightly on each slice. Then the work for each slice is calculated, the total work is approximated by adding together (a Riemann sum) the work for each slice, and, finally, a limit is taken to get a definite integral representing the total work. **There are so many possible variations in work problems that it is vital that you understand the process.**

The work done on an object by a constant force is the magnitude of the force applied to the object multiplied by the distance over which the force is applied: **work = (force)·(distance)**.

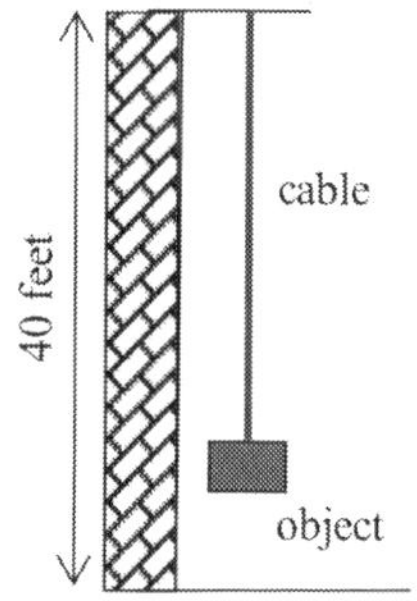

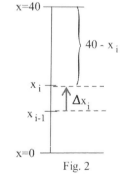

Fig. 1

Example 1: A 10 pound object is lifted 40 feet from the ground to the top of a building using a cable which weighs 1/2 pound per foot (Fig. 1). How much work is done?

Solution: This type of problem appeared in section 4.7, but it is a good example of the process of dividing the problem into pieces and analyzing each piece. We can partition the height of the building (Fig. 2). Then the work done to lift the object from the height x_i to the height x_{i+1} is the force applied times the distance moved:

force = (weight of the object) + (weight of the cable)
= (10 pounds) + (0.5 (length of hanging cable))
= $10 + 0.5(40 - x_i)$ pounds = $30 - 0.5x_i$ pounds

distance = $x_i - x_{i-1}$ feet = Δx_i feet

work = (force)(distance) = $\{ 30 - 0.5x_i \} \Delta x_i$ foot-pounds.

Total work $\approx \sum_{i=1}^{n} \{ \text{work on } i^{\text{th}} \text{ slice} \} = \sum_{i=1}^{n} \{ 30 - 0.5x_i \} \Delta x_i$ foot-pounds

$$\longrightarrow \int_0^{40} \{ 30 - 0.5x \} \, dx = (30x - 0.25x^2) \Big|_0^{40} = 800 \text{ foot-pounds.}$$

Fig. 2

Practice 1: How much work is done lifting a 130 pound injured person to the top of a 30 foot cliff using a stretcher weighing 10 pounds and a cable weighing 2 pounds per foot?

In the previous Example and Practice problem the distance moved on each part of the partition was always Δx, and the force was more complicated. In some of the following examples, the Δx is part of the force calculation. Analyze each problem.

Example 2: A cola glass in Fig. 3 has the dimensions given in Table 1. Approximately how much work do you do when you drink a cola glass full of water

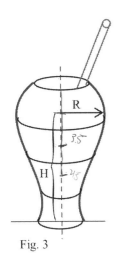

Fig. 3

(weight density = 62.5 pounds/ft^3 = 0.5787 ounces/in^3) by sucking it through a straw to a point 3 inches above the top edge of the glass?

Solution: The Table naturally partitions the water into 1 inch thick "slices" (Fig. 4). The work to move each slice is approximately the weight of the slice times the distance it is moved. We can use the radius at the bottom of each slice to approximate the volume and then the weight of the slice, and a point half way up each slice to calculate the distance the slice is moved.

Height above bottom of the glass (inches)	Inside radius (inches)
4	1.4
3	1.6
2	1.5
1	1.0
0	1.1

Table 1: Inside radius of a cola glass

top slice: force = weight = (volume)(density) ≈ $\pi(1.6 \text{ in})^2$(1 in)(0.5787 oz/in^3) ≈ 4.7 oz.

distance ≈ (distance from middle of slice to lips) = 3.5 in.

work ≈ (force)(distance) = (4.7 oz)(3.5 in) = 16.4 oz–in.

next slice: force = weight = (volume)(density) ≈ $\pi(1.5 \text{ in})^2$(1 in)(0.5787 oz/in^3) ≈ 4.1 oz.

distance ≈ (distance from middle of slice to lips) = 4.5 in

work ≈ (force)(distance) = (4.1 oz)(4.5 in) = 18.4 oz–in.

The work for the last two slices is (1.8 oz)(5.5 in) = 9.9 oz–in and (2.2 oz)(6.5 in) = 14.3 oz–in.

The total work is the sum of the work needed to raise each slice of water:

Total work ≈ (16.4 oz–in) + (18.4 oz–in) + (9.9 oz–in) + (14.3 oz–in) = 59 oz–in.

Practice 2: Approximate the total work needed to raise the water in Example 2 by using the **top** radius of the slice to approximate the weight and the midpoint of each slice to approximate the distance the slice is raised.

If we knew the radius of the glass at every height, then we could improve our approximation by taking thinner slices. In fact, if we knew the radius at every height we could have formed a Riemann sum, taken the limit of the Riemann sum as the thickness of the slices approached 0, and obtained a definite integral. In the next Example we do know the radius of the container at every height.

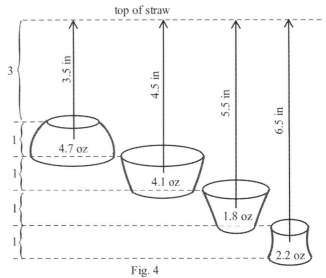

Fig. 4

Example 3: Find the work needed to raise the water in the cone in Fig. 5a to the top of the straw.

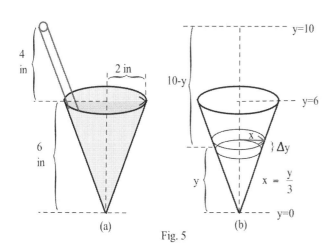

Fig. 5

Solution: We can label the cone (Fig. 5b), and partition the height of the cone to get slices of water. The work done raising the ith slice is the distance the slice is raised times the force needed to move it, the weight of the slice. For any c_i in the subinterval $[y_{i-1}, y_i]$, the slice is raised a distance of approximately $(10 - c_i)$ inches.

Each slice is approximately a right circular cylinder so its volume is $\pi(\text{radius})^2 \Delta y$. At the height y, the radius of the cylinder is $x = y/3$ so at the height c_i the radius is $c_i/3$. Then the force is

$$\text{force} = (\text{volume})(\text{density}) \approx \pi(\text{radius})^2(\Delta y_i)(0.5787 \text{ oz/in}^3) = \pi(c_i/3)^2 (\Delta y_i)(0.5787) \text{ ounces}.$$

The work to raise the ith slice $\approx \pi(c_i/3)^2(\Delta y_i)(0.5787)(10 - c_i)$ ounce–inches, and the total work is approximately $\sum_{i=1}^{n} \pi(c_i/3)^2(\Delta y_i)(0.5787)(10 - c_i)$. As the mesh of the partition approaches 0, the Riemann sum approaches the definite integral:

$$\text{total work} \approx \sum_{i=1}^{n} \pi(c_i/3)^2(\Delta y_i)(0.5787)(10 - c_i) \longrightarrow \int_0^6 \pi(y/3)^2(0.5787)(10 - y)\, dy.$$

$$\text{Total work} = \int_0^6 \pi(y/3)^2(0.5787)(10 - y)\, dy = \frac{0.5787\pi}{9} \int_0^6 10y^2 - y^3 \, dy$$

$$= \frac{0.5787\pi}{9} \left\{ \frac{10}{3}(6)^3 - \frac{1}{4}(6)^4 \right\} = 79.99 \text{ oz–in}$$

In this example, both the force and the distance were variables and both depended on the height of the slice above the bottom of the cone.

Practice 3: How much work is done in drinking just the top 3 inches of the water in Example 3?

Example 4: The trough in Fig. 6 is filled with a liquid weighing 70 pounds per cubic foot. How much work is done pumping the liquid over the wall next to the trough?

Solution: As before, we can partition the height of the trough to get slices of liquid. In order to form a Riemann sum for the total work, we need the weight of a typical slice (Fig. 7) and the distance it is raised.

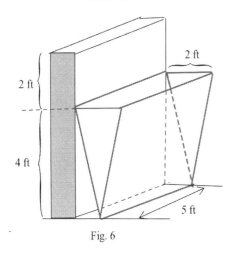

Fig. 6

slice at height y_i: weight = (volume)(density)
= (length)(width)(height)(70 pounds/ft^3)
= $(5 \text{ ft})(\frac{1}{2} y_i \text{ ft})(\Delta y_i \text{ ft})(70 \text{ pounds/ft}^3)$
= $175 y_i \Delta y_i$ pounds.

distance raised = $6 - y_i$ ft

work = $(175 y_i \Delta y_i \text{ pounds})(6 - y_i \text{ feet})$
= $175 y_i (6 - y_i) \Delta y_i$ foot–pounds.

The rest of the solution is straightforward and follows the pattern of the previous problems:

Total work $\approx \sum_{i=1}^{n} \{ \text{work to raise } i^{\text{th}} \text{ slice} \}$

$= \sum_{i=1}^{n} \{ 175 y_i (6 - y_i) \Delta y_i \text{ foot–pounds} \}$

$\longrightarrow \int_0^4 175 y(6 - y) \, dy$

$= 175(3y^2 - \frac{y^3}{3}) \Big|_0^4 = 4666.7$ foot–pounds.

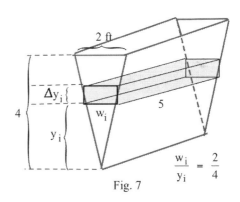

Fig. 7

$\frac{w_i}{y_i} = \frac{2}{4}$

"Raise the liquid" problems can be handled by partitioning the height of the container and then focusing your attention on **one typical slice**. If you can calculate the **weight** of that slice and the **distance** it is raised, the rest of the steps are straightforward: form a Riemann sum, form a definite integral, and evaluate the integral to get the total work.

Work Moving An Object In A Straight Line

Suppose we are pushing a box along a flat surface (Fig. 8a) which is smooth in places and rough in other places so at some places we only have to push lightly and in other places we have to push hard. If f(x) is the amount of force we need to use at location x, and we want to push the box along a straight line from x=a to x=b, then we can partition the interval [a, b] into pieces, $[a, x_1], [x_1, x_2], \ldots, [x_{n-1}, b]$ (Fig. 8b). The work to move the box along the ith piece from x_{i-1} to x_i is approximately

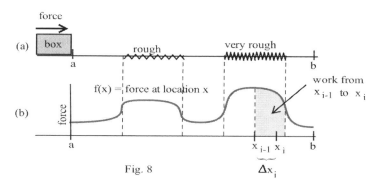

Fig. 8

(force)•(distance) ≈ $f(c_i) \cdot (x_i - x_{i-1}) = f(c_i) \Delta x_i$ for any c_i in the subinterval $[x_{i-1}, x_i]$.

The total work is the sum of the work along each piece, $\sum_{k=1}^{n} f(c_i) \Delta x_i$, a Riemann sum. As we take smaller and smaller subintervals (as the mesh of the partition approaches 0), the Riemann sum approaches the definite integral:

$$\sum_{k=1}^{n} f(c_i) \Delta x_i \quad \longrightarrow \quad \int_a^b f(x)\, dx = \text{total work} .$$

If an object starts at $x = a$ and is moved in a straight line to the location $x = b > a$ by applying a force of $f(x)$ at every location x between a and b,

then the total work done on the object is $\int_a^b f(x)\, dx$.

This has a simple geometric interpretation. If $f(x)$ is the force applied at the position x (Fig. 9), then the work done to move from position $x = a$ to position $x = b$ is the area under the graph of f from $x = a$ to $x = b$.

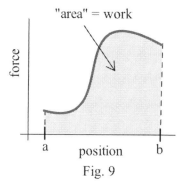

Fig. 9

5.3 More Work Applications

Example 5: Suppose a force of $7x$ pounds is required to stretch a spring (Fig. 10) x inches past its natural length. How much work will be done stretching the spring from its natural length ($x = 0$) to 5 inches beyond its natural length ($x = 5$ inches)?

Solution: Work $= \int_a^b f(x)\, dx = \int_0^5 7x\, dx = \left. \frac{7x^2}{2} \right|_0^5 = 87.5$ inch–pounds.

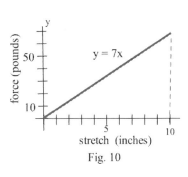

Fig. 10

This can also be don graphically (Fig. 11).

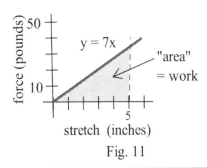

Fig. 11

Practice 4: How much work is done to stretch the spring in the previous example from 5 inches past its natural length to 10 inches past its natural length?

The spring example is an application of a physical principle discovered by the English physicist Robert Hooke (1635–1703), a contemporary of Newton.

Hooke's Law: The force needed to stretch or compress a spring x units from its natural length is proportional to the distance x : force $f(x) = kx$ for some constant k. (Fig. 12)

The "k" in Hooke's Law is called the "spring constant". It varies from spring to spring (depending on the materials and dimensions of the spring and even on the temperature), but is constant for each spring as long as the spring is not overextended or overcompressed. In fact, Hooke's law holds for most solid objects, at least for limited ranges of force:

"Nor is it observable in those bodies only, but in all other springy bodies whatsoever, whether metal, wood, stones, baked earth, hair, horns, silk, bones, sinews, glass and the like." (Hooke).

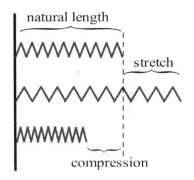

Fig. 12

Most bathroom scales use springs and are based on Hooke's Law for compressing a spring.

Example 6: A spring has a natural length of 43 centimeters, and a weight of 4 grams stretches it to a total length of 75 centimeters. How much work is done stretching the spring from a total length of 63 cm to a total length of 93 cm?

Solution: First we need to use the given information to find the value of k, the spring constant. A force of 4 g produces a stretch of 32 cm (total length of 75 cm minus the rest length of 43 cm). Substituting $x = 32$ cm and $f(x) = 4$ g into Hooke's Law, $f(x) = kx$, we have $4\text{ g} = k\,(32\text{ cm})$ so $k = \frac{4\text{ g}}{32\text{ cm}} = \frac{1\text{ g}}{8\text{ cm}} = .125$ g/cm .

The total length of 63 cm represents a stretch of 20 cm beyond the natural length, and the total length of 93 cm represents a 50 cm stretch. Then the work done is

$$\text{work} = \int_a^b f(x)\,dx = \int_{20}^{50} (.125)\,x\,dx = (.125) \cdot \frac{x^2}{2} \Big|_{20}^{50} = 131.25 \text{ g–cm.}$$

Practice 5: A spring has a natural length of 3 inches, and a force of 2 pounds stretches it to a total length of 8 inches. How much work is done stretching the spring from a total length of 5 inches to a total length of 10 inches?

Lifting a Payload: The problem of finding the work done lifting a payload from the surface of a moon (or any body with no atmosphere) is very similar. Suppose the moon has a radius of R miles and the payload weighs P pounds at the surface of the moon (at a distance of R miles from the center of the moon). When the payload is x miles from the center of the moon (x ≥ R), the gravitational attraction between the moon and the payload is proportional to the reciprocal of the square of the distance x between the centers of the moon and the payload:

required force = $f(x) = \dfrac{R^2 P}{x^2}$ pounds (Fig. 13).

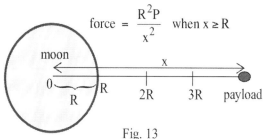

Fig. 13

The total amount of work done raising the payload from the surface (altitude is 0, so $x = R$) to an altitude of R ($x = R+R = 2R$) is

$$\text{work} = \int_a^b f(x)\,dx = \int_R^{2R} \frac{R^2 P}{x^2}\,dx = R^2 P\left(-\frac{1}{x}\right)\Big|_R^{2R} = R^2 P\left(-\frac{1}{2R}\right) - R^2 P\left(-\frac{1}{R}\right) = \frac{RP}{2} \text{ mile–pounds.}$$

Practice 6: How much work will be needed to raise the payload from the altitude R above the surface ($x = 2R$) to an altitude of 2R?

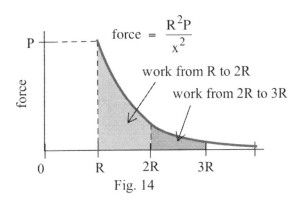

Fig. 14

The appropriate areas under the force graph (Fig. 14) illustrate why the work to raise the payload from $x = R$ to $x = 2R$ was so much larger than the work to raise it from $x = 2R$ to $x = 3R$. In fact, the work to raise the payload from $x = 2R$ to $x = 100R$ is $0.49RP$ which is still less that the $0.5RP$ needed to raise the payload from $x = R$ to $x = 2R$.

The real problem of lifting a payload is much more difficult because the rocket doing the lifting must also lift itself (more work) and the mass of the rocket will keep changing as it burns up fuel. Lifting a payload from a body with an atmosphere is even harder: there is friction from the atmosphere, and the frictional force depends on the density of the atmosphere (which varies with height), the speed of the rocket and the shape of the rocket. Life can get complicated.

PROBLEMS

1. A tank 4 feet long, 3 feet wide and 7 feet tall (Fig. 15) is filled with water which weighs 62.5 pounds per cubic foot. How much work is done pumping the water out over the top of the tank?

2. A tank 4 feet long, 3 feet wide and 6 feet tall is filled with a oil which weighs 60 pounds per cubic foot.
   (a) How much work is done pumping the oil over the top edge of the tank?
   (b) How much work is done pumping the 3 feet of oil of the top edge of the tank?

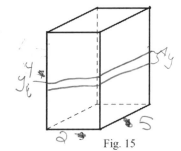
Fig. 15

3. A tank 5 feet long, 2 feet wide and 4 feet tall is filled with a oil which weighs 60 pounds per cubic foot.
   (a) How much work is done pumping all of the oil out over the top edge of the tank?
   (b) How much work is done pumping the top 36 cubic feet of oil out over the top edge of the tank?
   (c) How long does a 1 horsepower pump take to empty the tank over the top edge of the tank?
       (A 1 horsepower pump works at a rate of 33,000 foot–pounds per minute.) A 1/2 horsepower pump? Which pump does more work?

4. A cylindrical aquarium with radius 2 feet and height 5 feet (Fig. 16) is filled with salt water (65 pounds/ft^3).
   (a) How much work is done pumping all of the water over the top edge of the tank?
   (b) How much work is done pumping the water to a point 3 feet above the top edge of the tank?
   (c) How long does a 1 horsepower pump take to empty the tank over the top edge of the tank? A 1/2 horsepower pump? Which pump does more work?

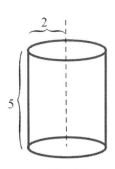
Fig. 16

5.3   More Work Applications   Contemporary Calculus   9

5. A cylindrical barrel with a radius of 1 foot and a height of 6 feet is filled with oil (60 pounds/ft^3).

   (a) How much work is done pumping all of the oil over the top edge of the barrel?

   (b) How much work is done pumping the top 1 foot of oil to a point 2 feet above the top of the barrel?

   (c) How long will it take a 1 horsepower pump to empty the top 3 feet of oil over the top edge of the barrel?

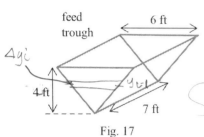

Fig. 17

6. An animal feed trough (Fig. 17) is filled with food weighing 80 pounds/ft^3 . How much work is done lifting all of the food over the top of the trough?

7. How much work is done lifting the top 1 foot of food over the top of the trough in Problem 6?

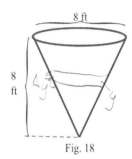

Fig. 18

8. The conical container in Fig. 18 is filled with loose grain which weighs 40 pounds/ft^3 .

   (a) How much work is done lifting all of the grain over the top of the cone?

   (b) lifting the top 2 feet of grain over the top of the cone?

9. If you and a friend share the work equally in emptying the conical container in Problem 8, what depth of grain should the first person leave for the second person to empty?

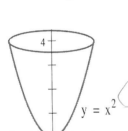

Fig. 19

10. The parabolic container in Fig. 19 is filled with water. (a) How much work is done pumping the water over the top of the tank? (b) to a point 3 feet above the top of the tank?

11. The parabolic container in Fig. 20 is filled with water.

   (a) How much work is done pumping the water over the top of the tank?

   (b) to a point 3 feet above the top of the tank?

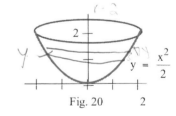

Fig. 20

12. The spherical tank in Fig. 21 is full of water. How much work is done lifting the water to the top of the tank?

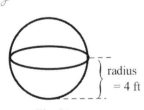

Fig. 21

13. There are two feet of water in the bottom of spherical tank in Fig. 21. How much work is done lifting the water to the top of the tank?

14. The student said, "I've got a shortcut for these tank problems, but it doesn't always work. I figure the weight of the liquid and multiply that by the distance I have to move the middle point in the water. It worked for the first 5 problems and then it didn't."

    (a) Does it really give the right answer for the first 5 problems?

    (b) How are the containers in the first 5 problems different from the others?

    (c) For which of the containers in Fig. 22 will the "shortcut" work?

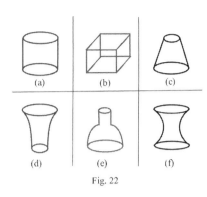
Fig. 22

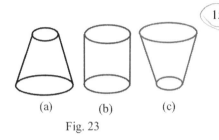
Fig. 23

15. All of the containers in Fig. 23 have the same height and hold the same volume of water. Which requires

    (a) the most work to empty?

    (b) the least work to empty?

    Explain how you reached your conclusions.

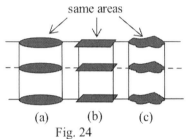
Fig. 24

16. All of the containers in Fig. 24 have the same height and at each height x they all have the same cross sectional area. Which requires

    (a) the most work to empty? (b) the least work to empty?

    Justify your conclusions.

17. Fig. 25 shows the force required to move a box along a rough surface. How much work is done pushing the box (a) from x = 0 to x = 5 feet? (b) from x = 3 to x = 5 feet?

18. How much work is done pushing the box in Fig. 25

    (a) from x = 3 to x = 7 feet? (b) from x = 0 to x = 7 feet?

19. A spring requires a force of $6x$ ounces to stretch it x inches past its natural length. How much work is done stretching the spring

    (a) from its natural length ($x = 0$) to 3 inches beyond its natural length?

    (b) from its natural length to 6 inches beyond its natural length?

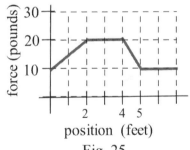
Fig. 25

20. A spring requires a force of $5x$ grams to compress it x cm. How much work is done compressing the spring (a) 7 cm from its natural length? (b) 10 cm from its natural length?

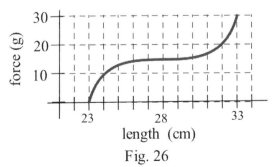
Fig. 26

21. Fig. 26 shows the force needed to stretch a material that does not obey Hooke's Law. Approximately how much work is done stretching it

    (a) from a total length of 23 cm to 33 cm?

    (b) from a total length of 28 cm to 33 cm?

22. Approximately how much work is done stretching the defective spring in the previous problem
    (a) from a total length of 23 cm to 26 cm? (b) from a total length of 30 cm to 35 cm?

23. A 3 pound object stretches a spring 5 inches. How much work is done stretching it 4 more inches?

24. A 2 pound fish stretches a spring 3 inches. How much work is done stretching it 3 more inches?

25. A payload weighs 100 pounds at the surface of an asteroid which has a radius of 300 miles. How much work is done lifting the payload from the asteroid's surface to an altitude of (a) 100 miles? (b) 200 miles? (c) 300 miles?

26. Calculate the amount of work required to lift **you** from the surface of the moon where your weight is approximately 1/6 what it is on earth to an altitude of 200 miles? (The moon's radius is approximately 1,080 miles.)

27. Calculate the amount of work required to lift **you** from the surface of the earth to an altitude (a) of 200 miles? (b) of 400 miles? (c) of 1,000,000 miles? (The earth's radius is approximately 4,000 miles.)

28. An object located at the origin repels **you** with a force inversely proportional to your distance from the object ($f(x) = -\frac{1}{kx}$). When you are 10 feet from the origin the repelling force is 0.1 pound. How much work is done as you move (a) from x = 20 to x = 10? (b) from x = 10 to x = 1? (c) from x = 1 to x = 0.1?

29. An object located at the origin repels you with a force inversely proportional to the square of your distance from the object ($f(x) = -\frac{1}{kx^2}$). When you are 10 feet from the origin the repelling force is 0.1 pound. How much work is done as you move (a) from x = 20 to x = 10? (b) from x = 10 to x = 1? (c) from x = 1 to x = 0.1?

30. The student said "I've got a 'work in a line' shortcut that always seems to work. I figure the average force and then multiply by the total distance. Will it always work?" (a) Will it? Justify your answer. (Hint: What is the formula for 'average force'?) (b) Is it a shortcut?

Work Along A Curved Path

Suppose the location of a moving object is defined parametrically as $x = x(t)$ and $y = y(t)$ for $a \le t \le b$, and the force is a function of t, $f = f(t)$. Then we can represent the work done moving along the path as a definite integral. Partition the time interval $[a, b]$ into short subintervals. For the interval $[t_{i-1}, t_i]$:

force $\approx f(c_i)$ for any c_i in $[t_{i-1}, t_i]$

distance moved $\approx \sqrt{(\Delta x_i)^2 + (\Delta y_i)^2} = \sqrt{(\Delta x_i/\Delta t_i)^2 + (\Delta y_i/\Delta t_i)^2} \; \Delta t_i$

work $\approx f(c_i) \sqrt{(\Delta x_i/\Delta t_i)^2 + (\Delta y_i/\Delta t_i)^2} \; \Delta t_i$

Total work $\approx \sum$ {work along each subinterval} $\approx \sum f(c_i) \sqrt{(\Delta x/\Delta t)^2 + (\Delta y/\Delta t)^2} \; \Delta t$

$$\int_{t=a}^{t=b} f(t) \sqrt{(dx/dt)^2 + (dy/dt)^2} \; dt = \text{total work along the path } (x(t), y(t)).$$

In problems 31 – 35, find the total work along the given parametric path. If necessary, approximate the value of the integral using your calculator. f is in pounds, x and y are in feet, t is in minutes.

31. $f(t) = t$. $x(t) = \cos(t), y(t) = \sin(t), 0 \le t \le 2\pi$.

32. $f(t) = t$. $x(t) = t, y(t) = t^2, 0 \le t \le 1$.

33. $f(t) = t$. $x(t) = t^2, y(t) = t, 0 \le t \le 1$.

34. $f(t) = \sin(t)$. $x(t) = 2t, y(t) = 3t, 0 \le t \le \pi$. (Fig. 27)

35. $f(t) = t$. $x(t) = \cos(t), y(t) = \sin(t), 0 \le t \le 2\pi$ (Fig. 28). (Can you find a geometric way to calculate the shaded area?)

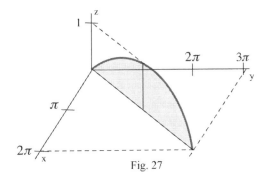

Fig. 27

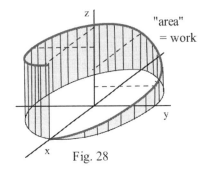

Fig. 28

Section 5.3 PRACTICE Answers

Practice 1: x_i = distance to top of cliff, Δx_i = distance lifted (length of "slice")

$$w_i = (130 + 10 + 2x_i)\, \Delta x_i = (140 + 2x_i)\, \Delta x_i\ .$$

$$\text{Total work} = \int_{x=0}^{x=30} (140 + 2x)\, dx = 140x + x^2 \Big|_0^{30}$$

$$= 4{,}200 + 900 = \mathbf{5{,}100\ \ foot\text{-}pounds}.$$

Practice 2: Total work $\approx \{\pi(1.4)^2(3.5) + \pi(1.6)^2(4.5) + \pi(1.5)^2(5.5) + \pi(1.0)^2(6.5)\} \cdot (0.5787)$

$$\approx \mathbf{67.73\ \ inch\text{-}ounces}.$$

Practice 3: Total work to drink top 3 inches in Example 3 is

$$\int_{y=3}^{y=6} \pi(y/3)^2(0.5787)(10 - y)\, dy = \frac{0.5787\pi}{9}\left\{\frac{10}{3}x^3 - \frac{1}{4}x^4\right\}\Big|_3^6$$

$$= \frac{0.5787\pi}{9}\{396 - 69.75\} \approx \mathbf{65.904\ inch\text{-}ounces}.$$

Practice 4: From Example 5 we know $f(x) = 7x$ so the total work "5 inches past its natural length to 10 inches past its natural length" is

$$\int_a^b f(x)\, dx = \int_5^{10} 7x\, dx = \frac{7x^2}{2}\Big|_5^{10} = \mathbf{262.5\ \ inch\text{--}pounds}.$$

Graphically, this total work is the area of the trapezoidal region bounded by $y = 7x$, the x–axis, and vertical lines at $x = 5$ and $x = 10$.

Practice 5: Be careful to do all calculations based on the amount of **stretch**, not just on the length.

$f(x) = kx$ and we are told that $f(5) = 2$ so $2 = 5k$ and $k = 2/5$. Then the total work is

$$\int_a^b f(x)\, dx = \int_2^7 \frac{2}{5}x\, dx = \frac{x^2}{5}\Big|_2^7 = \frac{45}{5} = \mathbf{9\ \ inch\text{--}pounds}.$$

Practice 6: $\text{work} = \int_a^b f(x)\, dx = \int_{2R}^{3R} \frac{R^2 P}{x^2}\, dx = R^2 P\left(-\frac{1}{x}\right)\Big|_{2R}^{3R}$

$$= R^2 P\left(-\frac{1}{3R}\right) - R^2 P\left(-\frac{1}{2R}\right) = \mathbf{\frac{RP}{6}\ \ mile\text{--}pounds}.$$

5.4 MOMENTS & CENTERS OF MASS

This section develops a method for finding the center of mass of a thin, flat shape — the point at which the shape will balance without tilting (Fig. 1). Centers of mass are important because in many applied situations an object behaves as though its entire mass is located at its center of mass. For example, the work done to pump the water in a tank to a higher point is the same as the work to move the center of mass of the water to the higher point (Fig. 2), a much easier problem, if we know the mass and the center of mass of the water. Also, volumes and surface areas of solids of revolution can be easy to calculate, if we know the center of mass of the region being revolved.

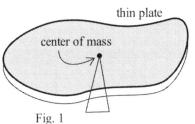

Fig. 1

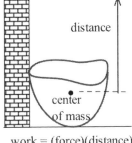

work = (force)(distance)
Fig. 2

POINT MASSES

Before looking for the centers of mass of complicated regions, we consider point masses and systems of point masses, first in one dimension and then in two dimensions.

Point Masses Along A Line

Two people with different masses can position themselves on a seesaw so that the seesaw balances (Fig. 3). The person on the right causes the seesaw to "want to turn" clockwise about the fulcrum, and the person on the left causes it to "want to turn" counterclockwise. If these two "tendencies" are equal, the seesaw will balance. A measure of this tendency to turn about the fulcrum is called the **moment** about the fulcrum of the system, and its magnitude is the mass multiplied by the distance from fulcrum.

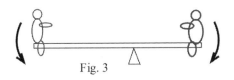

Fig. 3

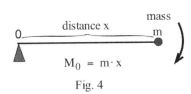

$M_0 = m \cdot x$
Fig. 4

In general, the **moment about the origin, M_0**, produced by a mass m at a location x is $m \cdot x$, the product of the mass and the "signed distance" of the mass from the origin (Fig. 4). For a system of masses $m_1, m_2, \ldots, m_n$ at locations $x_1, x_2, \ldots, x_n$, respectively,

$$M = \text{total mass of the system} = \sum_{i=1}^{n} m_i, \text{ and}$$

$$M_0 = \text{moment about the origin} = x_1 \cdot m_1 + x_2 \cdot m_2 + x_3 \cdot m_3 + \ldots + x_n \cdot m_n = \sum_{i=1}^{n} x_i \cdot m_i.$$

If the moment about the origin is positive then the system tends to rotate clockwise about the origin. If the moment about the origin is negative then the system tends to rotate counterclockwise about the origin. If the moment about the origin is zero, then the system does not tend to rotate in either direction about the origin; it balances on a fulcrum at the origin.

The **moment about the point p, M_p**, produced by a mass m at the location x is the signed distance of x from p times the mass m: $(x-p) \cdot m$. The moment about the point p produced by masses $m_1, m_2, \ldots, m_n$ at locations $x_1, x_2, \ldots, x_n$, respectively, is

$$M_p = \text{moment about the point } p = (x_1-p) \cdot m_1 + (x_2-p) \cdot m_2 + \ldots + (x_n-p) \cdot m_n = \sum_{i=1}^{n} (x_i-p) \cdot m_i .$$

The point at which the system balances is called the **center of mass** of the system and is written $\bar{x}$ (pronounced "x bar"). Since the system balances at $\bar{x}$, the moment about $\bar{x}$ must be zero. Using this fact and properties of summation, we can find a formula for $\bar{x}$.

$$0 = M_{\bar{x}} = \text{moment about } \bar{x} = \sum_{i=1}^{n} (x_i - \bar{x}) \cdot m_i = \sum_{i=1}^{n} (x_i m_i - \bar{x} m_i)$$

$$= \sum_{i=1}^{n} x_i m_i - \sum_{i=1}^{n} \bar{x} m_i = \left(\sum_{i=1}^{n} x_i m_i \right) - \bar{x} \left(\sum_{i=1}^{n} m_i \right) , \text{ since } \bar{x} \text{ is a constant.}$$

So $\left(\sum_{i=1}^{n} x_i m_i \right) = \bar{x} \left(\sum_{i=1}^{n} m_i \right)$, and $\bar{x} = \dfrac{\sum_{i=1}^{n} x_i m_i}{\sum_{i=1}^{n} m_i} = \dfrac{M_0}{M} = \dfrac{\text{moment about the origin}}{\text{total mass}}$.

The **center of mass** of a system of masses $m_1, m_2, \ldots, m_n$ at locations $x_1, x_2, \ldots, x_n$ is the point $\bar{x}$ at which the system balances. The moment of the system about $\bar{x}$ is zero.

$$\bar{x} = (\text{moment about the origin}) / (\text{total mass}) = M_0 / M = \sum_{i=1}^{n} x_i \cdot m_i \Bigg/ \sum_{i=1}^{n} m_i .$$

The single point mass with mass M located at $\bar{x}$, the center of mass of the system, produces the same moment about any point on the line as the whole system. For many purposes, the mass of the entire system can be thought of as "concentrated at $\bar{x}$."

Example 1: Find the center of mass of the first three point–masses given in Table 1.

Solution: $M = 2 + 3 + 1 = 6$. $M_0 = (2)(-3) + (3)(4) + (1)(6) = 12$.
$\bar{x} = M_0 / M = 12/6 = 2$.

The first three point–masses will balance on a fulcrum located at 2.

Practice 1: Find the center of mass of the last three point–masses given in Table 1.

Point Masses In The Plane

i	m_i	x_i
1	2	–3
2	3	4
3	1	6
4	5	–2
5	3	4

Table 1

The ideas of moments and centers of mass extend nicely from one dimension to a system of masses located at points in the plane. For a "knife edge" fulcrum located along the y–axis (Fig. 5), the moment of a mass m at the point (x, y) is the mass times the signed distance of the mass from the y–axis:

(mass)(signed distance from the y–axis) = m·x. This "tendency to rotate about the y–axis" is called the **moment about the y–axis**, written M_y: $M_y = m \cdot x$. Similarly, the mass M at the point (x, y) has a **moment about the x–axis** (Fig. 6): $M_x = m \cdot y$. For a system of masses m_i located at the points (x_i, y_i),

$$M = \text{total mass} = \sum_{i=1}^{n} m_i \qquad M_y = \text{moment about y–axis} = \sum_{i=1}^{n} m_i \cdot x_i$$

$$M_x = \text{moment about x–axis} = \sum_{i=1}^{n} m_i \cdot y_i$$

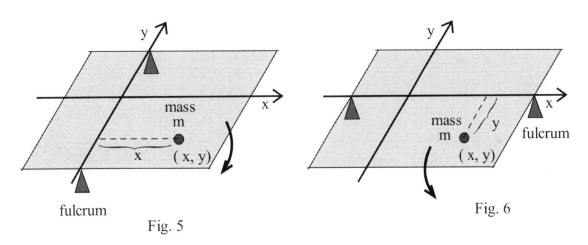

Fig. 5   Fig. 6

The total mass M of the system, located at the center of mass $(\bar{x}, \bar{y})$, has the same moment about any line as the entire system has about that line. For the moment about the y–axis, $M \cdot \bar{x} = M_y$ so $\bar{x} = M_y / M$. Similarly, for the moment about the x–axis, $M \cdot \bar{y} = M_x$ so $\bar{y} = M_x / M$.

	Point Masses:	**Along a Line**	**In the Plane**
masses:		$m_1, m_2, \ldots, m_n$	$m_1, m_2, \ldots, m_n$
locations:		$x_1, x_2, \ldots, x_n$	$(x_1, y_1), (x_2, y_2), \ldots, (x_n, y_n)$
total mass:		$M = \sum_{i=1}^{n} m_i$	$M = \sum_{i=1}^{n} m_i$
moments:		$M_0 = \sum_{i=1}^{n} m_i \cdot x_i$	$M_y = \sum_{i=1}^{n} m_i \cdot x_i \quad , \quad M_x = \sum_{i=1}^{n} m_i \cdot y_i$
center of mass:		$\bar{x} = M_0 / M$	$\bar{x} = M_y / M \quad , \quad \bar{y} = M_x / M$

Example 2: Find the center of mass of the first three point–masses in Table 2.

Solution: $M = 2 + 3 + 1 = 6$. $M_y = (2)(-3) + (3)(4) + (1)(6) = 12$.

$M_x = (2)(4) + (3)(-7) + (1)(-2) = -15$.

$\bar{x} = M_y / M = 12/6 = 2$. $\bar{y} = M_x / M = -15/6 = -2.5$.

The first three point–masses will balance at the point $(2, -2.5)$.

i	m_i	x_i	y_i
1	2	–3	4
2	3	4	–7
3	1	6	–2
4	5	–2	1
5	3	4	–6

Table 2

Practice 2: Find the center of mass of all five point–masses in Table 2.

CENTER OF MASS OF A REGION

When we move from discrete point masses to whole, continuous regions in the plane, we move from finite sums and arithmetic to limits of Riemann sums, definite integrals, and calculus. The following material extends the ideas and calculations from point masses to uniformly thin, flat plates that have a constant density given as mass per area such as "grams/cm^2" (Fig. 7). The center of mass of one of these plates is the point $(\bar{x}, \bar{y})$ at which the plate balances without tilting. It turns out that the center of mass $(\bar{x}, \bar{y})$ of such a plate depends only on the region of the plane covered by the plate and not on its density.

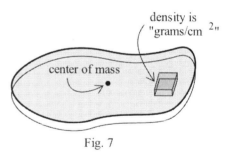

Fig. 7

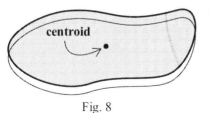

Fig. 8

The point $(\bar{x}, \bar{y})$ is also called the **centroid of the region**. (Fig. 8) In the following discussion, you should notice that each finite sum that appeared in the discussion of point masses has a counterpart for these thin plates in terms of integrals.

Rectangles

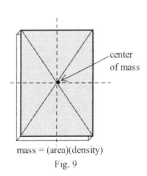

mass = (area)(density)
Fig. 9

The rectangle is the basic shape used to extend the point mass ideas to regions. The total mass of a rectangular plate is the area of the plate multiplied by the density constant: mass M = {area}{density}.

We assume that the center of mass of a thin, rectangular plate is located half way up and half way across the rectangle, at the point where the diagonals of the rectangle cross (Fig. 9). Then the moments of the rectangle can be found by treating the rectangle as a point with mass M located at the center of mass of the rectangle.

Example 3: Find the moments about the x–axis, y–axis, and the line x = 5 of the rectangular plate in Fig. 10.

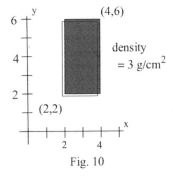

Fig. 10

Solution: The density of the plate is 3 g/cm^2 and the area of the plate is (2 cm)(4 cm) = 8 cm^2 so the total mass is M = (8 cm^2)(3 g/cm^2) = 24 g. The center of mass of the rectangular plate is $(\bar{x}, \bar{y}) = (3, 4)$.

The moment about the x–axis is the mass multiplied by the signed distance of the mass from the x–axis:

M_x = (24 g)·{signed distance of (3,4) from the x–axis} = (24 g)(4 cm) = 96 g–cm.

Similarly,

M_y = (24 g)·{signed distance of (3,4) from the y–axis} = (24 g)(3 cm) = 72 g–cm.

The moment about the line x = 5 is

(24 g)·{signed distance of (3,4) from the line x = 5} = (24 g)(2 cm) = 48 g–cm.

To find the moments and center of mass of a plate made up of several rectangular regions, just treat each of the rectangular pieces as a point mass concentrated at its center of mass. Then the plate is treated as a system of discrete point masses..

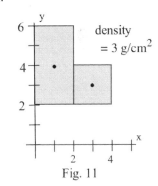

Fig. 11

The plate in Fig. 11 can be divided into two rectangular plates, one with mass 24 g and center of mass (1,4), and one with mass 12 g and center of mass (3,3). The total mass of the pair is M = 36 g, and the moments about the axes are

M_x = (24 g)(4 cm) + (12 g)(3 cm) = 132 g–cm, and

M_y = (24 g)(1 cm) + (12 g)(3 cm) = 60 g–cm.

Then $\bar{x} = M_y/M$ = (60 g–cm)/(36 g) = 5/3 cm and $\bar{y} = M_x/M$ = (132 g–cm)/(36 g) = 11/3 cm so the center of mass of the plate is at $(\bar{x}, \bar{y}) = (5/3, 11/3)$.

Practice 3: Find the center of mass of the region in Fig. 12.

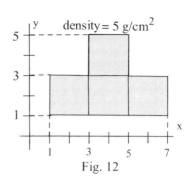

Fig. 12

To find the center of mass of a thin plate, we will "slice" the plate into narrow rectangular plates and treat the collection of rectangular plates as a system of point masses located at the centers of mass of the rectangles. The total mass and moments about the axes for the system of point masses will be Riemann sums. Then, by taking limits as the widths of the rectangles approach 0, we will obtain exact values for the mass and moments as definite integrals

$\bar{x}$ For A Region

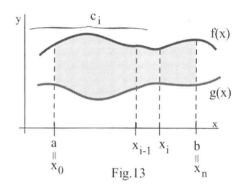

Fig. 13

Suppose $f(x) \geq g(x)$ on [a,b] and R is a plate on the region between the graphs of f and g for $a \leq x \leq b$ (Fig. 13). If the interval [a, b] is partitioned into subintervals $[x_{i-1}, x_i]$ and the point c_i is the midpoint of each subinterval, then the slice between vertical cuts at x_{i-1} and x_i is approximately rectangular and has mass approximately equal to

(area)(density) = (height)(width)(density) $\approx \{ f(c_i) - g(c_i) \}(x_{i-1} - x_i) k = \{ f(c_i) - g(c_i) \} (\Delta x_i) k$.

The mass of the whole plate is approximately

$$M = \sum \{f(c_i) - g(c_i)\} (\Delta x_i) k \longrightarrow k \int_a^b \{ f(x) - g(x) \} \, dx = k \cdot \{ \text{area of the region between f and g} \}.$$

The moment about the y–axis of each rectangular piece is

(distance from the y–axis to the center of mass of the piece)·(mass) = $c_i \cdot \{f(c_i) - g(c_i)\} (\Delta x_i) k$

so $\quad M_y = \sum c_i \{ f(c_i) - g(c_i) \} (\Delta x_i) k \longrightarrow k \int_a^b x \cdot \{ f(x) - g(x) \} \, dx$.

Then the center of mass of the plate is $\bar{x} = \dfrac{M_y}{M} = \dfrac{\int_a^b x \{ f(x) - g(x) \} \, dx}{\int_a^b f(x) - g(x) \, dx}$.

The density constant k is a factor of M_y and of M, so it has no effect on the value of $\bar{x}$. The value of $\bar{x}$ depends only on the shape and location of the region, and $\bar{x}$ is the x coordinate of the centroid of the region between the graphs of $y = f(x)$ and $y = g(x)$ for $a \le x \le b$.

If the bottom curve is the x–axis, then $g(x) = 0$, and the previous results simplify to

$$M = k \int_a^b f(x)\,dx, \quad M_y = k \int_a^b x \cdot f(x)\,dx, \quad \text{and} \quad \bar{x} = \frac{M_y}{M}.$$

Practice 4: Find the x–coordinate of the center of mass of the region between $f(x) = x^2$ and the x–axis for $0 \le x \le 2$. (In this case, $g(x) = 0$.)

$\bar{y}$ For A Region

To find $\bar{y}$, the y–coordinate of the center of mass of a plate R, we need to find M_x, the moment of the plate about the x–axis. When R is partitioned vertically (Fig. 14), the moment of each (very narrow) strip about the x–axis, M_x, is

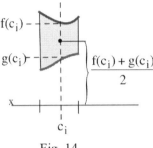

Fig. 14

(the signed distance from the x–axis to the center of mass of the strip)·(mass of strip).

Since each thin strip is approximately rectangular, the y–coordinate of the center of mass of each strip is approximately **half way** up the strip: $\bar{y}_i \approx \{f(c_i)+g(c_i)\}/2$. Then

M_x for the strip = (signed distance from the x–axis to the center of mass of the strip)·(mass of strip)

= (signed distance from x–axis)·(height of strip)·(width of strip)·(density constant)

$$= \frac{f(c_i)+g(c_i)}{2} \cdot (f(c_i) - g(c_i)) \cdot (\Delta x_i) \cdot k.$$

The total moment about the x–axis is $M_x = \sum_{i=1}^{n} \frac{f(c_i)+g(c_i)}{2} \cdot \{f(c_i) - g(c_i)\}(\Delta x_i)\,k$

$$\longrightarrow \quad k \int_a^b \frac{f(x)+g(x)}{2} \cdot \{f(x) - g(x)\}\,dx = \frac{k}{2} \int_a^b \{f^2(x) - g^2(x)\}\,dx.$$

If the lower curve is the x–axis, then $g(x) = 0$ and the formulas simplify to

$$M = k \int_a^b f(x)\,dx, \quad M_x = k \int_a^b \frac{f(x)}{2} \cdot f(x)\,dx = \frac{k}{2} \int_a^b f^2(x)\,dx, \quad \text{and} \quad \bar{y} = \frac{M_x}{M}.$$

Example 4: Find the y–coordinate of the centroid of the region between the x–axis and the top half of a circle of radius r (Fig. 15).

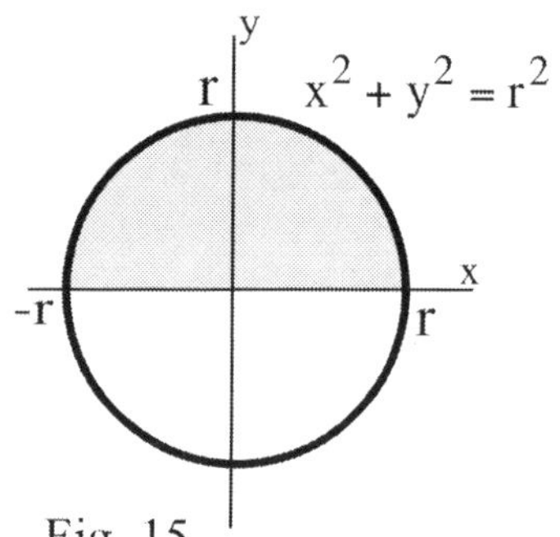

Solution: The equation of the circle is $x^2 + y^2 = r^2$ so $f(x) = y = \sqrt{r^2 - x^2}$.

$$M = k \int_a^b f(x)\, dx = k \int_{-r}^r \sqrt{r^2 - x^2}\, dx$$

$$= k \tfrac{1}{2}(\text{area of the circle of radius r}) = k \tfrac{1}{2}(\pi r^2) = \tfrac{1}{2} k\pi r^2 .$$

Fig. 15

The region is symmetric about the y–axis so $\bar{x} = 0$. The moment of the region about the x–axis is

$$M_x = \tfrac{k}{2} \int_a^b f^2(x)\, dx = \tfrac{k}{2} \int_{-r}^r (\sqrt{r^2 - x^2})^2\, dx = \tfrac{k}{2} \int_{-r}^r r^2 - x^2\, dx = \tfrac{k}{2} \left\{ r^2 x - \tfrac{x^3}{3} \right\} \Big|_{-r}^r = \tfrac{2}{3} k r^3 .$$

Finally, $\bar{y} = \dfrac{M_x}{M} = \dfrac{\tfrac{2}{3} k r^3}{\tfrac{1}{2} k \pi r^2} = \dfrac{4}{3} \dfrac{r}{\pi} \approx 0.4244\, r$.

Practice 5: Show that the centroid of a triangular region with vertices $(0,0), (0,h)$ and $(b,0)$ is $(\bar{x}, \bar{y}) = (b/3, h/3)$.

5.4 Moments & Centers of Mass   Contemporary Calculus

The integral formulas for moments are given below in a form useful for actually calculating moments of regions between the graphs of two functions, but it is also important that you understand the process used to derive the formulas.

	Point masses in the plane	**Region between f and g for $a \leq x \leq b$ ($f \geq g$)**
total mass:	$M = \sum_{i=1}^{n} m_i$	$M = \int_a^b \{area\}\{density\} = k \cdot \int_a^b f(x) - g(x) \, dx$
moments:	$M_y = \sum_{i=1}^{n} x_i \cdot m_i$	$M_y = \int_a^b \{\text{dist. of c.m. of slice to y-axis}\}\{mass\}$ $= k \cdot \int_a^b x \cdot \{f(x) - g(x)\} \, dx$
	$M_x = \sum_{i=1}^{n} y_i \cdot m_i$	$M_x = \int_a^b \{\text{dist. of c.m. of slice to x-axis}\}\{mass\}$ $= \frac{k}{2} \cdot \int_a^b \{f^2(x) - g^2(x)\} \, dx$
center of mass:	$\bar{x} = \frac{M_y}{M}$, $\bar{y} = \frac{M_x}{M}$	$\bar{x} = \frac{M_y}{M}$, $\bar{y} = \frac{M_x}{M}$

Example 5: Find the centroid of the region bounded between the graphs of $y = x$ and $y = x^2$ for $0 \leq x \leq 1$.

Solution: $M = k \cdot \int_0^1 (x - x^2) \, dx = \frac{k}{6}$, $M_y = k \cdot \int_0^1 x(x - x^2) \, dx = \frac{k}{12}$ and

$M_x = \frac{k}{2} \cdot \int_0^1 (x)^2 - (x^2)^2 \, dx = \frac{k}{15}$. Then $\bar{x} = M_y / M = 1/2$ and $\bar{y} = \frac{M_x}{M} = 2/5$.

Symmetry

Symmetry is a very powerful geometric concept that can simplify many mathematical and physical problems, including the task of finding centroids of regions. For some regions, we can use symmetry alone to determine the centroid.

Geometrically, a region R is **symmetric about a line L** if, when R is folded along the line L, each point of R on one side of the fold matches up with one point of R on the other side of the fold (Fig. 16).

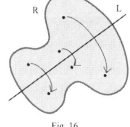

Fig. 16

Example 6: Sketch two lines of symmetry for each region in Fig. 17.

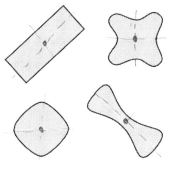

Fig. 17

Solution: The lines of symmetry are shown in Fig. 18. Every line through the center of the circular region is a line of symmetry.

A very useful fact about symmetric regions is that the centroid $(\bar{x}, \bar{y})$ of a symmetric region must lie on **every** line of symmetry of the region. If a region has two different lines of symmetry, then the centroid must lie on each of them, so the centroid is located at the point where the lines of symmetry intersect.

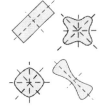

Fig. 18

Practice 6: Locate the centroid of each region in Fig. 17.

WRAP UP

The purpose of this section is to illustrate again the process of going from an applied problem, to an approximate solution as a Riemann sum, to an exact solution represented as a definite integral. The emphasis is on using calculus to solve an applied problem. However, centroids and centers of mass can themselves be used to solve other applied problems. If we know the center of mass of a region, then some work, volume of revolution, and surface area of revolution problems become simple. These applications of centroids and centers of mass are discussed very briefly in the "optional" section of the problems, and a physical method for determining the centroid of a region is described. Centers of mass of regions with variable density are discussed in a later chapter.

PROBLEMS

1. (a) Find the total mass and the center of mass for a system consisting of the 3 masses in Table 3.

   (b) Where should you locate a new object with mass 8 so the new system has its center of mass at $x = 5$?

   (c) How much mass should be located at $x = 10$ so the original system plus the new mass at $x = 10$ has is center of mass at $x = 6$?

m	x
2	4
5	2
5	6

   Table 3

2. (a) Find the total mass and the center of mass for a system consisting of the 4 masses in Table 4.

   (b) Where should you locate a new object with mass 10 so the new system has its center of mass at $x = 6$?

   (c) How much mass should be located at $x = 14$ so the original system plus the new mass at $x = 14$ has is center of mass at $x = 6$?

m	x
5	1
3	7
2	5
6	2

   Table 4

3. (a) Find the total mass and the center of mass for a system consisting of the 3 masses in Table 5.

   (b) Where should you locate a new object with mass 10 so the new system has its center of mass at $(5, 2)$?

m	x	y
2	4	3
5	2	4
5	6	2

   Table 5

4. (a) Find the total mass and the center of mass for a system consisting of the 4 masses in Table 6.

   (b) Where should you locate a new object with mass 12 so the new system has its center of mass at $(3, 5)$?

m	x	y
1	5	4
2	2	7
3	1	0
5	3	8

   Table 6

In problems 5 – 10, divide the flat plate in each Figure into rectangles and semicircles, calculate the mass, moments and centers of mass of each piece, and use those results to find the center of mass of the plate. Assume that the density of the plate is 1 g/cm^2. Plot the location of the center of mass for each shape. **(See Example 4 for the centroid of a semicircular region.)**

5. Fig. 19.   6. Fig. 20.   7. Fig. 21.   8. Fig. 22.   9. Fig. 23.   10. Fig. 24.

Fig. 19

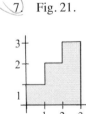
Fig. 20

Fig. 21

Fig. 22

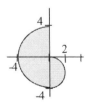
Fig. 23

Fig. 24

In problems 11 – 26, sketch the region bounded between the given functions on the interval and calculate the centroid of each region (use Simpson's rule with n = 20 if necessary). Plot the location of the centroid on your sketch of the region.

11. $y = x$ and the x–axis for $0 \leq x \leq 3$.

12. $y = x^2$ and the x–axis for $-2 \leq x \leq 2$.

13. $y = x^2$ and the line $y = 4$ for $-2 \leq x \leq 2$.

14. $y = \sin(x)$ and the x–axis for $0 \leq x \leq \pi$.

15. $y = 4 - x^2$ and the x–axis for $-2 \leq x \leq 2$.

16. $y = x^2$ and $y = x$ for $0 \leq x \leq 1$.

17. $y = 9 - x$ and $y = 3$ for $0 \leq x \leq 3$.

18. $y = \sqrt{1 - x^2}$ and the x–axis for $0 \leq x \leq 1$.

19. $y = \sqrt{x}$ and the x–axis for $0 \leq x \leq 9$.

20. $y = \ln(x)$ and the x–axis for $1 \leq x \leq e$.

21. $y = e^x$ and the line $y = e$ for $0 \leq x \leq 1$.

22. $y = x^2$ and the line $y = 2x$ for $0 \leq x \leq 2$.

23. An empty one foot square tin box (Fig. 25) weighs 10 pounds and its center of mass is 6 inches above the bottom of the box. When the box is **full** with 60 pounds of liquid, the center of mass of the box–liquid system is again 6 inches above the bottom. (a) Write the height of the center of mass of the box–liquid system as a function of the x, the height of liquid in the box.
(b) What height of liquid in the bottom of the box results in the box–liquid system having the lowest center of mass (and the greatest stability)?

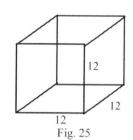
Fig. 25

24. The empty can in Fig. 26 weighs 1 ounce when empty and 13 ounces when full. Write the height of the center of mass of the can–liquid system as a function of the height of the liquid in the can.

25. The empty glass in Fig. 27 weighs 4 ounces when empty and 20 ounces when full. Write the height of the center of mass of the glass–liquid system as a function of the height of the liquid in the glass.

26. Give a **practical** set of directions someone could actually use to find the **height** of the center of gravity of their body with their arms at their sides (Fig. 28). How will the height of the center of gravity change if they lift their arms? (In a uniform gravitational field such as at the surface of the earth, the center of mass and center of gravity are at the same point.)

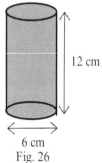
Fig. 26

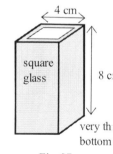
Fig. 27

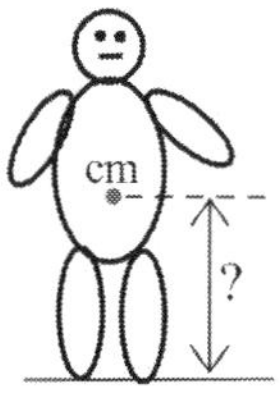

27. Try the following experiment. Stand straight with your back and heels against a wall. Slowly raise one leg, keeping it straight, in front of you. What happened? Why?

Fig. 28

28. Why can't two dancers stand in the position shown in Fig. 29?

29. Determine the centroid of your state.

30. (a) Sketch regions with exactly 2 lines of symmetry, exactly 3 lines of symmetry, and exactly 4 lines of symmetry.

    (b) If a shape has exactly two lines of symmetry, the lines can meet at right angles. Do they have to meet at right angles?

Fig. 29

Work

In a uniform gravitational field, the center of gravity of an object is at the same point as the center of mass of the object, and the work done to lift an object is the weight of the object multiplied by the distance that the center of gravity of the object is raised:

work = (total weight of object)·(distance the center of gravity of the object is raised).

In the high jump, this explains the effectiveness of the "Fosbury Flop", a technique in which the jumper assumes an inverted U position while going over the bar (Fig. 30). In this way, the jumper's body goes over the bar while the jumper's center of gravity goes under it, so a given amount of upward thrust produces a higher bar cleared.

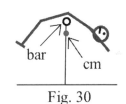

Fig. 30

If the center of gravity of an object is known, some work problems become easy.

31. A rectangular box is filled to a depth of 4 feet with 300 pounds of liquid. How much work is done pumping the liquid to a point 10 feet high? (How high is the center of gravity of the liquid, and how much must it be raised?)

32. A cylinder is filled to a depth of 2 feet with 40 pounds of liquid. How much work is done pumping the liquid to a point 7 feet high? (How high is the center of gravity of the liquid, and how much must it be raised?)

33. A sphere of radius 1 foot is filled with 250 pounds of liquid. How much work is done pumping the liquid to a point 3 feet above the top of the sphere? (Draw a picture.)

34. A sphere of radius 2 foot is filled with 2000 pounds of liquid. How much work is done pumping the liquid to a point 5 feet above the top of the sphere?

If the amount of work is already known, it can be used to find the height of the center of gravity.

Theorems of Pappus

When location of the center of mass of an object is known, the theorems of Pappus make some volume and surface area calculations very easy.

Volume of Revolution:

If a plane region with **area** A and centroid $(\bar{x}, \bar{y})$ is revolved around a line in the plane which does not go through the region (touching the boundary is alright),

then the **volume** swept out by one revolution is the area of the region times the distance traveled by the centroid (Fig. 31):

Volume about line L = $A \cdot 2\pi \cdot \{$ distance of $(\bar{x}, \bar{y})$ from the line L $\}$.

Volume about x–axis = $A \cdot 2\pi \cdot \bar{y}$.   Volume about y–axis = $A \cdot 2\pi \cdot \bar{x}$.

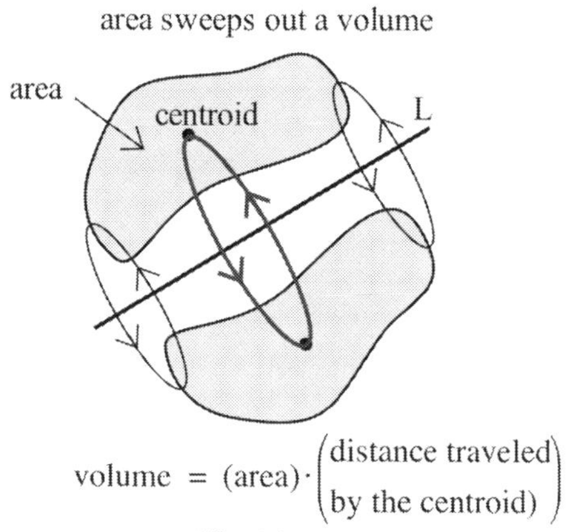

$$\text{volume} = (\text{area}) \cdot \begin{pmatrix} \text{distance traveled} \\ \text{by the centroid} \end{pmatrix}$$

Fig. 31

Surface Area of Revolution

If a plane region with **perimeter** P and centroid of the edge $(\bar{x}, \bar{y})$ is revolved around a line in the plane which does not go through the region (touching the boundary is alright),

then the **surface area** swept out by one revolution is the perimeter of the region times the distance traveled by the centroid (Fig. 32):

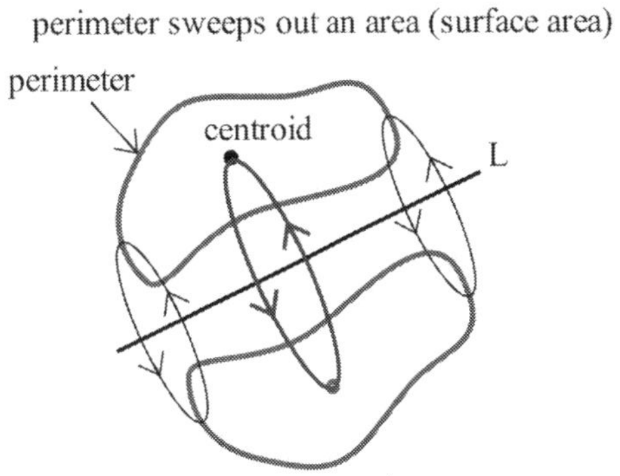

surface area = (perimeter)·(distance traveled by the centroid)

Fig. 32

Surface area about line L = P·2π·{ distance of $(\bar{x}, \bar{y})$ from the line L }.

Surface area about x–axis = P· 2π· $\bar{y}$.   Surface area about y–axis = P· 2π· $\bar{x}$.

35. The center of a square region with 2 foot sides is at the point (3,4). Use the Theorems of Pappus to find the volume and surface area swept out when the square is rotated (a) about the x–axis, (b) about the y–axis, and (c) about the horizontal line y = 6.

36. The lower left corner of a rectangular region with an 8 inch base and a 4 inch height is at the point (3,5). Use the Theorems of Pappus to find the volume and surface area swept out when the rectangle is rotated (a) about the x–axis, (b) about the y–axis, and (c) about the line y = x + 5.

37. The center of a circle with radius 2 feet is at the point (3,5). Use the Theorems of Pappus to find the volume and surface area swept out when the circular region is rotated (a) about the x–axis, (b) about the y–axis, and (c) about the vertical line x = 6.

38. The center of a circle (Fig. 33) with radius r is at the point (0, R). Use the Theorems of Pappus to find the volume and surface area swept out when the circular region is rotated about the x–axis.

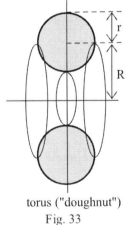

torus ("doughnut")
Fig. 33

39. Find the volumes and surface areas swept out when the rectangles in Fig. 34 are rotated about the line L. (Measurements are in feet.)

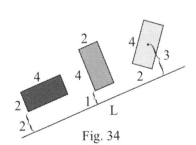

Fig. 34

Physically Approximating Centoids of Regions

The centroid of a region can be approximated experimentally, even if the region, such as a state or country, is not described by a formula.

Cut the shape out of a piece of some uniformly thick material such as paper and pin an edge to a wall. The shape will pivot about the pin until its center of mass is directly below the pin (Fig. 35) so the center of mass of the shape must lie directly below the pin, on the line connecting the pin with the center of mass of the earth. Repeat the process using a different point near the edge of the shape and a different line can be found. The center of mass also lies on the new line, so we can conclude that the centroid of the shape is located where the two lines intersect, the only point located on both lines (Fig. 36). It is a good idea to pick a third point near the edge and plot a third line. This line should also pass through the point of intersection of the other two lines.

Fig. 35

The "population center" of a region can be physically approximated by attaching weights proportional to the populations of the different areas and then repeating the "pin" process with this weighted model. The point on the new model where the lines intersect is the approximate "population center" of the region.

Fig. 36

Section 5.4   **PRACTICE Answers**

Practice 1: $M = 1 + 5 + 3 = 9$. $M_0 = (1)(6) + (5)(-2) + (3)(4) = 8$.

$\bar{x} = M_0 / M = 8/9$

The last three point–masses will balance on a fulcrum located at $x = 8/9$.

i	m_i	x_i
1	2	–3
2	3	4
3	1	6
4	5	–2
5	3	4

Table 1

Practice 2: $M = 2 + 3 + 1 + 5 + 3 = 14$.

$M_y = (2)(-3) + (3)(4) + (1)(6) + (5)(-2) + (3)(4) = 14$.

$M_x = (2)(4) + (3)(-7) + (1)(-2) + (5)(1) + (3)(-6) = -28$.

$\bar{x} = M_y / M = 14/14 = 1$. $\bar{y} = M_x / M = -28/14 = -2$.

The five point–masses balance at the point $(1, -2)$.

i	m_i	x_i	y_i
1	2	–3	4
2	3	4	–7
3	1	6	–2
4	5	–2	1
5	3	4	–6

Table 2

Practice 3: There are several ways to break the region in Fig. 12 into "easy" pieces — one way is to consider the four 2–by–2 cm squares.

The cm of each square is located at the center of the square (at (2,2), (4,2), (6,2), and (4,4)), and each square has mass $(4 \text{ cm}^2)(5 \text{ g/cm}^2) = 20$ g.

$M = 4(20 \text{ g}) = 80$ g.

$M_y = 2(20) + 4(20) + 6(20) + 4(20) = 320$ g·cm

$M_x = 2(20) + 2(20) + 2(20) + 4(20) = 200$ g·cm

Then $\bar{x} = M_y / M = \dfrac{320 \text{ g·cm}}{80 \text{ g}} = $ **4 cm** and

$\bar{y} = M_x / M = \dfrac{200 \text{ g·cm}}{80 \text{ g}} = $ **2.5 cm** .

The center of mass is (4, 2.5) .

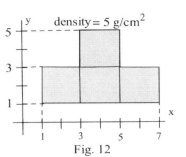

Fig. 12

Practice 4: $M = k \int_a^b f(x) \, dx = k \int_0^2 x^2 \, dx = \dfrac{8}{3} k$, $M_y = k \int_a^b x \cdot f(x) \, dx = k \int_0^2 x \cdot x^2 \, dx = 4k$.

Then $\bar{x} = M_y / M = \dfrac{4k}{\frac{8}{3}k} = 1.5$.

Practice 5: The triangular region is shown in Fig. 36 : $f(x) = h - \dfrac{h}{b} x$ for $0 \le x \le b$.

$M = k \int_a^b f(x) \, dx = k \int_0^b (h - \dfrac{h}{b} x) \, dx = k \{ hx - \dfrac{h}{b} \dfrac{1}{2} x^2 \} \Big|_0^b$

$= k\{ hb - \dfrac{h}{b} \dfrac{b^2}{2} \} = \dfrac{k}{2} hb$.

$M_y = k \int_a^b x \cdot f(x) \, dx = k \int_0^b x \cdot (h - \dfrac{h}{b} x) \, dx = k \{ h \dfrac{x^2}{2} - \dfrac{h}{b} \dfrac{x^3}{3} \} \Big|_0^b$

$= k\{ h \dfrac{b^2}{2} - \dfrac{h}{b} \dfrac{b^3}{3} \} = \dfrac{k}{6} hb^2$.

$M_x = \dfrac{k}{2} \int_a^b f^2(x) \, dx = \dfrac{k}{2} \int_0^b (h - \dfrac{h}{b} x)^2 \, dx = \dfrac{kh^2}{2} \{ x - \dfrac{2}{b} \dfrac{x^2}{2} + \dfrac{1}{b^2} \dfrac{x^3}{3} \} \Big|_0^b$

$= \dfrac{kh^2}{2} \{ b - b + \dfrac{b}{3} \} = \dfrac{k}{6} h^2 b^2$.

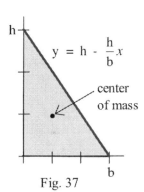

Fig. 37

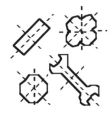

Fig. 18

Finally, $\bar{x} = \dfrac{M_y}{M} = \dfrac{\frac{k}{6} hb^2}{\frac{k}{2} hb} = \dfrac{b}{3}$ and $\bar{y} = \dfrac{M_x}{M} = \dfrac{\frac{k}{6} h^2 b^2}{\frac{k}{2} hb} = \dfrac{h}{3}$ so **cm = (b/3, h/3)** .

Practice 6: The centroid of each region in Fig. 18 is located at the point where the lines of symmetry intersect.

5.5 ADDITIONAL APPLICATIONS

This section **introduces** three additional applications of integrals and once again illustrates the process of going from a problem to a Riemann sum and on to a definite integral. A fourth application is included which does not follow this process. It uses the idea of "area" to try to model an election and to qualitatively understand why certain election outcomes occur.

The main point of this section is to show the power of definite integrals to solve a wide variety of applied problems. Each of these new applications is treated more briefly than those in the previous sections.

1. Liquid Pressures and Forces

The hydrostatic pressure on an immersed object is the density of the fluid times the depth of the object: **pressure = (density)(depth)**. The total hydrostatic force against an immersed object is the sum of the hydrostatic forces against each part of the object. If our entire object is at the same depth, we can determine the total hydrostatic force simply by multiplying the density times the depth times the area. If the unit of density is "pounds per cubic foot" and the depth is given in "feet," then the unit of pressure is "pounds per square foot," a measure of **force per area**. If a pressure, with the units "pounds per square foot," is multiplied by an area with the units "square feet," the result is a force, "pounds."

Example 1: Find the total hydrostatic force against the bottom of the aquarium shown in Fig. 1.

Solution: The density of water is 62.5 pounds/ft^3, so

total hydrostatic force = (density)·(depth)·(area)

= (62.5 pounds/ft^3)·(3 feet)·(2 square feet) = 375 pounds.

Finding the total hydrostatic force against the front of the aquarium is a very different problem because different parts of the front are at different depths and are subject to different pressures. To find the force against the front of the aquarium, we can partition it into thin horizontal slices (Fig. 2) and focus on one of them. Since the slice is very thin, every part of the ith slice is at almost the same depth so every part of the slice is subject to almost the same pressure. We can approximate the total hydrostatic force against the slice at the depth x_i as

(density)·(depth)·(area) = (62.5 pounds/ft^3)(x_i feet)(2 feet)(Δx_i feet)

= $125 x_i \cdot \Delta x_i$ pounds.

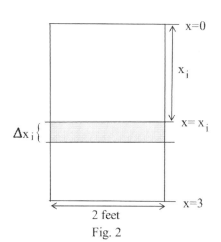
Fig. 2

The total hydrostatic force against the front is the sum of the forces against each slice,

$$\text{total hydrostatic force} \approx \sum 125 x_i \, \Delta x_i \, , \text{ a Riemann sum.}$$

The limit of the Riemann sum as the slices get thinner ($\Delta x \to 0$) is a definite integral:

$$\text{total hydrostatic force} \approx \sum 125 x_i \, \Delta x_i \longrightarrow \int_{x=0}^{3} 125 x \, dx = 62.5 x^2 \Big|_{x=0}^{3} = 562.5 \text{ pounds.}$$

Practice 1: Find the total hydrostatic force against one side of the aquarium.

Hydrostatic Force

If   the width of a slice of a horizontal object at depth x is $w(x)$ (Fig. 3)

then the total hydrostatic force against the object between

        depths a and b is

$$\text{total hydrostatic force} \approx \sum (\text{density}) \cdot (\text{depth}) \cdot w(x_i) \, \Delta x_i$$

$$\longrightarrow \int_{x=a}^{b} (\text{density}) \cdot x \cdot w(x) \, dx \, .$$

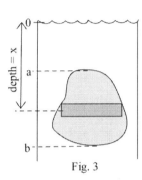

Fig. 3

Example 2: Find the total hydrostatic force against windows A and B in Fig. 4.

Solution: Window A: Using similar triangles,

$$\frac{w}{6-x} = \frac{3}{2} \quad \text{so} \quad w(x) = \frac{3}{2}(6-x) \, .$$

Then total hydrostatic force

$$= \int_{x=4}^{6} (\text{density}) \cdot x \cdot w(x) \, dx$$

$$= \int_{4}^{6} (60) \cdot x \cdot \frac{3}{2}(6-x) \, dx = 90 \left(3x^2 - \frac{x^3}{3} \right) \Big|_{4}^{6} = 840 \text{ pounds.}$$

Window B: The equation of the circle is $(5-x)^2 + (w/2)^2 = 1$ so $w(x) = 2\sqrt{1 - (5-x)^2}$. Then

$$\text{total hydrostatic force} = \int_{4}^{6} (60) \cdot x \cdot 2\sqrt{1 - (5-x)^2} \, dx \approx 938.1 \text{ pounds (using a calculator).}$$

Practice 2: Find the total hydrostatic force against windows C and D in Fig. 4.

Because the total force at even moderate depths is so large, the underwater windows at aquariums are made of thick glass or plastic and strongly secured to their frames. Similarly, the bottom of a dam is much thicker than the top in order to withstand the greater force against the bottom.

2. Kinetic Energy of a Rotating Object

> The **Kinetic Energy** (the energy of motion) of an object is defined to be half the mass of the object multiplied by the square of the velocity of the object: $\mathbf{KE = \frac{1}{2} m \cdot v^2}$.

The larger an object is or the faster it is moving, the greater its kinetic energy. If every part of the object has the same velocity, then it is easy to compute its kinetic energy. Sometimes, however, different parts of the object move with different velocities. For example, if an ice skater is spinning with an angular velocity of 2 revolutions per second, then her arms travel further in one second (have a greater linear velocity) when they are extended than when they are drawn in close to her body (Fig. 5). So the ice skater, spinning at 2 revolutions per second, has greater kinetic energy when her arms are extended. Similarly, the tip of a rotating propeller or of a swinging baseball bat has a greater has a greater linear velocity than other parts of the propeller or bat. If the units of mass are "grams" and the units of velocity are "centimeters per second," then the units of kinetic energy $KE = \frac{1}{2} m \cdot v^2$ are "ergs."

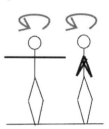

Fig. 5

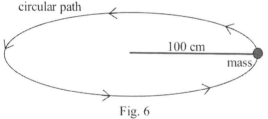

Fig. 6

Example 3: A point–mass of 1 gram at the end of a (massless) 100 centimeter long string is rotated at a rate of 2 revolutions per second (Fig. 6). (a) Find the kinetic energy of the point–mass. (b) Find its kinetic energy if the string is 200 centimeters long.

Solution: (a) In one second, the mass travels twice around a circle with radius 100 centimeters so it travels $2 \cdot (2\pi \cdot \text{radius}) = 400\pi$ centimeters. The velocity is $v = 400\pi$ cm/s, and

$$KE = \frac{1}{2} m \cdot v^2 = \frac{1}{2}(1 \text{ g}) \cdot (400\pi \text{ cm/s})^2 = 80,000 \, \pi^2 \text{ ergs}.$$

(b) If the string is 200 centimeters long, then the velocity is $2 \cdot (2\pi \cdot \text{radius})/\text{second} = 800\pi$ cm/s,

and $KE = \frac{1}{2} m \cdot v^2 = \frac{1}{2}(1 \text{ g}) \cdot (800\pi \text{ cm/sec})^2 = 320,000 \, \pi^2$ ergs.

When the length of the string doubles, the velocity is twice as large and the kinetic energy is 4 times as large.

Practice 3: A 1 gram point–mass at the end of a 2 meter (massless) string is rotated at a rate of 4 revolutions per second. Find the kinetic energy of the point mass.

If different parts of a rotating object are different distances from the axis of rotation, then those parts have different linear velocities, and it is more difficult to calculate the total kinetic energy of the object. By now the method should seem very familiar: partition the object into small pieces, approximate the kinetic energy of each piece, and add the kinetic energies of the small pieces (a Riemann sum) to approximate the total kinetic energy of the object. The limit of the Riemann sum as the pieces get smaller is a definite integral.

Example 4: The density of a narrow bar (Fig. 7) is 5 grams per meter of length. Find the kinetic energy of the 3 meter long bar when it is rotated at a rate of 2 revolutions per second.

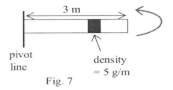

Fig. 7

Solution: If the length of the bar is partitioned (Fig. 8), the mass of the ith piece is

$$m_i \approx (\text{length}) \cdot (\text{density}) = (\Delta x_i \text{ meters})(5 \text{ grams/meter}) = 5 \Delta x_i \text{ grams}.$$

In one second the ith piece will make two revolutions and will travel approximately

$$2 \cdot (2\pi \cdot \text{radius}) = 400\pi \cdot x_i \text{ centimeters so } v_i \approx 400\pi x_i \text{ cm/sec}.$$

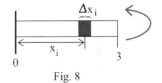

Fig. 8

The kinetic energy of the ith piece is

$$ke_i = \frac{1}{2} m_i \cdot v_i^2 = \frac{1}{2}(5 \Delta x_i \text{ grams}) \cdot (400\pi x_i \text{ cm/sec})^2$$

$$= 400{,}000 \pi^2 (x_i)^2 \Delta x_i \text{ ergs},$$

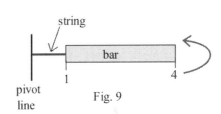

Fig. 9

and the total kinetic energy of the rotating bar is

$$KE = \sum ke_i = \sum 400{,}000\pi^2 (x_i)^2 \Delta x_i \longrightarrow \int_0^3 400{,}000\pi^2 \cdot x^2 \, dx$$

$$= 400{,}000 \, \pi^2 \frac{x^3}{3} \Big|_0^3 = 3{,}600{,}000 \, \pi^2 \text{ ergs}.$$

Practice 4: Find the kinetic energy of the bar in the previous example if it is rotated at 2 revolutions per second at the end of a 100 centimeter (massless) string (Fig. 9).

Example 5: Find the kinetic energy of the object in Fig. 10 when it rotates at 2 revolutions per second.

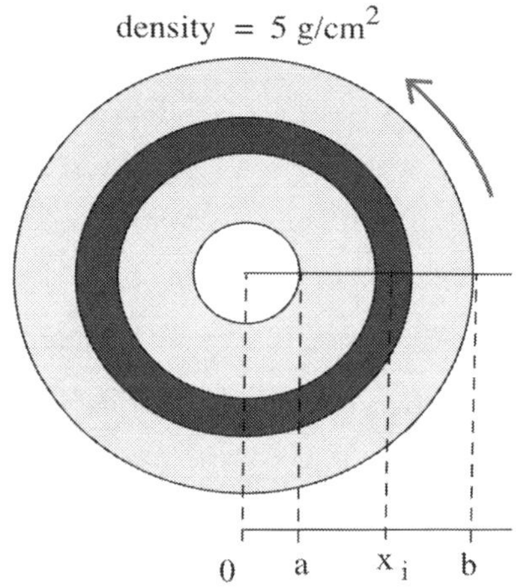

Solution: We can partition along one radial line, and form circular "slices." Then the "slice" between x_i and $x_i + \Delta x$ is a thin circular band with mass m_i = (volume)·(density) $\approx 2\pi x_i \Delta x_i \cdot h \cdot d$ Newtons. Each part of the "slice" is approximately x_i centimeters from the axis of rotation so each part has approximately the same velocity:
$v_i \approx$ (2 rev/sec)·($2\pi x_i$ centimeters/rev) = $4\pi \cdot x_i$ centimeters/sec.

The kinetic energy of the ith piece is

$$ke_i \approx = \frac{1}{2} m_i \cdot v_i^2 = \frac{1}{2}(2\pi \cdot x_i \Delta x_i \cdot h \cdot d) \cdot (4\pi \cdot x_i)^2 = 16\pi^3 \cdot h \cdot d \cdot (x_i)^3 \Delta x_i ,$$

so $KE = \sum ke_i = \sum 16\pi^3 \cdot h \cdot d \cdot (x_i)^3 \Delta x$

$$\longrightarrow \int_a^b 16\pi^3 \cdot h \cdot d \cdot x^3 \, dx = 16\pi^3 h d \int_a^b x^3 \, dx = 4\pi^3 h d \cdot (b^4 - a^4) .$$

Since b is raised to the 4th power, a small increase in the value of b leads to a large increase in the kinetic energy.

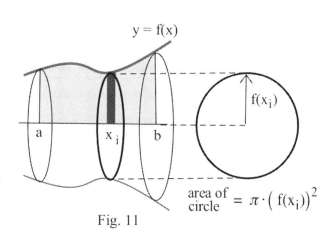

Fig. 10

3. Volumes of Revolution Using "Tubes" (Shells)

In Section 5.2 the "disk" method was used for finding the volume swept out when a region is revolved about a line (Fig. 11). To find the volume swept out when a region was revolved about the x–axis, we made cuts perpendicular to the x–axis so each slice was a "disk" with volume $\pi(\text{radius})^2 \cdot (\text{thickness})$. After adding the volumes of the slices together (a Riemann sum) and taking a limit as the thicknesses approached 0, we obtained a definite integral representation for the exact volume:

{ volume of revolution about the x–axis }
$$= \int_a^b \pi f^2(x) \, dx .$$

However, the disk method can be cumbersome if we want the volume when the region in the figure is revolved about the y–axis or some other vertical line. To revolve the region about the y–axis, the disk method requires that we represent the original equation $y = f(x)$ as a function of y: $x = g(y)$.

Fig. 11

area of circle $= \pi \cdot (f(x_i))^2$

Sometimes that is easy: if $y = 3x$ then $x = y/3$. But sometimes it is not easy at all: if $y = x + e^x$, then we can not solve for y as an elementary function of x. The "tube" method lets us use the original equation $y = f(x)$ to find the volume when the region is revolved about a vertical line.

We partition the x–axis to cut the region into thin, almost rectangular "slices." When the thin "slice" at x_i is revolved about a vertical line (Fig. 12a), the volume of the resulting "tube" can be approximated by cutting the wall of the tube and laying it out flat (Fig. 12b) to get a thin, solid rectangular box. The volume of the tube is approximately the same as the volume of the solid box:

$V_{tube} \approx V_{box}$ = (length)•(height)•(thickness) = **(2π•radius)•(height)•(Δx_i)** = $(2\pi x_i)•(f(x_i))\Delta x_i$.

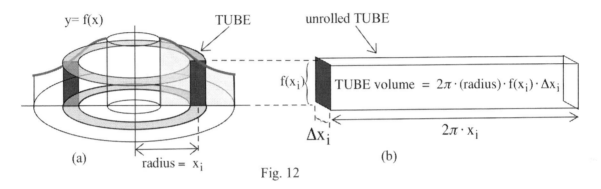

Fig. 12

The volume swept out when the whole region is revolved is the sum of the volumes of these "tubes", a Riemann sum. The limit of the Riemann sum is

$$\{ \text{volume of rotation about a vertical line} \} = \int_{x=a}^{b} 2\pi \cdot (\text{radius}) \cdot (\text{height}) \, dx .$$

The "tube" pattern for the volume of a region defined by a single function extends easily to regions between two functions.

Volume of Revolution Using "Tubes" (Shells)

If   region R is bounded between the functions $f(x) \geq g(x)$ for $0 \leq a \leq b$ (Fig. 13),

then   { volume obtained when R is revolved about a vertical line } = $\int_{x=a}^{b} 2\pi \cdot (\text{radius}) \cdot \{ f(x) - g(x) \} \, dx$.

Example 6: Find the volume when the region R in Fig. 14 is revolved about a vertical line.

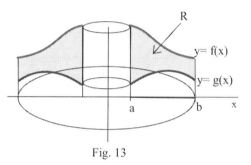

Fig. 13

Solution: We can partition the interval $[2, 4]$ on the x–axis to get thin slices of R. When the slice at x_i is revolved around the y–axis, a tube is swept out, and the volume if this i^{th} tube is

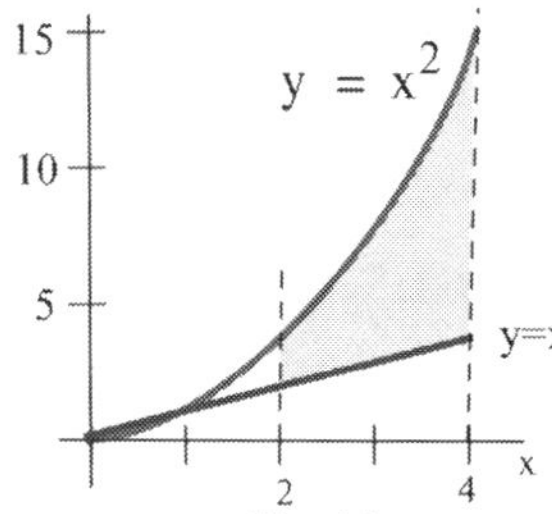

$$v_i \approx (2\pi \cdot \text{radius}) \cdot (\text{height}) \cdot (\text{thickness}) \approx 2\pi (x_i)(x_i^2 - x_i)(\Delta x_i) = 2\pi (x_i^3 - x_i^2) \Delta x_i$$

Fig. 14

The total volume is the sum of the volumes of the tubes:

$$V = \sum v_i = \sum 2\pi (x_i^3 - x_i^2) \Delta x_i \quad \text{(a Riemann sum)}$$

$$\longrightarrow \int_{2}^{4} 2\pi (x^3 - x^2) \, dx = 2\pi \left(\frac{x^4}{4} - \frac{x^3}{3} \right) \Big|_{2}^{4} = \frac{248 \pi}{3} \approx 259.7 .$$

Example 7: Write a definite integral to represent the volume swept out when the region in Fig. 15 is revolved about the vertical line $x = 4$.

Solution: $V = \int_{a}^{b} 2\pi \cdot (\text{radius}) \cdot (\text{height}) \, dx$

$= \int_{0}^{\pi} 2\pi \cdot (4 - x) \cdot \sin(x) \, dx$

≈ 30.526 (using calculator integrate command))

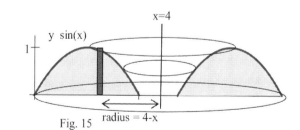

Fig. 15

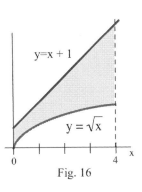

Fig. 16

Practice 5: Represent the volume when the region in Fig. 16 is revolved about the y–axis.

In theory, each method works for each volume of revolution problem. In practice, however, for any particular problem one of the methods is often easier to use than the other.

4. Areas & Elections

The previous applications used definite integrals to determine areas, volumes, pressures, and energies **exactly**. But exactness and numerical precision are not the same as "understanding," and sometimes we can gain insight and understanding simply by determining which of two areas or integrals is larger. One situation of this type involves elections.

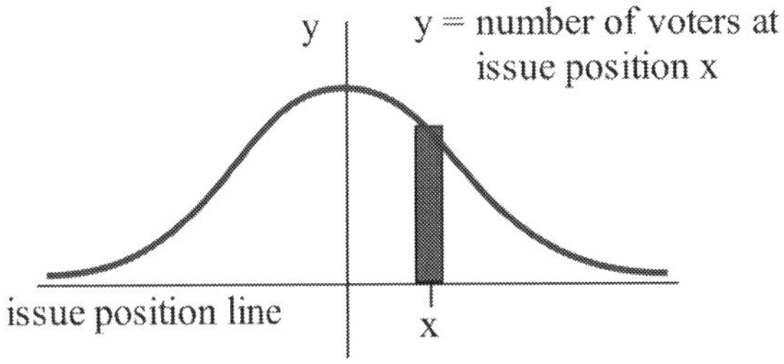

Fig. 17

Suppose the voters of a state have been surveyed about their positions on a **single** issue, and the distribution of voters who place themselves at each position on this issue is shown in Fig. 17. Also suppose that each voter votes for the candidate closest to that voter. If two candidates have taken the positions labeled A and B in Fig. 18, then a voter at position c votes for the candidate at A since A is closer to c than B is to c.

Fig. 18

Similarly, a voter at position d votes for the candidate at B. The total votes for the candidate at A in this election is the shaded area under the curve, and the candidate with the larger number of votes, the larger area, is the winner. In this illustration, the candidate at A wins.

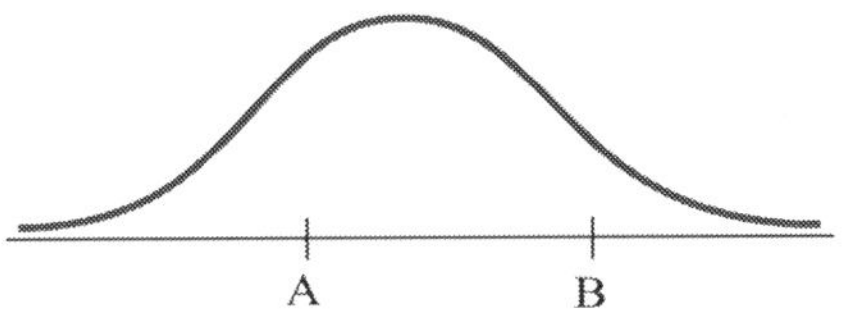

Fig. 19

Example 8: The distribution of voters on an issue is shown in Fig. 19. If these voters decide between candidates on the basis of that single issue, which candidate will win the election?

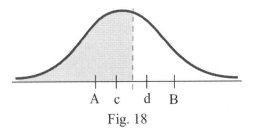

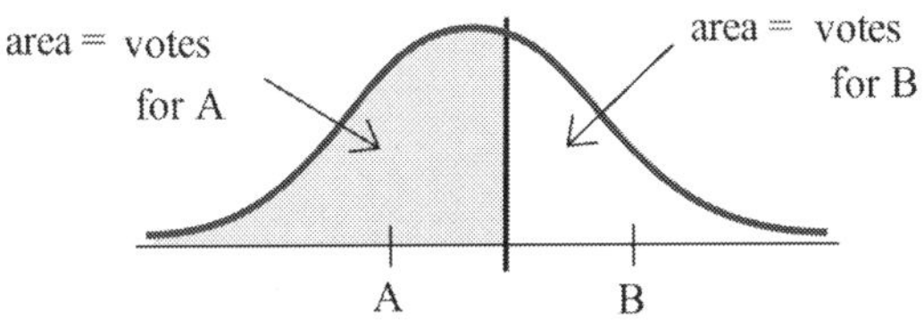

Fig. 20

Solution: Fig. 20 illustrates that A has a larger area (more votes) than B. A will win.

Practice 6: In an election between candidates with positions A and B in Fig. 21, who will win?

If voters behave as described and if the election is between **2 candidates**, then we can give the candidates some advice.

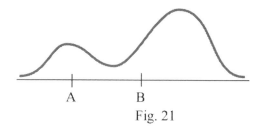
Fig. 21

The best position for a candidate is at the "median point," the location that divides the voters into two equal sized (equal area) groups so half of the voters are on one side of the median point and half are on the other side (Fig. 22). A candidate at the median point gets more votes than a candidate at any other point. (Why?)

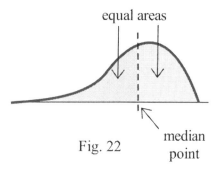
Fig. 22

If two candidates have positions on opposite sides of the median point (Fig. 23), then a candidate can get more votes by moving a bit toward the median point. This "move toward the middle ground commonly occurs in elections as candidates try to sell themselves as "moderates" and their opponents as "extremists."

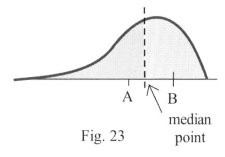
Fig. 23

If there are more than two candidates running in an election, then the situation changes dramatically, and a candidate at the median position, the unbeatable place in a 2–candidate election, can even get the fewest votes. If Fig. 24 is the distribution of voters on the single issue in the

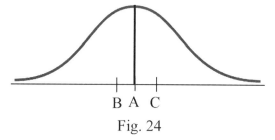
Fig. 24

election, then candidate A would beat B in an election just between A and B (Fig. 25a); and A would beat C in an election just between A and C (Fig. 25b). But in an election among all 3 candidates, A would get the fewest votes of the 3 candidates (Fig. 25c). This type of situation really does occur. It leads to the political saying about a primary election with lots of candidates and a general election between the final nominees of the two parties:

"extremists can win primaries, but moderates are elected to office."

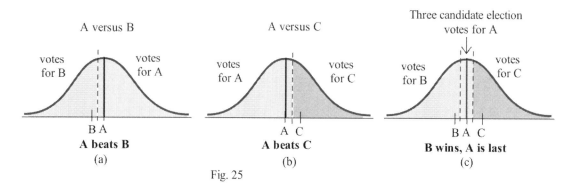
Fig. 25

The previous discussion of elections and areas is greatly oversimplified. Most elections involve several issues of different importance to different voters, and the views of the voters are seldom completely known before the election. Many candidates take "fuzzy" positions on issues. And it is not even certain that real voters vote for the "closest" candidate: perhaps they don't vote at all unless some candidate is "close enough" to their position. But the very simple model of elections can still help us understand how and why some things happen in elections. It is also a starting place for building more sophisticated models to help understand more complicated election situations and to test assumptions about how voters really do make voting decisions.

PROBLEMS

Liquid Pressure

For problems 1–5, assume that the liquid has density d.

1. Calculate the total force against windows A and B in Fig. 26

2. Calculate the total force against windows C and D in Fig. 26.

3. Calculate the total force against each end of the tank in Fig. 27. How does the total force against the ends of the tank change if the length of the tank is doubled?

4. Calculate the total force against the ends of the tank in Fig. 28.

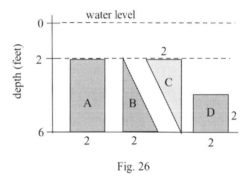

Fig. 26

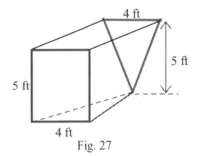

Fig. 27

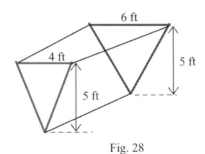

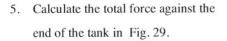

Fig. 28

5. Calculate the total force against the end of the tank in Fig. 29.

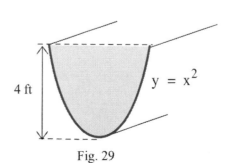

Fig. 29

6. The 3 tanks in Fig. 30 are all 6 feet tall and the top perimeter of each tank is 10 feet. Which tank has the greatest total force against its sides?

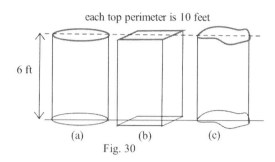
Fig. 30

7. The 3 tanks in Fig. 31 are all 6 feet tall and the cross sectional area of each tank is 16 square feet. Which tank has the greatest total force against its sides?

8. Calculate the total force against the bottom 2 feet of the sides of a 40 foot by 40 foot tank which is filled (a) to a depth of 30 feet with water and (b) to a depth of 35 feet.

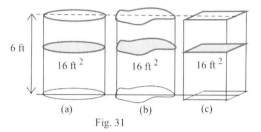
Fig. 31

9. Calculate the total force against the bottom 2 feet of the sides of a cylindrical tank with a radius of 20 feet which is filled (a) to a depth of 30 feet with water and (b) to a depth of 35 feet.

10. Calculate the total force against the sides and bottom of a 12 ounce can of cola (height = 12 cm, radius = 3 cm). (The density of cola is approximately the same as water.) Why is the total force more than 12 ounces?

Kinetic energy

11. Calculate the kinetic energy of a 20 gram object rotating at 3 revolutions per second at the end of (a) a 15 cm (massless) string, and (b) a 20 cm string.

12. One centimeter of a metal bar weighs 3 grams. Calculate the kinetic energy of a 50 centimeter bar which is rotating at a rate of 2 revolutions per second about one end.

13. One centimeter of a metal bar weighs 3 grams. Calculate the kinetic energy of a 50 centimeter bar which is rotating at a rate of 2 revolutions per second at the end of a 10 cm piece of string.

14. Calculate the kinetic energy of a 20 gram meter stick which is rotating at a rate of 1 revolution per second about one end.

15. Calculate the kinetic energy of a 20 gram meter stick if it is rotating at a rate of 1 revolution per second about its middle point.

16. A flat, circular plate is made from material which weighs 2 grams per cubic centimeter. The plate is 5 centimeters thick, has a radius of 30 centimeters and is rotating about its center at a rate of 2 revolutions per second. (a) Calculate its kinetic energy. (b) Find the radius of the plate that would have twice the kinetic energy of this plate? (Assume the density, thickness, and rotation rate are the same.)

17. Each "washer" in Fig. 32 is made from material weighing 1 gram per cubic centimeter, and each is rotating about its center at a rate of 3 revolutions per second. Calculate the kinetic energy of each washer.

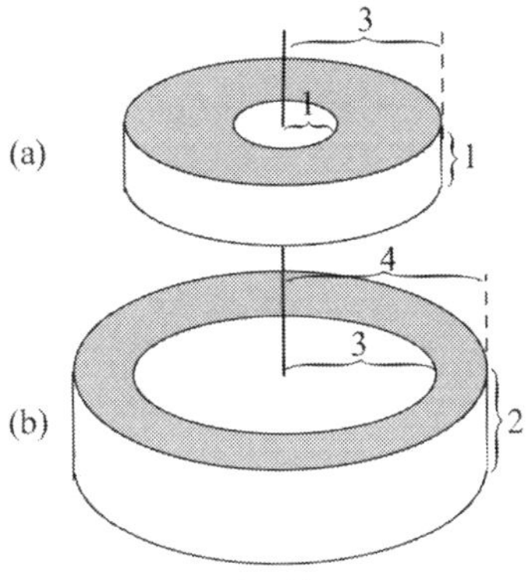

18. The rectangular plate is 1 cm thick, 10 cm long and 6 cm wide and is made of material which weighs 3 grams per cubic centimeter. Calculate the kinetic energy of of the plate if it is rotated at a rate of 2 revolutions per second
    (a) about its 10 cm side (Fig. 33) and
    (b) about its 6 cm side.

19. Calculate the kinetic energy of of the plate in Problem 18 if it is rotated at a rate of 2 revolutions per second about a vertical line
    (a) through the center of the plate (Fig. 34a) and
    (b) through the center of the plate (Fig. 34b).

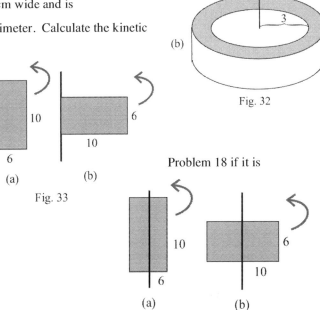

Fig. 32

Fig. 33

Fig. 34

Volumes by "Tubes"

In problems 20 – 25, sketch each region and calculate the volume swept out when the region is revolved about the specified vertical line.

20. The region between $y = 2x - x^2$ and the x–axis for $0 \le x \le 2$ is rotated about the y-axis.

21. The region between $y = \sqrt{1 - x^2}$ and the x–axis for $0 \le x \le 1$ is rotated about the y-axis.

22. The region between $y = \dfrac{1}{1 + x^2}$ and the x–axis of $0 \le x \le 3$ is rotated about the y-axis.

23. The region between $y = 2x$ and $y = x^2$ for $0 \le x \le 3$ is rotated about the $x = 4$ line.

24. The region between $y = x$ and $y = 2x$ for $1 \le x \le 3$ is rotated about the $x = 1$ line.

25. The region between $y = 1/x$ and $y = 1/3$ for $1 \le x \le 3$ is rotated about the $x = 5$ line.

In problems 26 – 30, write a definite integral representing the volume swept out when the region is revolved about the y–axis, and use a calculator to evaluate the integral.

26. The region between $y = e^x$ and $y = x$ for $0 \le x \le 2$.

27. The region between $y = \ln(x)$ and $y = x$ for $1 \le x \le 4$.

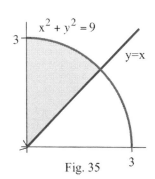

Fig. 35

28. The region between $y = x^2$ and $y = 6 - x$ for $1 \leq x \leq 4$.

29. The shaded region in Fig. 35.

30. The shaded region in Fig. 36.

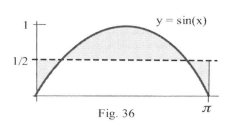
Fig. 36

Areas & Elections

31. For the voter distribution in Fig. 37, which candidates would the voters at positions a, b and c vote for?

32. For the voter distribution in Fig. 38, which candidates would the voters at positions a, b and c vote for?

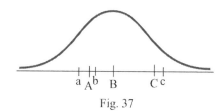
Fig. 37

33. Shade the region representing votes for candidate A in Fig. 39. Which candidate wins?

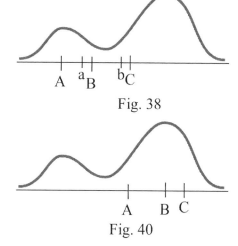
Fig. 38

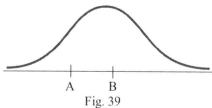
Fig. 39

34. Shade the region representing votes for candidate A in Fig. 40. Which candidate wins?

Fig. 40

35. In Fig. 41, (a) which candidate wins?
    (b) If candidate B withdraws before the election then which candidate will win?
    (c) If candidate B stays in the election, but C withdraws, then who will win?

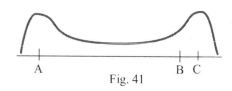
Fig. 41

36. In Fig. 42, (a) which candidate wins?
    (b) If candidate A withdraws before the election, which candidate wins?
    (c) If candidate B stays in the election, but C withdraws, then who wins?

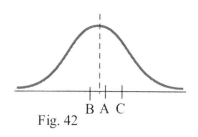
Fig. 42

37. In Fig. 43,

(a) if the election was only between A and B, who would win?

(b) If the election was only between A and C, who would win?

(c) If the election was among A, B, and C, who would win?

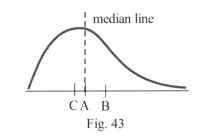

Fig. 43

38. In Fig. 44,

(a) if the election was only between A and B, who would win?

(b) If the election was only between A and C, who would win?

(c) If the election was among A, B, and C, who would win?

39. Sketch what a voter distribution might look like for a 2 issue election.

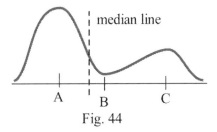

Fig. 44

Section 5.5 Practice Answers

Practice 1: The reasoning is the same as in Example 1 except that the width of the front is 2 feet but the width of the side is 1 foot. Then

$$(\text{density})\cdot(\text{depth})\cdot(\text{area}) = (62.5 \text{ pounds/ft}^3)(x_i \text{ feet})(1 \text{ foot})(\Delta x_i \text{ feet}) = 62.5\, x_i \cdot \Delta x_i \text{ pounds}$$

and

$$\text{hydrostatic force} \approx \sum 125 x_i\, \Delta x_i \longrightarrow \int_{x=0}^{3} 62.5 x\, dx = 31.25\, x^2 \Big|_{x=0}^{3} = 281.25 \text{ pounds}.$$

Practice 2: C: $\dfrac{w}{x-4} = \dfrac{3}{2}$ in Fig. 45 so $w = \dfrac{3}{2}(x-4)$. Then

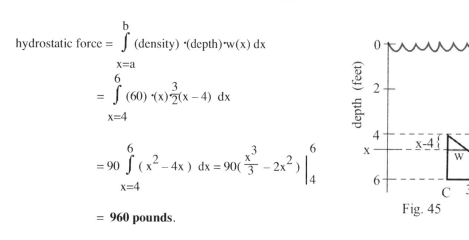
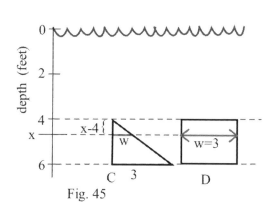

$$\text{hydrostatic force} = \int_{x=a}^{b} (\text{density})\cdot(\text{depth})\cdot w(x)\, dx$$

$$= \int_{x=4}^{6} (60)\cdot(x)\cdot\frac{3}{2}(x-4)\, dx$$

$$= 90 \int_{x=4}^{6} (x^2 - 4x)\, dx = 90\left(\frac{x^3}{3} - 2x^2\right)\Big|_{4}^{6}$$

$$= \mathbf{960 \text{ pounds}}.$$

Fig. 45

D: $w = 3$ at all depths so

$$\text{hydrostatic force} = \int_{x=a}^{b} (\text{density})\cdot(\text{depth})\cdot w(x)\, dx = \int_{x=4}^{6} (60)\cdot(x)\cdot(3)\, dx$$

$$= 180 \int_{x=4}^{6} x\, dx = 90\, x^2 \Big|_{4}^{6} = \mathbf{1800 \text{ pounds}}.$$

Windows A and C (with a flip) fit together to form window D, and it is encouraging that the sum of the total hydrostatic forces on A and C is the same as the total hydrostatic force on D.

Practice 3: The object travels $2\pi(\text{radius}) = 2\pi(2 \text{ meters}) = 4\pi$ meters in one revolution so in the 1 second it takes to make 4 revolutions it travels 16π meters: $v = 1600\pi$ cm/second.

$$KE = \frac{1}{2} m v^2 = \frac{1}{2}(1 \text{ g})\cdot(1600\pi \text{ cm/s})^2 = 1{,}280{,}000\pi^2 \text{ ergs} \approx 12{,}633{,}094 \text{ ergs}.$$

Practice 4: Since the bar and the number of revolutions per second are the same as in Example 4,

$$ke_i = \tfrac{1}{2} m_i \cdot v_i^2 = \tfrac{1}{2}(5 \, \Delta x_i \text{ grams}) \cdot (400\pi \, x_i \text{ cm/sec})^2 = 400{,}000\pi^2 (x_i)^2 \, \Delta x_i \text{ ergs}.$$

Then, since the bar is at the end of a 1 meter string, we integrate from $x = 1$ to $x = 1+3 = 4$:

$$KE = \sum ke_i = \sum 400{,}000\pi^2 (x_i)^2 \, \Delta x_i \quad \longrightarrow \quad \int_1^4 400{,}000\pi^2 \cdot x^2 \, dx$$

$$= 400{,}000\pi^2 \frac{x^3}{3} \Big|_1^4 = 8{,}400{,}000 \, \pi^2 \text{ ergs}.$$

Practice 5: { volume obtained when R is revolved about the y–axis } $= \int_{x=a}^{b} 2\pi \cdot x \cdot \{ f(x) - g(x) \} \, dx$

so volume $= \int_{x=0}^{4} 2\pi \cdot x \cdot \{ (x+1) - \sqrt{x} \} \, dx = 2\pi \int_{x=0}^{4} (x^2 + x - x^{3/2}) \, dx$

$$= 2\pi \left(\frac{x^3}{3} + \frac{x^2}{2} - \frac{2}{5} x^{5/2} \right) \Big|_0^4 \approx 2\pi(16.53) \approx 103.9 .$$

Practice 6: The shaded regions in Fig. 46 show the total votes for each candidate: **B wins**.

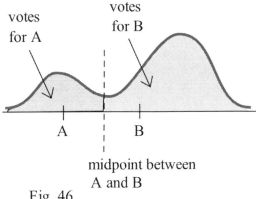

Fig. 46

PROBLEM ANSWERS Chapter Five

Section 5.1

1. volume = $(5 \cdot 6 \cdot 1) + (4 \cdot 4 \cdot 2) + (3 \cdot 3 \cdot 1) = 71$ 2. $v = 60 + 20 + 9 + 4 = 93$

3. $v = (\pi \cdot 4^2 \cdot 0.5) + (\pi \cdot 3^2 \cdot 1) + (\pi \cdot 1^2 \cdot 2) = 19\pi$ 5.   volume = $(9 \cdot 0.2) + (6 \cdot 0.2) + (2 \cdot 0.2) = 3.4$

7. $\int_0^3 \text{(base)(height)(thickness)} = \int_0^3 (2)(5-x)\, dx = 10x - x^2 \Big|_0^3 = (30-9) - (0) = 21 \text{ in}^3$

8. $\int_0^3 \text{(base)(height)(thickness)} = \int_0^3 (x)(x^2)\, dx = \int_0^3 x^3\, dx = \frac{1}{4} x^4 \Big|_0^3 = \frac{81}{4} \text{ in}^3 = 20.25 \text{ in}^3$

9. $\int_1^4 \frac{1}{2}\text{(base)(height)(thickness)} = \int_1^4 \frac{1}{2}(x+1)(x^{1/2})\, dx = \frac{1}{2} \int_1^4 x^{3/2} + x^{1/2}\, dx$

$= \frac{1}{2} \{ \frac{2}{5} x^{5/2} + \frac{2}{3} x^{3/2} \} \Big|_1^4 = \frac{1}{2} \{ \frac{64}{5} + \frac{16}{3} \} - \frac{1}{2} \{ \frac{2}{5} + \frac{2}{3} \} = \frac{128}{15} \text{ m}^3$

11. $\int_2^4 \pi(\text{radius})^2 (\text{thickness}) = \int_2^4 \pi(\frac{1}{2}(4-x))^2\, dx = \frac{\pi}{4} \int_2^4 16 - 8x + x^2\, dx$

$= \frac{\pi}{4}(16x - 4x^2 + \frac{1}{3} x^3) \Big|_2^4 = \frac{\pi}{4} \{ \frac{64}{3} \} - \frac{\pi}{4} \{ \frac{56}{3} \} = \frac{2\pi}{3} \text{ m}^3$

13. Solids A and B have equal volumes. Why?

14. $\int_0^3 \frac{1}{2}\text{(base)(height)(thickness)} = \int_0^3 \frac{1}{2}(x)(2x)\, dx = \int_0^3 x^2\, dx = (\frac{1}{3} x^3) \Big|_0^3 = 9$ cubic units.

15. $\int_1^4 \text{(side)(side)(thickness)} = \int_1^4 (\frac{1}{x})(\frac{1}{x})\, dx = \int_1^4 x^{-2}\, dx = -x^{-1} \Big|_1^4 = (\frac{-1}{4}) - (\frac{-1}{1}) = \frac{3}{4}$ cubic units.

16. $\int_0^4 \pi(\text{radius})^2 (\text{thickness}) = \int_0^4 \pi(2 - \sqrt{x})^2\, dx = \pi(4x - 4(\frac{2}{3}) x^{3/2} + \frac{1}{2} x^2) \Big|_0^4 = \frac{8}{3} \pi \approx 8.38$.

17. $\int_0^4 \pi(\text{radius})^2 (\text{thickness}) = \int_0^4 \pi(3 - \sqrt{x})^2\, dx = \pi \int_0^4 (9 - 6\sqrt{x} + x)\, dx$

$= \pi(9x - 6(\frac{2}{3}) x^{3/2} + \frac{1}{2} x^2) \Big|_0^4 = \pi(36 - 32 + 8) - \pi(0) = 12\pi \approx 37.7$ cubic units.

18. $\int_0^\pi \pi(\text{radius})^2 (\text{thickness}) = \int_0^\pi \pi(\frac{1}{2} - \sin(x))^2\, dx = \pi \int_0^\pi (\frac{1}{4} - \sin(x) + \sin^2(x))\, dx$

$= \pi \{ \frac{1}{4} x + \cos(x) + \frac{1}{2} x - \frac{1}{4} \sin(2x) \} \Big|_0^\pi$

$= \pi \{ \frac{\pi}{4} + (-1) + \frac{\pi}{2} - 0 \} - \pi \{ 0 + 1 + 0 - 0 \} = \frac{3}{4} \pi^2 - 2\pi \approx 1.12$ cubic units.

Odd Answers: Chapter Five Contemporary Calculus

19. $\int_0^2 \frac{1}{2}(\text{base})(\text{height})(\text{thickness}) = \int_0^2 \frac{1}{2}(x)(x^2)\,dx = \frac{1}{2}\int_0^2 x^3\,dx = \frac{1}{2}(\frac{1}{4}x^4)\Big|_0^2 = \frac{1}{8}(16) - \frac{1}{8}(0) = 2$ cubic units.

20. $\int_0^{\pi/2} 2\cdot\cos(x)\,dx = 2$ cubic units.

21. $\int_1^2 \pi(\text{radius})^2(\text{thickness}) = \int_1^2 \pi(x^2)^2\,dx - \int_1^2 \pi(1)^2\,dx = \pi\int_1^2 x^4 - 1\,dx = \pi(\frac{1}{5}x^5 - x)\Big|_1^2$

$= \pi(\frac{32}{5} - 2) - \pi(\frac{1}{5} - 1) = \frac{26}{5}\pi \approx 16.34$ cubic units.

22. $\frac{256}{3}\pi \approx 268.08$ cubic units.   23. $\frac{32}{3}\pi$ 25. 60π 27. $2Rr^2\pi^2$

29. on your own   31. (a) $H^2 L$ (b) $\frac{1}{3}H^2 L$ (c) ratio $= \frac{1}{3}$

33. (a) BL   (b) Volume $= \frac{1}{3}$ BL   (c) ratio $= \frac{1}{3}$

Section 5.2

1. $\sqrt{15^2 + 20^2} + \sqrt{(-24)^2 + 18^2} + \sqrt{12^2 + (-12)^2} \approx 71.97$ feet.

2. $71.97 + \sqrt{11^2 + 7^2} + \sqrt{(-14)^2 + (-17)^2} + \sqrt{(-10)^2 + (-9)^2} \approx 120.48$ feet.

3. $\sqrt{1^2 + 1^2} + \sqrt{1^2 + 2^2} \approx 3.65$

4. $\sqrt{1^2 + (1/2)^2} + \sqrt{1^2 + (1/6)^2} + \sqrt{1^2 + (1/12)^2} \approx 3.135$

5. (a) (0,1) to (2,5): $\sqrt{2^2 + 4^2} = \sqrt{20} = 2\sqrt{5}$

   (b) $y' = 2$: $L = \int_0^2 \sqrt{1 + (2)^2}\,dx = \sqrt{5}\int_0^2 1\,dx = \sqrt{5}\,x\Big|_0^2 = 2\sqrt{5}$ (same as in part (a))

7. (a) (2,1) to (5,-5): $\sqrt{3^2 + 6^2} = \sqrt{45} = 3\sqrt{5}$

   (b) $x' = 1, y' = -2$: $L = \int_0^3 \sqrt{1^2 + (-2)^2}\,dt = \sqrt{5}\int_0^3 1\,dt = \sqrt{5}\,t\Big|_0^3 = 3\sqrt{5}$

9. $y' = \sqrt{x} = x^{1/2}$: $L = \int_0^4 \sqrt{1 + (y')^2}\,dx = \int_0^4 \sqrt{1 + (\sqrt{x})^2}\,dx = \int_0^4 \sqrt{1 + x}\,dx$

   $= \frac{2}{3}(1+x)^{3/2}\Big|_0^4 = \frac{2}{3}(5)^{3/2} - \frac{2}{3}(1)^{3/2} \approx 6.787$

11. $y' = x^2 - \frac{1}{4x^2}$: $1 + (y')^2 = 1 + (x^4 - \frac{1}{2} + \frac{1}{16x^4}) = x^4 + \frac{1}{2} + \frac{1}{16x^4} = (x^2 + \frac{1}{4x^2})^2$

$$L = \int_1^5 \sqrt{1+(y')^2}\ dx = \int_1^5 \sqrt{x^2 + \frac{1}{4x^2}}\ dx = \int_1^5 \sqrt{x^2 + \frac{1}{4}x^{-2}}\ dx$$

$$= \frac{1}{3}x^3 - \frac{1}{4}x^{-1}\Big|_1^5 = \left(\frac{1}{3}(5)^3 - \frac{1}{4(5)}\right) - \left(\frac{1}{3}(1)^3 - \frac{1}{4(1)}\right) \approx 41.53$$

13. $y' = x^4 - \frac{1}{4}x^{-4}$: $1+(y')^2 = 1 + (x^8 - \frac{1}{2} + \frac{1}{16x^8}) = x^8 + \frac{1}{2} + \frac{1}{16x^8} = (x^4 + \frac{1}{4x^4})^2$

$$L = \int_1^5 \sqrt{1+(y')^2}\ dx = \int_1^5 x^4 + \frac{1}{4}x^{-4}\ dx = \frac{1}{5}x^5 - \frac{1}{12}x^{-3}\Big|_1^5$$

$$= \left(\frac{1}{5}(5)^5 - \frac{1}{12(5)^3}\right) - \left(\frac{1}{5}(1)^5 - \frac{1}{12(1)^3}\right) \approx 624.88$$

15. $y' = 2x$: $L = \int_0^1 \sqrt{1+(y')^2}\ dx = \int_0^1 \sqrt{1+(2x)^2}\ dx$

$$= \int_0^1 \sqrt{1+4x^2}\ dx \approx 1.479 \text{ (using calculator)}$$

16. $y' = 3x^2$: $L = \int_0^1 \sqrt{1+(y')^2}\ dx = \int_0^1 \sqrt{1+9x^4}\ dx \approx 1.548$ (using calculator)

17. $y' = \frac{1}{2}x^{-1/2}$: $L = \int_1^9 \sqrt{1+(y')^2}\ dx = \int_1^9 \sqrt{1+\frac{1}{4x}}\ dx \approx 8.268$ (using calculator)

18. $y' = \frac{1}{x}$: $L = \int_1^e \sqrt{1+(y')^2}\ dx = \int_1^e \sqrt{1+\frac{1}{x^2}}\ dx \approx 2.003$ (using calculator)

19. $y' = \cos(x)$: (a) $L = \int_0^{\pi/4} \sqrt{1+\cos^2(x)}\ dx \approx 1.058$ (using calculator)

    (b) $L = \int_{\pi/4}^{\pi/2} \sqrt{1+\cos^2(x)}\ dx \approx 0.852$ (using calculator)

20. $x' = -3\cdot\sin(t)$, $y' = 4\cdot\cos(t)$

$$L = \int_0^{2\pi} \sqrt{(-3\cdot\sin(t))^2 + (4\cdot\cos(t))^2}\ dt$$

$$= \int_0^{2\pi} \sqrt{9\cdot\sin^2(t) + 16\cdot\cos^2(t)}\ dt \approx 22.103 \text{ (using calculator)}$$

21. $x' = -5\cdot\sin(t)$, $y' = 2\cdot\cos(t)$

$$L = \int_0^{2\pi} \sqrt{25\cdot\sin^2(t) + 4\cdot\cos^2(t)}\ dt \approx 23.018 \text{ (using calculator)}$$

22. $x' = -t\cdot\sin(t) + \cos(t)$, $y' = t\cdot\cos(t) + \sin(t)$. Then

$(x')^2 + (y')^2 = t^2 + 1$ (check it) so $L = \int_0^{2\pi} \sqrt{1+t^2}\ dt \approx 21.256$ (using calculator)

23. same x' and y' as in #22. $L = \int_{10}^{20} \sqrt{1+t^2}\ dt \approx 150.346$ (using calculator)

24. $x' = R\cdot(1-\cos(t))$ and $y' = R\cdot\sin(t)$: $(x')^2 + (y')^2 = 2R^2\cdot(1-\cos(t))$

$$L = \int_0^{2\pi} \sqrt{2R^2\cdot(1-\cos(t))}\ dt = R\sqrt{2}\int_0^{2\pi}\sqrt{1-\cos(t)}\ dt \approx 8R\ (\text{Actually, } = 8R)$$

25. 1 mile = 5,280 feet at 2π feet per revolution so 1 mile = $\frac{5280\text{ ft}}{2\pi\text{ ft/rev}} \approx 840.338$ revolutions.
   From #24, total distance = (840.338 rev.)(8 feet/rev) $\approx$ 6722.705 feet ($\approx$ 1.27 miles).

26. 2 (Why?)

27. $y' = 2x$: total length = $\int_0^4 \sqrt{1+4x^2}\ dx \approx 16.8186$ (using calculator)

   (a) Find T so $\int_0^T \sqrt{1+4x^2}\ dx \approx \frac{1}{2}(16.8186) = 8.4093$.

   By experimenting on a calculator, $T \approx 2.77$.

   (b) Find A so $\int_0^A \sqrt{1+4x^2}\ dx \approx \frac{1}{3}(16.8186) = 5.6062$ and

   B so $\int_0^B \sqrt{1+4x^2}\ dx \approx \frac{2}{3}(16.8186) = 11.2124$.

   By experimenting on a calculator, $A \approx 2.22$ and $B \approx 3.23$.

28., 29. on your own.

30. (a) A: $SA_x = 2\pi(4)(4) = 32\pi$ B: $SA_x = 2\pi(2)(2) = 8\pi$
    (b) A: $SA_y = 2\pi(3)(4) = 24\pi$ B: $SA_y = 2\pi(6)(2) = 24\pi$

31. (a) A: $SA_x = 2\pi(3)(3) = 18\pi$ B: $SA_x = 2\pi(4)(\sqrt{3^2+4^2}) = 40\pi$
    (b) A: $SA_y = 2\pi(2)(3) = 12\pi$ B: $SA_y = 2\pi(6.5)(5) = 65\pi$

33. (a) about $y = 1$: A: $SA = 2\pi(2)(3) = 12\pi$ B: $SA = 2\pi(3)(5) = 30\pi$
    (b) about $x = -2$: A: $SA = 2\pi(4)(3) = 24\pi$ B: $SA = 2\pi(8.5)(5) = 85\pi$

35. The largest surface area occurs when the midpoint is farthest from the line of rotation, the x–axis, when $\theta = 90°$. Then SA = 2π(dist. to x–axis)(length) = $2\pi(6)(2) = 24\pi$.

37. $y' = 6x^2$: $SA_y = \int_0^1 2\pi x\sqrt{1+(y')^2}\ dx = \int_0^1 2\pi x\sqrt{1+(6x^2)^2}\ dx$

   $= \int_0^1 2\pi x\sqrt{1+36x^4}\ dx \approx 10.207$ (using calculator)

39. $y' = 4x$: $SA_x = \int_0^1 2\pi y \sqrt{1 + (y')^2}\, dx = \int_0^1 2\pi(2x^2)\sqrt{1 + 16x^2}\, dx \approx 13.306$ (using calculator)

41. $y' = 3x^2$: $SA_x = \int_0^2 2\pi y \sqrt{1 + (y')^2}\, dx = \int_0^2 2\pi(x^3)\sqrt{1 + 9x^4}\, dx \approx 203.046$ (using calculator)

43. $y' = 2x$: $SA_y = \int_0^2 2\pi x \sqrt{1 + (y')^2}\, dx = \int_0^2 2\pi x \sqrt{1 + 4x^2}\, dx \approx 36.177$ (using calculator)

or use a u–substitution with $u = 1 + 4x^2$: $SA_y = \int_{u=1}^{u=17} 2\pi u^{1/2} \frac{1}{8}\, du \approx 36.177$.

45–50. on your own

Section 5.3

1. Fig. 1. Weight = (volume)(density) = $(3 \cdot 4 \cdot \Delta y)(62.5)$,

   distance raised = $7 - y_i$, so Work = $(750 \Delta y)(7 - y_i)$.

   Total work = $\int_0^7 (7 - y)(750)\, dy = 750(7y - \frac{1}{2} y^2)\Big|_0^7 = 18,375$ ft–lbs.

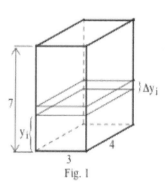

Fig. 1

3. Slice at height y: weight = $(5 \cdot 2 \cdot \Delta y)(60) = 600 \Delta y$ lb,

   distance raised = $4 - y$ ft.

   (a) Total work = $\int_0^4 (4 - y)(600)\, dy = 600(4y - \frac{1}{2} y^2)\Big|_0^4 = 4,800$ ft–lbs.

   (b) The top 36 cubic feet of water corresponds to dimensions of 5 ft. long, 2 ft wide, and $\frac{36}{(5)(2)} = 3.6$ ft. high.

   Weight of slice = $600 \Delta y$, distance raised = $4 - y$ (where y goes from 0.4 ft. to 4 ft.)

   Total work = $\int_{0.4}^4 (4 - y)(600)\, dy = 600(4y - \frac{1}{2} y^2)\Big|_{0.4}^4 = 3,888$ ft–lbs.

   (c) With 1 HP pump: it takes $\frac{4800}{33000}$ minutes ≈ 0.145 min ≈ 8.73 seconds to empty the tank.

   With 1/2 HP pump: $\frac{8.73}{1/2} = 17.46$ seconds.

   Both pumps do the same amount of work but in different times and thus have different power.

5. Slice at height y: weight = $(\pi \cdot r^2 \cdot \Delta y)(60) = (\pi \cdot 1^2 \cdot \Delta y)(60) = 60\pi \Delta y$ lb, distance raised = $6 - y$ ft.

   (a) Total work = $\int_0^6 (6 - y)(60\pi)\, dy = 60\pi(6y - \frac{1}{2} y^2)\Big|_0^6 = 60\pi(18) \approx 3,393$ ft–lbs.

   (b) weight of slice = $60\pi \Delta y$ lb, distance raised = $(2 + 6) - y = 8 - y$ ft.

   Total work = $\int_{6-1}^6 (8 - y)(60\pi)\, dy = 60\pi(8y - \frac{1}{2} y^2)\Big|_5^6 = 150\pi \approx 471.2$ ft–lbs.

   (c) First find the **work** needed to empty the top 3 feet; using integration of part (a) with new limits.

$$\text{work} = \int_3^6 (6-y)(60\pi)\, dy = 60\pi(6y - \frac{1}{2}y^2)\Big|_3^6 = 60\pi(4.5) \approx 848 \text{ ft-lbs}.$$

Knowing that 1 HP pump works at a rate of 33,000 ft–lbs/minute, then it takes

$$\frac{848}{33000} \approx 0.0257 \text{ minutes} \approx 1.54 \text{ seconds}.$$

7. Fig. 2. Use similar triangles: slice at height y has

   weight = (volume)(density) = $(7 \cdot \frac{3}{2} \cdot y \cdot \Delta y)(80) = 840y\,\Delta y$ and

   distance raised = $4 - y$.

   Total work = $\int_3^4 (4-y)(840y)\, dy = 840(2y^2 - \frac{1}{3}y^3)\Big|_3^4 = 1{,}400$ ft–lbs.

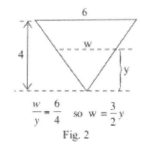

Fig. 2

8. Fig. 3. Each slice (perpendicular to the y–axis) is a disk with

   volume = $\pi(\text{radius})^2(\text{thickness}) = \pi(\frac{y}{2})^2 \Delta y$ so

   weight = (volume)(density) = $(\pi(\frac{y}{2})^2 \Delta y)(40) = 10\pi y^2 \Delta y$.

   (a) Distance to top = $8 - y$ so

   $$\text{Total work} = \int_0^8 (8-y)(10\pi y^2)\, dy = 10\pi \int_0^8 8y^2 - y^3\, dy$$

   $$= 10\pi(\frac{8}{3}y^3 - \frac{1}{4}y^4)\Big|_0^8 = \frac{10\pi}{12}\cdot 8^4 \approx 3{,}413.3\,\pi \approx 10{,}723 \text{ ft-lbs}.$$

   (b) The limits of integration are now 6 to 8:

   $$\text{work} = \int_6^8 (8-y)(10\pi y^2)\, dy = 10\pi \int_6^8 8y^2 - y^3\, dy$$

   $$= 10\pi(\frac{8}{3}y^3 - \frac{1}{4}y^4)\Big|_6^8 \approx 893.3\,\pi \approx 2{,}806 \text{ ft-lbs}.$$

9. Find H so $10\pi \int_0^H 8y^2 - y^3\, dy = \frac{1}{2}\{\text{total work done in 8(a)}\} = \frac{1}{2}\{3{,}413.3\,\pi\} \approx 1{,}706.6\pi$.

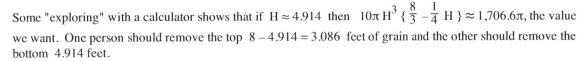

Fig. 3

$$10\pi \int_0^H 8y^2 - y^3\, dy = 10\pi\{\frac{8}{3}y^3 - \frac{1}{4}y^4\}\Big|_0^H = 10\pi H^3\{\frac{8}{3} - \frac{1}{4}H\}.$$

Some "exploring" with a calculator shows that if $H \approx 4.914$ then $10\pi H^3\{\frac{8}{3} - \frac{1}{4}H\} \approx 1{,}706.6\pi$, the value we want. One person should remove the top $8 - 4.914 = 3.086$ feet of grain and the other should remove the bottom 4.914 feet.

11. Fig. 4. Slice at height y:

weight = (volume)(density) = $(\pi r^2 \Delta y)(62.5)$

$= (\pi (\sqrt{2y})^2 \Delta y)(62.5) = 125\pi y \Delta y$.

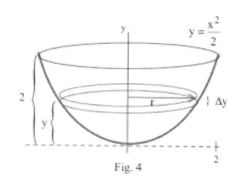

Fig. 4

(a) Distance raised = 2 – y so

work = $\int_0^2 (2-y)(125\pi y) \, dy = 125\pi \{ y^2 - \frac{1}{3} y^3 \} \Big|_0^2$

$= 125\pi (4 - \frac{8}{3}) \approx 523.6$ ft–lbs.

(b) The only change is the distance raised: distance = 5 – y.

work = $\int_0^2 (5-y)(125\pi y) \, dy = 125\pi \{ \frac{5}{2} y^2 - \frac{1}{3} y^3 \} \Big|_0^2$

$= 125\pi (10 - \frac{8}{3}) \approx 2,879.8$ ft–lbs.

13. Fig. 5. Slice at height y:

weight = (volume)(density) = $(\pi r^2 \Delta y)(62.5)$

$= (\pi (\sqrt{16 - y^2})^2 \Delta y)(62.5) = 62.5\pi (16 - y^2) \Delta y$.

distance = 4 – y.

work = $\int_{-4}^{-2} (4-y)(62.5\pi)(16-y^2) \, dy$

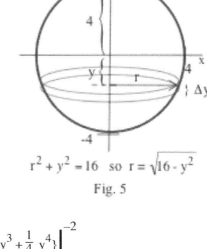

$r^2 + y^2 = 16$ so $r = \sqrt{16 - y^2}$

Fig. 5

$= 62.5\pi \int_{-4}^{-2} 64 - 16y - 4y^2 + y^3 \, dy = 62.5\pi \{ 64y - 8y^2 - \frac{4}{3} y^3 + \frac{1}{4} y^4 \} \Big|_{-4}^{-2}$

$\approx 62.5\pi (89.33) \approx 17,540$ ft–lbs.

15. { work for (a) } > { work for (b) } > { work for (c) }. For container (a), most of the water is raised only a small distance, but for container (a), most of the water is raised a larger distance.

17. Work = area under the curve of force vs. position.

(a) work = $(10)(5) + \frac{1}{2}(2)(10) + (2)(10) + \frac{1}{2}(1)(10) = 85$ ft–lbs.

(b) work = $(2)(10) + (1)(10) + \frac{1}{2}(1)(10) = 35$ ft–lbs.

19. F = kx so 6x = kx and k = 6.

(a) work = $\int_0^3 6x \, dx = 3x^2 \Big|_0^3 = 27$ in–oz.

(b) work = $\int_0^6 6x \, dx = 3x^2 \Big|_0^6 = 108$ in–oz.

21. Work = { area under graph of force vs. length }. Use the graph in Fig. 26 to approximate the area.

(a) From x = 23 to x = 33, area ≈ (33 – 23 cm) { avg. height of about 15 g } = 150 g–cm.

(b) From $x = 28$ to $x = 33$, area $\approx (33 - 28 \text{ cm})\{15 \text{ g}\} + \frac{1}{2}(33-30)\{15 \text{ g}\} = 97.5$ g–cm.

23. $F = kx$ so $3 = k \cdot 5$ and $k = 3/5$.

    (a) work = $\int_{5}^{5+4} \frac{3}{5} x \, dx = \frac{3}{10} x^2 \Big|_5^9 = 16.8$ in–lbs.

25. Work = $\int_a^b f(x) \, dx = \int_R^{R+h} \frac{R^2 P}{x^2} \, dx = \int_{300}^{300+h} \frac{(300)^2(100)}{x^2} \, dx = 9 \cdot 10^6 \left\{ \frac{-1}{x} \right\} \Big|_{300}^{300+h} = 9 \cdot 10^6 \{ \frac{1}{300} - \frac{1}{300+h} \}$

    (a) $h = 100$: work $= 9 \cdot 10^6 \{ \frac{1}{300} - \frac{1}{400} \} \approx 7,500$ mile–pounds.

    (b) $h = 200$: work $= 9 \cdot 10^6 \{ \frac{1}{300} - \frac{1}{500} \} \approx 12,000$ mile–pounds.

    (c) $h = 300$: work $= 9 \cdot 10^6 \{ \frac{1}{300} - \frac{1}{600} \} \approx 15,000$ mile–pounds.

27. Assume you weigh P pounds.

    Work $= \int_a^b f(x) \, dx = \int_R^{R+h} \frac{R^2 P}{x^2} \, dx = \int_{4000}^{4000+h} \frac{(4000)^2 P}{x^2} \, dx$

    $= 1.6 \cdot 10^7 P \left\{ \frac{-1}{x} \right\} \Big|_{4000}^{4000+h} = 1.6 \cdot 10^7 P \{ \frac{1}{4000} - \frac{1}{4000+h} \}$

    (a) $h = 200$: work $= 1.6 \cdot 10^7 P \{ \frac{1}{4000} - \frac{1}{4200} \}$ mile–pounds.

    (b) $h = 400$: work $= 1.6 \cdot 10^7 P \{ \frac{1}{4000} - \frac{1}{4400} \}$ mile–pounds.

    (c) $h = 10^6$: work $= 1.6 \cdot 10^7 P \{ \frac{1}{4000} - \frac{1}{4000+10^6} \}$ mile–pounds. What happens when h is really large?

29. $f(x) = \frac{-1}{kx^2}$ and we know that $f = 0.1$ when $x = 10$ so $0.1 = \frac{-1}{k10^2}$ and $k = -0.1$.

    work $= \int_a^b f(x) \, dx = \int_a^b \frac{-1}{(-0.1)x^2} \, dx = 10 \left\{ \frac{-1}{x} \right\} \Big|_a^b = 10 \{ \frac{1}{a} - \frac{1}{b} \}$.

    (a) from $x = 20$ to $x = 10$: work $= 10 \{ \frac{1}{10} - \frac{1}{20} \} = 0.5$ ft–lb.

    (b) from $x = 10$ to $x = 1$: work $= 10 \{ \frac{1}{1} - \frac{1}{10} \} = 9$ ft–lb.

    (c) from $x = 1$ to $x = 0.1$: work $= 10 \{ \frac{1}{0.1} - \frac{1}{1} \} = 90$ ft–lb.

31. Work $= \int_{t=a}^{t=b} f(t) \sqrt{(dx/dt)^2 + (dy/dt)^2} \, dt$

    $= \int_0^{2\pi} t \sqrt{\sin^2(t) + \cos^2(t)} \, dt = \int_0^{2\pi} t \, dt = \frac{1}{2} t^2 \Big|_0^{2\pi} = 2\pi^2$.

33. Work = $\int_{t=a}^{t=b} f(t)\sqrt{(dx/dt)^2 + (dy/dt)^2}\, dt = \int_0^1 t\sqrt{4t^2+1}\, dt$.

(Use the substitution $u = 4t^2 + 1$ to find an antiderivative, and then evaluate.)

35. "Unroll" the region to get a triangle with base = 2π(radius) = $2\pi(1) = 2\pi$ and height = $f(2\pi) = 2\pi$.

The "area" of the triangle is $\frac{1}{2}$(base)(height) = $\frac{1}{2}(2\pi)(2\pi) = 2\pi^2$. (This is the same as problem 31.)

Section 5.4

1. (a) $M = 2 + 5 + 5 = 12$. $M_o = (2)(4) + (5)(2) + (5)(6) = 48$. $\bar{x} = \frac{M_o}{M} = \frac{48}{12} = 4$.

    (b) new $\bar{x} = 5 = \frac{M_o}{M} = \frac{(2)(4) + (5)(2) + (5)(6) + (8)(?)}{2+5+5+8} = \frac{48 + 8(?)}{20}$ so $? = \frac{100-48}{8} = 6.5$.

    (c) $\bar{x} = 6 = \frac{M_o}{M} = \frac{48 + (?)(10)}{12 + ?}$ so $72 + 6(?) = 48 + 10(?)$ and $4(?) = 24$ so $? = 6$.

3. (a) $M = 2 + 5 + 5 = 12$. $\bar{x} = \frac{M_y}{M} = \frac{(2)(4) + (5)(2) + (5)(6)}{12} = \frac{48}{12} = 4$.

    $\bar{y} = \frac{M_x}{M} = \frac{(2)(3) + (5)(4) + (5)(2)}{12} = \frac{36}{12} = 3$.

    (b) new $\bar{x} = 5 = \frac{48 + 10(?)}{12 + 10}$ so $? = \frac{110-48}{10} = 6.2$.

    new $\bar{y} = 2 = \frac{36 + 10(?)}{12+10}$ so $? = \frac{44-36}{10} = 0.8$.

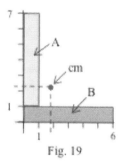
Fig. 19

5. See Fig. 19. A: mass = (1)(6)(density) = 6. Center of mass = (0.5, 4)

    B: mass = (1)(6)(density) = 6. Center of mass = (3, 0.5)

    Total mass = 6+6 = 12. $M_y = (6)(0.5) + (6)(3) = 21$. $M_x = (6)(4) + (6)(0.5) = 27$.

    $\bar{x} = \frac{M_y}{M} = \frac{21}{12} = 1.75$ and $\bar{y} = \frac{M_x}{M} = \frac{27}{12} = 2.25$.

    Notice that the center of mass is not "in" the region.

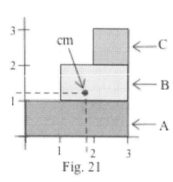
Fig. 21

7. See Fig. 21. A: mass = (1)(3) = 3. Center of mass = (1.5, 0.5)

    B: mass = (1)(2) = 2. Center of mass = (2, 1.5)

    C: mass = (1)(1) = 1. Center of mass = (2.5, 2.5)

    Total mass = 3+2+1 = 6. $M_y = (3)(1.5)+(2)(2)+(1)(2.5) = 11$.

    $M_x = (3)(0.5)+(2)(1.5)+(1)(2.5) = 7$.

    $\bar{x} = \frac{M_y}{M} = \frac{11}{6} \approx 1.83$. $\bar{y} = \frac{M_x}{M} = \frac{7}{6} \approx 1.17$.

9. See Fig. 23. A: mass = (8)(4) = 32. Center of mass = (2, 4)

B: mass $= 0.5\pi(2^2) \approx 6.28$.

Center of mass $\approx (2, 8+0.4244(2)) = (2, 8.85)$

C: mass $= 0.5\pi(4^2) \approx 25.13$.

Center of mass $\approx (4+0.4244(4), 4) = (5.70, 4)$

Total mass ≈ 63.416. $M_y \approx (32)(2)+(6.28)(2)+(25.13)(5.70) = 219.801$.

$M_x \approx (32)(4) + (6.28)(8.85) + (25.13)(4) = 284.098$.

$\bar{x} = \dfrac{M_y}{M} \approx \dfrac{219.801}{63.416} \approx 3.47$. $\bar{y} = \dfrac{M_x}{M} = \dfrac{284.098}{63.416} \approx 4.48$.

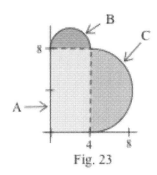

Fig. 23

11. Fig. 1. $\bar{x} = \dfrac{\int_0^3 x \cdot f(x)\, dx}{\int_0^3 f(x)\, dx} = \dfrac{\int_0^3 x \cdot x\, dx}{\int_0^3 x\, dx} = \dfrac{\left. \tfrac{1}{3}x^3 \right|_0^3}{\left. \tfrac{1}{2}x^2 \right|_0^3} = \dfrac{9}{4.5} = 2$

$\bar{y} = \dfrac{\tfrac{1}{2}\int_0^3 (f(x))^2\, dx}{\int_0^3 f(x)\, dx} = \dfrac{\tfrac{1}{2}\int_0^3 x^2\, dx}{\int_0^3 x\, dx} = \dfrac{\left. \tfrac{1}{2}\tfrac{1}{3}x^3 \right|_0^3}{\left. \tfrac{1}{2}x^2 \right|_0^3} = \dfrac{4.5}{4.5} = 1$

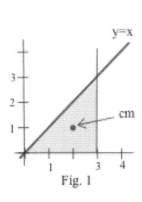

Fig. 1

13. Fig. 2. Because of symmetry about the y–axis, $\bar{x} = 0$. $\bar{y} = \dfrac{M_x}{M}$.

$M_x = \tfrac{1}{2} \int_a^b \{ f^2(x) - g^2(x) \}\, dx = \tfrac{1}{2} \int_{-2}^{2} \{ 4^2 - (x^2)^2 \}\, dx = \tfrac{1}{2} \{ 16x - \tfrac{1}{5} x^5 \} \Big|_{-2}^{2}$

$= \tfrac{1}{2} \{ (32-\tfrac{32}{5}) - (-32+\tfrac{32}{5}) \} \approx 25.6$

$M = \int_a^b \{ 4 - x^2 \}\, dx = \{ 4x - \tfrac{1}{3} x^3 \} \Big|_{-2}^{2} = \{ (8-\tfrac{8}{3}) - (-8+\tfrac{8}{3}) \} = \tfrac{32}{3} \approx 10.67$.

$\bar{y} = \dfrac{M_x}{M} \approx \dfrac{25.6}{10.67} = 2.4$

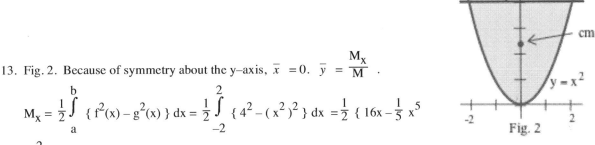

Fig. 2

15. Fig. 3. Because of symmetry about the y–axis, $\bar{x} = 0$. $\bar{y} = \dfrac{M_x}{M}$.

$$M_x = \frac{1}{2}\int_a^b \{f^2(x)\}\,dx = \frac{1}{2}\int_{-2}^{2} (4-x^2)^2\,dx = \frac{1}{2}\left\{16x - \frac{8}{3}x^3 + \frac{1}{5}x^5\right\}\Big|_{-2}^{2}$$

$$= \frac{1}{2}\left\{(32 - \frac{64}{3} + \frac{32}{5}) - (-32 + \frac{64}{3} - \frac{32}{5})\right\} \approx 17.07$$

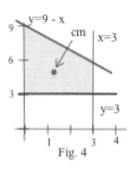
Fig. 3

$$M = \int_a^b \{4-x^2\}\,dx = \left\{4x - \frac{1}{3}x^3\right\}\Big|_{-2}^{2} = \left\{(8 - \frac{8}{3}) - (-8 + \frac{8}{3})\right\} = \frac{32}{3} \approx 10.67.$$

$$\bar{y} = \frac{M_x}{M} \approx \frac{17.07}{10.67} \approx 1.6$$

17. Fig. 4. $M = \displaystyle\int_0^3 f(x) - g(x)\,dx = \int_0^3 9 - x - 3\,dx = 6x - \frac{1}{2}x^2\Big|_0^3 = 13.5$

$$M_y = \int_0^3 x\cdot\{f(x)-g(x)\}\,dx = \int_0^3 x\cdot(9-x-3)\,dx = 3x^2 - \frac{1}{3}x^3\Big|_0^3 = 27 - 9 = 18.$$

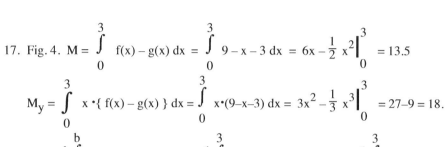
Fig. 4

$$M_x = \frac{1}{2}\int_a^b \{f^2(x) - g^2(x)\}\,dx = \frac{1}{2}\int_0^3 \{(9-x)^2 - (3)^2\}\,dx = \frac{1}{2}\int_0^3 \{81 - 18x + x^2 - 9\}\,dx$$

$$= \frac{1}{2}\left\{72x - 9x^2 + \frac{1}{3}x^3\right\}\Big|_0^3 = \frac{1}{2}\{216 - 81 + 9\} = 72.$$

$$\bar{x} = \frac{M_y}{M} = \frac{18}{13.5} \approx 1.33. \quad \bar{y} = \frac{M_x}{M} = \frac{72}{13.5} \approx 5.33.$$

19. Fig. 5. $M = \displaystyle\int_a^b f(x)\,dx = \int_0^9 \sqrt{x}\,dx = \frac{2}{3}x^{3/2}\Big|_0^9 = \frac{54}{3} \approx 18$

$$M_y = \int_0^9 x\cdot\{f(x)\}\,dx = \int_0^9 x\cdot(\sqrt{x})\,dx = \frac{2}{5}x^{5/2}\Big|_0^9 = \frac{486}{5} = 97.2$$

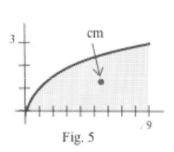
Fig. 5

$$M_x = \frac{1}{2}\int_a^b f^2(x)\,dx = \frac{1}{2}\int_0^9 (\sqrt{x})^2\,dx = \frac{1}{2}\frac{1}{2}x^2\Big|_0^9 = \frac{81}{4} = 20.25.$$

$$\bar{x} = \frac{M_y}{M} = \frac{97.2}{18} = 5.4. \quad \bar{y} = \frac{M_x}{M} = \frac{20.25}{18} \approx 1.125.$$

21. Fig. 6. $M = \displaystyle\int_a^b f(x)\,dx = \int_0^1 e - e^x\,dx = e\cdot x - e^x\Big|_0^1 = (e - e) - (0 - 1) = 1$

$$M_y = \int_0^1 x \cdot \{f(x)\} \, dx = \int_0^1 x \cdot (e - e^x) \, dx = \left. e \cdot \frac{1}{2} x^2 - e^x (x-1) \right|_0^1$$

$$= (\frac{e}{2} - 0) - (0 + 1) \approx 0.359$$

$$M_x = \frac{1}{2} \int_a^b \{ f^2(x) - g^2(x) \} \, dx = \frac{1}{2} \int_0^1 e^2 - e^{2x} \, dx$$

$$= \frac{1}{2} \{ e^2 \cdot x - \frac{1}{2} e^{2x} \} \bigg|_0^1 = \frac{1}{2} \{ e^2 - \frac{1}{2} e^2 \} - \frac{1}{2} \{ 0 - \frac{1}{2} \} \approx 2.097.$$

$$\bar{x} = \frac{M_y}{M} = \frac{0.359}{1} = 0.359. \quad \bar{y} = \frac{M_x}{M} = \frac{2.097}{1} = 2.097.$$

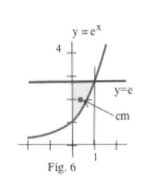

Fig. 6

23. (a) $h = \dfrac{M_0}{M} = \dfrac{(10)(6) + (\frac{x}{2})(\frac{x}{12})(60)}{10 + (\frac{x}{12})(60)} = \dfrac{60 + \frac{5}{2} x^2}{10 + 5x}$.

(b) Find minimum of h in part (a): calculate $\dfrac{dh}{dt}$, set $\dfrac{dh}{dt} = 0$ and solve for x to find critical points.

$$\frac{dh}{dt} = \frac{5x(10 + 5x) - 5(60 + \frac{5}{2} x^2)}{(10 + 5x)^2}.$$

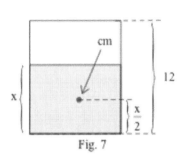

Fig. 7

If $\dfrac{dh}{dt} = 0$ then $5x(10 + 5x) - 5(60 + \frac{5}{2} x^2) = 0$ so $x = \dfrac{-4 \pm \sqrt{16 + 96}}{2} \approx 3.3$ inches.

25. Empty glass: weight = 4 oz., $\bar{y} = \frac{8}{2} = 4$ cm.

Full glass (liquid only): weight = 20–4 = 16 oz., $\bar{y} = 4$ cm, volume = (4 cm)(4 cm)(8 cm) = 128 cm^3.

$$\text{density of liquid} = \frac{\text{weight}}{\text{volume}} = \frac{16 \text{ oz.}}{128 \text{ cm}^3} = \frac{1}{8} \frac{\text{oz.}}{\text{cm}^3}$$

Partially full glass (liquid only): **x = height of liquid**, $\bar{y} = \frac{x}{2}$ cm,

$$\text{weight} = (4 \text{ cm})(4 \text{ cm})(x \text{ cm})(\frac{1}{8} \frac{\text{oz.}}{\text{cm}^3}) = 2x \text{ oz}$$

h = height of cm of glass containing x cm of liquid = $\dfrac{(4 \text{ oz})(4 \text{ cm}) + (2x \text{ oz})(\frac{x}{2} \text{ cm})}{4 \text{ oz} + 2x \text{ oz}} = \dfrac{16 + x^2}{4 + 2x}$ cm.

27. and 29. On your own.

31. Center of gravity is 2 feet above the ground and is raised to a height of 10 feet: distance c.g. raised = 8 feet.
work = (force)(distance) = (300)(8) = 2,400 ft–lb.

33. C.g. is raised 3 + 1 = 4 feet. Work = (250)(4) = 1,000 ft–lb.

35. (a) Volume about x–axis = A·2π·y = 2^2·2π·4 = 32π ft^3.
    Surface area about x–axis = P·2π·y = 8·2π·4 = 64π ft^2.

    (b) Volume about y–axis = A·2π·x = 4·2π·3 = 24π ft^3.
    Surface area about y–axis = P·2π·x = 8·2π·3 = 48π ft^2.

    (c) Volume about line = A·2π·(dist. of c.g. from line) = 4·2π·2 = 16π ft^3.
    Surface area about line = P·2π·(dist of c.g. from line) = 8·2π·2 = 32π ft^2.

37. (a) Volume about x–axis = A·2π·y = $(\pi 2^2)$·2π·5 = $40\pi^2$ ft^3.
    Surface area about x–axis = P·2π·y = 2π(2)·2π·5 = $40\pi^2$ ft^2.

    (b) Volume about y–axis = A·2π·x = 4π·2π·3 = $24\pi^2$ ft^3.
    Surface area about y–axis = P·2π·x = 4π·2π·3 = $24\pi^2$ ft^2.

    (c) Volume about line = A·2π·(dist. of c.g. from line) = 4π·2π·3 = $24\pi^2$ ft^3.
    Surface area about line = P·2π·(dist of c.g. from line) = 4π·2π·3 = $24\pi^2$ ft^2.

39. Each rectangle has area = 8 ft^2, perimeter = 12 ft., and centroid 3 ft. from the line of rotation.
    Volume of each = 2π(radius)(area) = 2π(3 ft)(8 ft^2) = 48π ft^3 ≈ 150.8 ft^3.
    Surface area of each = 2π(radius)(perimeter) = 2π(3 ft)(12 ft) = 72π ft^2 ≈ 226.2 ft^2.

Section 5.5

Liquid Pressure

1. A: $\int_{2}^{6} d \cdot x \cdot (2)\, dx = 2d \int_{2}^{6} x\, dx = d x^2 \Big|_{2}^{6} = 32d$.

   Fig. 1. B: $\int_{2}^{6} d \cdot x \cdot \frac{1}{2}(x-2)\, dx = \frac{d}{2}\int_{2}^{6} x^2 - 2x\, dx$

   $= \frac{d}{2}(\frac{1}{3}x^3 - x^2) \Big|_{2}^{6} = \frac{56}{3} d$.

2. Fig. 2. C: $\int_{2}^{6} d \cdot x \cdot \frac{1}{2}(6-x)\, dx = 13.33\, d$

   D: $\int_{4}^{6} d \cdot x \cdot (2)\, dx = 20\, d$.

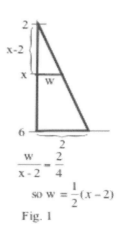

Fig. 1

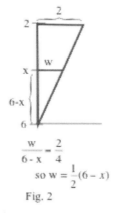

Fig. 2

Odd Answers: Chapter Five Contemporary Calculus 14

3. Rectangular end: $\int_0^5 d \cdot x \cdot (4) \, dx = 4d \int_0^5 x \, dx = 2d x^2 \Big|_0^5 = 50 d$.

Triangular end (Fig. 3): $\int_0^5 d \cdot x \cdot (\frac{4}{5})(5-x) \, dx = \frac{4}{5} d \int_0^4 5x - x^2 \, dx$

$$= \frac{4}{5} d \{ \frac{5}{2} x^2 - \frac{1}{3} x^3) \Big|_0^5 = \frac{50}{3} d .$$

The total force is unchanged if the length is doubled.

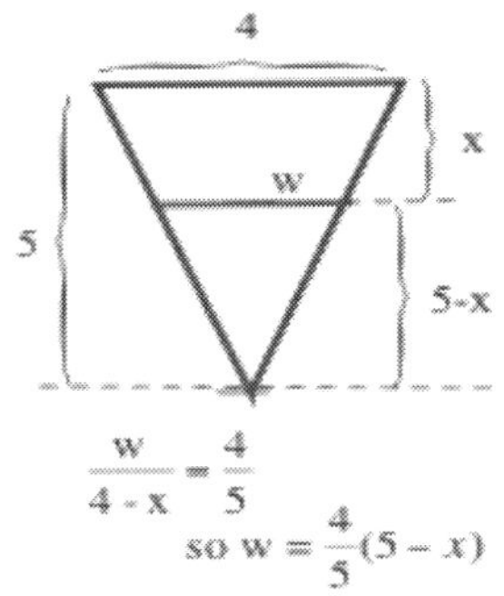

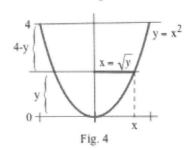

Fig. 3

5. Fig. 4. $\int_0^4 d \cdot (4-y) \cdot (2\sqrt{y}) \, dy = 2d \int_0^4 4y^{1/2} - y^{3/2} \, dy$

$$= 2d \{ 4(\frac{2}{3} y^{3/2}) - \frac{2}{5} y^{5/2} \} \Big|_0^4$$

$$= 2d \{ 4(\frac{2}{3}(8)) - \frac{2}{5}(32) \} = \frac{256}{15} d .$$

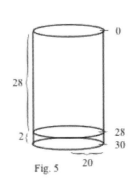

Fig. 4

6. All three have the same total force against their sides.
7. The one with the largest perimeter: not (a), probably (b)

9. Fig. 5. (a) $\int_{28}^{30} d \cdot x \cdot (40\pi) \, dx = 20\pi d \, x^2 \Big|_{28}^{30} = 20\pi d (116) = 2{,}320 \pi d$.

(b) $\int_{33}^{35} d \cdot x \cdot (40\pi) \, dx = 20\pi d \, x^2 \Big|_{33}^{35} = 20\pi d (136) = 2{,}720 \pi d$.

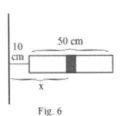

Fig. 5

Kinetic Energy

11. (a) M = 20 g. v = 3 rev/sec = 3(2π(15cm))/sec = 90π cm/sec.

$KE = \frac{1}{2} M v^2 = \frac{1}{2}(20 g)(90\pi \text{ cm/s})^2 = 81{,}000 \pi^2$ ergs ≈ 799,438 ergs.

(b) M = 20 g. v = 3 rev/sec = 3(2π(**20**cm))/sec = 120π cm/sec.

$KE = \frac{1}{2} M v^2 = \frac{1}{2}(20 g)(120\pi \text{ cm/s})^2 = 144{,}000 \pi^2$ ergs ≈ 1,421,223 ergs.

13. Fig. 6. $KE_i = \frac{1}{2}(3 \Delta x)(2 \cdot 2\pi x)^2 = 24\pi^2 x^2 \Delta x$

$KE = \int_{10}^{60} 24\pi^2 x^2 \, dx = 8\pi^2 x^3 \Big|_{10}^{60} = 1.72 \cdot 10^6 \pi^2$ ergs ≈ $1.698 \cdot 10^7$ ergs.

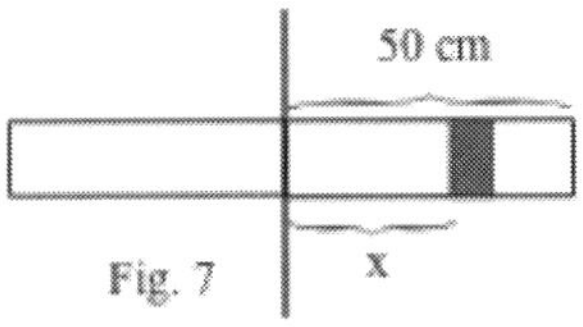

Fig. 6

15. Fig. 7. (20 g)/(100 cm) = 0.2 g/cm. $KE_i = \frac{1}{2}(0.2 \Delta x)(2\pi x)^2 = 0.4\pi^2 x^2 \Delta x$

$KE = \int_{50}^{0} 0.4\pi^2 x^2 \, dx = 0.4\pi^2 \frac{1}{3} x^3 \Big|_0^{50} = 16{,}667 \pi^2$ ergs ≈ 164,493 ergs.

Total KE = 2{ 16,667 π^2 ergs } ≈ 328,986 ergs.

Fig. 7

17. (a) $m_i = 2\pi r \, \Delta x = 2\pi x \, \Delta x$,

$v_i = 3$ rev/sec $= 3(2\pi$ radians $)/$sec $= 6\pi x$ radians/sec.

$KE_i = \frac{1}{2} m_i (v_i)^2 = \frac{1}{2}(2\pi x \, \Delta x)(6\pi x)^2 = 36 \pi^3 x^3 \Delta x$

$KE = \int_1^3 36 \pi^3 x^3 \, dx = 36 \pi^3 (\frac{1}{4}) x^4 \Big|_1^3 = 9 \pi^3 (3^4 - 1^4) = 720 \pi^3$

(b) $m_i = 4\pi x \, \Delta x$, $v_i = 3$ rev/sec $= 3(2\pi$ radians $)/$sec $= 6\pi x$ radians/sec.

$KE_i = \frac{1}{2} m_i (v_i)^2 = \frac{1}{2}(4\pi x \, \Delta x)(6\pi x)^2 = 72 \pi^3 x^3 \Delta x$

$KE = \int_3^4 72 \pi^3 x^3 \, dx = 72 \pi^3 (\frac{1}{4}) x^4 \Big|_3^4 = 18 \pi^3 (4^4 - 3^4) = 3{,}150 \pi^3$

19. density $= 3$ g/cm^3.

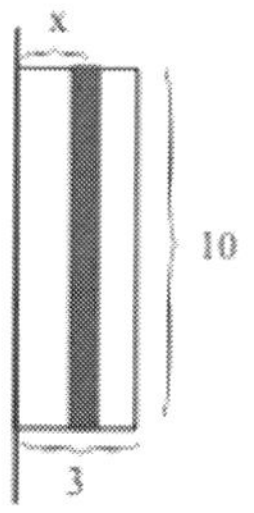

(a) Fig. 8. $m_i = $ (area)$(3$ g/cm$^3) = (10 \, \Delta x)(3$ g/cm$^3) = 30 \, \Delta x$.

$v_i = 2$ rev/sec $= 2(2\pi x$ radians $)/$sec $= 4\pi x$ radians/sec.

$KE_i = \frac{1}{2} m_i (v_i)^2 = \frac{1}{2}(30 \, \Delta x)(4\pi x)^2 = 240 \pi^2 x^2 \Delta x$

$KE = \int_0^3 240 \pi^2 x^2 \, dx = 80 \pi^2 x^3 \Big|_0^3 = 18 \pi^2 (27) = 2{,}160 \pi^2$

Total KE $= 2\{ 2{,}160 \pi^2 \} = 4{,}320 \pi^2$

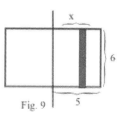

Fig. 8

(b) Fig. 9. $m_i = $ (area)$(3$ g/cm$^3) = (6 \, \Delta x)(3$ g/cm$^3) = 18 \, \Delta x$. $v_i = 4\pi x$ radians/sec.

$KE_i = \frac{1}{2} m_i (v_i)^2 = \frac{1}{2}(18 \, \Delta x)(4\pi x)^2 = 144 \pi^2 x^2 \Delta x$

$KE = \int_0^5 144 \pi^2 x^2 \, dx = 48 \pi^2 x^3 \Big|_0^5 = 48 \pi^2 (125) = 6{,}000 \pi^2$

Total KE $= 2\{ 6{,}000 \pi^2 \} = 12{,}000 \pi^2$

Fig. 9

Volumes using tubes: volume $= \int_a^b 2\pi(\text{radius})(\text{height})(\text{thickness})$

21. Fig. 10. $\int_a^b 2\pi(\text{radius})(\text{height})(\text{thickness})$

$= \int_0^1 2\pi(x)(\sqrt{1-x^2}) \, dx$ (put $u = 1 - x^2$, then $du = -2x \, dx$)

$= 2\pi \int_{u=1}^{u=0} u^{1/2} (\frac{-1}{2}) \, du = -\pi (\frac{3}{2}) u^{3/2} \Big|_{u=1}^{u=0}$

$= (-2\pi,3)(0) - (\frac{-2\pi}{3}(1)) = \frac{2}{3}\pi$.

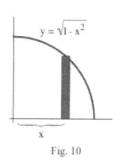

Fig. 10

Odd Answers: Chapter Five   Contemporary Calculus   16

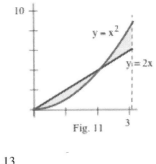

Fig. 11

23. Fig. 11. $\int_a^b 2\pi(\text{radius})(\text{height})(\text{thickness})$

$= \int_0^2 2\pi(4-x)(2x-x^2)\,dx + \int_2^3 2\pi(4-x)(x^2-2x)\,dx$

$= 2\pi\int_0^2 -x^3 + 6x^2 - 8x\,dx + 2\pi\int_2^3 x^3 - 6x^2 + 8x\,dx = 2\pi\{4\} + 2\pi\{\frac{7}{4}\} \approx 36.13$

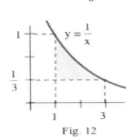

Fig. 12

25. Fig. 12. $\int_a^b 2\pi(\text{radius})(\text{height})(\text{thickness}) = \int_1^3 2\pi(5-x)(\frac{1}{x} - \frac{1}{3})\,dx$

$= 2\pi\int_1^3 \frac{5}{x} - \frac{8}{3} + \frac{x}{3}\,dx = 2\pi\{5\ln(x) - \frac{8}{3}x + \frac{1}{6}x^2\}\Big|_1^3$

$= 2\pi\{-4 + 5\ln(3)\} = 9.38$.

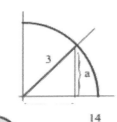

Fig. 13

27. Fig. 13. $\int_a^b 2\pi(\text{radius})(\text{height})(\text{thickness}) = \int_1^4 2\pi(x)(x - \ln(x))\,dx$

≈ 85.8 (Using calculator.)

29. Fig. 14. $\int_a^b 2\pi(\text{radius})(\text{height})(\text{thickness}) = \int_0^{3\sqrt{2}} 2\pi(x)(\sqrt{9-x^2} - x)\,dx$

= finish on your own.

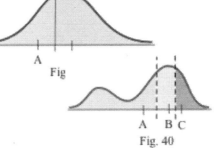

14

Voting

31. a votes for cand. A, b votes for cand. A, c votes for cand. C.

32. a votes for cand. B, b votes for cand. C, c votes for cand. C.

33. Fig. 39: B wins.   34. Fig. 40: A wins.

35. (a) Fig. 41a: A wins.

(b) Fig. 41b: C wins.

(c) Fig. 41c: B wins.

Fig. 40

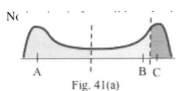
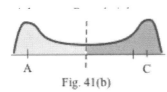
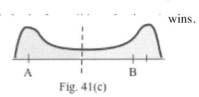

Fig. 41(a)   Fig. 41(b)   Fig. 41(c)

36. (a) B wins.   (b) B wins.   (c) A wins.   37. (a) A wins.   (b) A wins.   (c) looks like C wins

38. (a) B wins.   (b) A wins.   (c) looks like A wins   39. on your own

6.0 INTRODUCTION TO DIFFERENTIAL EQUATIONS

This chapter is an introduction to differential equations, a major field in applied and theoretical mathematics and a very useful one for engineers, scientists, and others who study changing phenomena. The physical laws of motion and heat and electricity can be written as differential equations. The growth of a population, the changing gene frequencies in that population, and the spread of a disease can be described by differential equations. Economic and social models use differential equations, and the earliest examples of "chaos" came from studying differential equations used for modeling atmospheric behavior. Some scientists even say that the main purpose of a calculus course should be to teach people to understand and solve differential equations.

The purpose of this chapter is to introduce some basic ideas, vocabulary, and techniques for differential equations and to explore additional applications. Applications in this chapter include

> exponential population growth and calculating how long it takes a population to double in size,
>
> radioactive decay and its use for dating ancient objects and detecting fraud,
>
> describing the motion of an object through a resisting medium, and
>
> chemical mixtures and rates of reaction.

More complicated differential equations and ways of using and solving them are discussed in Chapter 17 and in later courses.

Differential Equations

Algebraic equations contain constants and variables, and the solutions of an algebraic equation are typically numbers. For example, $x = 3$ and $x = -2$ are solutions of the algebraic equation $x^2 = x + 6$. Differential equations contain derivatives or differentials of functions. Solutions of differential equations are functions. The differential equation $y' = 3x^2$ has infinitely many solutions, and two of those solutions are the functions $y = x^3 + 2$ and $y = x^3 - 4$ (Fig. 1).

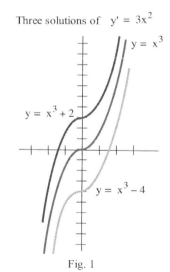

Fig. 1

You have already solved lots of differential equations: every time you found an antiderivative of a function $f(x)$, you solved the differential equation $y' = f(x)$ to get a solution y. You have also used differential equations in applications. Areas, volumes, work, and motion problems all involved integration and finding antiderivatives so they all used differential equations. The differential equation $y' = f(x)$, however, is just the beginning. Other applications generate different differential equations.

Checking Solutions of Differential Equations

Whether a differential equation is easy or difficult to solve, it is important to be able to check that a possible solution really satisfies the differential equation.

A possible solution of an algebraic equation can be checked by putting the solution into the equation to see if the result is true: $x = 3$ is a solution of $5x + 1 = 16$ since $5(3) + 1 = 16$ is true. Similarly, a solution of a differential equation can be checked by substituting the function and the appropriate derivatives into the equation to see if the result is true: $y = x^2$ is a solution of $xy' = 2y$ since $y' = 2x$ and $x(2x) = 2(x^2)$ is true.

Example 1: Check (a) that $y = x^2 + 5$ is a solution of $y'' + y = x^2 + 7$ and (b) that $y = x + 5/x$ is a solution of $y' + \frac{y}{x} = 2$.

Solution: (a) $y = x^2 + 5$ so $y' = 2x$ and $y'' = 2$. Substituting these functions for y and y'' into the differential equation $y'' + y = x^2 + 7$, we have $y'' + y = (2) + (x^2 + 5) = x^2 + 7$, so $y = x^2 + 5$ is a solution of the differential equation.

(b) $y = x + 5/x$ so $y' = 1 - 5/x^2$. Substituting these functions for y and y' in the differential equation $y' + \frac{y}{x} = 2$, we have

$$y' + \frac{y}{x} = (1 - 5/x^2) + \frac{1}{x}(x + 5/x) = 1 - 5/x^2 + 1 + 5/x^2 = 2,$$ the result we wanted to verify.

Practice 1: Check (a) that $y = 2x + 6$ is a solution of $y - 3y' = 2x$ and (b) that $y = e^{3x}$, $y = 5e^{3x}$, and $y = Ae^{3x}$ (A is a constant) are all solutions of $y'' - 2y' - 3y = 0$.

A solution of a differential equation with the initial condition $y(x_0) = y_0$ is a function that satisfies the differential equation and the initial condition. To check the solution of an initial value problem we must check that both the equation and the initial condition are satisfied.

Example 2: Which of the given functions is a solution of the initial value problem $y' = 3y$, $y(0) = 5$? (a) $y = e^{3x}$ (b) $y = 5e^{3x}$ (c) $y = -2e^{3x}$

Solution: All three functions satisfy the differential equation $y' = 3y$, but only one of the functions satisfies the initial condition that $y(0) = 5$. If $y = e^{3x}$, then $y(0) = e^{3(0)} = 1 \neq 5$ so $y = e^{3x}$ does not satisfy the initial condition (Fig. 2). If $y = 5e^{3x}$, then $y(0) = 5e^{3(0)} = 5$ satisfies the initial condition. If $y = -2e^{3x}$, then $y(0) = -2e^{3(0)} = -2 \neq 5$ so $y = e^{3x}$ does not satisfy the initial condition.

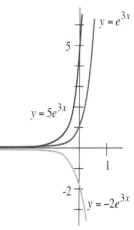

Fig. 2

Practice 2: Which function is a solution of the initial value problem $y'' = -9y$, $y(0) = 2$?

(a) $y = \sin(3x)$ (b) $y = 2\sin(3x)$ (c) $y = 2\cos(3x)$.

Finding the Value of the Constant

Differential equations usually have lots of solutions, a whole "family" of them, and each solution satisfies a different initial condition. To find which solution of a differential equation also satisfies a given initial condition $y(x_0) = y_0$, we can replace x and y in the solution function with the values x_0 and y_0 and algebraically solve for the value of the unknown constant.

Example 3: For every value of C, the function $y = Cx^2$ is a solution of $xy' = 2y$ (Fig. 3). Find the value of C so that $y(5) = 50$.

Solution: Substituting the initial condition $x = 5$ and $y = 50$ into the solution $y = Cx^2$, we have that $50 = C(5)^2$ so $C = 50/25 = 2$. The function $y = 2x^2$ satisfies both the differential equation and the initial condition.

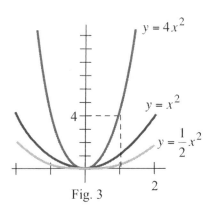

Fig. 3

Practice 3: For every value of C, the function $y = e^{2x} + C$ is a solution of $y' = 2e^{2x}$. Find the value of C so that $y(0) = 7$. (Be careful, $C \neq 7$.)

PROBLEMS

In problems 1 – 10, check that the function y is a solution of the given differential equation.

1. $y' + 3y = 6$. $y = e^{-3x} + 2$.

2. $y' - 2y = 8$. $y = e^{2x} - 4$.

3. $y'' - y' + y = x^2$. $y = x^2 + 2x$

4. $3y'' + y' + y = x^2 - 4x$. $y = x^2 - 6x$

5. $xy' - 3y = x^2$. $y = 7x^3 - x^2$

6. $xy'' - y' = 3$. $y = x^2 - 3x + 5$

7. $y' + y = e^x$. $y = \frac{1}{2}e^x + 2e^{-x}$

8. $y'' + 25y = 0$. $y = \sin(5x) + 2\cos(5x)$

9. $y' = -x/y$. $y = \sqrt{7 - x^2}$.

10. $y' = x - y$. $y = x - 1 + 2e^{-x}$.

In problems 11 – 20, check that the function y is a solution of the given initial value problem.

11. $y' = 6x^2 - 3$ and $y(1) = 2$. $y = 2x^3 - 3x + 3$.

12. $y' = 6x + 4$ and $y(2) = 3$. $y = 3x^2 + 4x - 17$.

13. $y' = 2\cos(2x)$ and $y(0) = 1$. $y = \sin(2x) + 1$.

14. $y' = 1 + 6\sin(2x)$ and $y(0) = 2$. $y = x - 3\cos(2x) + 5$.

15. $y' = 5y$ and $y(0) = 7$. $y = 7e^{5x}$.

16. $y' = -2y$ and $y(0) = 3$. $y = 3e^{-2x}$.

17. $xy' = -y$ and $y(1) = -4$. $y = -4/x$.

18. $yy' = -x$ and $y(0) = 3$. $y = \sqrt{9 - x^2}$.

19. $y' = 5/x$ and $y(e) = 3$. $y = 5\ln(x) - 2$.

20. $y' + y = e^x$ and $y(0) = 5$. $y = \frac{1}{2}e^x + \frac{9}{2}e^{-x}$.

In problems 21 – 30, a family of solutions of a differential equation is given. Find the value of the constant C so the solution satisfies the initial value condition.

21. $y' = 2x$ and $y(3) = 7$. $y = x^2 + C$.

22. $y' = 3x^2 - 5$ and $y(1) = 2$. $y = x^3 - 5x + C$.

23. $y' = 3y$ and $y(0) = 5$. $y = Ce^{3x}$.

24. $y' = -2y$ and $y(0) = 3$. $y = Ce^{-2x}$.

25. $y' = 6\cos(3x)$ and $y(0) = 4$. $y = 2\sin(3x) + C$.

26. $y' = 3 - 2\sin(2x)$ and $y(0) = 1$. $y = 3x + \cos(2x) + C$.

27. $y' = 1/x$ and $y(e) = 2$. $y = \ln(x) + C$.

28. $y' = 1/x^2$ and $y(1) = 3$. $y = -1/x + C$.

29. $y' = -y/x$ and $y(2) = 10$. $y = -C/x$.

30. $y' = -x/y$ and $y(3) = 4$. $y = \sqrt{C - x^2}$.

In problems 31 – 40, find the function y which satisfies each initial value problem.

31. $y' = 4x^2 - x$ and $y(1) = 7$.

32. $y' = x - e^x$ and $y(0) = 3$.

33. $y' = 3/x$ and $y(1) = 2$.

34. $xy' = 1$ and $y(e) = 7$.

35. $y' = 6e^{2x}$ and $y(0) = 1$.

36. $y' = 36(3x - 2)^2$ and $y(1) = 8$.

37. $y' = x \cdot \sin(x^2)$ and $y(0) = 3$.

38. $y' = 6/x^2$ and $y(1) = 2$.

39. $xy' = 6x^3 - 10x^2$ and $y(2) = 5$.

40. $x^2 y' = 6x^3 - 1$ and $y(1) = 10$.

41. Show that if $y = f(x)$ and $y = g(x)$ are solutions to $y' + 5y = 0$, then so are $y = 3 \cdot f(x)$, $y = 7 \cdot g(x)$, $y = f(x) + g(x)$, and $y = A \cdot f(x) + B \cdot g(x)$ for any constants A and B.

42. Show that if $y = f(x)$ and $y = g(x)$ are solutions to $y'' + 2y' - 3y = 0$, then so are $y = 3 \cdot f(x)$, $y = 7 \cdot g(x)$, $y = f(x) + g(x)$, and $y = A \cdot f(x) + B \cdot g(x)$ for any constants A and B.

43. Show that $y = \sin(x) + x$ and $y = \cos(x) + x$ are both solutions of $y'' + y = x$.
    Are $y = 3\{\sin(x) + x\}$ and $y = \{\sin(x) + x\} + \{\cos(x) + x\}$ solutions of $y'' + y = x$?

44. Show that $y = e^{3x} - 2$ and $y = 5e^{3x} - 2$ are both solutions of $y' - 3y = 6$.
    Are $y = 7\{e^{3x} - 2\}$ and $y = \{e^{3x} - 2\} + \{5e^{3x} - 2\}$ solutions of $y' - 3y = 6$?

45. $\frac{dy}{dt} = A - By$ (A and B positive constants) describes the concentration y of glucose in the blood at time t. Check that $y = \frac{A}{B} - C \cdot e^{-Bt}$ is a solution of the differential equation for every value of the constant C.

46. $\frac{dy}{dt} = Ay$ (A a constant) is used to model "exponential" growth and decay. Check that $y = C \cdot e^{At}$ is a solution of the differential equation for every value of the constant C.

47. $L\frac{dI}{dt} + RI = E$ (L, R, and E are positive constants) describes the current I in an electrical circuit. Show that $I = \frac{E}{R}(1 - e^{-Rt/L})$ is a solution of the differential equation.

48. $m \cdot y'' = -C \cdot y$ (C a positive constant) describes the position y of an object on a spring as it moves up and down. Show that $y = A \cdot \sin(wt) + B \cdot \cos(wt)$ with $w = \sqrt{C/m}$ is a solution of the differential equation for all values of the constants A and B.

Section 6.0   **PRACTICE Answers**

Practice 1:   (a) $y = 2x + 6$. $y' = 2$. $y - 3y' = (2x+6) - 3(2) = 2x$ (OK)

(b) $y = e^{3x}$. $y' = 3e^{3x}$. $y'' = 9e^{3x}$. $y'' - 2y' - 3y = 9e^{3x} - 2(3e^{3x}) - 3(e^{3x}) = 0$ (OK)

$y = 5e^{3x}$. $y' = 15e^{3x}$. $y'' = 45e^{3x}$. $y'' - 2y' - 3y = 45e^{3x} - 2(15e^{3x}) - 3(5e^{3x}) = 0$ (OK)

$y = Ae^{3x}$. $y' = 3Ae^{3x}$. $y'' = 9Ae^{3x}$. $y'' - 2y' - 3y = 9Ae^{3x} - 2(3Ae^{3x}) - 3(Ae^{3x}) = 0$ (OK)

Practice 2:   Want $y'' = -9y$, $y(0) = 2$.

(a) $y = \sin(3x)$, $y' = 3\cos(3x)$, $y'' = -9\sin(3x) = 9y$ (OK) but $y(0) = 2\sin(0) = 0 \neq 2$.

(b) $y = 2\sin(3x)$, $y' = 6\cos(3x)$, $y'' = -18\sin(3x) = 9y$ (OK) but $y(0) = 2\sin(0) = 0 \neq 2$.

(c) $y = 2\cos(3x)$. $y' = -6\sin(3x)$. $y'' = -18\cos(3x) = 9y$ (OK) and $y(0) = 2\cos(0) = 2$ (OK).

$y = 2\cos(3x)$ satisfies both the differential equation and the initial condition.

Practice 3: $y = e^{2x} + C$. $y' = 2e^{2x}$ (OK)

$7 = y(0) = e^{2 \cdot 0} + C = 1 + C$ so $C = 6$. Then $y = e^{2x} + 6$.

6.1 THE DIFFERENTIAL EQUATION y' = f(x)

This section introduces the basic concepts and vocabulary of differential equations as they apply to the familiar problem, $y' = f(x)$. The fundamental notions of a general solution of a differential equation, the particular solution of an initial value problem, and a direction field are introduced here and are used in later sections as we examine more complicated differential equations and their applications.

Solving y' = f(x)

The solution of the differential equation $y' = f(x)$ is the collection of all antiderivatives of f, $y = \int f(x)\,dx$. If $y = F(x)$ is one antiderivative of f, then we have essentially found all of the antiderivatives of f since every antiderivative of f has the form $F(x) + C$ for some value of the constant C. If F is an antiderivative of f, the collection of functions $F(x) + C$ is called the **general solution** of $y' = f(x)$. The general solution is a whole family of functions.

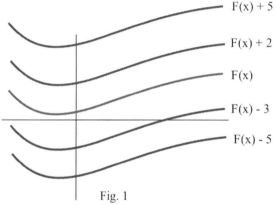

Fig. 1

Example 1: Find the general solution of the differential equation $y' = 2x + e^{3x}$.

Solution: $y = \int 2x + e^{3x}\,dx = x^2 + \frac{1}{3} e^{3x} + C$.

Practice 1: Find the general solutions of the differential equations $y' = x + \frac{3}{x+2}$ and $y' = \frac{6}{x^2 + 1}$.

Direction Fields

Geometrically, a derivative gives the slope of the tangent line to a curve, so the differential equation $y' = f(x)$ can be interpreted as a geometric condition: at each point (x,y) on the graph of y, the slope of the tangent line is $f(x)$. The differential equation $y' = 2x$ says that at each point (x,y) on the graph of y, the **slope** of the line tangent to the graph is 2x: if the point $(5,3)$ is on the graph of y then the slope of the tangent line there is $2 \cdot 5 = 10$. This information can be represented graphically as a **direction field** for $y' = 2x$. A direction field for $y' = 2x$ is a a collection of short line segments through a number of sample points (x, y) in the plane (Fig. 2), and the slope of the segment through (x,y) is 2x. Figures 3a and 3b show direction fields for the differential equations $y' = 3x^2$ and $y' = \cos(x)$. For a differential equation of the form $y' = f(x)$, the values of y' depend only on x so along any vertical line (for a fixed value of x) all the line segments have the same y', the same slope, and they are parallel (Fig. 4).

Fig. 2: Direction field for $y' = 2x$

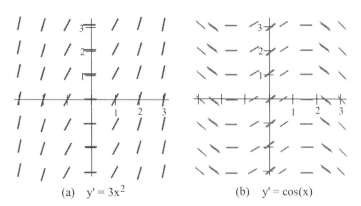
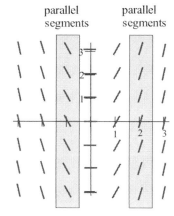

Fig. 3: Direction fields   Fig. 4: Direction field for $y' = f(x)$

If y' depends on both x and y, then the slopes of the line segments will depend on both x and y, and the slopes along a vertical line can vary. Fig. 5 shows a direction field for the differential equation $y' = x - y$ in which y' is a function of both x and y.

A **direction field** of a differential equation $y' = g(x,y)$ is a collection of short line segments with slopes $g(x,y)$ at the points (x, y).

Practice 2: Construct direction fields for (a) $y' = x + 1$ and (b) $y' = x + y$ by sketching a short line segment with slope y' at each point (x, y) with integer coordinates between -3 and 3.

If the function f in the differential equation $y' = f(x)$ is given graphically, an approximate direction field can be constructed.

Example 2: Construct a direction field for the differential equation $y' = f(x)$ for the f shown in Fig. 6.

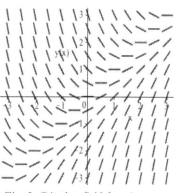

Fig. 5: Direction field for $y' = x - y$

Solution: If $x = 0$, then $y' = f(0) = 1$ so at every point on the vertical line $x = 0$ (the y–axis) the line segments of the direction field have slope $y' = 1$. Several short segments with slope 1 are shown along the y–axis in Fig. 7. Similarly, if $x = 1$, then $y' = f(1) = 0$ so the line segments of the direction field have slope $y' = 0$ at every point on the vertical line $x = 1$. The line segments along each vertical line are parallel.

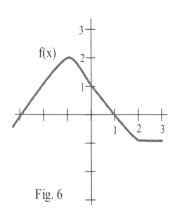

Fig. 6

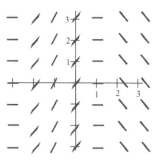

Fig. 7: Direction field for $y' = f(x)$

Practice 3: Construct a direction field for the differential equation $y' = f(x)$ for the f shown in Fig. 8.

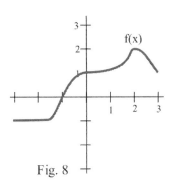

Fig. 8

Once we have a direction field for a differential equation, we can sketch several curves which have the appropriate tangent line slopes (Fig. 9). In that way we can see the shapes of the solutions even if we do not have formulas for them. These shapes can be useful for estimating which initial conditions lead to linear solutions or periodic solutions or solutions with other properties, and they can help us understand the behavior of machines and organisms.

Direction fields are tedious to plot by hand, but computers and calculators can plot them quickly. Programs are available for plotting direction fields on graphic calculators.

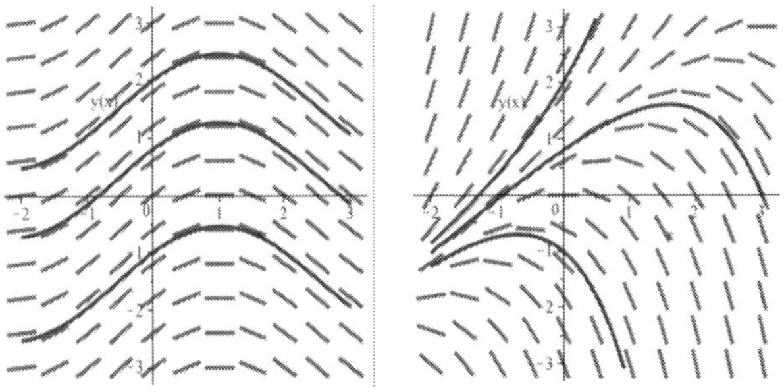

Initial Conditions and Particular Solutions

An **initial condition** $y(x_0) = y_0$ specifies that the solution y of the differential equation should go through the point (x_0, y_0) in the plane. To solve a differential equation with an initial condition, we typically use integration to find the general solution (the family of solutions which contains an arbitrary constant), and then we use algebra to find the value of the constant so the solution satisfies the initial condition. The member of the family that satisfies the initial condition is called a **particular solution**.

Fig. 9: Functions with the appropriate slopes for the two direction fields

Example 3: Solve the differential equation $y' = 2x$ with the initial condition $y(2) = 1$.

Solution: The general solution is $y = \int 2x \, dx = x^2 + C$. Substituting the values $x_0 = 2$ and $y_0 = 1$ into the general solution, $1 = (2)^2 + C$ so $C = -3$. Then the solution we want is $y = x^2 - 3$. (A quick check verifies that $y' = 2x$ and $y = 1$ when $x = 2$.) Fig. 10 shows the direction field $y' = 2x$ and the particular solution which goes through the point $(2,1)$, $y = x^2 - 3$. The solution of the differential equation which satisfies the initial condition $y(0) = -1$ is also shown.

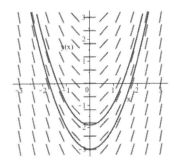

Fig. 10: Direction field $y'=2x$ and solutions y for $y(2)=1$ and $y(0)=-1$

Example 4: If a ball is tossed upward with a velocity of 100 feet/second, its height y at time t satisfies the differential equation $y' = 100 - 32t$. Sketch the direction field for y ($0 \le t \le 4$) and then sketch the solution that satisfies the initial condition that the ball is 200 feet high after 3 seconds.

Solution: $y = \int 100 - 32t \, dt = 100t - 16t^2 + C$.

When $t=3$, $y=200$ so $200 = 100(3) - 16(3)^2 + C = 156 + C$

and $C = 200 - 156 = 44$.

The particular function we want is $y = 100t - 16t^2 + 44$. The direction field and the particular solution are shown in Fig. 11.

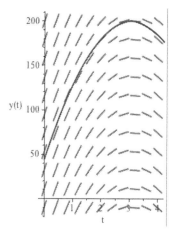

Fig. 11: Direction field $y' = 100 - 32t$ and solution with $y(3) = 200$

Practice 4: Find the solution of $y' = 9x^2 - 6\sin(2x) + e^x$ which goes through the point (0,6).

Example 5: A direction field for $y' = x - y$ is shown in Fig. 12. Sketch the three particular solutions of the differential equation $y' = x - y$ which satisfy the initial conditions $y(0) = 2$, $y(0) = -1$, and $y(1) = -2$.

Solution: The three particular solutions are shown in Fig. 12.

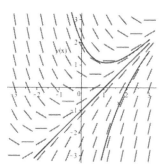

Fig. 12: Particular solutions of $y' = x - y$

PROBLEMS

In problems 1 – 6, the direction field of a differential equation is shown. Sketch the solutions which satisfy the given initial conditions.

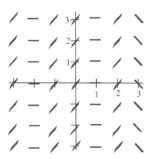

Fig. 13: Direction field for $y' = f(x)$

1. Fig. 13. The initial conditions are
   (a) $y(0) = 1$, (b) $y(1) = -2$, and (c) $y(1) = 3$.

2. Fig. 13. The initial conditions are
   (a) $y(0) = 2$, (b) $y(1) = -1$, and
   (c) $y(0) = -2$.

3. Fig. 14. The initial conditions are
   (a) $y(-2) = 1$, (b) $y(0) = 1$, and (c) $y(2) = 1$.

4. Fig. 14. The initial conditions are
   (a) $y(-2) = -1$, (b) $y(0) = -1$, and (c) $y(2) = -1$.

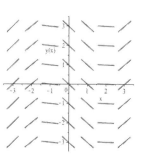

Fig. 14: Direction field for $y' = g(x)$

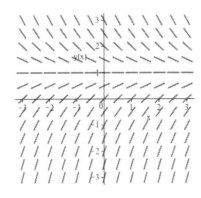

5. Fig. 15. The initial conditions are (a) $y(0) = -2$, (b) $y(0) = 0$, and (c) $y(0) = 2$. What happens to these three solutions for large values of x?

6. Fig. 15. The initial conditions are (a) $y(2) = -2$, (b) $y(2) = 0$, and (c) $y(2) = 2$. What happens to these three solutions for large values of x?

Fig. 15: Direction field for $y'=f(x)$

In problems 7 – 12, (i) sketch the direction field for each differential equation and (ii) without solving the differential equation, sketch the solutions that go through the points $(0,1)$ and $(2,0)$.

7. $y' = 2x$.

8. $y' = 2 - x$.

9. $y' = 2 + \sin(x)$.

10. $y' = e^x$.

11. $y' = 2x + y$.

12. $y' = 2x - y$.

In problems 13 – 18, (a) find the family of functions which solve each differential equation, and (b) find the particular member of the family that goes through the given point.

13. $y' = 2x - 3$ and $y(1) = 4$.

14. $y' = 1 - 2x$ and $y(2) = -3$.

15. $y' = e^x + \cos(x)$ and $y(0) = 7$.

16. $y' = \sin(2x) - \cos(x)$ and $y(0) = -5$.

17. $y' = \dfrac{6}{2x+1} + \sqrt{x}$ and $y(1) = 4$.

18. $y' = e^x/(1 + e^x)$ and $y(0) = 0$.

Problems 19 and 20 concern the direction field (Fig. 16) that comes from a differential equation called the logistic equation ($y' = y(1 - y)$, that is used to model the growth of a population in an enviroment with renewable but limited resources. It is also used to describe the spread of a rumor or disease through a population.

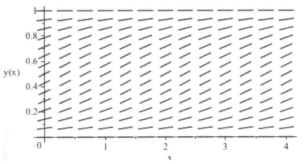

Fig. 16: Direction field for a Logistic Equation $y' = y*(1 - y)$

19. Sketch the solution that satisfies the initial condition $P(0) = 0.1$. What letter of the alphabet does this solution look like?

20. Sketch several solutions that have different initial values for $P(0)$. What appears to happen to all of these solutions after a "long time" (for large values of x)?

In problems 21 and 22, the figures show the direction of surface flow at different locations along a river. Sketch the paths small corks will follow if they are put into the river at the dots in each figure. (Since the magnitude and the direction of flow are indicated, each diagram is called a **vector field**.) Notice that corks that start close to each other can drift far apart and corks that start far apart can drift close.

21. Fig. 17 .   22. Fig. 18 .

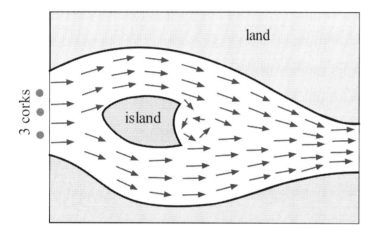

Fig. 17: Surface flow along a river

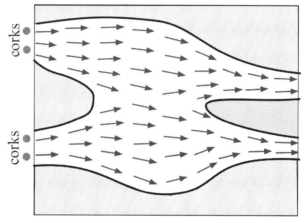

Fg. 18: Surface flow along a river

MAPLE

Many of the direction fields in this section were created using a computer program called MAPLE.
The commands below were used to create Fig. 12

with(DEtools):   *This loads a library needed for the DEplott command.*

DEplot(diff(y(x),x) = x - y(x), y(x), x=-3..3, y=3..3, dirgrid=[11,11], color=red, arrows=line, thickness=2,
       [[y(0)=-1], [y(0)=2],[y(2)=0]], linecolor=blue);

Plots the direction field for y ' = x – y
for -3 ≤ x ≤ 3 and -3 ≤ y ≤ 3
using an 11 by 11 grid
of red arrows with thickness 2.
   (arrows can be set equal to "small", "medium", "large" or "line,"

It includes the solution curves for the initial conditions y(0)=-1, y(0)=2 and y(2)=0 in blue.

Section 6.1 PRACTICE Answers

Practice 1: (a) $y' = x + \dfrac{3}{x+2}$. $y = \int x + \dfrac{3}{x+2}\,dx = \dfrac{1}{2}x^2 + 3\cdot \ln|x+2| + C$.

(b) $y' = \dfrac{6}{x^2+1}$. $y = \int \dfrac{6}{x^2+1}\,dx = 6\cdot \arctan(x) + C$.

Practice 2: (a)

x	y' = x + 1
−3	−2
−2	−1
−1	0
0	1
1	2
2	3
3	4

(See Fig. 19)

y' does not depend on the value of y.

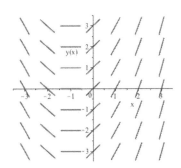

Fig. 19: Direction field for $y' = x + 1$

(b) $y' = x + y$ (See Fig. 20)

y = −3		y = −2		y = −1	
x	y' = x−3	x	y' = x−2	x	y' = x−1
−3	−6	−3	−5	−3	−4
−2	−5	−2	−4	−2	−3
−1	−4	−1	−3	−1	−2
0	−3	0	−2	0	−1
1	−2	1	−1	1	0
2	−1	2	0	2	1
3	0	3	1	3	2

Fig. 20: Direction field for $y' = x + y$

Practice 3: The direction field for the differential equation $y' = f(x)$ for f (given in Fig. 8) is shown in Fig. 21.

Practice 4: $y' = 9x^2 - 6\sin(2x) + e^x$ and $y(0) = 6$.

$y = \int 9x^2 - 6\sin(2x) + e^x\,dx$

$= 3x^3 + 3\cos(2x) + e^x + C$.

$6 = y(0) = 3(0)^3 + 3\cos(2\cdot 0) + e^0 + C = 0 + 3 + 1 + C$

so $C = 6 - 3 - 1 = 2$ and

$y = 3x^3 + 3\cos(2x) + e^x + 2$.

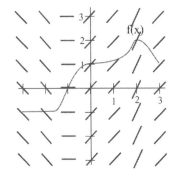

Fig. 21: Directio field for Practice 3

6.2 SEPARABLE DIFFERENTIAL EQUATIONS

In the previous section, y' was a function of x alone and the slopes of the line segments of the direction field did not depend on the y-coordinate of the location of the line segment. In many situations, however, y' depends on both x and y, for example, $y' = xy$ (Fig. 1) or $y' = x + y$ (Fig. 2). This section emphasizes how to solve differential equations in which the variables can be "separated," and the next section examines several applications of these "separable" differential equations.

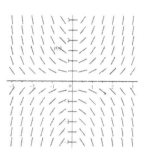

Fig. 1: $y' = x*y$

Definition: A differential equation is called **separable** if the variables can be separated algebraically so that the equation has the form
{function of **y** alone}·y' = {function of **x** alone}: $g(y) \cdot y' = f(x)$.

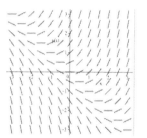

Fig. 2: $y' = x + y$

Example 1: "Separate the variables" by writing each differential equation in the form $g(y) \cdot y' = f(x)$ (x, y > 0).

(a) $y' = xy$ (b) $xy' = \frac{y+1}{x}$ (c) $y' = \frac{1 + \sin(x)}{y^2 + y}$ (d) $y' = y$

Solution: (a) Dividing each side of $y' = xy$ by y, we have $\frac{1}{y} \cdot y' = x$ so $g(y) = \frac{1}{y}$ and $f(x) = x$.

(b) Dividing each side by $x(y+1)$, we have $\frac{1}{y+1} \cdot y' = \frac{1}{x^2}$ so $g(y) = \frac{1}{y+1}$ and $f(x) = \frac{1}{x^2}$.

(c) Multiplying each side by $y^2 + y$, $(y^2 + y) \cdot y' = 1 + \sin(x)$ so $g(y) = y^2 + y$ and $f(x) = 1+\sin(x)$.

(d) Dividing each side by y, $\frac{1}{y} \cdot y' = 1$ so $g(y) = \frac{1}{y}$ and $f(x) = 1$.

The differential equations (a) – (d) are separable.

Practice 1: Show these differential equations are separable by rewriting them in the form $g(y) \cdot y' = f(x)$.

(a) $y' = x^3 (y - 5)$ (y > 5) (b) $y' = \frac{3}{2x + x \cdot \sin(y+2)}$ (x > 0)

Many differential equations can not be written in the form $g(y) \cdot y' = f(x)$; they are not separable. For example, $y' = x + y$ and $y' = \sin(xy) + x$ are not separable. Techniques for solving some of these nonseparable equations are discussed in Chapter 17.

Solving Separable Differential Equations

The steps to solve a separable differential equation are straightforward:
- use algebra to separate the variables,
- put the equation into an equivalent form with differentials, and
- integrate each side of the equation.

Example 2: Find the general solution of $\frac{1}{x} y' = \frac{x}{2y}$ ($x, y > 0$).

Solution: By multiplying each side by $2xy$, this differential equation can be written as $2y\, y' = x^2$, so it is separable and can be put into differential form:

$$2y\, y' = x^2 \quad \text{so} \quad 2y \frac{dy}{dx} = x^2 \quad \text{and} \quad 2y\, dy = x^2\, dx.$$

Integrating each side, $\int 2y\, dy = \int x^2\, dx$, so $y^2 = \frac{1}{3} x^3 + C$, an implicit form of the general solution.

(Each antiderivative has an integration constant, $y^2 + C_1 = \frac{1}{3} x^3 + C_2$, but C_1 can be moved to the right side of the equation, combined with C_2, and the final result expressed using only a single constant, $C = C_2 - C_1$. Then $y^2 = x^3/3 + C$.)

Finally, solving for y, we have $y = \pm \sqrt{\frac{1}{3} x^3 + C}$, the explicit form of the general solution.

Steps for solving a separable equation $g(y) \cdot y' = f(x)$:

(a) Rewrite in differential form: $\quad g(y)\, dy = f(x)\, dx$

(b) Integrate each side: $\quad \int g(y)\, dy = \int f(x)\, dx$

(c) Find antiderivatives of g and f: $\quad G(y) = F(x) + C \quad$ ($G' = g$ and $F' = f$)

(d) If an initial value (x_0, y_0) is given, put the values for x_0 and y_0 into F and G and solve for C.

(e) If possible, explicitly solve for y.

Example 3: Find the solution of $y' = \frac{6x + 1}{2y}$ ($y > 0$) which satisfies $y(2) = 3$.

Solution: This differential equation can be written as $2y\, y' = 6x + 1$ so it is separable and can be written using differentials:

$$2y\, y' = 6x + 1 \quad \text{so} \quad 2y\frac{dy}{dx} = 6x + 1 \quad \text{and} \quad 2y\, dy = (6x + 1)\, dx$$

Integrating each side, $\int 2y\, dy = \int 6x + 1\, dx$ so $y^2 = 3x^2 + x + C$.

In an initial value problem, it is usually safest to solve for C immediately after finding the antiderivatives. Putting $x = 2$ and $y = 3$ into the general solution $y^2 = 3x^2 + x + C$,

$$(3)^2 = 3(2)^2 + (2) + C \quad \text{so} \quad 9 = 12 + 2 + C \quad \text{and} \quad C = -5.$$

Then $y^2 = 3x^2 + x - 5$ or $y = \pm\sqrt{3x^2 + x - 5}$. Since the point $(2, 3)$ is on the top half of the circle, we use only the + value of the square root for y: $y = +\sqrt{3x^2 + x - 5}$.

The general solution of $y' = \dfrac{6x + 1}{2y}$ is $y^2 = 3x^2 + x + C$ or $y = \sqrt{3x^2 + x + C}$.

The particular solution which satisfies the initial condition $y(2) = 3$ is $y^2 = 3x^2 + x - 5$ or $y = \sqrt{3x^2 + x - 5}$.

Practice 2: Find the general solution of $y' = \dfrac{1 - \sin(x)}{3y^2}$ and the particular solution through $(0, 2)$.

Sometimes algebra is the hardest part of the problem, and logarithms are often involved.

Example 4: Solve $x\, y' = y + 3$ assuming $x \neq 0$ and $y \neq -3$.

Solution: Putting the problem into the form $g(y) \cdot y' = f(x)$: $\dfrac{1}{y+3}\, y' = \dfrac{1}{x}$.

Rewriting this in differential form and integrating, we have that

$$\frac{1}{y+3}\, dy = \frac{1}{x}\, dx \quad \text{and} \quad \int \frac{1}{y+3}\, dy = \int \frac{1}{x}\, dx \quad \text{so} \quad \ln|y + 3| = \ln|x| + C,$$

an implicit form of the general solution. To explicitly solve for y, recall that $e^{\ln(a)} = a$. Then

$$e^{\ln|y+3|} = e^{\ln|x| + C} = e^{\ln|x|} \cdot e^C \quad \text{so} \quad |y + 3| = |x| \cdot e^C \quad \text{and} \quad y + 3 = \pm x \cdot e^C$$

Replacing the complicated constant $\pm e^C$ with A and subtracting 3 from each side, we have

$y = Ax - 3$, an explicit form of the general solution.

Two Special Cases: $y' = ky$ and $y' = k(y - a)$

The separable differential equations $y' = ky$ and $y' = k(y - a)$ are relatively simple, but they describe a wealth of important situations, including population growth, radioactive decay, drug testing, heating and cooling. The two differential equations are solved here and some of their applications are explored in section 6.3.

The Differential Equation $y' = ky$:

The differential equation $y' = ky$ describes a function y whose rate of change is proportional to the value of y. Fig. 3 shows direction fields for $y' = 1y$ (growth) and $y' = -2y$ (decay). The differential equation $y' = ky$ models the behavior of populations (the number of babies born is proportional to the number of people in the population), radioactive decay (the number of atoms which decay is proportional to the number of atoms present), the absorption of some medicines by our bodies, and many other situations. The solutions of $y' = ky$ will help us determine how long it takes a population to double in size, how old some prehistoric artifacts are, and even how often some medicines should be taken in order to maintain a safe and effective concentration of medicine in our bodies.

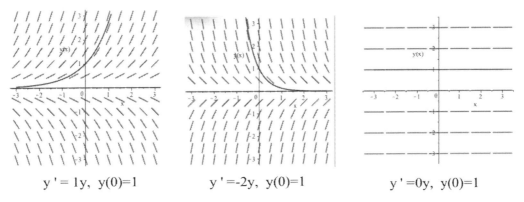

$\quad\quad y' = 1y,\ y(0)=1 \quad\quad\quad y' = -2y,\ y(0)=1 \quad\quad\quad y' = 0y,\ y(0)=1$

Fig. 3: Direction fields for $y' = ky$

> If $y' = ky$ ($y > 0$), then $y = y(0) \cdot e^{kx}$.

Proof: $y' = ky$ can be rewritten as $\frac{1}{y} y' = k$ so it is a separable differential equation can be written using differentials as

$\frac{1}{y} dy = k\, dx$. Then $\int \frac{1}{y} dy = \int k\, dx$ so $\ln(y) = kx + C$.

When $x = 0$, $y = y(0)$ so $\ln(y(0)) = k \cdot 0 + C$ and $C = \ln(y(0))$. Then

$\ln(y) = kx + \ln(y(0))$, so $e^{\ln(y)} = e^{kx + \ln(y(0))} = e^{kx} \cdot e^{\ln(y(0))}$ and $y = y(0) \cdot e^{kx}$.

Practice 3: The population of a town is 7,000 people and it is growing at a rate so $P' = 0.08 \cdot P$ people/year. Write an equation for the population of the town t years from now and use the equation to estimate the towns population in 10 years.

The Differential Equation $y' = k(y - a)$:

The differential equation $y' = k(y - a)$ describes a function y whose rate of change is proportional to the difference of y and the number a. Figure 4 shows the direction fields for $y' = 1(y - a) = y - a$ and $y' = -1(y - a) = a - y$. In the first case, the solution curves are "repelled" by the horizontal line $y = a$, and in the second case they are "attracted" by the line. The differential equation $y' = k(y - a)$ models the changing temperature of a cup of tea (the rate of cooling is proportional to the difference in temperature of the tea and the surrounding air) and the changing pressure within a balloon (the rate of pressure change is proportional to the difference in pressure between the inside and outside of the balloon). The solutions of the differential equation will help us determine how long it takes the hot tea to cool (or cold tea to warm up) to any given temperature and how long it takes a slowly leaking balloon (or tire) to lose half of its air.

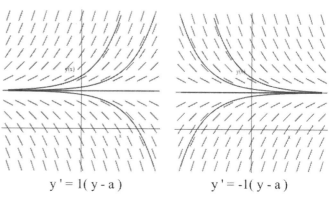

$y' = 1(y - a)$ $y' = -1(y - a)$

Fig. 4: Direction fields for $y' = k(y - a)$

> If $y' = k(y - a)$, then $y - a = (y_0 - a) \cdot e^{kx}$.

Proof: Since the differentiable equation $y' = k(y - a)$ is separable, we can separate the variables, integrate, and solve for y. The equation can be written as

$$\frac{1}{y - a} \, dy = k \, dx. \text{ Then } \int \frac{1}{y - a} \, dy = \int k \, dx \text{ so } \ln(y - a) = k \cdot x + C.$$

When $x = 0, y = y_0$ so $\ln(y_0 - a) = k \cdot 0 + C$ and $C = \ln(y_0 - a)$. Then

$$\ln(y - a) = k \cdot x + \ln(y_0 - a) \text{ so } e^{\ln(y - a)} = e^{k \cdot x + \ln(y_0 - a)} = e^{k \cdot x} e^{\ln(y_0 - a)} = (y_0 - a) \cdot e^{k \cdot x}$$

and $y - a = (y_0 - a) \cdot e^{kx}$.

Practice 4: When a pan of $90°C$ water ($T_0 = 90$) is placed in a $70°C$ room ($a = 70$), the rate at which the water cools is $T' = -0.15(T - 70)$ degrees per minute (Fig. 5). Write a formula for the temperature T of the water t minutes after it is placed in the room and use the equation to estimate the temperature of the water after 5 minutes, 10 minutes, and 15 minutes.

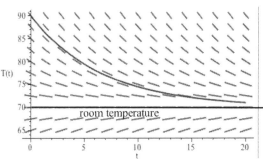

Fig. 5: Temperature of warm water in a cool room

The solutions of these two differential equations are used in applied problems in Section 6.3.

PROBLEMS

1. Fig. 6 shows the direction field of the separable differential equation $y' = 2xy$. Sketch the solutions of the differential equation which satisfy the initial conditions $y(0) = 3$, $y(0) = 5$, and $y(1) = 2$.

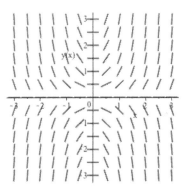

Fig. 6: Direction field for $y' = 2xy$

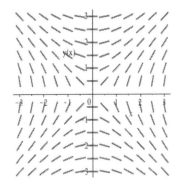

2. Fig. 7 shows the direction field of the separable differential equation $y' = x/y$. Sketch the solutions of the differential equation which satisfy the initial conditions $y(0) = 3$, $y(0) = 5$, and $y(1) = 2$.

Fig. 7: Direction field for $y' = x/y$

In problems 3 – 10, (a) separate the variables and rewrite the differential equation in the form $g(y) \cdot y' = f(x)$, and (b) solve the resulting differential equation. (Assume that x and y are restricted so that division by zero does not occur.)

3. $y' = 2xy$
4. $y' = x/y$
5. $(1 + x^2) \cdot y' = 3$
6. $xy' = y + 3$

7. $y' \cos(x) = e^y$
8. $y' = x^2 y + 3y$
9. $y' = 4y$
10. $y' = 5(2 - y)$

In problems 11 – 18, solve the initial value separable differential equations.

11. $y' = 2xy$ for $y(0) = 3, y(0) = 5$, and $y(1) = 2$. 12. $y' = x/y$ for $y(0) = 3, y(0) = 5$, and $y(1) = 2$.

13. $y' = 3y$ for $y(0) = 4, y(0) = 7$, and $y(1) = 3$. 14. $y' = -2y$ for $y(0) = 4, y(0) = 7$, and $y(1) = 3$.

15. $y' = 5(2 - y)$ for $y(0) = 5$ and $y(0) = -3$. 16. $y' = 7(1 - y)$ for $y(0) = 4$ and $y(0) = -2$.

17. $(1 + x^2) \cdot y' = 3$ for $y(1) = 4$ and $y(0) = 2$. 18. $xy' = y + 3$ for $y(1) = 20$. Can $y(0) = 2$?

19. The rate of growth of a population $P(t)$ which starts with 3,000 people and increases by 4% per year (Fig. 8) is $P'(t) = 0.0392 \cdot P(t)$. Solve the differential equation and use the solution to estimate the population in 20 years.

20. The rate of growth of a population $P(t)$ which starts with 5,000 people and increases by 3% per year is $P'(t) = 0.0296 \cdot P(t)$. Solve the differential equation and use the solution to estimate the population in 20 years.

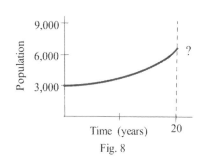
Fig. 8

21. The rate of decay of a piece of carbon–14 in a piece of material containing 3 grams of carbon–14 is $C'(t) = (-0.00012) \cdot C(t)$ where $C(t)$ is the number of grams present after t years (Fig. 9). Solve the differential equation and use the solution to estimate the amount of carbon–14 present after 10,000 years.

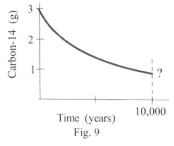
Fig. 9

22. The rate of decay of iodine–131 is $I'(t) = -0.086 \cdot I(t)$ where $I(t)$ is the number of grams present after t days. Solve the differential equation. If we start with 5 grams of iodine–131, how much will be present after 2 hours and after 10 hours? (First find a formula for $I(t)$.)

23. The rate of temperature change of a bowl of soup in a 25°C room is $T' = -0.12(T - 25)$ where T is the temperature of the soup after t minutes. If the soup originally is 80°C, find a formula for T and use it to estimate the temperature of the soup after 5 minutes (Fig. 10).

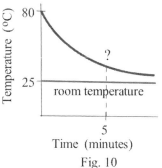
Fig. 10

24. When the switch is closed in an electrical circuit with a constant voltage source of 9 volts, a resistance of 2 ohms and an inductance of 3 henries, the rate of change of the current i (in amperes) is described by the differential equation $3\frac{di}{dt} + 2i = 9$ where t is the time in seconds. Solve the separable differential equation for i.

Section 6.2 **PRACTICE Answers**

Practice 1: (a) $y' = x^3(y-5)$ so $\frac{1}{y-5} y' = x^3$. In the pattern, $g(y) = \frac{1}{y-5}$ and $f(x) = x^3$.

(b) $y' = \dfrac{3}{2x + x \cdot \sin(y+2)}$ so $(2 + \sin(y+2)) \cdot y' = \frac{3}{x}$. $g(y) = 2 + \sin(y+2)$ and $f(x) = \frac{3}{x}$.

Practice 2: $y' = \dfrac{1 - \sin(x)}{3y^2}$ so $3y^2 \cdot y' = 1 - \sin(x)$ and $3y^2 \, dy = (1 - \sin(x)) \, dx$. Then

General solution: $\int 3y^2 \, dy = \int (1 - \sin(x)) \, dx$ so $y^3 = x + \cos(x) + C$.

Particular solution $(0,2)$: $(2)^3 = 0 + \cos(0) + C$ so $8 = 1 + C$ and $C = 7$.
$$y^3 = x + \cos(x) + 7.$$

Practice 3: ($y' = ky$ so $y = y(0) \cdot e^{kx}$.) $P' = 0.08 \cdot P$ so $P(t) = P(0) \cdot e^{0.08t}$ with $P(0) = 7{,}000$.
$\mathbf{P(t) = 7{,}000 \cdot e^{0.08t}}$. $P(10) = 7{,}000 \cdot e^{0.08(10)} = 7{,}000 \cdot e^{0.8} \approx 7{,}000(2.22554) = \mathbf{15{,}579.}$

Practice 4: ($y' = k(y - a)$ so $y - a = \{y_0 - a\} \cdot e^{kx}$)

$T' = -0.15(T - 70)$. $k = -0.15$, $a = 70$ and $T_0 = 90$. Then

$T - 70 = (T_0 - 70)e^{-0.15t} = (90 - 70)e^{-0.15t} = 20e^{-0.15t}$ and $\mathbf{T = 70 + 20e^{-0.15t}}$.

When $t = 5$, $T = 70 + 20e^{-0.15t} = 70 + 20e^{-0.15(5)} = 70 + 20e^{-0.75} \approx 70 + 20(0.472) \approx 79.4°$.

When $t = 10$, $T = 70 + 20e^{-0.15t} = 70 + 20e^{-0.15(10)} = 70 + 20e^{-1.5} \approx 70 + 20(0.223) \approx 74.5°$.

When $t = 15$, $T = 70 + 20e^{-0.15t} = 70 + 20e^{-0.15(15)} = 70 + 20e^{-2.25} \approx 70 + 20(0.105) = 72.1°$.

After a "long" time, the temperature will be very close to (and slightly above) $70°$.

6.3 GROWTH, DECAY, AND COOLING

A population of people, a chunk of radioactive material, money in the bank, and a cup of hot soup all share a common trait. In each situation, the rate at which an amount is changing is proportional to the amount:

- the number of births per year is proportional to the number of people in the population

- the number of atoms per hour that release a particle is proportional to the number of atoms present

- the number of dollars of interest per year is proportional to the amount of money in the bank account

- the number of degrees the soup cools per minute is proportional to the temperature difference between the soup and the air.

All of these situations can be modeled with separable differential equations we solved in Section 6.2. In fact, the first three can be modeled with the same differential equation: $y' = ky$. The cooling soup uses $y' = k(y-a)$. In this section our focus is on using the equations and their solutions to answer questions about applied problems. The applications here all involve the rate of change of some quantity with respect to time and the notation is usually changed so the independent variable is time t (instead of x) and the dependent quantity is $f(t)$ (instead of y). The statement $y' = ky$ then becomes $f'(t) = k \cdot f(t)$, and the solution $y = y_0 \cdot e^{kx}$ becomes $f(t) = f(0) \cdot e^{kt}$.

Theorem: If the rate of change of f is proportional to the value of f, $f'(t) = k \cdot f(t)$,

then $f(t) = f(0) \cdot e^{kt}$.

When k is positive, $f(t) = f(0) \cdot e^{kt}$ represents **exponential growth**, and k is called the **growth constant**. When k is negative, $f(t) = f(0) \cdot e^{kt}$ represents **exponential decay**, and k is called the **decay constant**. Fig. 1 shows the graphs of $f(t) = e^{kt}$ for several values of k.

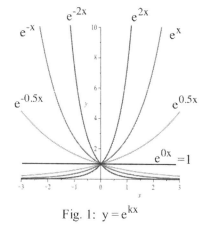

Fig. 1: $y = e^{kx}$

Exponential Growth

When the initial population $f(0)$ and the growth constant k are known, we can write an equation for $f(t)$, the population at any time t, and use it to answer questions about the population.

Example 1: The number of bacteria on a petri plate t hours after the experiment starts is $2000 \cdot e^{0.0488t}$.

    (a) How many bacteria are on the plate after 1 hour? 2 hours?

    (b) What is the percentage growth of the population from t=0 to t=1? From t=1 to t=2?

    (c) How many hours does it take for the population to reach 3000? To double?

Solution: The population after t hours is $f(t) = 2000 \cdot e^{0.0488t}$ (Fig. 2)

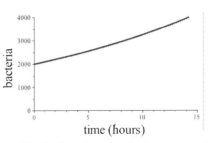

Fig. 2: Growing population of bacteria

(a) $f(1) = 2000 \cdot e^{0.0488(1)} = 2000 \cdot e^{0.0488} \approx 2000(1.0500) = 2100$.

$f(2) = 2000 \cdot e^{0.0488(2)} = 2000 \cdot e^{0.0976} \approx 2000(1.1025) = 2205$.

(b) Percentage growth from t=0 to t=1 is

$\frac{f(1) - f(0)}{f(0)} \cdot 100 = \frac{2100 - 2000}{2000} \cdot 100 = (0.05)(100) = 5\%$.

Percentage growth from t=1 to t=2 is $\frac{f(2) - f(1)}{f(1)} \cdot 100 = \frac{2205 - 2100}{2100} \cdot 100 = (0.05)(100) = 5\%$.

During the first hour, the population grows by 100 and during the second hour it grows by 105, but the percentage growth during each hour is a constant 5 %.

(c) We can find the value of t so $3000 = f(t) = 2000 \cdot e^{0.0488t}$ by dividing each side by 2000:

$1.5 = e^{0.0488t}$

taking logarithms to get t out of the exponent: $\ln(1.5) = \ln(e^{0.0488t}) = 0.0488t \ln(e) = .0488\, t$

and dividing by 0.0488 to solve for t: $t = \frac{1}{0.0488} \ln(1.5) \approx \frac{1}{0.0488} (0.4055) \approx 8.31$ hours.

Since the original population is 2000, the doubled population is 4000. We can find the value of t so that $4000 = f(t) = 2000 \cdot e^{0.0488t}$ by dividing each side by 2000 and taking logarithms: $\ln(2) = \ln(e^{0.0488t}) = 0.0488t \ln(e) = 0.0488\, t$. Then $t = \frac{\ln(2)}{0.0488} \approx \frac{0.693}{0.0488} \approx 14.2$ hours.

The bacteria population will double every 14.2 hours, the **doubling time** for this population.

Practice 1: Use $f(t) = 2000 \cdot e^{0.0488t}$ from Example 1. (a) What is the population when t=5?
(b) How long until the population is 5000? (c) How long until the population triples?

If the value of the growth constant k is not given, usually our first step is to use the given information to find it. Once we know the population at two different times, we can find k.

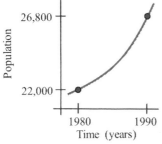

Fig. 3: Growing community

Example 2: The population of a community was 22,000 in 1980 and 26,800 in 1990. Assuming that the community maintains the same rate of exponential growth, (Fig. 3)

(a) what is a formula for the population t years after 1980?

(b) what is the annual percentage rate of growth of the community?

Solution: Let t represent the number of years since 1980, so 1980 corresponds to t=0 and 1990 corresponds to t=10. Then $f(0) = 22{,}000$, $f(10) = 26{,}800$, and $f(t) = f(0) \cdot e^{kt} = 22{,}000 \cdot e^{kt}$.

(a) To find a formula for f(t), we can use the 1990 (t=10) population to find the value for k:

$$26{,}800 = f(10) = 22{,}000 \cdot e^{k(10)} \text{ so } 1.218 = e^{k(10)} \text{ and}$$

$k = \frac{1}{10} \ln(1.218) \approx \frac{1}{10}(0.197) = 0.0197$. Then $\mathbf{f(t) = 22{,}000 \cdot e^{(0.0197)t}}$.

(b) $f(0) = 22{,}000$ and $f(1) = 22{,}000 \cdot e^{(0.0197)1} \approx 22{,}000(1.01989) = 22{,}437.58$ so the annual percentage increase was $\frac{f(1) - f(0)}{f(0)} \cdot 100 = \frac{437.58}{22000} \cdot 100 \approx \mathbf{1.989\ \%}$.

Practice 2: An experiment was begun by releasing 12,000 free neutrons into a material, and 2 seconds later, the material contained 18,000 free neutrons. Assuming the number of free neutrons grows exponentially, (a) determine a formula for the number present t seconds after the beginning of the experiment, and (b) determine how long it takes for the number of free neutrons to double.

Compound interest is another example of exponential growth.

Example 3: How long does it take $1000 to double at an effective annual rate of return of 5%? 10%? (This assumes that the yield is computed and compounded continuously.)

Solution: Let f(t) be the amount of money after t years. Then $f(0) = 1000$ and $f(t) = 1000 \cdot e^{kt}$.

5%: After 1 year, the investment will be $\$1000 + (.05)(\$1000) = \$1050$ so $f(1) = 1050 = 1000 \cdot e^{k \cdot 1}$. Solving for k, $1.05 = e^k$ so $k = \ln(1.05)$ and $f(t) = 1000 \cdot e^{\ln(1.05)t}$. Solving $2000 = 1000 \cdot e^{\ln(1.05)t}$ for t gives $t = \frac{\ln(2)}{\ln(1.05)} \approx \mathbf{14.2\ years}$.

10%: After 1 year the investment will be $1100. Then $k = \ln(1.10)$ so $f(t) = 1000 e^{\ln(1.10)t}$ and the doubling time is $t = \frac{\ln(2)}{\ln(1.10)} \approx \mathbf{7.27\ years}$.

Practice 3: How long does it take an investment to double if the rate of return is 12%?

When we know the growth constant k, the doubling time is simple to find. If $f(t) = f(0) \cdot e^{kt}$ then the doubling time is the time t so that $2f(0) = f(t) = f(0) \cdot e^{kt}$. Then

$$2 = e^{kt} \text{ and } \ln(2) = kt \text{ so } t = \frac{\ln(2)}{k}.$$

Doubling Time: If $f(t) = f(0) \cdot e^{kt}$, then the doubling time is $t = \frac{\ln(2)}{k}$. (Fig. 4)

An important aspect of exponential growth is that the **doubling time depends only on the growth constant k** and not on the population or the starting time. The previous Example and Practice problem illustrate the basis for a "rule" used in business:

Rule of 72: An investment with an annual rate of return of R% takes about $\frac{72}{R}$ years to double in value.

Table 1 shows the exact values for doubling times obtained using exponential growth with those obtained using the Rule of 72. The Rule of 72 gives good approximations and is easy to use. Problem 12 asks you to show why this "rule" works, and problem 13 asks you to find a "rule" for an investment to triple in value.

	Doubling Time (years)	
Rate of return (%)	Exact	Rule of 72
4	17.7	18.0
5	14.2	14.4
6	11.9	12.0
7	10.2	10.3
9	8.0	8.0
10	7.3	7.2
12	6.1	6.0
20	3.8	3.6

Table 1: Time for an investment to double in value

Exponential Decay

Exponential decay occurs when the rate of loss of something is proportional to the amount present. One example of exponential decay is radioactive decay: the number of atoms of a radioactive substance that "decay" (split into nonradioactive atoms and release p proportional to the number of radioactive atoms present. Exponentia quickly some medicines are absorbed from the bloodstream and even Exponential decay calculations are similar to those for growth, but th about "**half–life**", the time for half of the material to decay or be abso Table 2 shows the half–lives of some isotopes.

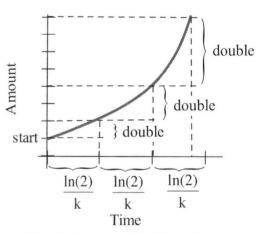

Fig. 4: Constant doubling times

strontium–90	29 years
argon–39	265 years
carbon–14	5700 years
plutonium–239	24,400 years
uranium–238	4.51×10^5 years
uranium–234	2.47×10^9 years

Table 2: Half–lives of some isotopes

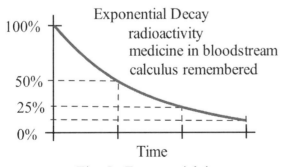

Fig. 5: Exponential decay

Example 4: You started with 10 g of radioactive Q, but after 6 days of decay there were only 3 g left (Fig. 6).

(a) Find a formula for the amount of Q present after t days.

(b) What is the half–life of Q?

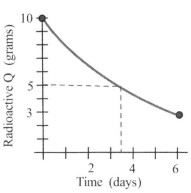

Fig. 6: Radioactive decay

Solution: Let f(t) be the amount of Q present after t days. Then $f(t) = f(0) \cdot e^{kt} = 10 \cdot e^{kt}$.

(a) $3 = f(6) = 10 \cdot e^{k6}$ so $0.3 = e^{6k}$ and $\ln(0.3) = 6k$. Then $k = \frac{1}{6} \ln(0.3) \approx -0.2007$ and $\mathbf{f(t) = 10 \cdot e^{(-0.2007)t}}$.

(b) The half–life is the time required for half of the material to decay, so we need to solve

$5 = 10 \cdot e^{(-0.2007)t}$ for t. Dividing by 10 and then taking logarithms,

$1/2 = e^{(-0.2007)t}$ and $\ln(1/2) = (-0.2007)t$ so $t = \frac{\ln(0.5)}{-0.2007} \approx 3.45$ days.

Carbon–14 Dating: If the half life of a substance is known and we know how much of the substance is present in a sample now, we can determine how much was present at some past time or determine how long ago the sample contained a particular amount of the substance. Radioactive carbon–14 with a half–life of about 5700 years is used in this way to estimate how long ago plants and animals lived. When a plant is alive it continually exchanges carbon–14 and ordinary carbon with the atmosphere so the ratio of carbon-14 to nonradioactive carbon stays relatively constant. But once the plant dies, this exchange stops. The ordinary carbon remains in the material, but the carbon–14 decays so the ratio of carbon–14 to ordinary carbon decreases at a known rate. By measuring the ratio of carbon–14 to ordinary carbon in a sample of plant tissue, scientists can determine how long ago the plant died and obtain an estimate for the age of the sample.

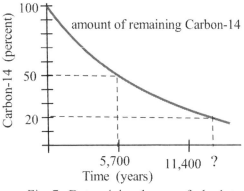

Fig. 7: Determining the age of a basket

Example 5: The amount of carbon–14 in plant fiber of a woven basket is 20% of the amount present in a living plant. Estimate the age of the basket. (Fig. 7)

Solution: Let f(t) represent the amount of carbon–14 in a sample that is t years old. Since we know the half–life is 5700 years, then

$f(5700) = f(0) \cdot e^{k \cdot 5700} = \frac{1}{2} f(0)$ so $e^{k \cdot 5700} = \frac{1}{2}$.

Solving for k,

$5700 k = \ln(1/2)$ and $k = \frac{\ln(.5)}{5700} \approx -0.0001216$ so $f(t) = f(0) \cdot e^{(-0.0001216)t}$.

Since 20% of the carbon–14 remains in our sample, we want the value of t so that

$0.20 \cdot f(0) = f(t) = f(0) \cdot e^{(-0.0001216)t}$.

Dividing by f(0), taking logarithms, and solving for t, we get $t = \frac{\ln(0.2)}{-0.0001216} \approx 13{,}235$ years.

The basket was made from a plant that died about 13,200 years ago. (Does that mean the basket was made about 13,200 years ago?) This dating method is very sensitive to small changes in the measured amount of carbon–14.

When the decay constant k is known, the half–life is simple to find. If $f(t) = f(0) \cdot e^{kt}$ then the half–life is the time t so that $\frac{1}{2} f(0) = f(t) = f(0) \cdot e^{kt}$. Solving for t, we have $t = \frac{\ln(1/2)}{k}$.

Half–life: If $f(t) = f(0) \cdot e^{kt}$, then the half–life is $t = \frac{\ln(1/2)}{k}$. (Fig. 8)

The half–life depends only on the decay constant k and not on the amount of material we have. If the half–life is known, then $k = \frac{\ln(1/2)}{\text{half-life}}$.

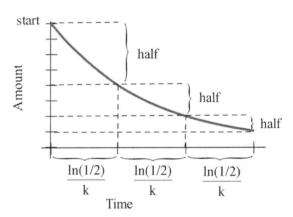

Fig. 8: Constant halfing times

Practice 4: The half–life of an isotope is 8 days. Write a formula for the amount of the isotope present t days after you began with 10 mg.

The rate at which many medicines are absorbed from the blood is proportional to the concentration of the medicine in the blood: the higher the concentration in the blood, the faster it is absorbed from the blood.

Example 6: Suppose medicine M has an absorption (decay) constant of –0.17 (determined experimentally), and that the lowest concentration of M that is "effective" is 0.3 mg/l (milligrams of M per liter of blood). If a patient who has 8 liters of blood is injected with 20 mg of M, how long will the M be effective?

Solution: Since the patient is starting with 20 mg of M in 8 liters of blood, the initial concentration is 20 mg/8 l = 2.5 mg/l . Then the amount of M at time t hours is $f(t) = 2.5 e^{-0.17t}$, and we want to find t so that f(t) = 0.3 mg/l : $0.3 = 2.5 e^{-0.17t}$ so $t = \frac{1}{-0.17} \ln(0.3/2.5) \approx 12.5$ hours.

Many medicines have a "safe and effective" range of concentrations (Fig. 9), and the goal of a schedule for taking the medicine is to keep the concentration near the middle of that range. Taking doses too close together in time can result in an overdose (Fig. 10), and taking them too far apart is eventually ineffective.

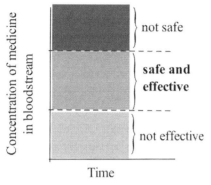

Fig. 9 : Safe and effective region

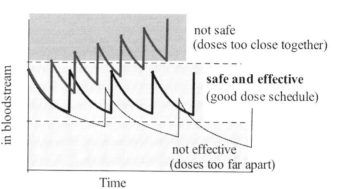

Fig. 10 : Safe and effective dosing schedule

Newton's Law of Cooling/Warming

Some rates of change depend on **how far** a value is from a fixed value. The rate at which a hot cup of soup cools (or a cool cup of milk warms up) is proportional to the difference in temperature between the soup and the surrounding air. This principle is called Newton's Law of Cooling/Warming.

Newton's Law of Cooling/Warming

If $f(t)$ is the temperature at time t of an object in an atmosphere with temperature a,

then the rate of change of f is proportional to the difference between f and a, $f'(t) = k\{f(t)-a\}$,

and $f(t) = a + \{f(0) - a\} \cdot e^{kt}$.

The statement that the rate of change is proportional to the difference, $f'(t) = k\{f(t) - a\}$, is a result from physics. The differential equation is separable, and was solved in the last section. Figure 11 shows some functions that have different initial values and that satisfy the differential equation $f'(t) = f(t) - 5$.

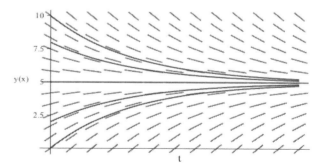

Fig. 11: Solutions of $f'(t) = (-1)*(f(t) - 5)$

6.3 Growth, Decay, and Cooling

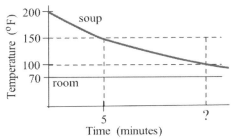
Fig. 12: Cooling cup of soup

Example 7: A cup of hot soup is in a room with a temperature of 70° F. When first poured, the soup was 200° F, and 5 minutes later it was 150° F (Fig. 12).

(a) Find an equation for the temperature f of the soup at any time t.

(b) How long does it take the 200° F soup to cool to 100° F?

(c) What will the temperature of the soup be after a "long" time?

Solution: (a) In this example, $a = 70°$ F and $f(0) = 200°$ F so $f'(t) = k\{ f(t) - 70 \}$ and $f(t) = 70 + \{200 - 70\} \cdot e^{kt} = 70 + 130 \cdot e^{kt}$. We can use the information that $f(5) = 150°$ F to find the value of k and an equation for f(t):

$150 = f(5) = 70 + 130 \cdot e^{k5}$ so $k = \frac{1}{5} \cdot \ln(80/130) \approx -0.0971$ and **$f(t) = 70 + 130 \cdot e^{(-0.0971)t}$**.

(b) We want to find the t so that $f(t) = 100$. Using the result from part (a),

$100 = f(t) = 70 + 130 \cdot e^{(-0.0971)t}$ so $30 = 130 \cdot e^{(-0.0971)t}$ and $t = \frac{\ln(30/130)}{-0.0971} \approx 15.1$ minutes.

(c) "After a long time" means for very large values of t.

$$\lim_{t \to \infty} 70 + 130 \cdot e^{(-0.0971)t} = \lim_{t \to \infty} 70 + \frac{130}{e^{(0.0971)t}} \longrightarrow 70 + 0 = 70° F.$$

Eventually, the soup will cool down to (almost) the temperature of the room.

PROBLEMS

1. Fig. 13 shows the growth of a city over several decades. How long did it take the city to double in population from 10,000 to 20,000? How long did it take to double from 15,000 to 30,000? What is the approximate doubling time of this population?

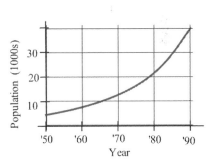
Fig. 13: Population growth of a city

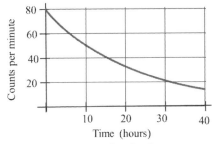
Fig. 14: Radioactive decay

2. Fig. 14 shows the counts per minute from a piece of radioactive material. How long did it take the counts to decrease from 80 per minute to 40? From 60 to 30? From 40 to 20? What is the half–life of this material?

3. The population of a community in 1970 was 48,000 people and in 1990 it was 64,000 people. (a) Write a formula for the population of the community t years after 1970. (b) Estimate the population in the year 2000. (c) Approximately when will the population be 100,000? (d) What is the doubling time of the population of this community?

4. The population of a community in 1970 was 40,000 people and in 1990 it was 60,000 people. (a) Write a formula for the population of the community t years after 1970. (b) Estimate the population in the year 2000. (c) Approximately when will the population be 100,000? (d) What is the doubling time of the population of this community?

5. You have found a terrific investment which pays at an effective annual rate of 15%. (a) Use the Rule of 72 and the exponential growth model to calculate how long it will take a $5,000 investment to double. (b) How long will it take for the investment to triple?

6. You have $3,000 invested for 10 years at an effective annual rate of 7.5% and a friend has the same amount invested at an effective annual interest rate of 7.75%. Your friend will get back how much more money than you at the end of (a) 10 years? (b) 20 years?

7. Find a formula for the population of the city in Fig. 13.

8. (Without using calculus.) Each bacterium of a certain species splits into two bacteria at the end of each minute. If we start with a few bacteria in a bowl at 3 pm and the bowl is full of bacteria at 4:30 pm, when was the bowl half full?

9. The newscaster said that the population of the world is now doubling every 20 years. What annual rate of growth results in a 20 year doubling time?

10. Group A has a population of 150,000 and a growth rate of 4%. Group B has a population of 100,000 and a growth rate of 7%. In how many years will the two groups be the same size? (Fig. 15)

11. Group A has a population of 600,000 and a growth rate of 3%. Group B has a population of 400,000 and a growth rate of 6%. In how many years will the two groups be the same size?

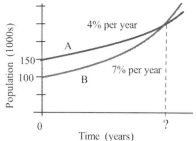

Fig. 15: Two growing populations

12. Derive the "Rule of 72." For an investment with an annual rate of return of R%, show that the value of the growth constant is $k = \ln(1 + R/100)$ so the doubling time is $\ln(2)/\ln(1 + R/100)$. Calculate the values of k for R between 5 and 15, and observe that for these values of R the exact doubling time $\frac{\ln(2)}{\ln(1 + R/100)}$ is approximately equal to $\frac{72}{R}$.

13. Develop a "Rule of M" for the **tripling** time of a investment. Find a value for M so M/R is a good approximation of the time its takes an investment with a rate of return of R % to triple in value. Assume that R is between 5 and 15.

14. The unregulated population of fish in a certain lake grows by 30% per year under optimum conditions, and the result of a fish census is that there are approximately 20,000 fish in the lake. How many fish can be harvested (Fig. 16) at the end of the year in order to maintain a stable population from year to year? (This is an example of calculating the yield for a "renewable resource." In practice, the calculations are more sophisticated and also take into account the distribution of species, ages and genders.)

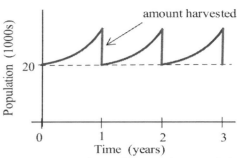

Fig. 16: Harvesting a fish growing population

15. The annual growth constant for the population of snails is $k = 0.14$, and currently we have 8,000 snails.
    (a) Graph the snail population for 20 months if we harvest 2,000 snails at the end of every 2 months.
    (b) Graph the snail population for 20 months if we harvest 3,000 snails at the end of every 2 months.
    (c) How many snails can we harvest every 2 months in order to maintain a stable population?

16. An exponential growth function $f(t) = A e^{kt}$ has a constant doubling time, but there are functions with constant doubling times which are not exponential. (a) Show that the exponential function $f(t) = 2^t = e^{\ln(2)t}$ has a constant doubling time of 1. (Show that $f(t+1) = 2f(t)$.) (b) Graph the function $g(t) = 2^t(1 + A \cdot \sin(2\pi t))$ for $A = 0.5$ and 1.5. Show that g has a constant doubling time 1 for every choice of A.

17. We started an experiment with 10 grams of a radioactive material and 14 days later there were 2 grams left. (a) Find an equation for the amount of material remaining t days after the beginning of the experiment. (b) Find the half–life of the material. (c) How long after the beginning of the experiment will there be 0.7 grams of the material left?

18. We start with 8 mg of a radioactive substance and 10 days later determine that there is 6.3 mg of the substance left. (a) Find an equation for the amount left t days after the start. (b) Find the half–life of the material. (c) How long after the start will there be one milligram of the substance left?

19. The Geiger counter initially recorded 187 counts per minute from a radioactive material, but 2 days later the count was down to 143 counts per minute. (a) What is the half–life of the material? (b) When will the count be down to 20 counts per minute? (The count per minute is proportional to the amount of radioactive material present.)

20. The initial measurement from a radioactive material was 540 counts per minute, and a week later it was 500 counts per minute.
    (a) What is the half–life of the material?
    (b) When will the count be down to 100 counts per minute?

21. Determine an equation for the counts per minute for the radioactive material A in Fig. 17.

22. Determine an equation for the counts per minute for the radioactive material B in Fig. 17.

23. A friend is considering purchasing a letter reputedly written by Isaac Newton (1642–1727), but an analysis of the paper shows that it contains 97.5% of the proportion of carbon–14 present in new paper. Can we be certain the letter is a forgery? If the paper is the right age, can we be certain the letter is genuine?

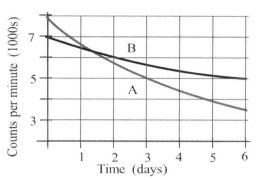

Fig. 17: Radioactive decay of two materials

24. For several centuries the Shroud of Turin was widely believed to be the shroud of Jesus. Three independent laboratories in England, Switzerland, and the United States used carbon–14 dating on a few square centimeters of the cloth, and in 1988 they reported that the Shroud of Turin was probably made in the early 1300s and certainly after 1200 A.D. (a) If the Shroud was made in 1300 A.D., what percentage of the original carbon–14 was still present in 1988? (b) If the Shroud was made in 30 A.D., what percentage of the original carbon–14 was still present in 1988? (Science 21, October 1988, Vol. 242, p. 378)

25. Half of a particular medicine is used up by the body every 6 hours, and the medicine is not effective if the concentration in the blood is less than 10 mg/l. If an ill person is given an initial dose of medicine to raise the concentration to 30 mg/l, how long will the medicine be effective?

26. A particular illegal substance has a half–life of 12 hours, and it can be detected in concentrations as low as 0.002 mg/l in the blood. (a) If a person has an initial concentration of the substance of 15 mg/l in the blood, how long can it be detected? (b) If the detection test is improved by a factor of 100 so it can detect a concentration of 0.00002 mg/l, how long can an initial concentration of 15 mg/l be detected?

27. A doctor gave a patient 9 mg of a medicine which has half–life of 15 hours in the body. How much of the medicine does the patient need to take **every 8 hours** in order to maintain between 6 and 9 mg of the medicine in the body all of the time? (Fig. 18)

28. Each layer of a dark film transmits 40% of the light that strikes it. (a) How many layers are needed for an eye shield to transmit 10% of the light? (b) How many layers are needed to transmit 2% of the light?

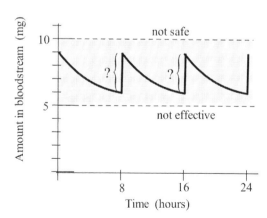

Fig. 18 : Finding a safe and effective dose

29. A region has been contaminated with radioactive iodine–131 to a level 5 times the safe level. How long will it take until the area is safe?

30. A region has been contaminated with radioactive strontium–90 to a level 5 times the safe level. How long will it take until the area is safe?

31. The population of a country is 4 million and is growing at 5% per year. Currently the country has 10 million acres of forests which are being cut down (and not replanted) at a rate of 300,000 acres per year. (a) Find an equation for the number of acres of forest per person.

    (b) How fast is the number of acres of forest per person changing?

    (c) If the population and harvest rates remain constant, in approximately how many years will there be one acre of forest per person?

32. When a pan of hot (200° F) water is removed from the stove in a 70° F kitchen, it takes 4 minutes to cool to a temperature of 150° F.

    (a) Find an equation for the temperature of the water t minutes after it is removed from the stove.

    (b) When will the water be 100° F?   (c) When will the water be 80° F?

    (d) When will be water be 60° F?

33. When the pan of 200° F water is taken outside on a cool (40° F) day, it only takes 4 minutes to cool to 150 F.

    (a) Find an equation for the temperature of the water t minutes after it is removed from the stove.

    (b) When will the water be 100° F?   (c) When will the water be 80° F?

    (d) When will be water be 60° F?

34. When a pitcher of orange juice is taken out of a 40° F refrigerator in a 70° F kitchen, it takes 5 minutes to warm up to 60° F.

    (a) Find an equation for the temperature of the juice t minutes after it is removed from the refrigerator.

    (b) How long does it take to warm up 50° F?

    (c) How long does it take to warm up to 65° F?

Section 6.3   PRACTICE Answers

Practice 1: $f(t) = 2000 \cdot e^{0.0488t}$.

(a) When $t = 5$, $f(5) = 2000 \cdot e^{0.0488(5)} = 2000 \cdot e^{0.244} \approx 2000(1.276) = \mathbf{2{,}552}$.

(b) $f(t) = 5000$: $5000 = 2000 \cdot e^{0.0488t}$ so $\frac{5000}{2000} = \frac{5}{2} = e^{0.0488t}$.

Taking the natural log of each side, $\ln(5/2) = 0.0488t$ and $t = \frac{1}{0.0488} \ln(5/2) \approx \mathbf{18.78}$.

(c) $f(0) = 2{,}000$. Triple $= 6{,}000$. $6000 = 2000 \cdot e^{0.0488t}$ so $\frac{6000}{2000} = 3 = e^{0.0488t}$.

Taking the natural log of each side, $\ln(3) = 0.0488t$ and $t = \frac{1}{0.0488} \ln(3) \approx \mathbf{22.51}$.

Practice 2: $f(0) = 12{,}000$ and $f(2) = 18{,}000$.

(a) $f(t) = 12{,}000 \cdot e^{kt}$. $18{,}000 = 12{,}000 \cdot e^{k(2)}$ so $\frac{18000}{12000} = 1.5 = e^{2k}$.

Taking the natural log of each side, $\ln(1.5) = 2k$ and $k = \frac{1}{2}\ln(1.5) \approx 0.2027$.

$f(t) = 12{,}000 \cdot e^{(0.5\ln(1.5))t} \approx 12{,}000 \cdot e^{0.2027t}$.

(b) Double $= 2(12{,}000) = 24{,}000$. $24{,}000 = 12{,}000 \cdot e^{0.2027t}$ so $\frac{24000}{12000} = 2 = e^{0.2027t}$.

Then $\ln(2) = 0.2027t$ so $t = \frac{1}{0.2027}\ln(2) \approx \mathbf{3.42}$.

Practice 3:   After 1 year, each \$1 investment will be $\$1 + (.12)(\$1) = \$1.12$ so $f(1) = 1.12 = 1 \cdot e^{k \cdot 1}$.

Solving for k, $1.12 = e^k$ so $k = \ln(1.12)$ and $f(t) = e^{\ln(1.12)t}$.

Solving $2 = e^{\ln(1.12)t}$ for t gives $t = \frac{\ln(2)}{\ln(1.12)} \approx \mathbf{6.12 \text{ years}}$.

Practice 4: $f(t) = f(0) \cdot e^{kt} = 10 \cdot e^{kt}$ with $k = \frac{\ln(1/2)}{\text{half life}} = \frac{\ln(1/2)}{8} \approx -0.0866$.

$f(t) \approx \mathbf{10 \cdot e^{-0.0866t}}$.

Differential equations in "literature":

From the murder mystery, The Calculus of Murder by Erik Rosenthal, St. Martin's Press, 1986:

"Maybe we could do some calculations before you call. From what you said, the rate of absorption would be proportional to the amount present and inversely proportional to the content of the stomach."

"Daniel, speak English."

"The more poison, the faster the rate of absorption: the greater the content of the stomach, the slower the . . . "

"Got it. Sounds right. So?"

"There's probably a constant of absorbency well known for arsenic and any given set of conditions. **It's a simple differential equation**."

"You're kidding."

The detective, a calculus teacher (!), then goes on to solve the differential equation $y' = cy$, to find the absorption constant c, and to figure out "whodunit".

PROBLEM ANSWERS Chapter Six

Section 6.0

1. $y = e^{-3x} + 2$. $y' = -3e^{-3x}$ so $y' + 3y = (-3e^{-3x}) + 3(e^{-3x} + 2) = -3e^{-3x} + 3e^{-3x} + 6 = 6$

3. $y = x^2 + 2x$. $y' = 2x + 2$. $y'' = 2$. $y'' - y' + y = (2) - (2x + 2) + (x^2 + 2x) = x^2$.

5. $y = 7x^3 - x^2$. $y' = 21x^2 - 2x$.
   $x \cdot y' - 3y = x(21x^2 - 2x) - 3(7x^3 - x^2) = 21x^3 - 2x^2 - 21x^3 + 3x^2 = x^2$.

7. $y = \frac{1}{2}e^x + 2e^{-x}$. $y' = \frac{1}{2}e^x - 2e^{-x}$. $y' + y = (\frac{1}{2}e^x - 2e^{-x}) + (\frac{1}{2}e^x + 2e^{-x}) = e^x$.

9. $y = (7 - x^2)^{1/2}$. $y' = \frac{1}{2}(7 - x^2)^{-1/2}(-2x) = \frac{-x}{\sqrt{7 - x^2}} = \frac{-x}{y}$.

11. $y = 2x^3 - 3x + 3$. $y(1) = 2(1)^3 - 3(1) + 3 = 2$ (OK). $y' = 6x^2 - 3$ (OK).

12. $y = 3x^2 + 4x - 17$. $y(2) = 3(2)^2 + 4(2) - 17 = 3$ (OK). $y' = 6x + 4$ (OK).

13. $y = \sin(2x) + 1$. $y(0) = \sin(0) = 1 = 1$ (OK). $y' = 2 \cdot \cos(2x)$ (OK).

15. $y = 7e^{5x}$. $y(0) = 7e^{5(0)} = 7$ (OK). $y' = 7e^{5x} \cdot 5 = 5y$ (OK).

17. $y = \frac{-4}{x}$. $y(1) = \frac{-4}{1} = -4$ (OK). $x \cdot y' = x \cdot (\frac{4}{x^2}) = \frac{4}{x} = -(\frac{-4}{x}) = -y$ (OK).

19. $y = 5 \cdot \ln(x) - 2$. $y(e) = 5 \cdot \ln(e) - 2 = 5 \cdot 1 - 2 = 3$ (OK). $y' = 5 \cdot \frac{1}{x}$ (OK).

21. $7 = y(3) = (3)^2 + C$ so $7 = 9 + C$ and $C = -2$.

23. $5 = y(0) = Ce^{3(0)} = C \cdot 1$ so $C = 5$.

25. $4 = y(0) = 2\sin(3 \cdot 0) + C = 0 + C$ so $C = 4$.

27. $2 = y(e) = \ln(e) + C = 1 + C$ so $C = 1$.

29. $10 = y(2) = \frac{-C}{2}$ so $C = -20$.

31. $y = \int 4x^2 - x \, dx = \frac{4}{3}x^3 - \frac{1}{2}x^2 + C$. $7 = y(1) = \frac{4}{3} - \frac{1}{2} + C$ so $C = \frac{37}{6}$.
    $y = \frac{4}{3}x^3 - \frac{1}{2}x^2 + \frac{37}{6}$.

33. $y = \int \frac{3}{x} \, dx = 3 \cdot \ln|x| + C$. $2 = y(1) = 3 \cdot \ln(1) + C = 0 + C$ so $C = 2$. **$y = 3 \cdot \ln|x| + 2$**.

35. $y = \int 6e^{2x} \, dx = 3e^{2x} + C$. $1 = y(0) = 3e^{2(0)} + C = 3 + C$ so $C = -2$. $\mathbf{y = 3e^{2x} - 2}$.

37. $y = \int x \sin(x^2) \, dx = -\frac{1}{2} \cos(x^2) + C$. $3 = y(0) = -\frac{1}{2} \cos(0) + C = \frac{-1}{2} + C$ so $C = 3.5$.
$\mathbf{y = -\frac{1}{2} \cos(x^2) + 3.5}$.

39. $y' = \frac{1}{x}(6x^3 - 10x^2) = 6x^2 - 10x$. $y = \int 6x^2 - 10x \, dx = 2x^3 - 5x^2 + C$.
$5 = y(2) = 2(2)^3 - 5(2)^2 + C = 16 - 20 + C$ so $C = 9$. $\mathbf{y = 2x^3 - 5x^2 + 9}$.

41. Know: $f'(x) + 5f(x) = 0$. $g'(x) + 5g(x) = 0$.
  (a) $y = 3 \cdot f(x)$. $y' + 5y = 3 \cdot f'(x) + 5 \cdot 3 \cdot f(x) = 3 \cdot \{ f'(x) + 5 \cdot f(x) \} = 3 \cdot \{0\} = 0$.
  (b) $y = 7 \cdot g(x)$. $y' + 5y = 7 \cdot g'(x) + 5 \cdot 7 \cdot g(x) = 7 \cdot \{ f'(x) + 5 \cdot g(x) \} = 7 \cdot \{0\} = 0$.
  (c) $y = f(x) + g(x)$. $y' + 5y = f'(x) + g'(x) + 5 \cdot f(x) + 5 \cdot g(x)$
  $= f'(x) + 5 \cdot f(x) + g'(x) + 5 \cdot g(x) = 0 + 0 = 0$.
  (d) $y = A \cdot f(x) + B \cdot g(x)$. $y' + 5y = A \cdot f'(x) + B \cdot g'(x) + 5 \cdot A \cdot f(x) + 5 \cdot B \cdot g(x)$
  $= A \cdot \{ f'(x) + 5 \cdot f(x) \} + B \cdot \{ g'(x) + 5 \cdot g(x) \} = 0 + 0 = 0$.

43. $y'' + y = x$.
  (a) $y = \sin(x) + x$. $y' = \cos(x) + 1$. $y''(x) = -\sin(x)$. $y'' + y = (-\sin(x)) + (\sin(x) + x) = x$. (OK)
  (b) $y = \cos(x) + x$. $y' = -\sin(x) + 1$. $y''(x) = -\cos(x)$. $y'' + y = (-\cos(x)) + (\cos(x) + x) = x$. (OK)
  (c) No. For $y = 3\{ \sin(x) + x \}$, $y'' + y = 3x$. For $y = \{ \sin(x) + x \} + \{ \cos(x) + x \}$, $y'' + y = 2x$.

45. $y = \frac{A}{B} - C \cdot e^{-Bt}$. $\frac{dy}{dt} = 0 - C \cdot \{ -B \cdot e^{-Bt} \} = BC \cdot e^{-Bt}$.

$A - By = A - B\{ \frac{A}{B} - C \cdot e^{-Bt} \} = A - A + BC \cdot e^{-Bt} = BC \cdot e^{-Bt}$. Therefore, $\frac{dy}{dt} = A - By$.

47. $I = \frac{E}{R} \{ 1 - e^{-Rt/L} \}$. $\frac{dI}{dT} = \frac{E}{R} \{ 0 - \frac{-R}{L} e^{-Rt/L} \} = \frac{E}{L} e^{-Rt/L}$.

$L \cdot \frac{dI}{dt} + R \cdot I = L \cdot \{ \frac{E}{L} e^{-Rt/L} \} + R \cdot \{ \frac{E}{R} (1 - e^{-Rt/L}) \} = E \cdot e^{-Rt/L} + E \cdot \{ 1 - e^{-Rt/L} \} = E$ (yes!)

Section 6.1

1. See Fig. 1.

3. See Fig. 2.

5. See Fig. 3. (d) All of the solutions seem to approach the horizontal line $y = 1$.

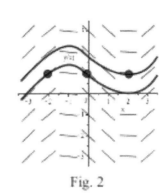
Fig. 2

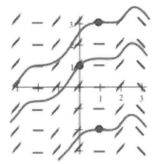
Fig. 1: Direction field for $y' = f(x)$

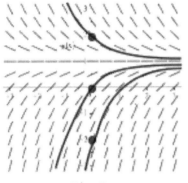
Fig. 3

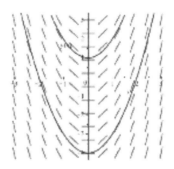
Fig. 4: Direction field for $y' = 2x$

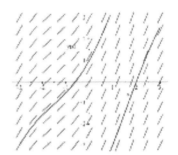
Fig. 5: Direction field for $y' = 2 + \sin(x)$

7. $y' = 2x$. See Fig. 4.

9. $y' = 2 + \sin(x)$. See Fig. 5.

11. $y' = 2x + y$. See Fig. 6.

13. $y' = 2x - 3$. (a) Family $y = \int 2x - 3 \, dx = x^2 - 3x + C$.

(b) $y(1) = 4$: $4 = (1)^2 - 3(1) + C$ so $C = 6$. Particualr solution is $\mathbf{y = x^2 - 3x + 6}$.

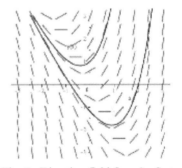
Fig. 6: Direction field for $y' = 2x+y$

15. $y' = e^x + \cos(x)$.

(a) Family $y = \int e^x + \cos(x) \, dx = e^x + \sin(x) + C$.

(b) $y(0) = 7$: $7 = e^0 + \sin(0) + C$ so $C = 6$. Particualr solution is $\mathbf{y = e^x + \sin(x) + 6}$.

17. $y' = \frac{6}{2x+1} + x^{1/2}$. (a) Family $y = \int \frac{6}{2x+1} + x^{1/2} \, dx = 3 \cdot \ln|2x+1| + \frac{2}{3} x^{3/2} + C$.

(b) $y(1) = 4$: $4 = 3 \cdot \ln|2 \cdot 1 + 1| + \frac{2}{3}(1)^{3/2} + C$ so $C = \frac{10}{3} - 3 \cdot \ln(3)$.

Particualr solution is $\mathbf{y = 3 \cdot \ln|2x+1| + \frac{2}{3} x^{3/2} + \{ \frac{10}{3} - 3 \cdot \ln(3) \}}$.

19. See Fig. 7. The letter "S."

20. After a "long time," all of the solutions approach the horizontal asymptote P = 100.

21. See Fig. 8. These are approximate paths. Yours can be somewhat different.

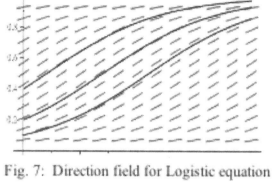
Fig. 7: Direction field for Logistic equation

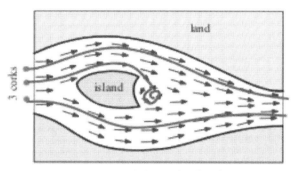
Fig. 8: Possible paths for the corks

Section 6.2

1. See Fig. 1

3. $\frac{1}{y} y' = 2x$. $\int \frac{1}{y} dy = \int 2x\, dx$ so $\ln|y| = x^2 + C$.

5. $y' = \frac{3}{1+x^2}$. $\int 1\, dy = \int \frac{3}{1+x^2} dx$

so $y = 3\cdot\arctan(x) + C$.

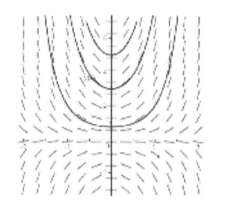
Fig. 1: Direction field for $y' = 2xy$

7. $\frac{1}{e^y} \cdot y' = \frac{1}{\cos(x)}$. $e^{-y} \cdot y' = \sec(x)$. $\int e^{-y} dy = \int \sec(x)\, dx$

so $-e^{-y} = \ln|\sec(x) + \tan(x)| + C$.

9. $y' = 4y$. $\frac{1}{y} \cdot y' = 4$. $\int \frac{1}{y} dy = \int 4\, dx$ so $\ln|y| = 4x + C$.

11. $\frac{1}{y} y' = 2x$. $\int \frac{1}{y} dy = \int 2x\, dx$ so $\ln|y| = x^2 + C$.

$y(0) = 3$: $\ln(3) = (0)^2 + C$ so $C = \ln(3)$. $\ln|y| = x^2 + \ln(3)$. $y = e^{x^2 + \ln(3)} = e^{x^2} \cdot e^{\ln(3)} = 3e^{x^2}$.

$y(0) = 5$: $\ln(5) = (0)^2 + C$ so $C = \ln(5)$. $\ln|y| = x^2 + \ln(5)$. $y = e^{x^2 + \ln(5)} = e^{x^2} \cdot e^{\ln(5)} = 5e^{x^2}$.

$y(1) = 2$: $\ln(2) = (1)^2 + C$ so $C = \ln(2) - 1$. $\ln|y| = x^2 + \ln(2) - 1$. $y = e^{x^2 + \ln(2) - 1} = e^{x^2 - 1} \cdot e^{\ln(2)} = 2e^{x^2 - 1}$.

13. $y' = 3y$. $\int \frac{1}{y} dy = \int 3 dx$ so **ln| y | = 3x + C**.

    $y(0) = 4$: $\ln(4) = 3(0) + C$ so $C = \ln(4)$. $\ln| y | = 3x + \ln(4)$.

    $y(0) = 7$: $\ln(7) = 3(0) + C$ so $C = \ln(7)$. $\ln| y | = 3x + \ln(7)$.

    $y(1) = 3$: $\ln(3) = 3(1) + C$ so $C = \ln(3) - 3$. $\ln| y | = 3x + \ln(3) - 3$.

15. $y' = 5(2-y)$. $\int \frac{1}{2-y} dy = \int 5 dx$ so **–ln|2–y| = 5x + C**.

    $y(0) = 5$: $-\ln|2-5| = 5(0) + C$ so $C = -\ln(3)$. $-\ln|2-y| = 5x - \ln(3)$.

    $y(0) = -3$: $-\ln|2-(-3)| = 5(0) + C$ so $C = -\ln(5)$. $-\ln|2-y| = 5x - \ln(5)$.

17. $y' = \frac{3}{1+x^2}$. $\int 1 dy = \int \frac{3}{1+x^2} dx$ so **y = 3·arctan(x) + C**.

    $y(1) = 4$: $4 = 3 \cdot \arctan(1) + C$ so $C = 4 - 3 \cdot \arctan(1) \approx 1.644$. $y = 3 \cdot \arctan(x) + 1.644$.

    $y(0) = 2$: $2 = 3 \cdot \arctan(0) + C$ so $C = 2$. $y = 3 \cdot \arctan(x) + 2$.

19. $P(t) = 3{,}000 \cdot e^{0.0392t}$. $P(20) = 3{,}000 \cdot e^{0.0392(20)} = 3{,}000 \cdot e^{0.784} \approx 3{,}000 \cdot (2.190) = 6{,}570$ people.

21. $C(t) = 3 \cdot e^{-0.00012t}$. $C(10{,}000) = 3 \cdot e^{(-0.00012)(10000)} = 3 \cdot e^{-1.2} \approx 3 \cdot (0.301) = 0.903$.

23. $T - a = (T_0 - a) \cdot e^{kt}$. $T - 25 = (80-25) \cdot e^{-0.12t}$.

    $T(5) = 25 + 55 \cdot e^{-0.12(5)} = 25 + 55 \cdot e^{-0.6} \approx 25 + 55 \cdot (0.5488) = 55.184$ °C

Section 6.3

1. Use Fig. 13 in text. Years from population of 10,000 to 20,000 approximately $1979 - 1965 = 14$.

    Years from population of 15,000 to 30,000 approximately $1987 - 1972 = 16$.

    Approximate doubling time: around 15 years.

3. (a) $f(t)$ = number of people t years after 1970.

        $f(t) = f(0) \cdot e^{kt}$, $f(0) = 48{,}000$ and $f(20) = 64{,}000$. $f(t) = 48{,}000 \cdot e^{kt}$.

        $64{,}000 = f(20) = 48{,}000 \cdot e^{k(20)}$ so $\frac{64000}{48000} = e^{k(20)}$ and $\ln(4/3) = 20k$.

        Then $k = \frac{1}{20} \ln(4/3) \approx 0.014$ and **f(t) = 48,000·e^(0.014)t**.

    (b) The year 2000 is $t = 30$. $f(30) = 48{,}000 \cdot e^{(0.014)(30)} \approx 73{,}054$ people.

    (c) Solve $100{,}000 = f(t) = 48{,}000 \cdot e^{(0.014)t}$ to get $t = \frac{\ln(100/48)}{0.014} \approx 52.4$ years.

    (d) Solve $96{,}000 = f(t) = 48{,}000 \cdot e^{(0.014)t}$ to get $t = \frac{\ln(2)}{0.014} \approx 49.5$ years.

5. (a) Rule of 72: Doubling time $\approx \frac{72}{15} = 4.8$ years.

   Exponential growth model: doubling time $= \frac{\ln(2)}{\ln(1.15)} \approx 4.96$ years.

   (b) After 1 year the value of the investment is $5,000 + (0.15)(5,000) = 5750$. Then $5750 = f(1) = 5000e^{k(1)}$ and $k = \ln(5750/5000) = \ln(1.15)$: $f(t) = 5000 \cdot e^{t \cdot \ln(1.15)}$.

   To find tripling time, solve $15,000 = 5,000 e^{t \cdot \ln(1.15)}$ for t.

   $\ln(3) = t \cdot \ln(1.15)$ so $t = \frac{\ln(3)}{\ln(1.15)} \approx 7.86$.

7. $f(t) = f(0) \cdot e^{kt}$. Put $t = 0$ for 1950, and $t = 10$ for 1960. Then $f(0) \approx 5,000$ and $f(10) \approx 7,500$ so $7500 = 5000 \cdot e^{k(10)}$ and $k = \frac{\ln(7500/5000)}{10} \approx 0.041$. $f(t) = 5000 \cdot e^{(0.041) \cdot t}$.

9. Doubling time $= \frac{\ln(2)}{k} = 20$ so $k = \frac{\ln(2)}{20} \approx 0.035$. Then $f(t) = f(0) \cdot e^{(0.035) \cdot t}$.

   Annual percentage rate of growth $= \frac{f(1) - f(0)}{f(0)}(100) = \frac{f(0) \cdot e^{(0.035) \cdot 1} - f(0)}{f(0)}(100) = \frac{e^{(0.035) \cdot 1} - 1}{1}(100) \approx 3.53$.

11. For group A, $P_a(t) = 600,000 \cdot e^{kt}$ with $k = \ln(1.03) \approx 0.0296$ so $P_a(t) = 600,000 \cdot e^{(0.0296)t}$.

    For group B, $P_b(t) = 400,000 \cdot e^{kt}$ with $k = \ln(1.06) \approx 0.0583$ so $P_b(t) = 400,000 \cdot e^{(0.0583)t}$.

    Set $P_a(t) = P_b(t)$ and solve for t: $600,000 \cdot e^{(0.0296)t} = 400,000 \cdot e^{(0.0583)t}$.

    (There are several ways to solve for t. This is one way.) Take natural logs of each side,

    $\ln(600,000) + (0.0296)t = \ln(400,000) + (0.0583)t$ so $\ln(600,000) - \ln(400,000) = \{0.0583 - 0.0296\} t$

    and $t = \frac{\ln(1.5)}{0.0287} \approx 14.13$ years.

    If they continue growing at their current rates, the two groups will be the same size in about 14.5 years.

13. When $R = 5$, the tripling time is $t = \frac{\ln(3)}{\ln(1.05)} \approx 22.5 = \frac{113}{R}$.

    When $R = 10$, the tripling time is $t = \frac{\ln(3)}{\ln(1.10)} \approx 11.53 \approx \frac{115}{R}$.

    When $R = 15$, the tripling time is $t = \frac{\ln(3)}{\ln(1.15)} \approx 7.86 \approx \frac{118}{R}$.

    For $5 \leq R \leq 15$, a good approximation of the tripling time is $\frac{115}{R}$.

15. $f(0) = 8{,}000$ snails and $k = 0.14$ so $f(t) = 8{,}000 \cdot e^{0.14 \cdot t}$.

   (a) Harvest 2,000 every 2 months: after 2 months $f(t) = 8000 \cdot e^{0.14 \cdot (2/12)} \approx 8{,}188$. After harvest = 6,188.

   after 4 months, population = $6{,}188 \cdot e^{0.14 \cdot (2/12)} \approx 6{,}334$. After harvest = 4,334.

   after 6 months, population = $4{,}334 \cdot e^{0.14 \cdot (2/12)} \approx 4{,}436$. After harvest = 2,436.

   after 8 months, population = $2{,}436 \cdot e^{0.14 \cdot (2/12)} \approx 2{,}494$. After harvest = 494.

   after 10 months, population = $494 \cdot e^{0.14 \cdot (2/12)} \approx 506$. After harvest, no snails remain..

   (b) No snails remain after the 6 month harvest. (5,177; 2,291; 0)

   (c) After one 2 month period, the population is $8000 \cdot e^{0.14 \cdot (2/12)} \approx 8{,}188$ so the population **growth** is 188 snails. We can harvest 188 snails every two months and maintain a stable population (between 8,000 and 8,188).

17. $f(0) = 10, f(14) = 2$, and $f(t) = f(0) \cdot e^{kt} = 10 \cdot e^{kt}$.

   (a) $2 = f(14) = 10 \cdot e^{k(14)}$ so $k = \frac{1}{14} \ln(\frac{2}{10}) \approx -0.115$. $f(t) = 10 \cdot e^{(-0.115)t}$.

   (b) half life = $\frac{\ln(1/2)}{k} \approx \frac{-0.693}{-0.115} \approx 6$ days.

   (c) $f(t) = 0.7 = 10 \cdot e^{(-0.115)t}$ so $t = \frac{1}{-0.115} \ln(0.07) \approx 23$ days.

19. (a) $f(t) = f(0) \cdot e^{kt}$ so $143 = 187 \cdot e^{k(2)}$ and $k = \frac{1}{2} \ln(143/187) \approx -0.134$.

   half life = $\frac{\ln(1/2)}{k} \approx 5.17$ days.

   (b) $20 = 187 \cdot e^{(-0.134)t}$ so $t = \frac{\ln(20/187)}{-0.134} \approx 16.7$ days.

21. From Fig. 17, $f(0) = 8$ and $f(6) = 3.5$ for material A. $3.5 = f(6) = f(0) \cdot e^{k(6)} = 8 \cdot e^{k(6)}$ so $k = \frac{1}{6} \ln(3.5/8) \approx -0.138$. $f(t) = 8 \cdot e^{(-0.138)t}$ counts (1000s) per minute on day t.

23. Carbon–14 has a half life of 5,700 years so $5{,}700 = \frac{\ln(1/2)}{k}$ and $k = \frac{\ln(1/2)}{5700} \approx -0.00012$.

   The amount present is 97.5% of new paper so $97.5 = 100 \cdot e^{(-0.00012)(t)}$ and $t = \frac{\ln(0.975)}{-0.00012} \approx 211$ years.

   The paper is about 211 years old, but Newton died in 1727, more than 211 years ago. The letter is a fake.

25. Half life = 6 hours so $6 = \frac{\ln(1/2)}{k}$ and $k \approx -0.116$. $f(t) = f(0) \cdot e^{kt} = 30 \cdot e^{(-0.116)t}$.

   Find t so $30 \cdot e^{(-0.116)t} \geq 10$: $(-0.116)t \geq \ln(10/30)$ so $t \leq \frac{-1.009}{-0.116} \approx 9.47$ hours.

   After about 9.5 hours, the concentration of medicine is no longer effective.

27. Half life = 15 hours so $k = \frac{\ln(1/2)}{15} \approx -0.046$. $f(t) = f(0) \cdot e^{kt} = 9 \cdot e^{(-0.046)t}$.

$f(8) = 9 \cdot e^{(-0.046)8} \approx 6.23$ mg which is a "decay" of $9 - 6.23 = 2.77$ mg during the 8 hours.

Taking a 2.77 mg dose every 8 hours keeps the medicine level in the safe and effective range over a long period of time.

29. Iodine–131 has a half life of 8.07 days so $k = \frac{\ln(1/2)}{8.07} \approx -0.086$. $f(t) = f(0) \cdot e^{(-0.086)t}$.

We have 5A present and want to find the time until the level is A:

$A = 5A \cdot e^{(-0.086)t}$ so $0.2 = e^{(-0.086)t}$ and $t = \frac{\ln(0.2)}{-0.086} \approx 18.7$ days

31. Population: $P(0) = 4$ million, $P(1) = (1.05)(4$ million$)$ so $f(1) = f(0) \cdot (1.05) = f(0) \cdot e^{k(1)}$. Then $k = \ln(1.05) \approx 0.049$ and $f(t) = (4{,}000{,}000) \cdot e^{(0.049)t}$.

Forest: $f(t) = 10{,}000{,}000 - (300{,}000) \cdot t$ acres remain after t years.

The entire forest will be gone in 33.3 years.

(a) acres per person $= \dfrac{10000000 - 300000 \cdot t}{4000000 \cdot e^{(0.049)t}} = \dfrac{100 - 3 \cdot t}{40 \cdot e^{(0.049)t}}$.

(b) rate of change = D(acres per person) $= \dfrac{40 \cdot e^{(0.049)t}(-3) - (100 - 3 \cdot t) \cdot 40 \cdot e^{(0.049)t} \cdot (0.049)}{\{40 \cdot e^{(0.049)t}\}^2}$

$= \dfrac{(-3) - (100 - 3 \cdot t)(0.049)}{40 \cdot e^{(0.049)t}} = \dfrac{-7.9 + 0.147t}{40 \cdot e^{(0.049)t}}$

(c) Solve $g(t) = \dfrac{100 - 3 \cdot t}{40 \cdot e^{(0.049)t}} = 1$ to get $t \approx 10.75$. (Used a graphic calculator to find $g(t) = 1$.)

33. (a) $f(t) = 40 + (200 - 40) \cdot e^{kt} = 40 + 160 \cdot e^{kt}$.

$150 = f(4) = 40 + 160 \cdot e^{k(4)}$ so $k = \frac{1}{4} \ln(110/160) \approx -0.094$. Then $f(t) = 40 + 160 \cdot e^{(-0.094)t}$.

(b) $100 = 40 + 160 \cdot e^{(-0.094)t}$ so $t = \dfrac{\ln(60/160)}{-0.094} \approx 10.4$ minutes.

(c) $80 = 40 + 160 \cdot e^{(-0.094)t}$ so $t = \dfrac{\ln(40/160)}{-0.094} \approx 14.7$ minutes.

(d) $60 = 40 + 160 \cdot e^{(-0.094)t}$ so $t = \dfrac{\ln(20/160)}{-0.094} \approx 22.1$ minutes.

CHAPTER 7: TRANSCENDENTAL FUNCTIONS

Introduction

In the previous chapters we saw how to calculate and use the derivatives and integrals of many of the most important and common functions in mathematics and applications. These common functions include polynomials, exponential functions, logarithms, the six trigonometric functions, and various combinations of them. There are, however, other important and useful functions, and we examine several of them in this chapter. The central purpose of this chapter is to extend the ideas and applications we have already seen to additional functions, primarily the inverse trigonometric functions.

The chapter begins with a short introduction to one–to–one functions. Section 7.1 is an examination of inverse functions and some of their properties. Section 7.2 introduces the inverses of the trigonometric functions, and Section 7.3 shows how to calculate and use the derivatives of the inverse trigonometric functions.

We have already been using a most important pair of one–to–one functions, e^x and $\ln(x)$. We have also used the fact that each of them "undoes" the effect of the other in order to solve equations such as $3 = 5e^{2x}$ and $8 = 2 \cdot \ln(x)$. **The functions e^x and $\ln(x)$ are inverses of each other, and you should keep these functions and their graphs in mind as we discuss general one–to–one and inverse functions.** Section 7.4 (optional) is a calculus–based presentation of the exponential and logarithm functions, and it includes verifications of a number of properties of the exponential and logarithm functions that we have already used, such as the multiplication law for logarithms: $\ln(a \cdot b) = \ln(a) + \ln(b)$.

One–to–One Functions

In earlier courses you saw that some equations have only one solution (for example, $5 - 2x = 3$ and $x^3 = 8$), some have two solutions ($x^2 + 3 = 7$), and some even have an infinite number of solutions ($\sin(x) = 0.8$). The graphs of the functions $y = 5 - 2x, y = x^3, y = x^2 + 3$, and $y = \sin(x)$ and the solutions of the equations are shown in Fig. 1. The functions f whose equations $f(x) = k$ have only one solution **for each value of k** (each outcome k comes from only one input x) are particularly common in applications, and they have a number of useful mathematical properties. The remainder of this section focuses on those functions and looks at some of their properties.

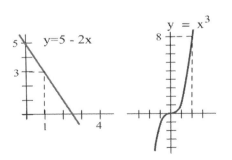
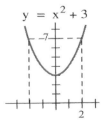
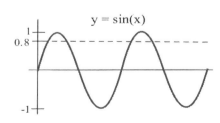

Fig. 1

Example 1: How many solutions does each of the following equations have?

(a) $f(x) = 0$ for $f(x) = x(x - 4)$ (b) $g(x) = 3$ for g given by Table 1

(c) $h(x) = 4$ for h given by the graph in Fig. 2. (d) $f(x) = k$ for $f(x) = e^x$.

Solution: (a) Two. $x(4 - x) = 0$ if $x = 0$ or $x = 4$. (b) One. $g(x) = 3$ if $x = 2$.

(c) Two. $h(x) = 4$ if $x = 1$ or $x = 5$.

(d) One, $x = \ln(k)$, if $k > 0$. None if $k \leq 0$.

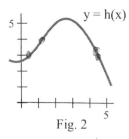

Fig. 2

x	g(x)
0	5
1	7
2	3
3	5
4	0
5	7

Table 1

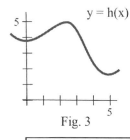

Fig. 3

Practice 1: How many solutions does each of the following equations have?

(a) $f(x) = 4$ for $f(x) = x(4 - x)$ (b) $g(x) = 7$ for g given by Table 1

(c) $h(x) = 3$ for h given by the graph in Fig. 3.

(d) $f(x) = 5$ for $f(x) = \ln(x)$

Horizontal Line Test for One–to–one (Definition of One–to–one)

A function is **one–to–one** if each **horizontal** line intersects the graph of the function at most once.

Equivalently, a function $y = f(x)$ is one–to–one if two **distinct** x–values always result in two **distinct** y–values: $a \neq b$ implies $f(a) \neq f(b)$. This immediately tells us that every strictly increasing function is one–to–one, and every strictly decreasing function is one–to–one. (Why?)

For any function, if we know the input x–value, we can calculate the output y–value, but an output may have come from any of several different inputs. With a one–to–one function, each output y–value comes from only one input x–value.

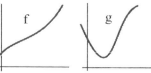

Fig. 4

Example 2: (a) Which functions in Fig. 4 are one–to–one?

(b) Which functions in Table 2 are one–to–one?

Solution: (a) In Fig. 4, functions f and h are one–to–one.

Function g is not one–to–one.

(b) In Table 2, function h is one–to–one. Functions f and g are not one–to–one.

x	f(x)	g(x)	h(x)
0	5	7	2
1	2	3	-1
2	3	0	5
3	5	1	4
4	0	6	3
5	1	3	0

Table 2

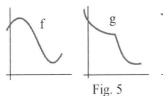

Fig. 5

Practice 2:

(a) Which functions in Fig. 5 are one–to–one?

(b) Which functions in Table 3 are one–to–one?

x	f(x)	g(x)	h(x)
0	4	2	2
1	2	3	−5
2	−2	0	1
3	5	4	14
4	3	6	3
5	1	7	1

Table 3

Example 3: Let $f(x) = 2x + 1$ (Fig. 6). Find the values of x so that

(a) $f(x) = 9$ and (b) $f(x) = a$.

(c) Solve $f(y) = x$ for y. ($f(y) = 2y + 1$)

Solution: (a) $9 = f(x) = 2x + 1$ so $8 = 2x$ and $x = 8/2 = 4$. (b) $a = 2x + 1$ so $2x = a − 1$ and $x = (a − 1)/2$.

(c) $x = f(y) = 2y + 1$ so $2y = x − 1$ and $y = (x − 1)/2$.

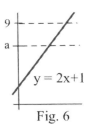

Fig. 6

Practice 3: Let $f(x) = 3x − 5$. Find values of x so (a) $f(x) = 7$ and (b) $f(x) = a$. (c) Solve $f(y) = x$ for y.

Practice 4: Show that exponential growth, $f(x) = e^{3x}$, and exponential decay, $g(x) = e^{-2x}$, are one–to–one.

PROBLEMS

In problems 1 – 4, state whether the given functions are one–to–one.

1. $f(x) = 3x − 5$, $y = 3 − x$, g(x) given by Table 4, and h(x) given by the graph in Fig. 7.

x	g(x)
0	3
1	4
2	5
3	2
4	4

Table 4

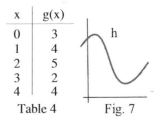

Fig. 7

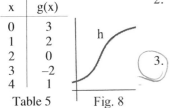

x	g(x)
0	3
1	2
2	0
3	−2
4	1

Table 5   Fig. 8

2. $f(x) = x/4$, $y = x^2 + 3$, g(x) given by Table 5, and h(x) given by the graph in Fig. 8.

3. $f(x) = \sin(x)$, $y = e^x − 2$, g(x) given by Table 6, and h(x) given by the graph in Fig. 9.

x	g(x)
0	−1
1	5
2	3
3	1
4	0

Table 6   Fig. 9

4. $f(x) = 17$, $y = x^3 − 1$, g(x) given by Table 7, and h(x) given by the graph in Fig. 10.

5. Are Social Security numbers one–to–one? Telephone numbers?

6. What would it mean if the scores on a calculus test were one–to–one?

7. What is the legal/social term for one–to–one in marriage?

x	g(x)
0	2
1	5
2	4
3	1
4	2

Table 7   Fig. 10

8. The function given below represents "y is married to x."

   (a) Is this f a function? (b) Is f one–to–one? (c) Is P breaking the law?

   (d) Is A breaking the law?

x	A	B	C	D
y	P	Q	P	R

9. How many places can a one–to–one function touch the x–axis?

10. Can a **continuous** one–to–one function have the values given below Explain why it is possible or why it is not possible?

x	1	3	5
f(x)	2	7	3

11. The graph of $f(x) = x - 2 \cdot INT(x)$ for $-2 \le x \le 3$ is given in Fig. 11.

    (a) Is f a one–to–one function?

    (b) Is f an increasing function? a decreasing function?

12. Is **every** linear function $f(x) = ax + b$ one–to–one?

13. Show that the function $f(x) = \ln(x)$ is one–to–one for $x > 0$.

14. Show that the function $f(x) = e^x$ is one–to–one.

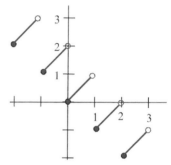

Fig. 11: $y = x - 2 \cdot INT(x)$

In problems 15 – 18, rules are given for encoding a 6 letter alphabet. For each problem:

(a) Is the encoding rule a function? (b) Is the encoding rule one–to–one? (c) Encode the word "bad."

(d) Write a table for decoding the encoded letters and use it to decode your answer to part (c).

(e) Graph the encoding rule and the decoding rule. (Fig. 12 shows the graphs for the code in problem 15). How are the encoding and decoding graphs related?

15. | original letter | a | b | c | d | e | f |
    |---|---|---|---|---|---|---|
    | encoded letter | d | c | f | e | b | a |

16. | original letter | a | b | c | d | e | f |
    |---|---|---|---|---|---|---|
    | encoded letter | b | d | f | b | a | c |

17. | original letter | a | b | c | d | e | f |
    |---|---|---|---|---|---|---|
    | encoded letter | d | f | e | a | c | b |

    How does your decoding rule compare with the encoding rule?

    What happens if you encode a word and then encode the encoded word: for example, Encode(Encode("bad")) = ?

18. | original letter | a | b | c | d | e | f |
    |---|---|---|---|---|---|---|
    | ncoded letter | e | a | f | c | b | d |

    What happens if you apply this coding rule three times:

    Encode(Encode(Encode("bad"))) = ?

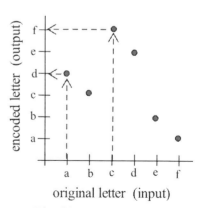

Fig. 12: Graph of a code

Section 7.0 PRACTICE Answers

Practice 1:
(a) One. Solve $x(4-x) = 4$ to get $x = 2$.
(b) At least two, $x = 1$ and $x = 5$.
(c) At least one, $x \approx 4$.
(d) Exactly one. Solve $5 = \ln(x)$ to get $x = e^5 \approx 148.4$.

Practice 2:
(a) Only g is 1–1.
(b) From the values in the table, f anf g are one–to–one.
(However, if f and g are continuous on [0,5], then neither of them is one–to–one. Why?)

Practice 3:
(a) $3x - 5 = 7$ so $x = 4$.
(b) $3x - 5 = a$ so $x = \dfrac{a+5}{3}$.
(c) $f(x) = 3x - 5$ so $f(y) = 3y - 5$. $f(y) = x$ means that $3y - 5 = x$ and $y = \dfrac{x+5}{3}$.

Practice 4: $f(x) = e^{3x}$. $f'(x) = 3 \cdot e^{3x} > 0$ so f is increasing, one–to–one, and has an inverse.
$g(x) = e^{-2x}$. $g'(x) = -2 \cdot e^{-2x} < 0$ so g is decreasing, one–to–one, and has an inverse.

7.1 INVERSE FUNCTIONS

One–to–one functions are important because their equations, $f(x) = k$, have (at most) a single solution. One-to-one functions are also important because they are the functions that can be uniquely "undone:" if f is a one–to–one function, then there is **another function** g which "undoes" the effect of f so $g(f(x)) = x$, the original input (Fig. 1). The function g which "undoes" the effect of f is called the inverse function of f or simply the **inverse of f**. If f is a function for encoding a message, then the inverse of f is the function that decodes an encoded message to get back to the original message. The functions e^x and $\ln(x)$ "undo" the effects of each other: $\ln(e^x) = x$ and $e^{\ln(x)} = x$ (for $x > 0$). The functions e^x and $\ln(x)$ are inverses of each other.

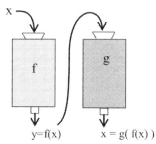

Fig. 1: g "undoes" the effect of f

In this section we examine some of the properties of inverse functions and see how to find the inverse of a function which is given by a table of data, a graph, or a formula.

Definition: If f and g are functions which satisfy $g(f(x)) = x$ and $f(g(x)) = x$

then g is the **inverse of f**, f is the **inverse of g**, and

f and g are a **pair of inverse functions**.

The **inverse function of f** is written f^{-1} (pronounced "eff inverse").
(Be careful. f^{-1} does **not** mean $1/f$.)

Example 1: A function f is given by Table 1. Evaluate $f^{-1}(0)$ and $f^{-1}(1)$.

Solution: For every x, $f^{-1}(f(x)) = x$ so the value of $f^{-1}(0)$ is the solution x of the equation $f(x) = 0$. The solution we want is $x = 2$, and we can check that $f^{-1}(0) = f^{-1}(f(2)) = 2$.
The value of $f^{-1}(1)$ is the solution of the equation $f(x) = 1$, and, from the table, we can see that solution is $x = 3$. We can check that $f^{-1}(1) = f^{-1}(f(3)) = 3$.
Similarly, $f^{-1}(2) = 1$ and $f^{-1}(3) = 0$. These results are shown in Table 2.

x	f(x)
0	3
1	2
2	0
3	1

Table 1

You should notice that if the ordered pair (a,b) is in the table for f, then the reversed pair (b,a) is in the table for f^{-1}.

x	$f^{-1}(x)$
0	2
1	3
2	1
3	0

Table 2

Practice 1: The function g is given by Table 3. Write a table of values for g^{-1}.

The method of interchanging the coordinates of a point on the graph (or in the table of values) of f to get a point on the graph (or in the table of values) of f^{-1} provides a efficient way of graphing f^{-1}.

x	g(x)
0	2
1	1
2	3
3	4
4	0

Table 3

> **Theorem:** If the point (a,b) is on the graph of f,
> then the point (b,a) is on the graph of f^{-1}.

Proof: If (a,b) is on the graph of f, then $b = f(a)$. Since b and $f(a)$ are equal, $f^{-1}(b)$ and $f^{-1}(f(a))$ are also equal, and $f^{-1}(b) = f^{-1}(f(a)) = a$ (by the definition of inverse functions) so the point (b,a) is on the graph of f^{-1}.

Graphically, when the coordinates of a point (a,b) are interchanged to get a new point (b,a), the new point is symmetric about the line $y = x$ to the original point (Fig. 2). If you put a spot of wet ink at the point (a,b) on a piece of paper and fold the paper along the line $y = x$, there will be new spot of ink at the point (b,a). Fig. 3 illustrates another graphical method for finding the location of the point (b,a).

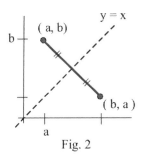

Fig. 2

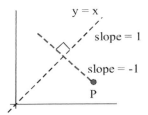
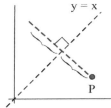
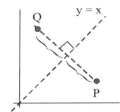

1. Draw the line $y=x$ and a line through P with slope -1

2. Measure the distance from P to the line $y=x$ and move that distance on the onther side of $y=x$

3. Plot the point Q at the new location

Fig. 3: Determining the location of a point symmetric to P about the line $y=x$

> **Corollary:** The graphs of f and f^{-1} are symmetric about the line $y = x$.

Example 2: The graph of f is shown in Fig. 4. Sketch the graph of f^{-1}.

Solution: Imagine the graph of f is drawn with wet ink, and fold the xy–plane along the line $y = x$. When the plane is unfolded, the new graph is f^{-1}. (Fig. 5)

Another method proceeds point–by–point. Pick several points (a,b) on the graph of f and plot the symmetric points (b,a). Use the new (b,a) points as a guide for sketching the graph of f^{-1}.

Fig. 4: Graph of f

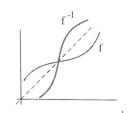

Fig. 5: Graphs of f and f^{-1}

Practice 2: The graph of g is shown in Fig. 6. Sketch the graph of g^{-1}. What happens to points on the graph of g that lie on the line $y = x$, points of the form (a, a)?

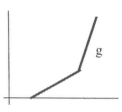

Fig. 6: Graph of g

Finding A Formula for $f^{-1}(x)$

When f is given by a table of values, a table of values for f^{-1} can be made by interchanging the values of x and y in the table for f as in Example 1. When f is given by a graph, a graph of f^{-1} can be made by reflecting the graph of f about the line $y = x$ as in Example 2. When f is given by a formula, we can try to find a formula for f^{-1}.

Example 3: The steps for wrapping a gift are (1) put gift in box, (2) put on paper, and (3) put on ribbon. What are the steps for opening the gift, the inverse of the wrapping operation?

Solution: (i) remove ribbon (undo step 3), (ii) remove paper (undo step 2), and (iii) remove gift from box (undo step 1). (Then show happiness and say thank you.) The point of this trivial example is to point out that the first unwrapping step undoes the last wrapping step, the second unwrapping step undoes the second–to–last wrapping step, . . . , and the last unwrapping step undoes the first wrapping step. This pattern holds for more numerical functions and their inverses too.

Example 4: The steps to evaluate $f(x) = 9x + 6$ are (1) multiply by 9 and (2) add 6. Write the steps, in words, for the inverse of this function, and then translate the verbal steps for the inverse into a formula for the inverse function.

Solution: (i) subtract 6 (undo step 2) and (ii) divide by 9 (undo step 1). $f^{-1}(x) = \dfrac{x-6}{9}$.

The following algorithm is a recipe for finding a formula for the inverse function.

Finding A Formula For f^{-1}:

Start with a formula for f: $y = f(x)$.

Interchange the roles of x and y: $x = f(y)$.

Solve $x = f(y)$ for y.

The resulting formula for y is the inverse of f: $y = f^{-1}(x)$.

The "interchange" and "solve" steps in the algorithm effectively undo the original operations in reverse order.

Example 5: Find formulas for the inverses of $f(x) = \dfrac{7x - 5}{4}$ and $g(x) = 2e^{5x}$.

Solution: Starting with $y = f(x) = \frac{7x-5}{4}$ and interchanging the roles of x and y, we have $x = \frac{7y-5}{4}$.

Solving for y, we get $4x = 7y - 5$, so $7y = 4x + 5$ and, finally, $y = \frac{4x+5}{7} = f^{-1}(x)$.

Starting with $y = g(x) = 2e^{5x}$ and interchanging the roles of x and y, we have $x = g(y) = 2e^{5y}$.

Solving for y, we get $\frac{x}{2} = e^{5y}$, so $\ln(\frac{x}{2}) = \ln(e^{5y}) = 5y$, and $y = \frac{1}{5} \cdot \ln(\frac{x}{2}) = g^{-1}(x)$.

Practice 3: Find formulas for the inverses of $f(x) = 2x - 5$, $g(x) = \frac{2x-1}{x+7}$, and $h(x) = 2 + \ln(3x)$.

Sometimes it is easy to "solve $x = f(y)$ for y," but not always. When we try to find a formula for the inverse of $y = f(x) = x + e^x$, the first step is easy: starting with $y = x + e^x$ and interchanging the roles of x and y, we have $x = y + e^y$. Then, unfortunately, there is no way to algebraically solve the equation $x = y + e^y$ explicitly for y. The function $y = x + e^x$ has an inverse function, but we cannot find an explicit formula for the inverse.

Which Functions Have Inverse Functions?

We have seen how to find the inverse function for some functions given by tables of values, by graphs, and by formulas, but there are functions which do not have inverse functions. The only way a graph and its reflection about the line $y = x$ can both be function graphs (f and f^{-1} are both functions) is if the graph satisfies both the Vertical Line Test (so f is a function) and the Horizontal Line Test (so f^{-1} is a function). This is stated more formally in the next theorem, although we do not prove it.

> Theorem: The function f has an inverse function if and only if f is one–to–one.

Two useful corollaries follow from this theorem.
  Corollary 1: If f is strictly increasing or is strictly decreasing, then f has an inverse function.
  Corollary 2: If $f'(x) > 0$ for all x or $f'(x) < 0$ for all x, then f has an inverse function.

Slopes of Inverse Functions

When a function f has an inverse, the symmetry of the graphs of f and f^{-1} also gives information about slopes and derivatives.

Example 6: Suppose the points P = (1,2) and Q = (3,6) are on the graph of f.

(a) Sketch the line through P and Q, and find the slope of the line through P and Q.

(b) Find the reflected points P* and Q* on the graph of f^{-1}, sketch the line segment through P* and Q*, and find the slope of the line through P* and Q*.

Solution: (a) The slope through P and Q is $m = \frac{6-2}{3-1} = 2$.

The graphs are shown in Fig. 7.

(b) The reflected points, obtained by interchanging the first and second coordinates of each point on f, are P* = (2,1) and Q* = (6,3). The slope of the line though P* and Q* is

$$\frac{3-1}{6-2} = \frac{1}{2} = \frac{1}{\text{slope of the segment through P and Q}}.$$

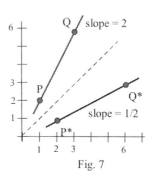

Fig. 7

In general,

if P = (a,b) and Q = (x,y) are points on the graph of f (Fig. 8)

then the reflected points P* = (b,a) and Q* = (y,x) are on the graph of f^{-1}, and

$$\{\text{slope of the segment P*Q*}\} = \frac{1}{\text{slope of the segment PQ}}.$$

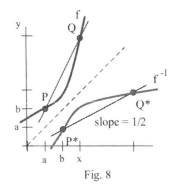

Fig. 8

Since the slope of a tangent line is the limit of slopes of secant lines, there is a similar relationship between the slope of the tangent line to f at the point (a,b) and slope of the tangent line to f^{-1} at the point (b,a). In Fig. 9, let the point Q* approach the point P* along the graph of f^{-1}. Then

$$(f^{-1})'(b) = \lim_{Q^* \to P^*} \{\text{slope of the segment P*Q*}\}$$

$$= \lim_{Q \to P} \frac{1}{\text{slope of the segment PQ}} = \frac{1}{f'(a)}.$$

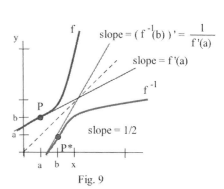

Fig. 9

Derivative of the Inverse Function

If $b = f(a)$, and f is differentiable at the point (a,b), and $f'(a) \neq 0$,

then $a = f^{-1}(b)$, and f^{-1} is differentiable at the point (b,a),

and $(f^{-1})'(b) = \dfrac{1}{f'(a)}$.

Example 7: The point $(3, 1.1)$ is on the graph of $f(x) = \ln(x)$ and $f'(3) = 1/3$. Let g be the inverse function of f. Give one point on the graph of g and evaluate g' at that point.

Solution: The point $(1.1, 3)$ is on the graph of g, and $g'(1.1) = \dfrac{1}{f'(3)} = \dfrac{1}{1/3} = 3$. In fact, the inverse of the natural logarithm is the exponential function e^x, and we can check that $e^{1.1} \approx 3$.

Example 8: In Table 4, fill in the values of $f^{-1}(x)$ and $(f^{-1})'(x)$ for $x = 0$ and 1.

Solution: $f(3) = 0$ so $f^{-1}(0) = 3$. $(f^{-1})'(0) = \dfrac{1}{f'(3)} = \dfrac{1}{2}$.

$f(2) = 1$ so $f^{-1}(1) = 2$. $(f^{-1})'(1) = \dfrac{1}{f'(2)} = \dfrac{1}{-1} = -1$.

x	f(x)	f'(x)	$f^{-1}(x)$	$(f^{-1})'(x)$
0	2	3	3	
1	3	−2		
2	1	−1		
3	0	2		

Table 4

Practice 4: In Table 4, fill in the values of $f^{-1}(x)$ and $(f^{-1})'(x)$ for $x = 2$ and 3.

PROBLEMS

1. The values of f and f' are given in Table 5. Complete the columns for f^{-1} and $(f^{-1})'$.

x	f(x)	f'(x)	$f^{-1}(x)$	$(f^{-1})'(x)$
1	3	−3		
2	1	2		
3	2	3		

Table 5

2. The values of g and g' are given in Table 6. Complete the columns for g^{-1} and $(g^{-1})'$.

x	g(x)	g'(x)	$g^{-1}(x)$	$(g^{-1})'(x)$
1	2	−2		
2	1	4		
3	3	2		

Table 6

3. The values of h and h' are given in Table 7. Complete the columns for h^{-1} and $(h^{-1})'$.

x	h(x)	h'(x)	$h^{-1}(x)$	$(h^{-1})'(x)$
1	2	2		
2	3	−2		
3	1	0		

Table 7

4. The values of w and w' are given in Table 8. Complete the columns for w^{-1} and $(w^{-1})'$.

5. Fig. 10 shows the graph of f. Sketch the graph of f^{-1}.

x	w(x)	w'(x)	$w^{-1}(x)$	$(w^{-1})'(x)$
1	1	2		
2	3	0		
3	2	5		

Table 8

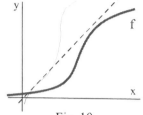

Fig. 10

6. Fig. 11 shows the graph of g. Sketch the graph of g^{-1}.

7. If the graphs of f and f^{-1} intersect at the point (a, b), how are a and b related?

8. If the graph f intersects the line y = x at x = a, does the graph of f^{-1} intersect y = x? If so, where?

9. The steps to evaluate the function $f(x) = \frac{7x-5}{4}$ are (1) multiply by 7, (2) subtract 5, and (3) divide by 4. Write the steps, in words, for the inverse of this function, and then translate the verbal steps for the inverse into a formula for the inverse function.

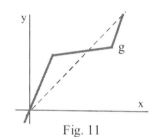

Fig. 11

10. Find a formula for the inverse function of f(x) = 3x – 2. Verify that $f^{-1}(f(5)) = 5$ and $f(f^{-1}(2)) = 2$.

11. Find a formula for the inverse function of g(x) = 2x + 1. Verify that $g^{-1}(g(1)) = 1$ and $g(g^{-1}(7)) = 7$.

12. Find a formula for the inverse function of $h(x) = 2e^{3x}$. Verify that $h^{-1}(h(0)) = 0$.

13. Find a formula for the inverse function of w(x) = 5 + ln(x). Verify that $w^{-1}(w(1)) = 1$.

14. If the graph of f goes through the point (2, 5) and f '(2) = 3, then the graph of f^{-1} goes through the point (__ , __) and $(f^{-1})'(5) =$ _____ ?

15. If the graph of f goes through the point (1, 3) and f '(1) > 0, then the graph of f^{-1} goes through the point (__ , __). What can be said about $(f^{-1})'(3)$?

16. If f(6) = 2 and f '(6) < 0, then the graph of f^{-1} goes through the point (__ , __). What can be said about $(f^{-1})'(2)$?

17. If f '(x) > 0 for all values of x, what can be said about $(f^{-1})'(x)$? What does this mean about the graphs of f and f^{-1} ?

18. If f '(x) < 0 for all values of x, what can be said about $(f^{-1})'(x)$? What does this mean about the graphs of f and f^{-1} ?

19. Find a linear function f(x) = ax + b so the graphs of f and f^{-1} are parallel and do not intersect.

20. Does f(x) = 3 + sin(x) have an inverse function? Justify your answer.

21. Does f(x) = 3x + sin(x) have an inverse function? Justify your answer.

22. For which positive integers n is $f(x) = x^n$ one–to–one? Justify your answer.

23. Some functions are their own inverses. For which **four** of these functions does $f^{-1}(x) = f(x)$?

   (a) f(x) = 1/x   (b) $f(x) = \frac{x+1}{x-1}$ (c) $f(x) = \frac{3x-5}{7x-3}$ (d) f(x) = x + a (a≠0)   (e) $f(x) = \frac{ax+b}{cx-a}$

Reflections on Folding

The symmetry of the graphs of a function and its inverse about the line y = x make it easy to sketch the graph of the inverse function from the graph of the function: we just fold the graph paper along the line y = x, and the graph of f^{-1} is the "folded" image of f. The simple idea of obtaining a new image of something by folding along a line can enable us to quickly "see" solutions to some otherwise difficult problems.

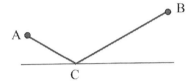

Fig. 12: Find the location for C so AC+CB is minimum

The minimum problem in Fig. 12 of finding the shortest path from town A to town B with an intermediate stop at the river is moderately difficult to solve using derivatives. Geometrically, it is quite easy (Fig. 13):

- obtain the point B* by folding the image of B across the river line
- connect A and B* with a straight line (the shortest path connecting A and B*)
- fold the C to B* line back across the river to obtain the A to C to B solution.

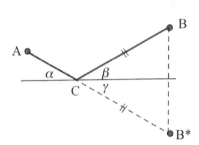

CB=CB* so AC+CB = AC+AB*
The straight line gives the shortest distance from A to B*

Fig. 13

As an almost free bonus, we see that, for the minimum path, the angle of incidence at the river equals the angle of reflection from the river.

24. Devise an algorithm using "folding" to find the where at the bottom edge of the billiards table you should aim ball A in Fig. 14 so that ball A will hit ball B after bouncing off the bottom edge of the table. (Assume that the angle of incidence equals the angle of reflection.)

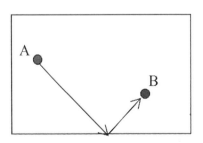

Fig. 14: Billiard balls

25. Devise an algorithm using "folding" to sketch the shortest path in Fig. 15 from town A to town B which includes a stop at the river and at the road. (One fold may not be enough.)

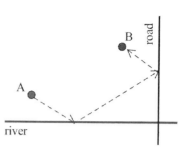

Fig. 15: Stops at a river and a road

26. Devise an algorithm using "folding" to find where you should aim ball A at the bottom edge of the billiards table in Fig. 16 so that ball A will hit ball B after bouncing off the bottom edge and the right edge of the table. (Assume that the angle of incidence equals the angle of reflection. Unfortunately, in a real game of billiards, the ball picks up a spin, "English", when it bounces off the first bank, and then on the second bounce the angle of incidence does not equal the angle of reflection.)

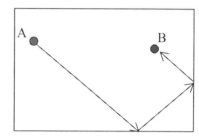

Fig. 16: Two bounces

The "folding" idea can even be useful if the path is not a straight line.

27. Fig. 17 shows the parabolic path of a thrown ball. If the ball bounces off the tall vertical wall in the Fig. 18, where will it land (hit the ground)? (Assume that the angle of incidence equals the angle of reflection and that the ball does not lose energy during the bounce.)

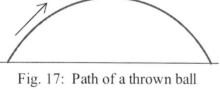

Fig. 17: Path of a thrown ball

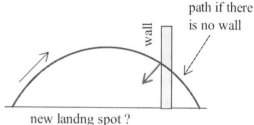

Fig. 18: Path of deflected ball

Sometimes "unfolding" a problem is useful too.

28. A spider and a fly are located at opposite corners of the cube as shown in Fig. 19. Sketch the shortest path the spider can take along the surface of the cube to reach the fly.

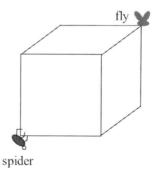

Fig. 19: Find the shortest path from the spider to the fly along the surface of the cube

Section 7.1 PRACTICE Answers

Practice 1: See Fig. 1.

x	g(x)	x	$g^{-1}(x)$
0	2	0	4
1	1	1	1
2	3	2	0
3	4	3	2
4	0	4	3

Fig. 1: Tables for g and g^{-1}

Practice 2: See Fig. 2. If (a,a) is on the graph of g, then (a,a) is also on the graph of g^{-1}.

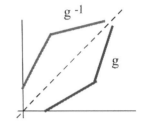

Fig. 20: Graphs of g and g^{-1}

Practice 3:

(a) $y = f(x) = 2x - 5$.
Find f^{-1}: $x = 2y - 5$ so $y = \frac{x+5}{2}$. $f^{-1}(x) = \frac{x+5}{2}$.

(b) $y = g(x) = \frac{2x-1}{x+7}$.
Find g^{-1}: $x = \frac{2y-1}{y+7}$ so $x(y+7) = 2y - 1$. Then $y(x-2) = -7x - 1$ so
$y = \frac{-7x-1}{x-2} = \frac{7x+1}{2-x}$. $g^{-1}(x) = \frac{7x+1}{2-x}$.

(c) $y = h(x) = 2 + \ln(3x)$:
Find h^{-1}: $x = 2 + \ln(3y)$ so $x - 2 = \ln(3y)$ and $e^{x-2} = e^{\ln(3y)}$.
Then $3y = e^{x-2}$ and $y = \frac{1}{3} e^{x-2}$. $h^{-1}(x) = \frac{1}{3} e^{x-2}$.

Practice 4:

x	f(x)	f '(x)	$f^{-1}(x)$	$(f^{-1})'(x)$
0	2	3	3	$\frac{1}{f'(3)} = \frac{1}{2}$
1	3	–2	2	$\frac{1}{f'(2)} = -1$
2	1	–1	0	$\frac{1}{f'(0)} = \frac{1}{3}$
3	0	2	1	$\frac{1}{f'(1)} = \frac{1}{-2}$

Fig. 3: Completed Table 4

7.2 INVERSE TRIGONOMETRIC FUNCTIONS

Section 7.2 is an introduction to the inverse trigonometric functions, their properties, and their graphs. The discussion focuses on the properties and techniques needed for derivatives and integrals. We emphasize the inverse sine and inverse tangent functions, the two inverse trigonometric functions most used in applications. This section provides the background and foundation for the calculus and applications of the inverse trigonometric functions in the next section.

Inverse Sine: Solving $k = \sin(x)$ for x

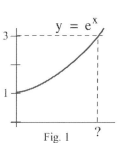
Fig. 1

It is straightforward to solve the equation $3 = e^x$ (Fig. 1): simply take the logarithm, the inverse of the exponential function e^x, of each side of the equation to get $\ln(3) = \ln(e^x) = x$. Since the function $f(x) = e^x$ is one–to–one, the equation $3 = e^x$ has only the one solution $x = \ln(3) \approx 1.1$.

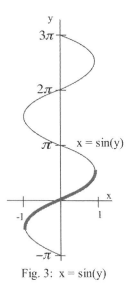
Fig. 3: $x = \sin(y)$

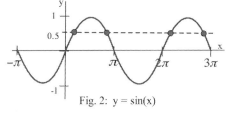
Fig. 2: $y = \sin(x)$

The solution of the equation $0.5 = \sin(x)$ (Fig. 2) presents more difficulties. As Fig. 2 illustrates, the function $f(x) = \sin(x)$ is not one–to–one, and it's reflected graph in Fig. 3 is not the graph of a function. However, sometimes it is important to "undo" the sine function, and we can do so by restricting its domain to the interval $[-\pi/2, \pi/2]$. For $-\pi/2 \le x \le \pi/2$, the function $f(x) = \sin(x)$ is one–to–one and has an inverse function, and the graph of the inverse function (Fig. 4) is the reflection about the line $y = x$ of the (restricted) graph of $y = \sin(x)$.

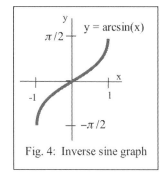
Fig. 4: Inverse sine graph

This inverse of the (restricted) sine function is written as $\sin^{-1}(x)$ or **arcsin(x)**. Most calculators use keys labeled $\sin^{-1}$ or INV SIN for this function.

Definition of Inverse Sine:

For $-1 \le x \le 1$ and $-\pi/2 \le y \le \pi/2$,

$y = \arcsin(x)$ if $x = \sin(y)$.

The domain of $f(x) = \arcsin(x)$ is $[-1, 1]$. The range of $f(x) = \arcsin(x)$ is $[-\pi/2, \pi/2]$.

The (restricted) sine function and the arcsin function are inverses of each other:

> For $-1 \leq x \leq 1$, $\sin(\arcsin(x)) = x$.
>
> For $-\pi/2 \leq y \leq \pi/2$, $\arcsin(\sin(y)) = y$.

Note: The notations $\sin^{-1}(x)$ and **arcsin(x)** are both commonly used in mathematical writings, and we use both. The symbol $\sin^{-1}(x)$ should **never** be used to represent $1/\sin(x)$. Represent $1/\sin(x)$ as $\csc(x)$ or, if you really want to use a negative exponent, as $(\sin(x))^{-1}$.

Note: The name "arcsine" comes from the unit circle definition of the sine function. On the unit circle (Fig. 5), if θ is the length of the **arc** whose sine is x, then $\sin(\theta) = x$ and $\theta = \mathbf{arcsin}(x)$.

Using the right triangle definition of sine (Fig. 6), θ represents an angle whose sine is x.

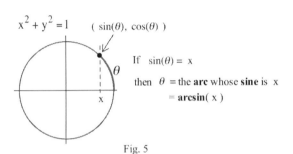

Fig. 5

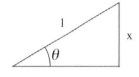

Fig. 6: $\sin(\theta) = x$, $\theta = \arcsin(x)$

Right Triangles and Arcsine

For the right triangle in Fig. 7, $\sin(\theta) = \dfrac{\text{opposite}}{\text{hypotenuse}} = \dfrac{3}{5}$ so $\theta = \arcsin(3/5)$. It is possible to evaluate other trigonometric functions, such as cosine and tangent, of an arcsine without explicitly solving for the value of the arcsine. For example,

$$\cos(\arcsin(3/5)) = \cos(\theta) = \dfrac{\text{adjacent}}{\text{hypotenuse}} = \dfrac{4}{5} \text{ and}$$

$$\tan(\arcsin(3/5)) = \tan(\theta) = \dfrac{\text{opposite}}{\text{adjacent}} = \dfrac{3}{4}.$$

$\sin(\theta) = \dfrac{3}{5}$

$\theta = \arcsin\left(\dfrac{3}{5}\right)$

($\theta \approx 0.6435$)

Fig. 7

Once we know the sides of the right triangle, the values of the other trigonometric functions can be evaluated using their standard right triangle definitions:

$\sin(\theta) = \dfrac{\text{opposite}}{\text{hypotenuse}}$ $\cos(\theta) = \dfrac{\text{adjacent}}{\text{hypotenuse}}$ $\tan(\theta) = \dfrac{\text{opposite}}{\text{adjacent}}$

$\csc(\theta) = \dfrac{1}{\sin(\theta)} = \dfrac{\text{hypotenuse}}{\text{opposite}}$ $\sec(\theta) = \dfrac{1}{\cos(\theta)} = \dfrac{\text{hypotenuse}}{\text{adjacent}}$ $\cot(\theta) = \dfrac{1}{\tan(\theta)} = \dfrac{\text{adjacent}}{\text{opposite}}$

If the angle θ is given as the arcsine of a number, but the sides of a right triangle are not given, we can construct our own triangle with the given angle: select values for the opposite side and hypotenuse so the ratio $\dfrac{\text{opposite}}{\text{hypotenuse}}$ is the value whose arcsin we want: arcsin(opposite/hypotenuse). The length of the other side can be calculated using the Pythagorean Theorem.

Example 1: Determine the lengths of the sides of a right triangle so one angle is $\theta = \arcsin(\frac{5}{13})$. Use the triangle to determine the values of $\tan(\arcsin(\frac{5}{13}))$ and $\csc(\arcsin(\frac{5}{13}))$.

Solution: We want the sine, the ratio $\dfrac{\text{opposite}}{\text{hypotenuse}}$, of θ to be 5/13 so we can choose the opposite side to be 5 and the hypotenuse to be 13 (Fig. 8). Then $\sin(\theta) = 5/13$, the value we want. Using the Pythagorean Theorem, the length of the adjacent side is $\sqrt{13^2 - 5^2} = 12$.

Finally, $\tan(\theta) = \tan(\arcsin(\frac{5}{13})) = \dfrac{\text{opposite}}{\text{adjacent}} = \dfrac{5}{12}$ and

$\csc(\theta) = \csc(\arcsin(\frac{5}{13})) = \dfrac{1}{\sin(\arcsin(\frac{5}{13}))} = \dfrac{13}{5}$.

Fig. 8: $\sin(\theta) = \dfrac{5}{13}$

Any choice of values for the opposite side and the hypotenuse is fine as long as the ratio of the opposite to the hypotenuse is 5/13 .

Practice 1: Determine the lengths of the sides of a right triangle so one angle is $\theta = \arcsin(\frac{6}{11})$. Use the triangle to determine the values of $\tan(\arcsin(\frac{6}{11}))$, $\csc(\arcsin(\frac{6}{11}))$, and $\cos(\arcsin(\frac{6}{11}))$.

Example 2: Determine the lengths of the sides of a right triangle so one angle is $\theta = \arcsin(x)$. Use the triangle to determine the values of $\tan(\arcsin(x))$ and $\cos(\arcsin(x))$.

Solution: We want the sine, the ratio $\dfrac{\text{opposite}}{\text{hypotenuse}}$, of θ to be x so we can choose the opposite side to be x and the hypotenuse to be 1 (Fig. 9). Then $\sin(\theta) = x/1 = x$, and, using the Pythagorean Theorem, the length of the adjacent side is $\sqrt{1 - x^2}$.

$\tan(\arcsin(x)) = \dfrac{\text{opposite}}{\text{adjacent}} = \dfrac{x}{\sqrt{1-x^2}}$ and

$\cos(\arcsin(x)) = \dfrac{\text{adjacent}}{\text{hypotenuse}} = \dfrac{\sqrt{1-x^2}}{1} = \sqrt{1-x^2}$.

Fig. 9: $\theta = \arcsin(x)$

Practice 2: Use the triangle in Example 2 to help evaluate $\sec(\arcsin(x))$ and $\csc(\arcsin(x))$.

Inverse Tangent: Solving k = tan(x) for x

The equation $0.5 = \tan(x)$ (Fig. 10) has many solutions; the function $f(x) = \tan(x)$ is not one–to–one; and it's reflected graph in Fig. 11 is not the graph of a function. However, if the domain of the tangent function is restricted to the interval $(-\pi/2, \pi/2)$, then $f(x) = \tan(x)$ is one–to–one and has an inverse function. The graph of the inverse tangent function (Fig. 12) is the reflection about the line $y = x$ of the (restricted) graph of $y = \tan(x)$.

This inverse tangent function is written as $\tan^{-1}(x)$ or **arctan**(x).

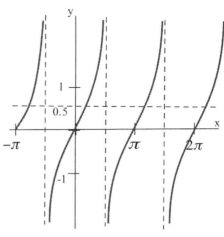

Fig. 10: y = tan(x)

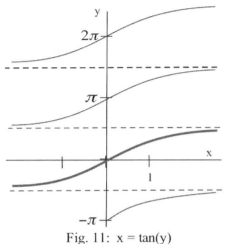

Fig. 11: x = tan(y)

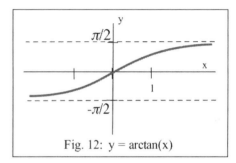

Fig. 12: y = arctan(x)

Note: The symbol $\tan^{-1}(x)$ should **never** be used to represent $\frac{1}{\tan(x)}$.

Definition of Inverse Tangent:

For all x and for $-\pi/2 < y \leq \pi/2$,

$y = \arctan(x)$ if $x = \tan(y)$.

The domain of $f(x) = \arctan(x)$ is all real numbers. The range of $f(x) = \arctan(x)$ is $(-\pi/2, \pi/2)$.

The (restricted) tangent function and the arctan function are inverses of each other.

For all values of x, $\tan(\arctan(x)) = x$.

For $-\pi/2 < y < \pi/2$, $\arctan(\tan(y)) = y$.

Arctan(x) is the length of the arc on the unit circle whose tangent is x. Arctan(x) is the angle whose tangent is x: $\tan(\arctan(x)) = x$.

Right Triangles and Arctan

For the right triangle in Fig. 13, $\tan(\theta) = \frac{\text{opposite}}{\text{adjacent}} = \frac{3}{2}$ so $\theta = \arctan(3/2)$.

Then $\sin(\arctan(3/2)) = \sin(\theta) = \frac{\text{opposite}}{\text{hypotenuse}} = \frac{3}{\sqrt{13}} \approx 0.83$ and

$\cot(\arctan(3/2)) = \frac{1}{\tan(\arctan(3/2))} = \frac{1}{3/2} = \frac{2}{3} \approx 0.67$.

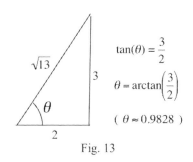

Fig. 13

Practice 3: Determine the lengths of the sides of a right triangle so one angle is $\theta = \arcsin(\frac{3}{5})$. Use the triangle to determine the values of $\tan(\arcsin(\frac{3}{5}))$, $\csc(\arcsin(\frac{3}{5}))$, and $\cos(\arcsin(\frac{3}{5}))$.

Example 3: On a wall 8 feet in front of you, the lower edge of a 5 foot tall painting is 2 feet above your eye level (Fig. 14). Represent your viewing angle θ using arctangents.

Solution: The viewing angle α to the bottom of the painting satisfies $\tan(\alpha) = \text{opposite/adjacent} = 2/8$, so $\alpha = \arctan(1/4)$. Similarly, the angle β to the top of the painting satisfies $\tan(\beta) = 7/8$ so $\beta = \arctan(7/8)$. Finally, the viewing angle θ for the painting is

$\theta = \beta - \alpha = \arctan(7/8) - \arctan(1/4) \approx 0.719 - 0.245 = 0.474$, or about $27°$.

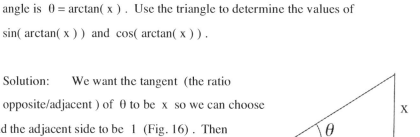

θ = viewing angle
Fig. 14: Viewing a picture

Practice 4: Determine the scoring angle for the soccer player in Fig. 15.

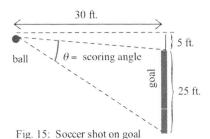

Fig. 15: Soccer shot on goal

Example 4: Determine the lengths of the sides of a right triangle so one angle is $\theta = \arctan(x)$. Use the triangle to determine the values of $\sin(\arctan(x))$ and $\cos(\arctan(x))$.

Solution: We want the tangent (the ratio opposite/adjacent) of θ to be x so we can choose the opposite side to be x and the adjacent side to be 1 (Fig. 16). Then $\tan(\theta) = x/1 = x$, and, using the Pythagorean Theorem, the length of the hypotenuse is $\sqrt{1+x^2}$. Finally,

Fig. 16: $\theta = \arctan(x)$

$\sin(\arctan(x)) = \frac{\text{opposite}}{\text{hypotenuse}} = \frac{x}{\sqrt{1+x^2}}$ and $\cos(\arctan(x)) = \frac{\text{adjacent}}{\text{hypotenuse}} = \frac{1}{\sqrt{1+x^2}}$.

Practice 5: Use the triangle in Fig. 16 to help evaluate $\sec(\arctan(x))$ and $\cot(\arctan(x))$.

Inverse Secant: Solving k = sec(x) for x

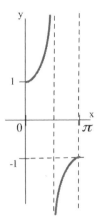

Fig. 18: y = sec(x)

The equation $2 = \sec(x)$ (Fig. 17) has many solutions, but we can create an inverse **function** for secant in the same way we did for sine and tangent, by suitably restricting the domain of the secant function so it is one–to–one. Fig. 18 shows the restriction $0 \le x \le \pi$ and $x \ne \pi/2$ which results in a one–to–one function which has an inverse. The graph of the inverse function (Fig. 19) is the reflection about the line $y = x$ of the (restricted) graph of $y = \sec(x)$.

This inverse secant function is written as $\sec^{-1}(x)$ or **arcsec**(x).

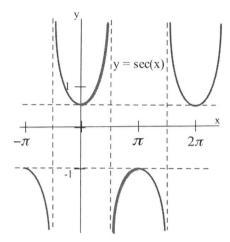

Fig. 17: y = sec(x)

Definition of Inverse Secant:

For $|x| \ge 1$ and for $0 \le y \le \pi$ with $y \ne \pi/2$,

$y = \text{arcsec}(x)$ if $x = \sec(y)$.

The domain of $f(x) = \text{arcsec}(x)$ is all x with $|x| \ge 1$.
The range of $f(x) = \text{arcsec}(x)$ is $[0, \pi/2)$ and $(\pi/2, \pi]$.

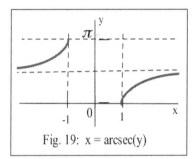

Fig. 19: x = arcsec(y)

The (restricted) secant function and the arcsec function are inverses of each other.

For all values of x with $|x| \ge 1$, $\sec(\text{arcsec}(x)) = x$.

For all values of y with $0 \le y \le \pi$ and $y \ne \pi/2$, $\text{arcsec}(\sec(y)) = y$.

Note: There are alternate ways to restrict secant to get a one–to–one function, and they lead to slightly different definitions of the inverse secant. We chose to use this restriction because it seems more "natural" than the alternatives, it is easier to evaluate on a calculator, and it is the most commonly used.

Example 5: Evaluate tan(arcsec(x)).

Solution: We want the secant (the ratio hypotenuse/adjacent) of θ to be x so we can choose the hypotenuse to be x and the adjacent side to be 1 (Fig. 20). Then $\sec(\theta) = x/1 = x$. Using the Pythagorean Theorem,

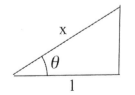

Fig. 20: θ = arcsec(x)

the length of the opposite side is $\sqrt{x^2-1}$, so

$$\tan(\operatorname{arcsec}(x)) = \frac{\text{opposite}}{\text{adjacent}} = \frac{\sqrt{x^2-1}}{1} = \sqrt{x^2-1}.$$

Practice 6: Evaluate $\sin(\operatorname{arcsec}(x))$ and $\cot(\operatorname{arcsec}(x))$.

The Other Inverse Trigonometric Functions

The inverse tangent and inverse sine functions are by far the most commonly used of the six inverse trigonometric functions. In particular, you will often encounter the arctangent function when you integrate rational functions such as $1/(1+x^2)$ and $(4x+7)/(x^2+6x+10)$. The inverse secant function is used less often.

The other three inverse trigonometric functions ($\cos^{-1}$, $\cot^{-1}$, and $\csc^{-1}$) can be defined as the inverses of a suitably restricted parts of cosine, cotangent, and cosecant, respectively. The graph of each inverse function is the reflection about the line $y = x$ of the graph of the restricted trigonometric function.

Inverse Trigonometric Functions, Calculators, and Computers

Most scientific calculators only have keys for $\sin^{-1}$, $\cos^{-1}$, and $\tan^{-1}$ (or INVerse SIN, COS, TAN), but those are enough to enable us to calculate the values of the other three inverse trigonometric functions by using the following identities.

If $x \neq 0$ and x is in the domains of the functions, then

$$\operatorname{arccot}(x) = \arctan(1/x), \quad \operatorname{arcsec}(x) = \arccos(1/x), \quad \operatorname{arccsc}(x) = \arcsin(1/x)$$

And some programming languages only have a single inverse trigonometric function, the arctangent function ATN(X), but even that is enough to enable us to evaluate the other five inverse trigonometric functions. Formulas for evaluating each inverse trigonometric function just in terms of ATN are given in an Appendix after the Practice Answers as are several additional identities involving the inverse trigonometric functions.

PROBLEMS

1. (a) List the three smallest positive angles θ that are solutions of the equation $\sin(\theta) = 1$.
   (b) Evaluate $\arcsin(1)$ and $\operatorname{arccsc}(1)$.

2. (a) List the three smallest positive angles θ that are solutions of the equation $\tan(\theta) = 1$.
   (b) Evaluate $\arctan(1)$ and $\operatorname{arccot}(1)$.

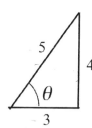

Fig. 21

3. Find the values of x **between 1 and 7** so (a) $\sin(x) = 0.3$, (b) $\sin(x) = -0.4$, and (c) $\sin(x) = 0.5$.

4. Find the values of x **between 5 and 12** so (a) $\sin(x) = 0.8$, and (b) $\sin(x) = -0.9$.

5. Find the values of x **between 2 and 7** so (a) $\tan(x) = 3.2$, and (b) $\tan(x) = -0.2$.

6. Find the values of x **between 1 and 5** so (a) $\tan(x) = 8$, and (b) $\tan(x) = -3$.

7. In Fig. 21, angle θ is (a) the arcsine of what number? (b) the arctangent of what number? (c) the arcsecant of what number? (d) the arccosine of what number?

Fig. 22

8. In Fig. 22, angle θ is (a) the arcsine of what number? (b) the arctangent of what number? (c) the arcsecant of what number? (d) the arccosine of what number?

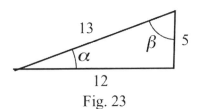

Fig. 23

9. For the angle α in Fig. 23, evaluate (a) $\sin(\alpha)$, (b) $\tan(\alpha)$, (c) $\sec(\alpha)$, and (d) $\cos(\alpha)$.

10. For the angle β in Fig. 23, evaluate (a) $\sin(\beta)$, (b) $\tan(\beta)$, (c) $\sec(\beta)$, and (d) $\cos(\beta)$.

11. For $\theta = \sin^{-1}(2/7)$, find the exact values of (a) $\tan(\theta)$, (b) $\cos(\theta)$, (c) $\csc(\theta)$, and (d) $\cot(\theta)$.

12. For $\theta = \tan^{-1}(9/2)$, find the exact values of (a) $\sin(\theta)$, (b) $\cos(\theta)$, (c) $\csc(\theta)$, and (d) $\cot(\theta)$.

13. For $\theta = \cos^{-1}(1/5)$, find the exact values of (a) $\tan(\theta)$, (b) $\sin(\theta)$, (c) $\csc(\theta)$, and (d) $\cot(\theta)$.

14. For $\theta = \sin^{-1}(a/b)$ with $0 < a < b$, find the exact values of (a) $\tan(\theta)$, (b) $\cos(\theta)$, (c) $\csc(\theta)$, and (d) $\cot(\theta)$.

15. For $\theta = \tan^{-1}(a/b)$ with $0 < a < b$, find the exact values of (a) $\tan(\theta)$, (b) $\sin(\theta)$, (c) $\cos(\theta)$, and (d) $\cot(\theta)$.

16. For $\theta = \tan^{-1}(x)$, find the exact values of (a) $\sin(\theta)$, (b) $\cos(\theta)$, (c) $\sec(\theta)$, and (d) $\cot(\theta)$.

17. Find the exact values of (a) $\sin(\cos^{-1}(x))$, (b) $\cos(\sin^{-1}(x))$, and (c) $\sec(\cos^{-1}(x))$.

18. Find the exact values of (a) $\tan(\cos^{-1}(x))$, (b) $\cos(\tan^{-1}(x))$, and (c) $\sec(\sin^{-1}(x))$.

19. (a) Does $\arcsin(1) + \arcsin(1) = \arcsin(2)$?
    (b) Does $\arccos(1) + \arccos(1) = \arccos(2)$?

20. (a) What is the viewing angle for the tunnel sign in Fig. 24? (b) Use arctangents to describe the viewing angle when the observer is x feet from the entrance of the tunnel.

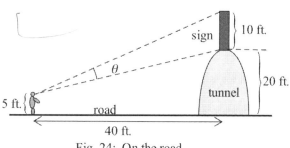
Fig. 24: On the road

21. (a) What is the viewing angle for the chalk board in Fig. 25 ? (b) Use arctangents to describe the viewing angle when the student is x feet from the front wall.

22. Graph $y = \arcsin(2x)$ and $y = \arctan(2x)$.

23. Graph $y = \arcsin(x/2)$ and $y = \arctan(x/2)$.

24. Which curve is longer, $y = \sin(x)$ from $x = 0$ to π or $y = \arcsin(x)$ from $x = -1$ to 1 ?

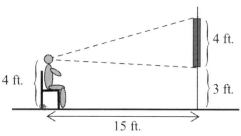
Fig. 25: In the classroom

For problems 25 – 28, $\frac{d\theta}{dt} = 12$ when $\theta = 1.3$, and θ and h are related by the given formula. Find $\frac{dh}{dt}$.

25. $\sin(\theta) = h/20$.
26. $\tan(\theta) = h/50$.
27. $\cos(\theta) = 3h + 20$.
28. $3 + \tan(\theta) = 7h$.

For problems 29 – 32, $\frac{dh}{dt} = 4$ when $\theta = 1.3$, and θ and h are related by the given formula. Find $\frac{d\theta}{dt}$.

29. $\sin(\theta) = h/38$.
30. $\tan(\theta) = h/40$.
31. $\cos(\theta) = 7h - 23$.
32. $\tan(\theta) = h^2$.

33. You are observing a rocket launch from a point 4000 feet from the launch pad (Fig. 26). When the observation angle is $\pi/3$, the angle is increasing at $\pi/12$ feet per second. How fast is the rocket traveling? (Hint: θ and h are functions of t. $\tan(\theta) = \frac{h}{4000}$ so $\frac{d}{dt}\tan(\theta) = \frac{d}{dt}(\frac{h}{4000})$. Then solve for $\frac{dh}{dt}$.)

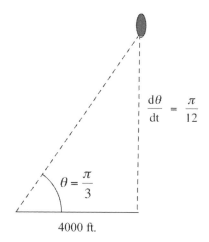
Fig. 26: Rocket launch

34. You are observing a rocket launch from a point 3000 feet from the launch pad. You heard on the radio the when the rocket is 5000 feet high its velocity is 100 feet per second. (a) What is the angle of elevation of the rocket when it is 5000 feet up? (b) How fast is the angle of elevation increasing when the rocket is 5000 feet up? (See the previous hint, but now solve for $d\theta/dt$.)

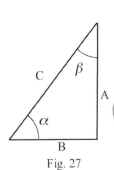

Fig. 27

35. In Fig. 27, (a) angle α is arcsine of what number?

    (b) angle β is the arccosine of what number?

    (c) For positive numbers A and B, evaluate arcsin(A/C) + arccos(A/C).

36. In Fig. 27, (a) angle α is arctangent of what number?

    (b) angle β is the arccotangent of what number?

    (c) For positive numbers A and B, evaluate arctan(A/B) + arccot(A/B).

37. In Fig. 27, (a) angle α is arcsecant of what number?

    (b) angle β is the arccosecant of what number?

    (c) For positive numbers A and B, evaluate arcsec(C/B) + arccsc(C/B).

38. Describe the pattern of your results for the previous three problems.

39. For the angle θ in Fig. 28, (a) θ is the arctangent of what number? (b) θ is arccotangent of what number?

40. For the angle θ in Fig. 28, (a) θ is the arcsine of what number?

    (b) θ is arccosecant of what number?

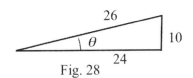

Fig. 28

41. For the angle θ in Fig. 28, (a) θ is the arccosine of what number?

    (b) θ is arcsecant of what number?

42. Describe the pattern of your results for the previous three problems.

In problems 43 – 51, use your calculator and the appropriate identities to evaluate the functions.

43. $\sec^{-1}(3)$ 44. $\sec^{-1}(-2)$ 45. $\sec^{-1}(-1)$ 46. $\cos^{-1}(0.5)$ 47. $\cos^{-1}(-0.5)$

48. $\cos^{-1}(1)$ 49. $\cot^{-1}(1)$ 50. $\cot^{-1}(0.5)$ 51. $\cot^{-1}(-3)$

52. For the triangle in Fig. 29, (a) θ = arctan(_____). (b) θ = arccot(_____).

    (c) arccot (_____) = arctan(_____).

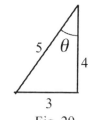

Fig. 29

53. For the triangle in Fig. 30, (a) θ = arcsin(_____). (b) θ = arccos(_____).

    (c) arccos(_____) = arcsin(_____).

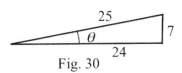

Fig. 30

$\tan(\arcsin(x)) = \dfrac{x}{\sqrt{1-x^2}}$ so $\arcsin(x) = \arctan\left(\dfrac{x}{\sqrt{1-x^2}}\right)$. Imitate this reasoning in problems 54 and 55.

54. Evaluate $\tan(\text{arccot}(x))$ and use the result to find a formula for $\text{arccot}(x)$ in terms of arctangent.

55. Evaluate $\tan(\text{arcsec}(x))$ and use the result to find a formula for $\text{arcsec}(x)$ in terms of arctangent.

56. Let $a = \arctan(x)$ and $b = \arctan(y)$. Use the identity $\tan(a+b) = \dfrac{\tan(a)+\tan(b)}{1-\tan(a)\cdot\tan(b)}$ to show that

$$\arctan(x) + \arctan(y) = \arctan\left(\dfrac{x+y}{1-xy}\right).$$

Section 7.2   PRACTICE Answers

Practice 1: See Fig. 31. $\sin(\theta) = \dfrac{\text{OPP}}{\text{HYP}} = \dfrac{6}{11}$.

$\tan(\arcsin(\tfrac{6}{11})) = \tan(\theta) = \dfrac{\text{OPP}}{\text{ADJ}} = \dfrac{6}{\sqrt{85}}$.

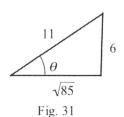
Fig. 31

$\csc(\arcsin(\tfrac{6}{11})) = \dfrac{1}{\sin(\arcsin(\tfrac{6}{11}))} = \dfrac{1}{6/11} = \dfrac{11}{6}$. $\quad \cos(\arcsin(\tfrac{6}{11})) = \cos(\theta) = \dfrac{\text{ADJ}}{\text{HYP}} = \dfrac{\sqrt{85}}{11}$.

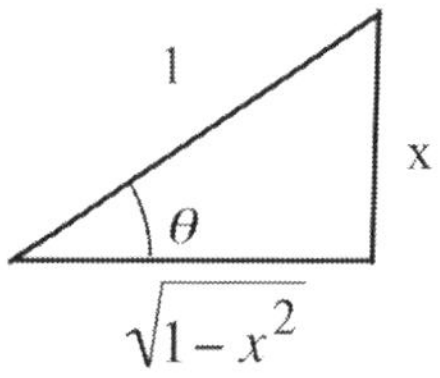

Practice 2: See Fig. 32: OPP = x, HYP = 1, ADJ = $\sqrt{1-x^2}$.

$\sec(\arcsin(x)) = \sec(\theta) = \dfrac{1}{\cos(\theta)} = \dfrac{1}{\text{ADJ}/\text{HYP}} = \dfrac{\text{HYP}}{\text{ADJ}} = \dfrac{1}{\sqrt{1-x^2}}$.

$\csc(\arcsin(x)) = \dfrac{1}{\sin(\arcsin(x))} = \dfrac{1}{x}$.

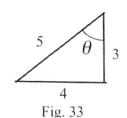
Fig. 32

Practice 3: See Fig. 33. $\sin(\theta) = \dfrac{\text{OPP}}{\text{HYP}} = \dfrac{3}{5}$. $\tan(\arcsin(\tfrac{3}{5})) = \tan(\theta) = \dfrac{\text{OPP}}{\text{ADJ}} = \dfrac{3}{4}$.

$\csc(\arcsin(\tfrac{3}{5})) = \dfrac{1}{\sin(\arcsin(\tfrac{3}{5}))} = \dfrac{1}{3/5} = \dfrac{5}{3}$.

$\cos(\arcsin(\tfrac{3}{5})) = \cos(\theta) = \dfrac{\text{ADJ}}{\text{HYP}} = \dfrac{4}{5}$.

Fig. 33

Practice 4: See Fig. 34. $\tan(\alpha) = \frac{5}{30}$ so $\alpha = \arctan(\frac{5}{30}) \approx 0.165$ (or $\approx 9.46°$).

$\tan(\alpha + \theta) = \frac{30}{30} = 1$ so $\alpha + \theta = \arctan(1) \approx 0.785$ (or $45°$).

Finally, $\theta = (\alpha + \theta) - (\alpha) \approx 0.785 - 0.165 = 0.62$
(or $\theta \approx 45° - 9.46° = 35.54°$).

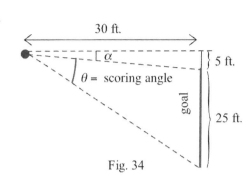

Fig. 34

Practice 5: See Fig. 35. $\tan(\theta) = \frac{x}{1} = x$ so $\theta = \arctan(x)$.

$\sec(\arctan(x)) = \sec(\theta) = \frac{1}{\cos(\theta)} = \frac{1}{\text{ADJ/HYP}}$

$= \frac{\text{HYP}}{\text{ADJ}} = \frac{\sqrt{1+x^2}}{1} = \sqrt{1+x^2}$.

$\cot(\arctan(x)) = \cot(\theta) = \frac{\text{ADJ}}{\text{HYP}} = \frac{1}{x}$.

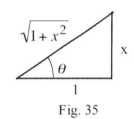

Fig. 35

Practice 6: See Fig. 20 in text. $\sin(\text{arcsec}(x)) = \frac{\sqrt{x^2-1}}{x}$. $\cot(\text{arcsec}(x)) = \frac{1}{\sqrt{x^2-1}}$

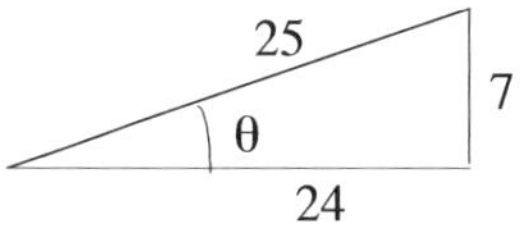

Fig. 30

Appendix: Identities for Inverse Trigonometric Functions

Theorem: $\operatorname{arccot}(x) = \arctan(1/x)$, $\operatorname{arcsec}(x) = \arccos(1/x)$, $\operatorname{arccsc}(x) = \arcsin(1/x)$.

Proof: $\tan(\operatorname{arccot}(x)) = \dfrac{1}{\cot(\operatorname{arccot}(x))} = \dfrac{1}{x}$ so, by taking the arctan of each side of the equation,

$\arctan\{\tan(\operatorname{arccot}(x))\} = \arctan\{1/x\}$ and $\operatorname{arccot}(x)) = \arctan\{1/x\}$.

Similarly, $\cos(\operatorname{arcsec}(x)) = \dfrac{1}{\sec(\operatorname{arcsec}(x))} = \dfrac{1}{x}$ so $\operatorname{arcsec}(x) = \arccos(1/x)$, and

$\sin(\operatorname{arccsc}(x)) = \dfrac{1}{\csc(\operatorname{arccsc}(x))} = \dfrac{1}{x}$ so $\operatorname{arccsc}(x) = \arcsin(1/x)$.

Theorem: For $0 \le x$, (a) $\arcsin(x) + \arccos(x) = \pi/2$, (b) $\arctan(x) + \operatorname{arccot}(x) = \pi/2$, and
(c) $\operatorname{arcsec}(x) + \operatorname{arccsc}(x) = \pi/2$.

Partial proof: If A and B are the complementary angles in a right triangle ($A + B = \pi/2$), then $\sin(A) = \cos(B)$. Let $x = \sin(A) = \cos(B)$. Then $A = \arcsin(x)$ and $B = \arccos(x)$ so the fact that $A + B = \pi/2$ becomes $\arcsin(x) + \arccos(x) = \pi/2$.
(This "partial proof" assumed that A and B were nonnegative acute angles, so the proof is only valid if x is the sine or cosine of such angles, that is, that $0 \le x \le 1$. In fact, (a) is true for $-1 \le x \le 1$.)

If $A + B = \pi/2$, then $\tan(A) = \cot(B)$. Let $x = \tan(A) = \cot(B)$. Then $A = \arctan(x)$ and $B = \operatorname{arccot}(x)$ so $A + B = \pi/2$ implies that $\arctan(x) + \operatorname{arccot}(x) = \pi/2$. (This "partial proof" is valid for $0 \le x$.)

Theorem: $\arcsin(-x) = -\arcsin(x)$, $\arctan(-x) = -\arctan(x)$, $\operatorname{arcsec}(-x) = \pi - \operatorname{arcsec}(x)$ (if $x \ge 1$).

Most versions of the programming language BASIC only have a single inverse trigonometric function, the arctangent function $\operatorname{ATN}(X)$, but that is enough to enable us to evaluate the other five inverse trigonometric functions.

$\arcsin(x) = \operatorname{ATN}\left(\dfrac{X}{\sqrt{1-X^2}}\right)$ $\qquad$ $\arccos(x) = \pi/2 - \arcsin(x) = \pi/2 - \operatorname{ATN}\left(\dfrac{X}{\sqrt{1-X^2}}\right)$

$\operatorname{arccot}(x) = \operatorname{ATN}(1/X)$

$\operatorname{arcsec}(x) = \operatorname{ATN}(\sqrt{X^2-1})$ $\qquad$ $\operatorname{arccsc}(x) = \pi/2 - \operatorname{arcsec}(x) = \pi/2 - \operatorname{ATN}(\sqrt{X^2-1})$

7.3 CALCULUS WITH THE INVERSE TRIGONOMETRIC FUNCTIONS

The three previous sections introduced the ideas of one–to–one functions and inverse functions and used those ideas to define arcsine, arctangent, and the other inverse trigonometric functions. Section 7.3 presents the **calculus** of inverse trigonometric functions. In this section we obtain derivative formulas for the inverse trigonometric functions and the associated antiderivatives. The applications we consider are both classical and sporting.

Derivative Formulas for the Inverse Trigonometric Functions

Derivative Formulas

(1) $D(\arcsin(x)) = \dfrac{1}{\sqrt{1-x^2}}$ (for $|x| < 1$)   (4) $D(\arccos(x)) = -\dfrac{1}{\sqrt{1-x^2}}$ (for $|x| < 1$)

(2) $D(\arctan(x)) = \dfrac{1}{1+x^2}$ (for all x)   (5) $D(\text{arccot}(x)) = -\dfrac{1}{1+x^2}$ (for all x)

(3) $D(\text{arcsec}(x)) = \dfrac{1}{|x|\sqrt{x^2-1}}$ (for $|x| > 1$)   (6) $D(\text{arccsc}(x)) = -\dfrac{1}{|x|\sqrt{x^2-1}}$ (for $|x| > 1$)

The proof of each of these differentiation formulas follows from what we already know about the derivatives of the trigonometric functions and the Chain Rule for Derivatives. Formula (2) is the most commonly used of these formulas, and it is proved below. The proofs of formulas (1), (4), and (5) are very similar and are left as problems. The proof of formula (3) is slightly more complicated and is included in an Appendix after the problems.

Proof of formula (2): The proof relies on two results from previous sections, that

$$D(\tan(f(x))) = \sec^2(f(x)) \cdot D(f(x))$$ (using the Chain Rule) and that $\tan(\arctan(x)) = x$.

Differentiating each side of the equation $\tan(\arctan(x)) = x$, we have

$$D(\tan(\arctan(x))) = D(x) = 1.$$

Evaluating each derivative in the last equation,

$$D(\tan(\arctan(x))) = \sec^2(\arctan(x)) \cdot D(\arctan(x)) \text{ and } D(x) = 1 \text{ so}$$

$$\sec^2(\arctan(x)) \cdot D(\arctan(x)) = 1.$$

Finally, we can divide each side by $\sec^2(\arctan(x))$ to get

$$D(\arctan(x)) = \frac{1}{\sec^2(\arctan(x))} = \frac{1}{\sec(\arctan(x)) \cdot \sec(\arctan(x))}$$

$$= \frac{1}{\sqrt{1+x^2}\sqrt{1+x^2}} = \frac{1}{1+x^2}.$$

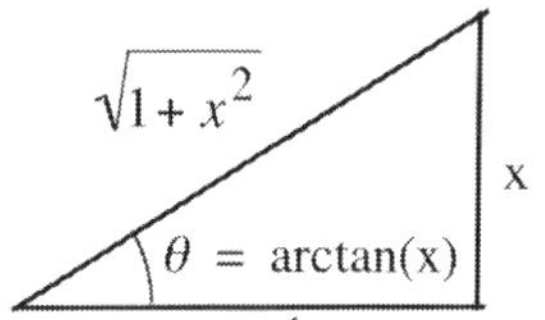

Fig. 1

Example 1: Calculate $D(\arcsin(e^x))$, $D(\arctan(x-3))$, $D(\arctan^3(5x))$, and $D(\ln(\arcsin(x)))$.

Solution: Each of the functions to be differentiated is a composition, so we need to use the Chain Rule.

$$D(\arcsin(e^x)) = \frac{1}{\sqrt{1-(e^x)^2}} \; D(e^x) = \frac{e^x}{\sqrt{1-e^{2x}}}.$$

$$D(\arctan(x-3)) = \frac{1}{1+(x-3)^2} \; D(x-3) = \frac{1}{1+(x-3)^2} = \frac{1}{x^2-6x+10}.$$

$$D(\arctan^3(5x)) = 3\arctan^2(5x) \; D(\arctan(5x)) = 3\arctan^2(5x) \cdot \frac{1}{1+(5x)^2} \cdot 5.$$

$$D(\ln(\arcsin(x))) = \frac{1}{\arcsin(x)} \; D(\arcsin(x)) = \frac{1}{\arcsin(x)} \cdot \frac{1}{\sqrt{1-x^2}}.$$

Practice 1: Calculate $D(\arcsin(5x))$, $D(\arctan(x+2))$, $D(\arcsec(7x))$, and $D(e^{\arctan(7x)})$.

A Classic Application

Mathematics is the study of patterns, and one of the pleasures of mathematics is that the same pattern can appear in unexpected places. The version of the classical Museum Problem below was first posed in 1471 by the mathematician Johannes Muller and is one of the oldest known maximization problems.

Museum Problem: The lower edge of a 5 foot painting is 4 feet above your eye level (Fig. 2). At what distance should you stand from the wall so your viewing angle of the painting is maximum?

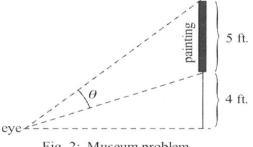

Fig. 2: Museum problem

On a typical autumn weekend, however, a lot more people would rather be watching or playing a football or soccer game than visiting a museum or solving a calculus problem about a painting. But the pattern of the Museum Problem even appears in football and soccer, sports not invented until hundreds of years after the original problem was posed and solved.

Since we also want to examine the Museum Problem in other contexts, let's solve the general version.

Example 2: General Museum Problem.

The lower edge of a H foot painting is A feet above your eye level (Fig. 3). At what distance x should you stand from the painting so the viewing angle is maximum?

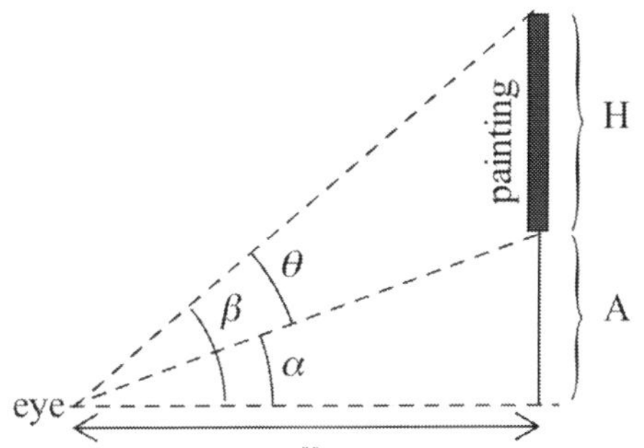

Solution: Let $B = A + H$. Then $\tan(\alpha) = \frac{A}{x}$ and $\tan(\beta) = \frac{B}{x}$ so $\alpha = \arctan(A/x)$ and $\beta = \arctan(B/x)$. The viewing angle is $\theta = \beta - \alpha = \arctan(B/x) - \arctan(A/x)$. We can maximize θ by calculating the derivative $\frac{d\theta}{dx}$ and finding where the derivative is zero. Since $\theta = \arctan(B/x) - \arctan(A/x)$,

Fig. 3: General museum problem

$$\frac{d\theta}{dx} = D(\arctan(B/x)) - D(\arctan(A/x))$$

$$= \frac{1}{1 + \left(\frac{B}{x}\right)^2}\left(-\frac{B}{x^2}\right) - \frac{1}{1 + \left(\frac{A}{x}\right)^2}\left(-\frac{A}{x^2}\right) = \frac{-B}{x^2 + B^2} + \frac{A}{x^2 + A^2} .$$

Setting $\frac{d\theta}{dx} = 0$ and solving for x, we have $x = \sqrt{AB} = \sqrt{A(A+H)}$. (We can disregard the endpoints since we clearly do not have a maximum viewing angle with our noses pressed against the wall or from infinity far away from the wall.)

Now the Original Museum Problem and the Football and Soccer versions below are straightforward.

In our original Museum Problem, $A = 4$ and $H = 5$, so the maximum viewing angle occurs when $x = \sqrt{4(4+5)} = 6$ feet. The maximum angle is $\theta = \arctan(9/6) - \arctan(4/6) \approx 0.983 - 0.588 \approx .395$ or about $22.6°$.

Practice 2: Football. A kicker is attempting a field goal by kicking the football between the goal posts

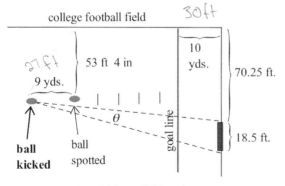

Fig. 4: Kicking a field goal

(Fig. 4). At what distance from the goal line should the ball be spotted so the kicker has the largest angle for making the field goal? (Assume that the ball is "spotted" on a "hash mark" that is 53 feet 4 inches from the edge of the field and is actually kicked from a point about 9 yards further from the goal line than where the ball is spotted.)

Practice 3: Soccer. Kelcey is bringing the ball down the middle of the soccer field toward the 25 foot wide goal which is defended by a goalie (Fig. 5). The goalie is positioned in the center of the goal and can stop a shot that is within four feet of the center of the goalie. At what distance from the goal should Kelcey shoot so the scoring angle is maximum?

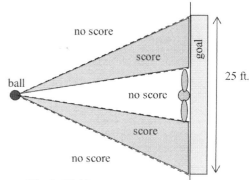

Fig. 5: Kicking a soccer goal

Antiderivative Formulas

Despite the Museum Problem and its sporting variations, the primary use of the inverse trigonometric functions in calculus is their use as antiderivatives. Each of the six differentiation formulas at the beginning of this section gives us an integral formula, but there are only three essentially different patterns:

$$\int \frac{1}{\sqrt{1-x^2}}\, dx = \arcsin(x) + C \quad \text{(for } |x| < 1\text{)}$$

$$\int \frac{1}{1+x^2}\, dx = \arctan(x) + C \quad \text{(for all } x\text{)}$$

$$\int \frac{1}{|x|\sqrt{x^2-1}}\, dx = \text{arcsec}(x) + C. \quad \text{(for } |x| > 1\text{)}$$

Most of the integrals we need are variations of the basic patterns, and usually we have to transform the integrand so that it exactly matches one of the basic patterns.

Example 3: Evaluate $\int \frac{1}{16+x^2}\, dx$.

Solution: We can transform this integrand into the arctangent pattern by factoring 16 from the denominator and changing to the variable $u = x/4$:

$$\frac{1}{16+x^2} = \frac{1}{16} \cdot \frac{1}{1+\frac{x^2}{16}} = \frac{1}{16} \cdot \frac{1}{1+(x/4)^2} = \frac{1}{16} \cdot \frac{1}{1+u^2}.$$

If $u = x/4$ then $du = \frac{1}{4}\, dx$ and **$dx = 4\, du$** so

$$\int \frac{1}{16+x^2}\, dx = \int \frac{1}{16} \cdot \frac{1}{1+u^2} \cdot 4\, du$$

$$= \frac{1}{4}\int \frac{1}{1+u^2}\, du = \frac{1}{4}\arctan(u) + C = \frac{1}{4}\arctan\left(\frac{x}{4}\right) + C.$$

Practice 4: Evaluate $\int \dfrac{1}{1+9x^2}\,dx$ and $\int \dfrac{1}{\sqrt{25-x^2}}\,dx$.

The most common integrands contain patterns with the forms $a^2 - x^2$, $a^2 + x^2$, and $x^2 - a^2$ where a is constant, and it is worthwhile to have general integral patterns for these forms.

$$\int \dfrac{1}{\sqrt{a^2 - x^2}}\,dx = \arcsin\left(\dfrac{x}{a}\right) + C \qquad (\text{for } |x| < |a|)$$

$$\int \dfrac{1}{a^2 + x^2}\,dx = \dfrac{1}{a}\arctan\left(\dfrac{x}{a}\right) + C \qquad (\text{for all } x \text{ and for } a \neq 0)$$

$$\int \dfrac{1}{|x|\sqrt{x^2 - a^2}}\,dx = \dfrac{1}{a}\operatorname{arcsec}\left(\dfrac{x}{a}\right) + C \qquad (\text{for all } |x| > |a| > 0)$$

These general formulas can be derived by factoring the a^2 out of the pattern and making a suitable change of variable. The final results can be checked by differentiating. **The arctan pattern is, by far, the most commonly needed.** The arcsin pattern appears occasionally, and the arcsec pattern only rarely.

Example 4: Derive the general formula for $\int \dfrac{1}{\sqrt{a^2 - x^2}}\,dx$ from the formula for $\int \dfrac{1}{\sqrt{1-x^2}}\,dx$.

Solution: We can algebraically transform the $a^2 + x^2$ pattern into an $1 + u^2$ pattern for an appropriate u:

$$\dfrac{1}{\sqrt{a^2 - x^2}} = \dfrac{1}{\sqrt{a^2(1 - x^2/a^2)}} = \dfrac{1}{a}\dfrac{1}{\sqrt{1 - (x/a)^2}}.$$

If we put $u = x/a$, then $du = 1/a\,dx$ and $dx = a\,du$ so

$$\int \dfrac{1}{\sqrt{a^2 - x^2}}\,dx = \dfrac{1}{a}\int \dfrac{1}{\sqrt{1 - (x/a)^2}}\,dx = \dfrac{1}{a}\int \dfrac{1}{\sqrt{1 - u^2}} \cdot a\,du$$

$$= \int \dfrac{1}{\sqrt{1 - u^2}}\,du = \arcsin(u) + C = \arcsin\left(\dfrac{x}{a}\right) + C.$$

Practice 5: Verify that the derivative of $\dfrac{1}{a} \cdot \arctan\left(\dfrac{x}{a}\right)$ is $\dfrac{1}{a^2 + x^2}$.

Example 5: Evaluate $\int \dfrac{1}{\sqrt{5 - x^2}}\,dx$ and $\int_1^3 \dfrac{1}{5 + x^2}\,dx$.

Solution: The constant a does not have to be an integer, so we can take $a^2 = 5$ and $a = \sqrt{5}$. Then

$$\int \frac{1}{\sqrt{5-x^2}} \, dx = \arcsin\left(\frac{x}{\sqrt{5}}\right) + C, \text{ and}$$

$$\int_1^3 \frac{1}{5+x^2} \, dx = \frac{1}{\sqrt{5}} \arctan\left(\frac{x}{\sqrt{5}}\right) \Big|_1^3 = \frac{1}{\sqrt{5}} \arctan\left(\frac{3}{\sqrt{5}}\right) - \frac{1}{\sqrt{5}} \arctan\left(\frac{1}{\sqrt{5}}\right) \approx 0.228.$$

The easiest way to integrate some rational functions is to split the original integrand into two pieces.

Example 6: Evaluate $\int \frac{6x+7}{25+x^2} \, dx$.

Solution: This integrand splits nicely into the sum of two other functions that can be easily integrated:

$$\int \frac{6x+7}{25+x^2} \, dx = \int \frac{6x}{25+x^2} \, dx + \int \frac{7}{25+x^2} \, dx.$$

The integral of $\frac{6x}{25+x^2}$ can be evaluated by changing the variable to $u = 25+x^2$ and $du = 2x \, dx$.

Then $6x \, dx = 3 \, du$ and $\int \frac{6x}{25+x^2} \, dx = \int \frac{3}{u} \, du = 3 \ln|u| + C = 3 \ln(25+x^2) + C$.

The integral of $\frac{7}{25+x^2}$ matches the arctangent pattern with $a = 5$:

$$\int \frac{7}{25+x^2} \, dx = \frac{7}{5} \arctan\left(\frac{x}{5}\right) + C.$$

Finally, $\int \frac{6x+7}{25+x^2} \, dx = \int \frac{6x}{25+x^2} \, dx + \int \frac{7}{25+x^2} \, dx = 3 \ln(25+x^2) + \frac{7}{5} \arctan\left(\frac{x}{5}\right) + C.$

The antiderivative of a linear function divided by an irreducible quadratic commonly involves a logarithm and an arctangent.

Practice 6: Evaluate $\int \frac{4x+3}{x^2+7} \, dx$.

PROBLEMS

In problems 1 – 15, calculate the derivatives.

1. $D(\arcsin(3x))$
2. $D(\arctan(7x))$
3. $D(\arctan(x+5))$

4. $D(\arcsin(x/2))$
5. $D(\arctan(\sqrt{x}))$
6. $D(\text{arcsec}(x^2))$

7. $D(\ln(\arctan(x)))$
8. $D(\sqrt{\arcsin(x)})$
9. $D((\text{arcsec}(x))^3)$

10. $D(\arctan(5/x))$ 11. $D(\arctan(\ln(x)))$ 12. $D(\arcsin(x+2))$

13. $D(e^x \cdot \arctan(2x))$ 14. $D\left(\dfrac{\arcsin(x)}{\arccos(x)}\right)$ 15. $D(\arcsin(x) + \arccos(x))$

16. $D(x \cdot \arctan(x))$ 17. $D\left(\dfrac{1}{\arcsin(x)}\right)$ 18. $D((1 + \operatorname{arcsec}(x))^3)$

19. $D(\sin(3 + \arctan(x)))$ 20. $D(\tan(x) \cdot \arctan(x))$ 21. $D(x \cdot \arctan(\tfrac{1}{x}))$

22. $\displaystyle\int \dfrac{7}{\sqrt{9-x^2}}\,dx$ 23. $\displaystyle\int_0^1 \dfrac{3}{x^2+25}\,dx$ 24. $\displaystyle\int_5^7 \dfrac{5}{x\sqrt{x^2-16}}\,dx$

25. $\displaystyle\int \dfrac{9}{\sqrt{49-x^2}}\,dx$ 26. $\displaystyle\int_1^4 \dfrac{2}{7+x^2}\,dx$ 27. $\displaystyle\int_6^{10} \dfrac{3}{x\sqrt{x^2-25}}\,dx$

28. $\displaystyle\int \dfrac{7}{(x-5)^2+9}\,dx$ 29. $\displaystyle\int_{-1}^1 \dfrac{e^x}{1+e^{2x}}\,dx$ 30. $\displaystyle\int_1^e \dfrac{1}{x} \cdot \dfrac{3}{1+(\ln(x))^2}\,dx$

31. $\displaystyle\int \dfrac{\cos(x)}{\sqrt{9-\sin^2(x)}}\,dx$ 32. $\displaystyle\int \dfrac{8x}{16+x^2}\,dx$ 33. $\displaystyle\int \dfrac{6x}{9+x^4}\,dx$

34. (a) Use arctangents to describe the viewing angle for the sign in Fig. 6 when the observer is x feet from the entrance to the tunnel. (b) At what distance x is the viewing angle maximized?

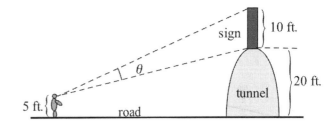

Fig. 6: Looking at a sign

35. (a) Use arctangents to describe the viewing angle for the chalk board in Fig. 7 when the student is x feet from the front wall. (b) At what distance x is the viewing angle maximized?

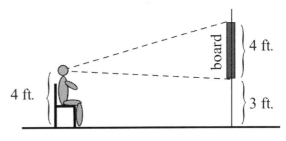

Fig. 7: In class

In problems 36 – 39, find the function y which satisfies the differential equation and goes through the given point.

36. $\dfrac{dy}{dx} = \dfrac{1}{y(1+x^2)}$ and the point (0,4). 37. $\dfrac{dy}{dx} = \dfrac{y}{\sqrt{1-x^2}}$ and $y(0) = e$.

38. $y' \cdot \sqrt{16 - x^2} = y$ and $y(4) = 1$. 39. $\dfrac{dy}{dx} = \dfrac{y^2}{9+x^2}$ and the point (1,2).

40. Prove differentiation formula (1): $D(\arcsin(x)) = \dfrac{1}{\sqrt{1-x^2}}$ (for $|x| < 1$).

41. Prove differentiation formula (4): $D(\arccos(x)) = -\dfrac{1}{\sqrt{1-x^2}}$ (for $|x| < 1$).

42. Prove differentiation formula (5): $D(\text{arccot}(x)) = -\dfrac{1}{1+x^2}$ (for all x).

43. Let $A(x) = \displaystyle\int_0^x \dfrac{1}{1+t^2}\, dt$, the area between the curve $y = \dfrac{1}{1+t^2}$ and the t–axis between $t = 0$ and x.

   (a) Evaluate $A(0), A(1), A(10)$. (b) Evaluate $\lim_{x \to \infty} A(x)$. (c) Find $\dfrac{d\,A(x)}{dx}$.

   (d) Is $A(x)$ an increasing, decreasing or neither? (e) Evaluate $\lim_{x \to \infty} A'(x)$.

44. Find area between the curve $y = \dfrac{1}{1+x^2}$ and the x–axis (a) from $x = -10$ to 10 and

   (b) from $x = -A$ to A. (c) Find the area under the **whole** curve. (Calculate the limit of your answer in part (b) as $A \to \infty$.)

45. $\displaystyle\int \dfrac{8x - 5}{x^2 + 9}\, dx$ 46. $\displaystyle\int \dfrac{1 - 4x}{x^2 + 1}\, dx$ 47. $\displaystyle\int \dfrac{7x + 3}{x^2 + 10}\, dx$ 48. $\displaystyle\int \dfrac{x + 5}{x^2 + 9}$

Problems 49 – 53 illustrate how we can sometimes decompose a difficult integral into easier ones.

49. (a) $\displaystyle\int \dfrac{8}{x^2 + 6x + 10}\, dx$ (hint: $x^2 + 6x + 10 = (x+3)^2 + 1$. Try $u = x + 3$.)

   (b) $\displaystyle\int \dfrac{4x + 12}{x^2 + 6x + 10}\, dx$ (hint: Try $u = x^2 + 6x + 10$. Then $(2x + 6)\,dx = 2\,du$.)

   (c) $\displaystyle\int \dfrac{4x + 20}{x^2 + 6x + 10}\, dx$ (hint: $\dfrac{4x + 20}{x^2 + 6x + 10} = \dfrac{4x + 12}{x^2 + 6x + 10} + \dfrac{8}{x^2 + 6x + 10}$

50. (a) $\int \frac{7}{x^2 + 4x + 5} \, dx$ (b) $\int \frac{12x + 24}{x^2 + 4x + 5} \, dx$ (c) $\int \frac{12x + 31}{x^2 + 4x + 5} \, dx$

51. $\int \frac{6x + 15}{x^2 + 4x + 20} \, dx$ 52. $\int \frac{2x + 5}{x^2 - 4x + 13} \, dx$

Section 7.3   **PRACTICE Answers**

Practice 1: $D(\arcsin(5x)) = \frac{1}{\sqrt{1 - (5x)^2}} \cdot 5 = \frac{5}{\sqrt{1 - 25x^2}}$.

$D(\arctan(x+2)) = \frac{1}{1 + (x+2)^2} = \frac{1}{x^2 + 4x + 5}$. $D(\text{arcsec}(7x)) = \frac{1}{|7x|\sqrt{(7x)^2 - 1}} \cdot 7 = \frac{1}{|x|\sqrt{49x^2 - 1}}$.

$D(e^{\arctan(7x)}) = e^{\arctan(7x)} D(\arctan(7x)) = e^{\arctan(7x)} \cdot \frac{7}{1 + (7x)^2} = e^{\arctan(7x)} \cdot \frac{7}{1 + 49x^2}$.

Practice 2: Football: A = 16.9 ft. H = 18.5 ft. (see Fig. 8) so $x = \sqrt{A(A+H)} = \sqrt{598.26} \approx 24.46$ feet from the back edge of the end zone. Unfortunately, that is still more than 5 feet into the endzone. Our mathematical analysis shows that to maximize the angle for kicking a field goal, the ball should be placed 5 feet into the end zone, a touchdown! If the ball is placed on a hash mark at the goal line, then the kicking distance is 57 feet (10 yards for the width of the endzone plus 9 yards that the ball is hiked) and the scoring angle for the kicker is

$\theta = \arctan(35.4/57) - \arctan(16.9/57) \approx 15.3^\circ$. It is somewhat interesting to see how the scoring angle $\theta = \arctan(35.4/x) - \arctan(16.9/x)$ changes with the distance x (feet), and also to compare the scoring angle for balls placed on the hash mark with those placed in the center of the field.

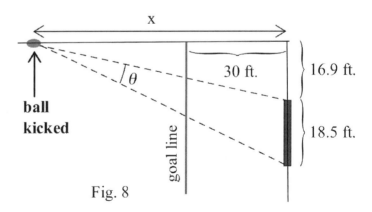

Fig. 8

Practice 3: Soccer: See Fig. 9. A = 4 ft. and H = 8.5 ft. so $x = \sqrt{A(A+H)} = \sqrt{50} \approx 7.1$ ft.

From 7.1 feet, the scoring angle on one side of the goalie is $\theta = \arctan(12.5/7.1) - \arctan(4/7.1) \approx 31°$.

For comparison, the scoring angles $\theta = \arctan(12.5/x) - \arctan(4/x)$ are given for some other distances x from the goal.

x	θ
5	29.5°
7.1	31.0°
10	29.5°
15	24.9°
20	20.7°
30	15.0°
40	11.6°

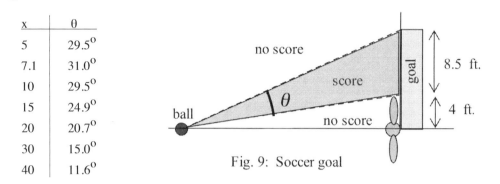

Fig. 9: Soccer goal

Practice 4: $\int \dfrac{1}{1+9x^2} dx = \int \dfrac{1}{1+(3x)^2} dx$. Put $u = 3x$. Then $du = 3\,dx$ and $dx = \dfrac{1}{3} du$.

$$\int \dfrac{1}{1+(3x)^2} dx = \int \dfrac{1}{1+(u)^2} \cdot \dfrac{1}{3} du = \dfrac{1}{3} \arctan(u) + C = \dfrac{1}{3} \arctan(3x) + C.$$

$$\int \dfrac{1}{\sqrt{25-x^2}} dx = \int \dfrac{1}{5} \dfrac{1}{\sqrt{1-(x/5)^2}} dx.$$ Put $u = \dfrac{x}{5}$. Then $du = \dfrac{1}{5} dx$ and $dx = 5\,du$.

$$\int \dfrac{1}{5}\dfrac{1}{\sqrt{1-(x/5)^2}} dx = \int \dfrac{1}{5}\dfrac{1}{\sqrt{1-(u)^2}} \cdot 5\,du = \arcsin(u) + C = \arcsin\left(\dfrac{x}{5}\right) + C.$$

Practice 5: $D\left(\dfrac{1}{a} \arctan\left(\dfrac{x}{a}\right)\right) = \dfrac{1}{a} \cdot \dfrac{1}{1+(x/a)^2} \cdot \dfrac{1}{a} = \dfrac{1}{a^2} \cdot \dfrac{1}{1+(x/a)^2} = \dfrac{1}{a^2+x^2}$.

Practice 6: $\int \dfrac{4x+3}{x^2+7} dx = \int \dfrac{4x}{x^2+7} dx + \int \dfrac{3}{x^2+7} dx$.

For the first integral on the right, put $u = x^2 + 7$. Then $du = 2x\,dx$ and $2\,du = 4x\,dx$. Then

$$\int \dfrac{4x}{x^2+7} dx = \int \dfrac{2}{u} du = 2\cdot\ln|u| + C = 2\cdot\ln(x^2+7) + C.$$

The second integral on the right matches the "arctan" pattern;

$$\int \dfrac{3}{x^2+7} dx = \dfrac{3}{\sqrt{7}} \arctan\left(\dfrac{x}{\sqrt{7}}\right) + C.$$

Therefore, $\int \dfrac{4x+3}{x^2+7} dx = 2\cdot\ln(x^2+7) + \dfrac{3}{\sqrt{7}} \arctan\left(\dfrac{x}{\sqrt{7}}\right) + C.$

Appendix: Proofs of Some Derivative Formulas

Proof of formula (1):   The proof relies on two results from previous sections, that

$\mathbf{D}(\sin(f(x))) = \cos(f(x)) \cdot \mathbf{D}(f(x))$ (the Chain Rule) and that $\sin(\arcsin(x)) = x$ (for $|x| \leq 1$).

Putting these two results together and differentiating both sides of $\sin(\arcsin(x)) = x$, we get

$$\mathbf{D}(\sin(\arcsin(x))) = \mathbf{D}(x).$$

Evaluating the derivatives,

$$\mathbf{D}(\sin(\arcsin(x))) = \cos(\arcsin(x)) \cdot \mathbf{D}(\arcsin(x)) \text{ and } \mathbf{D}(x) = 1$$

so $\qquad \cos(\arcsin(x)) \cdot \mathbf{D}(\arcsin(x)) = 1.$

Dividing each side by $\cos(\arcsin(x))$ and using the fact that $\cos(\arcsin(x)) = \sqrt{1-x^2}$, we have

$$\mathbf{D}(\arcsin(x)) = \frac{1}{\cos(\arcsin(x))} = \frac{1}{\sqrt{1-x^2}}.$$

The derivative of $\arcsin(x)$ is defined only when $1 - x^2 > 0$, or equivalently, if $|x| < 1$.

Proof of formula (3):   The proof relies on two results from previous sections, that

$$\mathbf{D}(\sec(f(x))) = \sec(f(x)) \cdot \tan(f(x)) \cdot \mathbf{D}(f(x)) \text{ (Chain Rule) and } \sec(\text{arcsec}(x)) = x.$$

Differentiating each side of $\sec(\text{arcsec}(x)) = x$, we have

$$\mathbf{D}(\sec(\text{arcsec}(x))) = \mathbf{D}(x) = 1.$$

Evaluating each derivative, we have

$$\mathbf{D}(\sec(\text{arcsec}(x))) = \sec(\text{arcsec}(x)) \cdot \tan(\text{arcsec}(x)) \cdot \mathbf{D}(\arctan(x)) = 1$$

so $\qquad \mathbf{D}(\text{arcsec}(x)) = \dfrac{1}{\sec(\text{arcsec}(x)) \cdot \tan(\text{arcsec}(x))} = \dfrac{1}{x \cdot \tan(\text{arcsec}(x))}.$

To evaluate $\tan(\text{arcsec}(x))$, we can use the identity $\tan^2(\theta) = \sec^2(\theta) - 1$ with $\theta = \text{arcsec}(x)$. Then $\tan^2(\text{arcsec}(x)) = \sec^2(\text{arcsec}(x)) - 1 = x^2 - 1$. Now, however, there is a slight difficulty: is $\tan(\text{arcsec}(x)) = +\sqrt{x^2-1}$ or is $\tan(\text{arcsec}(x)) = -\sqrt{x^2-1}$? It is clear from the graph of arcsec(x) (Fig. 8) that $\mathbf{D}(\text{arcsec}(x))$ is positive everywhere it is defined, so we need to choose the sign of the square root to guarantee that our calculated value for $\mathbf{D}(\text{arcsec}(x))$ is positive. An easier way to guarantee that the derivative is positive is simply to always use the positive square root and to take the absolute value of x. Then

$$\mathbf{D}(\text{arcsec}(x)) = \frac{1}{|x|\sqrt{x^2-1}} > 0.$$

The derivative of arcsec(x) is defined only when $x^2 - 1 > 0$, or equivalently, if $|x| > 1$.

PROBLEM ANSWERS Chapter Seven

Section 7.0

1. f is one–to–one (1–1), y is 1–1, g is not 1–1, h is not 1–1.

3. f is not 1–1, y is 1–1, g is 1–1, h is not 1–1.

5. I think SS numbers are supposed to be 1–1: each person should have one SS number, and each SS number should be assigned to only one person.

   Telephone numbers are not 1–1: the members of my family all have the same telephone number.

7. Monogamy.

9. The x–axis is a horizontal line so a 1–1 function can touch the x–axis in at most one place.

11. (a) Yes. It passes the horizontal line test.

    (b) f(0) > f(1) so f is NOT increasing. f(0) < f(0.7) so f is NOT decreasing.

13. $f(x) = \ln(x)$. $f'(x) = \frac{1}{x} > 0$ for $x > 0$.

    Assume $a > b > 0$ and $f(a) = f(b)$. Then by Rolle's theorem there is a point c between a and b so that $f'(c) = 0$. But $f'(c) = 0$ contradicts the fact that $f'(x) > 0$ for all $x > 0$, so our assumption that $f(a) = f(b)$ must be false. Therefore, $f(a) \neq f(b)$ and f is 1–1.

15. (a) Yes. (b) Yes. (c) cde

    (d) a b c d e f   (e) See Fig. 1.
        f e b a d c

17. (a) Yes. (b) Yes. (c) fda

    (d) a b c d e f   (e) See Fig. 2.
        d f e a c b

    (f) Same.

    (g) ENCODE(ENCODE(word)) = word

        ENCODE(ENCODE(bad)) = ENCODE(fda) = bad

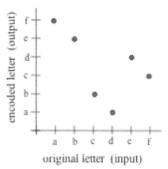

Fig. 1: Graph For Prob. 15

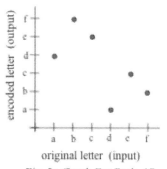

Fig. 2: Graph For Prob. 17

Section 7.1

1. See Fig. 1.

3. See Fig. 2.

5. See Fig. 3.

Fig. 1: Completed Table 5

x	f(x)	f '(x)	$f^{-1}(x)$	$(f^{-1})'(x)$
1	3	–3	2	$\frac{1}{f'(2)} = \frac{1}{2}$
2	1	2	3	$\frac{1}{f'(3)} = \frac{1}{3}$
3	2	3	1	$\frac{1}{f'(1)} = \frac{1}{-3}$

x	h(x)	h '(x)	$h^{-1}(x)$	$(h^{-1})'(x)$
1	2	2	3	$\frac{1}{h'(3)}$ undefined
2	3	–2	1	$\frac{1}{h'(1)} = \frac{1}{2}$
3	1	0	2	$\frac{1}{h'(2)} = \frac{1}{-2}$

Fig. 2: Completed Table 7

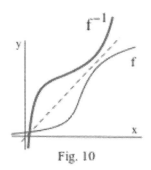

Fig. 10

7. If (a, b) is on the graphs of both f and f^{-1}, it is tempting, **but wrong**, to conclude that a = b. For example, if f(x) = 1 – x, then $f^{-1}(x) = 1 - x$ and the points (0,1), (1,0), and (3,–2) all lie on both graphs, as do infinitely many other points.

9. f: (i) multiply by 7, (ii) subtract 5, and (iii) divide by 4.
   f^{-1}: (i) multiply by 4, (ii) add 5, and (iii) divide by 7. $f^{-1}(x) = \frac{4x+5}{7}$.

11. y = g(x) = 2x + 1. Interchange x and y: x = 2y + 1 so $y = \frac{x-1}{2}$ and $g^{-1}(x) = \frac{x-1}{2}$.
    Checking: $g^{-1}(g(1)) = g^{-1}(2(1)+1) = g^{-1}(3) = \frac{(3)-1}{2} = 1$. (OK)
    $g(g^{-1}(1)) = g(\frac{(7)-1}{2}) = g(3) = 2(3) + 1 = 7$. (OK)

13. y = w(x) = 5 + ln(x). Interchange: x = 5 + ln(y) so x – 5 = ln(y) and $e^{x-5} = e^{\ln(y)} = y$.
    $w^{-1}(x) = e^{x-5}$. Check: $w^{-1}(w(1)) = w^{-1}(5 + \ln(1)) = w^{-1}(5) = e^{5-(5)} = e^0 = 1$. (OK)

15. f^{-1} goes through the point (3,1). $(f^{-1})'(3) = \frac{1}{f'(f^{-1}(x))} = \frac{1}{f'(1)} > 0$.

17. f '(x) > 0 for all x means f is an increasing function.
    $(f^{-1})'(x) = \frac{1}{f'(f^{-1}(x))} = \frac{1}{f'(\text{number})} > 0$ so f^{-1} is also an increasing function.

19. Suppose $f(x) = x + b$ $(b \neq 0)$. Interchange: $x = y + b$ so $y = x - b$ and $f^{-1}(x) = x - b$.

f and f^{-1} both have slope 1 so their graphs are parallel lines. The functions differ by a constant, $2b$, so the graphs do not intersect.

21. $f(x) = 3x + \sin(x)$. $f'(x) = 3 + \cos(x) > 0$ for all x so f is an increasing function and has an inverse.

23. (a) $f^{-1}(x) = \frac{1}{x} = f(x)$. (b) $f^{-1}(x) = f(x)$ (c) $f^{-1}(x) = f(x)$

(d) $f^{-1}(x) \neq f(x)$ unless $a = 0$ (e) $f^{-1}(x) = f(x)$

25. See Fig. 4.

(a) unfold along the road line to get B*

(b) unfold along the river line to get A*

Then the straight line from A* to B* is the shortest path

from A* to B*.

Get the final path by refolding the A* to B* line along the

river and along the road.

This method also works when the river and road are not perpendicular to each other. Try a couple of your own design.

27. Fold along the wall: the ball will land the same distance from the wall on your side as the distance it would have landed from the wall on the far side. See Fig. 5.

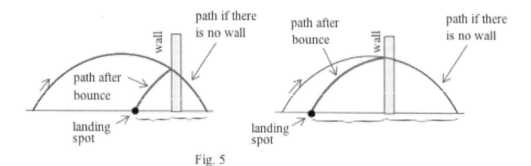

Fig. 5

Section 7.2

1. (a) $\sin(\theta) = 1$ so $\theta = \frac{\pi}{2} + 2\pi k$. When $k=0, \theta = \frac{\pi}{2}$. When $k=1, \theta = \frac{5\pi}{2}$. When $k=2, \theta = \frac{9\pi}{2}$.

    (b) $\arcsin() = \frac{\pi}{2}$. $\arcsin(1) = \arcsin(\frac{1}{1}) = \frac{\pi}{2}$.

3. (a) $\sin(x) = 0.3$ when $x \approx 0.304, x \approx \pi - 0.304 = 2.84$, and $x \approx 2\pi + 0.304 = 6.587$.

    (b) $\sin(x) = -0.4$ when $x \approx 3.553$ and $x \approx 5.872$.

    (c) $\sin(x) = 0.5$ when $x \approx 0.534, x \approx \pi - 0.534 \approx 2.618$, and $x = 2\pi + 0.534 \approx 6.807$.

5. (a) $\tan(x) = 3.2$ when $x \approx 1.268$ and $x \approx \pi + 1.268 \approx 4.410$.

    (b) $\tan(x) = -0.2$ when $x \approx 2.944$ and $x \approx \pi + 2.944 \approx 6.086$.

7. (a) $\sin(\theta) = \frac{4}{5}$ so $\theta = \arcsin(\frac{4}{5})$. (b) $\tan(\theta) = \frac{4}{3}$ so $\theta = \arctan(\frac{4}{3})$.

    (c) $\sec(\theta) = \frac{1}{\cos(\theta)} = \frac{1}{3/5} = \frac{5}{3}$ so $\theta = \text{arcsec}(\frac{5}{3})$.

    (d) $\cos(\theta) = \frac{3}{5}$ so $\theta = \arccos(\frac{3}{5})$.

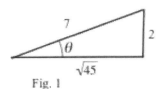

Fig. 1

9. (a) $\sin(\alpha) = \frac{5}{13}$ (b) $\tan(\alpha) = \frac{5}{12}$ (c) $\sec(\alpha) = \frac{1}{\cos(\alpha)} = \frac{13}{12}$ (d) $\cos(\alpha) = \frac{12}{13}$.

11. See Fig. 1. $\theta = \arcsin(\frac{2}{7})$.

    (a) $\tan(\theta) = \frac{2}{\sqrt{45}}$. (b) $\cos(\theta) = \frac{\sqrt{45}}{7}$ (c) $\csc(\theta) = \frac{1}{\sin(\theta)} = \frac{7}{2}$ (d) $\cot(\theta) = \frac{\sqrt{45}}{2}$.

13. See Fig. 2. $\theta = \arccos(\frac{1}{5})$.

    (a) $\tan(\theta) = \frac{\sqrt{24}}{1}$ (b) $\sin(\theta) = \frac{\sqrt{24}}{5}$ (c) $\csc(\theta) = \frac{1}{\sin(\theta)} = \frac{5}{\sqrt{24}}$ (d) $\cot(\theta) = \frac{1}{\sqrt{24}}$.

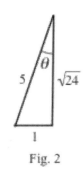

Fig. 2

15. See Fig. 3. $\theta = \arctan(\frac{a}{b})$.

    (a) $\tan(\theta) = \frac{a}{b}$ (b) $\sin(\theta) = \frac{a}{\sqrt{a^2+b^2}}$ (c) $\cos(\theta) = \frac{b}{\sqrt{a^2+b^2}}$ (d) $\cot(\theta) = \frac{b}{a}$.

17. $\sin^2(\theta) + \cos^2(\theta) = 1$ so $\sin(\theta) = \sqrt{1 - \cos^2(\theta)}$ and $\cos(\theta) = \sqrt{1 - \sin^2(\theta)}$.

    (a) $\sin(\arccos(x)) = \sqrt{1-x^2}$ (b) $\cos(\arcsin(x)) = \sqrt{1-x^2}$ (c) $\sec(\arccos(x)) = \frac{1}{x}$.

19. (a) No, $\arcsin(2)$ is not defined. $\arcsin(1) + \arcsin(1) = \frac{\pi}{2} + \frac{\pi}{2} = \pi$.

    (b) No, $\arccos(2)$ is not defined. $\arccos(1) + \arccos(1) = 0 + 0 = 0$.

21. See Fig. 4. Viewing angle $\theta = \alpha + \beta$.

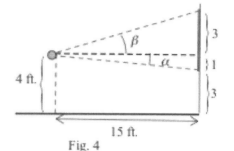

Fig. 4

(a) $\alpha = \arctan(\frac{1}{15}) \approx 0.067$ ($\approx 3.81°$)

$\beta = \arctan(\frac{3}{15}) \approx 0.197$ ($\approx 11.31°$)

so $\theta \approx 0.067 + 0.197 = 0.264$ ($\approx 15.12°$).

(b) In general, if x is the distance in feet, then $\alpha = \arctan(\frac{1}{x})$ and $\beta = \arctan(\frac{3}{x})$ so $\theta = \arctan(\frac{1}{x}) + \arctan(\frac{3}{x})$.

23. (a) See Fig. 5. $y = \arcsin(\frac{x}{2})$. (b) See Fig. 6. $y = \arctan(\frac{x}{2})$.

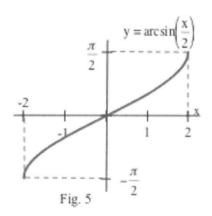

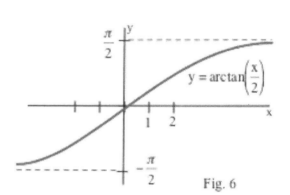

25. $\sin(\theta) = \frac{h}{20}$ so $\frac{d}{dt}\{\sin(\theta)\} = \cos(\theta) \cdot \frac{d\theta}{dt}$. Also, $\frac{d}{dt}\{\sin(\theta)\} = \frac{d}{dt}\{\frac{h}{20}\} = \frac{1}{20} \cdot \frac{dh}{dt}$.

Together, $\frac{1}{20} \cdot \frac{dh}{dt} = \cos(\theta) \cdot \frac{d\theta}{dt}$ so $\frac{dh}{dt} = 20 \cdot \cos(\theta) \cdot \frac{d\theta}{dt}$.

If $\theta = 1.3$ and $\frac{d\theta}{dt} = 12$, then $\frac{dh}{dt} = 20 \cdot \cos(\theta) \cdot \frac{d\theta}{dt} = 20 \cdot \cos(1.3) \cdot (12) \approx 64.20$.

27. $\cos(\theta) = 3h + 20$ so $\frac{d}{dt}\{\cos(\theta)\} = -\sin(\theta) \cdot \frac{d\theta}{dt}$. Also, $\frac{d}{dt}\{\cos(\theta)\} = \frac{d}{dt}\{3h + 20\} = 3 \cdot \frac{dh}{dt}$.

Together, $3 \cdot \frac{dh}{dt} = -\sin(\theta) \cdot \frac{d\theta}{dt}$ so $\frac{dh}{dt} = \frac{-1}{3} \cdot \sin(\theta) \cdot \frac{d\theta}{dt}$.

If $\theta = 1.3$ and $\frac{d\theta}{dt} = 12$, then $\frac{dh}{dt} = \frac{-1}{3} \cdot \sin(\theta) \cdot \frac{d\theta}{dt} \frac{-1}{3} \cdot \sin(1.3) \cdot (12) \approx -3.854$.

29. $\sin(\theta) = \dfrac{h}{38}$ so $\dfrac{d}{dt}\{\sin(\theta)\} = \cos(\theta)\cdot\dfrac{d\theta}{dt}$. Also, $\dfrac{d}{dt}\{\sin(\theta)\} = \dfrac{d}{dt}\{\dfrac{h}{38}\} = \dfrac{1}{38}\cdot\dfrac{dh}{dt}$.

Together, $\dfrac{1}{38}\cdot\dfrac{dh}{dt} = \cos(\theta)\cdot\dfrac{d\theta}{dt}$ so $\dfrac{d\theta}{dt} = \dfrac{1}{38\cdot\cos(\theta)}\cdot\dfrac{dh}{dt}$.

If $\theta = 1.3$ and $\dfrac{dh}{dt} = 4$, then $\dfrac{d\theta}{dt} = \dfrac{1}{38\cdot\cos(\theta)}\cdot\dfrac{dh}{dt} = \dfrac{1}{38\cdot\cos(1.3)}\cdot(4) \approx 0.394$.

31. $\cos(\theta) = 7h - 23$ so $\dfrac{d}{dt}\{\cos(\theta)\} = -\sin(\theta)\cdot\dfrac{d\theta}{dt}$. Also, $\dfrac{d}{dt}\{\cos(\theta)\} = \dfrac{d}{dt}\{7h - 23\} = 7\cdot\dfrac{dh}{dt}$.

Together, $7\cdot\dfrac{dh}{dt} = -\sin(\theta)\cdot\dfrac{d\theta}{dt}$ so $\dfrac{d\theta}{dt} = \dfrac{7}{-\sin(\theta)}\cdot\dfrac{dh}{dt}$.

If $\theta = 1.3$ and $\dfrac{dh}{dt} = 4$, then $\dfrac{d\theta}{dt} = \dfrac{7}{-\sin(\theta)}\cdot\dfrac{dh}{dt} = \dfrac{7}{-\sin(1.3)}\cdot(4) \approx -29.059$.

33. $\tan(\theta) = \dfrac{h}{4000}$ so $\dfrac{d}{dt}\{\tan(\theta)\} = \sec^2(\theta)\cdot\dfrac{d\theta}{dt}$. Also, $\dfrac{d}{dt}\{\tan(\theta)\} = \dfrac{d}{dt}\{\dfrac{h}{4000}\} = \dfrac{1}{4000}\cdot\dfrac{dh}{dt}$.

Together, $\dfrac{1}{4000}\cdot\dfrac{dh}{dt} = \sec^2(\theta)\cdot\dfrac{d\theta}{dt}$ so $\dfrac{dh}{dt} = 4000\cdot\sec^2(\theta)\cdot\dfrac{d\theta}{dt}$.

If $\theta = \dfrac{\pi}{3}$ and $\dfrac{d\theta}{dt} = \dfrac{\pi}{12}$, then $\dfrac{dh}{dt} = 4000\cdot\sec^2(\theta)\cdot\dfrac{d\theta}{dt} = 4000\cdot\sec^2(\dfrac{\pi}{3})\cdot(\dfrac{\pi}{12}) \approx 4{,}188.8$ ft/sec.

35. See Fig. 7. (a) $\sin(\alpha) = \dfrac{A}{C}$ so $\alpha = \arcsin(\dfrac{A}{C})$

(b) $\cos(\beta) = \dfrac{A}{C}$ so $\beta = \arccos(\dfrac{A}{C})$

(c) $\arcsin(\dfrac{A}{C}) + \arccos(\dfrac{A}{C}) = \alpha + \beta = \dfrac{\pi}{2}$ (or 90°)

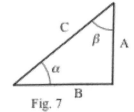

Fig. 7

37. See Fig. 7. (a) $\sec(\alpha) = \dfrac{1}{\cos(\alpha)} = \dfrac{C}{B}$ so $\alpha = \text{arcsec}(\dfrac{C}{B})$.

(b) $\csc(\beta) = \dfrac{1}{\sin(\beta)} = \dfrac{C}{B}$ so $\beta = \text{arccsc}(\dfrac{C}{B})$.

(c) $\text{arcsec}(\dfrac{C}{B}) + \text{arccsc}(\dfrac{C}{B}) = \alpha + \beta = \dfrac{\pi}{2}$ (or 90°)

39. See Fig. 8. (a) $\tan(\theta) = \dfrac{10}{24} = \dfrac{5}{12}$ so $\theta = \arctan(\dfrac{5}{12})$ (b) $\cot(\theta) = \dfrac{12}{5}$ so $\theta = \text{arccot}(\dfrac{12}{5})$

41. See Fig. 8. (a) $\sec(\theta) = \dfrac{1}{\cos(\theta)} = \dfrac{26}{24} = \dfrac{13}{12}$ so $\theta = \text{arcsec}(\dfrac{13}{12})$

(b) $\csc(\theta) = \dfrac{1}{\sin(\theta)} = \dfrac{26}{10} = \dfrac{13}{5}$ so $\theta = \text{arccsc}(\dfrac{13}{5})$.

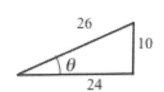

Fig. 8

43. $\text{arcsec}(3) = \arccos(\dfrac{1}{3}) \approx 1.231$ (or 70.53°)

45. $\operatorname{arcsec}(-1) = \arccos(\frac{1}{-1}) = \pi$ (or $180°$)

47. $\arccos(-0.5) = \frac{2\pi}{3} \approx 2.094$ (or $120°$)

49. $\operatorname{arccot}(1) = \arctan(\frac{1}{1}) = \frac{\pi}{4} \approx 0.785$ (or $45°$)

51. $\operatorname{arccot}(-3) = \arctan(\frac{1}{-3}) \approx 0.322$ (or $-18.43°$)

53. See Fig. 9. (a) $\sin(\theta) = \frac{7}{25}$ so $\theta = \arcsin(\frac{7}{25})$.

    (b) $\cos(\theta) = \frac{24}{25}$ so $\theta = \arccos(\frac{24}{25})$

    (c) $\arcsin(\frac{7}{25}) = \arccos(\frac{24}{25})$

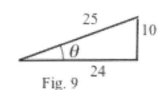

Fig. 9

55. If $\operatorname{arcsec}(x) = \alpha$ then $\sec(\alpha) = x$ so $\frac{1}{\cos(\alpha)} = x$ and $\cos(\alpha) = \frac{1}{x}$.

    Since $\sin(\alpha) = \sqrt{1 - \cos^2(\alpha)} = \sqrt{1 - \frac{1}{x^2}} = \frac{\sqrt{x^2-1}}{x}$, then

    $\tan(\alpha) = \frac{\sin(\alpha)}{\cos(\alpha)} = \frac{(\sqrt{x^2-1})/x}{1/x} = \sqrt{x^2-1}$. $\tan(\alpha) = \tan\{\operatorname{arcsec}(x)\} = \sqrt{x^2-1}$.

    Finally, taking arctan of each side of $\tan\{\operatorname{arcsec}(x)\} = \sqrt{x^2-1}$, we get that
    $\operatorname{arcsec}(x) = \arctan(\sqrt{x^2-1})$. (There are other ways to verify this result.)

Section 7.3

1. $D(\arcsin(3x)) = \frac{1}{\sqrt{1-(3x)^2}} \cdot D(3x) = \frac{3}{\sqrt{1-9x^2}}$. 2. $\frac{7}{1+49x^2}$

3. $D(\arctan(x+5)) = \frac{1}{1+(x+5)^2} \cdot D(x+5) = \frac{1}{x^2+10x+26}$. 4. $\frac{1}{\sqrt{4-x^2}}$

5. $D(\arctan(\sqrt{x})) = \frac{1}{1+(\sqrt{x})^2} \cdot D(\sqrt{x}) = \frac{1}{1+x} \cdot \frac{1}{2\sqrt{x}}$. 6. $\frac{2}{x\sqrt{x^4-1}}$

7. $D(\ln(\arctan(x))) = \frac{1}{\arctan(x)} \cdot D(\arctan(x)) = \frac{1}{\arctan(x)} \cdot \frac{1}{1+x^2}$. 8. $\frac{1}{2\sqrt{\arcsin(x)}} \cdot \frac{1}{\sqrt{1-x^2}}$

9. $D((\operatorname{arcsec}(x))^3) = 3(\operatorname{arcsec}(x))^2 \cdot D(\operatorname{arcsec}(x)) = 3(\operatorname{arcsec}(x))^2 \cdot \frac{1}{|x|\sqrt{x^2-1}}$. 10. $\frac{-5}{25+x^2}$

11. $D(\arctan(\ln(x))) = \frac{1}{1+(\ln(x))^2} \cdot D(\ln(x)) = \frac{1}{1+(\ln(x))^2} \cdot \frac{1}{x}$ 12. $\frac{1}{\sqrt{-x^2-4x-3}}$

13. $D(e^x \cdot \arctan(2x)) = e^x \cdot D(\arctan(2x)) + \arctan(x) \cdot D(e^x) = e^x \cdot \frac{1}{1+4x^2} \cdot (2) + \arctan(2x) \cdot e^x$.

14. $\dfrac{\arccos(x) \cdot \dfrac{1}{\sqrt{1-x^2}} - \arcsin(x) \cdot \dfrac{-1}{\sqrt{1-x^2}}}{(\arccos(x))^2}$

15. $\mathbf{D}(\arcsin(x) + \arccos(x)) = \dfrac{1}{\sqrt{1-x^2}} + \dfrac{-1}{\sqrt{1-x^2}} = 0$.   16. $x \cdot \dfrac{1}{1+x^2} + \arctan(x)$

17. $\mathbf{D}\left(\dfrac{1}{\arcsin(x)}\right) = \mathbf{D}((\arcsin(x))^{-1}) = -1 \cdot (\arcsin(x))^{-2} \cdot \mathbf{D}(\arcsin(x)) = \dfrac{-1}{(\arcsin(x))^2} \cdot \dfrac{1}{\sqrt{1-x^2}}$.

   (You could also use the quotient rule.)

18. $3 \cdot (1 + \text{arcsec}(x))^2 \dfrac{1}{|x|\sqrt{x^2-1}}$.

19. $\mathbf{D}(\sin(3 + \arctan(x))) = \cos(3 + \arctan(x)) \cdot \mathbf{D}(3 + \arctan(x)) = \cos(3 + \arctan(x)) \cdot \dfrac{1}{1+x^2}$.

20. $\tan(x) \cdot \dfrac{1}{1+x^2} + \arctan(x) \cdot \sec^2(x)$.

21. $\mathbf{D}\left(x \cdot \arctan\left(\dfrac{1}{x}\right)\right) = x \cdot \dfrac{1}{1+(1/x)^2} \cdot \mathbf{D}\left(\dfrac{1}{x}\right) + \arctan\left(\dfrac{1}{x}\right) \cdot \mathbf{D}(x)$

    $= x \cdot \dfrac{1}{1+(1/x)^2}\left(\dfrac{-1}{x^2}\right) + \arctan\left(\dfrac{1}{x}\right) = \dfrac{-x}{x^2+1} + \arctan\left(\dfrac{1}{x}\right)$.

22. $7 \cdot \arcsin\left(\dfrac{x}{3}\right) + C$

23. $\displaystyle\int_0^1 \dfrac{3}{x^2+25}\,dx = \dfrac{3}{25}\int_0^1 \dfrac{1}{(x/5)^2+1}\,dx$ Put $u = \dfrac{x}{5}$. Then $du = \dfrac{1}{5}dx$ and $dx = 5\,du$.

    $= \dfrac{3}{25}\displaystyle\int \dfrac{1}{(u)^2+1}\,5\,du = \dfrac{3}{5}\arctan(u) = \dfrac{3}{5}\arctan\left(\dfrac{x}{5}\right)\Big|_0^1 = \dfrac{3}{5}\arctan\left(\dfrac{1}{5}\right) - \dfrac{3}{5}\arctan\left(\dfrac{0}{5}\right) \approx 0.118$.

24. $\dfrac{5}{4} \cdot \text{arcsec}\left(\dfrac{x}{4}\right)\Big|_5^7 = \dfrac{5}{4} \cdot \text{arcsec}\left(\dfrac{7}{4}\right) - \dfrac{5}{4} \cdot \text{arcsec}\left(\dfrac{5}{4}\right)$

25. $\displaystyle\int \dfrac{9}{\sqrt{49-x^2}}\,dx = \dfrac{9}{7}\int \dfrac{1}{\sqrt{1-(x/7)^2}}\,dx$ Put $u = \dfrac{x}{7}$. Then $du = \dfrac{1}{7}dx$ and $dx = 7\,du$.

    $= \dfrac{9}{7}\displaystyle\int \dfrac{1}{\sqrt{1-(u)^2}}\,7\,du = 9\arcsin(u) + C = 9\arcsin\left(\dfrac{x}{7}\right) + C$.

26. $\dfrac{2}{\sqrt{7}} \cdot \arctan\left(\dfrac{x}{\sqrt{7}}\right)\Big|_1^4 = \dfrac{2}{\sqrt{7}} \cdot \arctan\left(\dfrac{4}{\sqrt{7}}\right) - \dfrac{2}{\sqrt{7}} \cdot \arctan\left(\dfrac{4}{\sqrt{7}}\right) \approx 0.472$

27. $\dfrac{3}{25} \displaystyle\int_{6}^{10} \dfrac{1}{\dfrac{x}{5}\sqrt{\left(\dfrac{x}{5}\right)^2 - 1}} \, dx$ Put $u = \dfrac{x}{5}$. Then $du = \dfrac{1}{5} dx$ and $dx = 5\, du$.

$\qquad = \dfrac{3}{25} \displaystyle\int \dfrac{1}{u\sqrt{(u)^2 - 1}} \, 5\, du = \dfrac{3}{5} \operatorname{arcsec}(u) = \dfrac{3}{5} \operatorname{arcsec}\left(\dfrac{x}{5}\right) \bigg|_{6}^{10}$

$\qquad = \dfrac{3}{5} \operatorname{arcsec}\left(\dfrac{10}{5}\right) - \dfrac{3}{5} \operatorname{arcsec}\left(\dfrac{6}{5}\right) = \dfrac{3}{5} \arccos\left(\dfrac{5}{10}\right) - \dfrac{3}{5} \arccos\left(\dfrac{5}{6}\right) \approx 0.277$.

28. $\dfrac{7}{3} \arctan\left(\dfrac{x-5}{3}\right) + C$. (Put $u = x - 5$)

29. $\displaystyle\int_{-1}^{1} \dfrac{e^x}{1 + e^{2x}} \, dx$. Put $u = e^x$. Then $du = e^x \, dx$.

$\qquad = \displaystyle\int \dfrac{1}{1 + (u)^2} \, du = \arctan(u) = \arctan(e^x) \bigg|_{-1}^{1} = \arctan(e^1) - \arctan(e^{-1}) \approx 0.866$.

30. $3 \cdot \arctan(\ln(e)) - 3 \cdot \arctan(\ln(1)) \approx 2.356$. (Put $u = \ln(x)$)

31. $\displaystyle\int \dfrac{\cos(x)}{\sqrt{9 - \sin^2(x)}} \, dx$ Put $u = \sin(x)$. Then $du = \cos(x) \, dx$.

$\qquad = \displaystyle\int \dfrac{1}{\sqrt{1 - u^2}} \, du = \arcsin\left(\dfrac{u}{3}\right) + C = \arcsin\left(\dfrac{1}{3} \sin(x)\right) + C$.

32. $\displaystyle\int \dfrac{8x}{16 + x^2} \, dx$ Put $u = 16 + x^2$. Then $du = 2x \, dx$ and $4\, du = 8x\, dx$.

$\qquad = \displaystyle\int \dfrac{1}{u} \, 4\, du = 4 \ln(u) + C = 4 \cdot \ln(16 + x^2) + C$.

33. $\displaystyle\int \dfrac{6x}{9 + x^4} \, dx$ Put $u = x^2$. Then $du = 2x \, dx$ and $3\, du = 6x\, dx$.

$\qquad \displaystyle\int \dfrac{1}{9 + u^2} \, 3\, du = 3 \cdot \dfrac{1}{3} \cdot \arctan\left(\dfrac{u}{3}\right) + C = \arctan\left(\dfrac{1}{3} x^2\right) + C$.

35. (a) See Section 7.2, Problem 19: $\theta = \arctan\left(\dfrac{1}{x}\right) + \arctan\left(\dfrac{3}{x}\right)$.

(b) $\dfrac{d\theta}{dx} = \dfrac{1}{1 + (1/x)^2} \cdot \dfrac{-1}{x^2} + \dfrac{1}{1 + (3/x)^2} \cdot \dfrac{-3}{x^2} = \dfrac{-1}{x^2 + 1} + \dfrac{-3}{x^2 + 9}$ which is never equal to 0, so

the angle θ is maximized at an endpoint: θ is maximum when $x = 0$. When $x = 0$, $\theta = \pi$.

37. Separable. $\frac{1}{y}\frac{dy}{dx} = \frac{1}{\sqrt{1-x^2}}$ so $\int \frac{1}{y} dy = \int \frac{1}{\sqrt{1-x^2}} dx$. Then $\ln(y) = \arcsin(x) + C$.

Using the initial condition $y(0) = e$: $\ln(e) = \arcsin(0) + C$ so $1 = 0 + C$ and $C = 1$.

$\ln(y) = \arcsin(x) + 1$ so (e raised to each side) $e^{\ln(y)} = e^{\arcsin(x) + 1}$ and $y = e^{\arcsin(x)} e^1$.

38. Similar to 37: $y = e^{\{\arcsin(x/4) - \pi/2\}} = \text{EXP}(\arcsin(\frac{x}{4}) - \frac{\pi}{2})$.

39. Separable: $\frac{1}{y^2}\frac{dy}{dx} = \frac{1}{9+x^2}$ so $\int \frac{1}{y^2} dy = \int \frac{1}{9+x^2} dx$ and $\frac{-1}{y} = \frac{1}{3}\arctan(\frac{x}{3}) + C$.

Using the initial condition $y(1) = 2$: $\frac{-1}{2} = \frac{1}{3}\arctan(\frac{1}{3}) + C$ so $C = \frac{-1}{2} - \frac{1}{3}\arctan(\frac{1}{3})$. Then

$\frac{-1}{y} = \frac{1}{3}\arctan(\frac{x}{3}) + \frac{-1}{2} - \frac{1}{3}\arctan(\frac{1}{3})$ and $y = \frac{-1}{\frac{1}{3}\arctan(\frac{x}{3}) + \frac{-1}{2} - \frac{1}{3}\arctan(\frac{1}{3})}$.

41. $\cos(\arccos(x)) = x$ so $D(\cos(\arccos(x))) = D(x)$ and $-\sin(\arccos(x)) \cdot D(\arccos(x)) = 1$. Then

$D(\arccos(x)) = \frac{-1}{\sin(\arccos(x))} = \frac{-1}{\sqrt{1-x^2}}$.

(Use the triangle with HYP = 1, ADJ = x, and OPP = $\sqrt{1-x^2}$ to verify that $\sin(\arccos(x)) = \sqrt{1-x^2}$.)

43. $A(x) = \int_0^x \frac{1}{1+t^2} dt = \arctan(x)$.

(a) $A(0) = \arctan(0) = 0$ (b) $A(1) = \arctan(1) = \pi/4 \approx 0.785$ (c) $A(10) = \arctan(10) \approx 1.471$.

(b) $\lim_{x \to \infty} A(x) = \lim_{x \to \infty} \arctan(x) = \frac{\pi}{2}$.

(c) $\frac{d}{dx} A(x) = \frac{d}{dx} \arctan(x) = \frac{1}{1+x^2} > 0$. (We can also get $A'(x)$ using the Fundamental Thm. of Calculus.)

(d) Since $A'(x) > 0$ for all x, $A(x)$ is an INCREASING function.

44. (a) $2 \cdot \arctan(10) \approx 2.942$. (b) $2(\frac{\pi}{2}) = \pi$.

45. $\int \frac{8x-5}{x^2+9} dx = \int \frac{8x}{x^2+9} dx - \int \frac{5}{x^2+9} dx = 4 \cdot \ln(x^2+9) - \frac{5}{3} \cdot \arctan(\frac{x}{3}) + C$.

46. $\arctan(x) - 2 \cdot \ln(x^2+1) + C$.

47. $\int \frac{7x+3}{x^2+10} dx = \int \frac{7x}{x^2+10} dx + \int \frac{3}{x^2+10} dx = \frac{7}{2} \cdot \ln(x^2+10) + \frac{3}{\sqrt{10}} \cdot \arctan(\frac{x}{\sqrt{10}}) + C$.

48. $\frac{1}{2} \cdot \ln(x^2+9) + \frac{5}{3} \cdot \arctan(\frac{x}{3}) + C$.

49. (a) Put $u = x + 3$, $du = dx$. $\int \frac{8}{(x+3)^2 + 1} dx = \int \frac{8}{u^2 + 1} du = 8 \cdot \arctan(u) + C = 8 \cdot \arctan(x+3) + C$.

(b) Put $u = x^2 + 6x + 10$. Then $du = (2x + 6) dx$ and $2 du = (4x + 12) dx$.

$\int \frac{4x + 12}{x^2 + 6x + 10} dx = \int \frac{2}{u} du = 2 \cdot \ln|u| + C = 2 \cdot \ln(x^2 + 6x + 10) + C$.

(c) final result = (answer in part a) + (answer in part b)

50. (a) $7 \cdot \arctan(x+2) + C$ (b) $6 \cdot \ln(x^2 + 4x + 5) + C$

(c) (answer in part a) + (answer in part b) = $7 \cdot \arctan(x+2) + 6 \cdot \ln(x^2 + 4x + 5) + C$

51. $\int \frac{6x + 15}{x^2 + 4x + 20} dx$ Put $u = x^2 + 4x + 20$. Then $du = (2x + 4) dx$ and $3 du = (6x + 12) dx$.

$= \int \frac{6x + 12}{x^2 + 4x + 20} dx + \int \frac{3}{x^2 + 4x + 20} dx = \{a\} + \{b\}$

$\{a\} = \int \frac{6x + 12}{x^2 + 4x + 20} dx = \int \frac{1}{u} 3 du = 3 \cdot \ln|u| = 3 \cdot \ln(x^2 + 4x + 20)$

$\{b\} = \int \frac{3}{(x+2)^2 + 16} dx = \int \frac{3}{(u)^2 + 16} dx = \frac{3}{4} \arctan(\frac{x+2}{4}) + C$ (Put $u = x+2$ and $du = dx$)

Total integral = $\{a\} + \{b\}$ = $3 \cdot \ln(x^2 + 4x + 20) + \frac{3}{4} \arctan(\frac{x+2}{4}) + C$.

52. $\ln(x^2 - 4x + 13) + 3 \cdot \arctan(\frac{x-2}{3}) + C$.

CHAPTER 8: IMPROPER INTEGRALS and INTEGRATION TECHNIQUES

Introduction

In previous sections we examined a variety of applications which require integrals, and we found antiderivatives of many important groups of functions: polynomials, some rational functions, the trigonometric functions, the logarithm functions and some exponential functions. With this information, it is easy to find antiderivatives of their sums and differences, but finding antiderivatives of their products, quotients, and compositions can still be quite difficult. This chapter introduces several techniques of integration which greatly expand the number and variety of functions you can integrate. The first section, however, does not discuss a technique for finding antiderivatives.

- Section 8.1 describes how to evaluate an **Improper Integral**, the integral of a function over an infinitely long interval or over a finite interval when the function is not bounded at one endpoint of the interval.

The rest of the chapter is devoted to finding antiderivatives, and the overall theme **transformation**, how to transform a new type of integrand into one we can integrate immediately or into one we can find in the tables at the end of the book. Our goal is to change the pattern of some new function or combination of functions into a pattern we recognize or can find in the tables.

- Section 8.2 **reviews** some of the most common patterns we have already encountered, and it emphasizes the powerful technique of **substitution**. When the substitution technique works, it is among the easiest and quickest to use.

The remaining four sections present additional techniques for finding antiderivatives. Each of the new techniques is very useful for finding antiderivatives of particular patterns and combinations of functions. There are more integration techniques besides the four presented here, but these four are the ones most commonly needed.

- Section 8.3 introduces a technique called **Integration By Parts** which is particularly useful for finding antiderivatives of products of functions. Integration by parts is the technique used to derive many of the integration formulas in the tables.

- Section 8.4 introduces an algebraic technique called **Partial Fraction Decomposition** for transforming difficult rational functions into sums of easier rational functions which can then be integrated using previous integration techniques.

- Section 8.5 introduces a technique called **Trigonometric Substitution** which is particularly useful for integrands containing sums and differences of squares of the forms $x^2 + a^2$, $x^2 - a^2$, and $a^2 - x^2$.

- Section 8.6 considers a variety of ways in which trigonometric identities and transformations can be used to find antiderivatives of some combinations of trigonometric functions.

Unfortunately, some functions simply do not have antiderivatives which are elementary combinations (sums, differences, products, quotients, roots, and compositions) of polynomials, rational functions, trigonometric, logarithmic and exponential functions, and none of the integration methods of this chapter will find their antiderivatives.

Historically, the integration techniques in this chapter and tables of antiderivatives were very important for people who needed to apply calculus and solve differential equations. Recently we have gained additional tools, computers and even calculators that can calculate the antiderivatives of many (but not all) functions. These electronic aides, like earlier tables of antiderivatives, can remove some computational difficulties on the way to an answer, but it is still up to you to understand and set up the problems and to interpret and use the answers. Now, perhaps more than ever, it is important that you master the concepts of calculus and understand how these concepts are related and are used. **The computer may help you get an answer once the problem has been understood and formulated in mathematical terms, but it is your understanding of the concepts that will enable you to formulate the problems so computers can help.**

Although the computational techniques in this chapter are less important than they were several years ago, they still contribute to understanding calculus and to recognizing patterns of functions and their derivatives and antiderivatives.

8.1 Improper Integrals

Our original development of the definite integral $\int_a^b f(x)\,dx$ used Riemann sums and assumed that

- the length of the interval of integration [a, b] was finite and
- that f(x) was defined and bounded at every point of the interval [a, b] (including the endpoints).

Sometimes, however, we need the value of an integral which does not satisfy one or both of these assumptions. In this section we extend the ideas of the definite integral to evaluate two types of **improper definite integrals**:

(1) the length of the interval of integration is not finite

(2) the integrand function is not bounded at a point of the interval of integration.

Example 1: Represent each area as an improper definite integral.

(a) The area of the infinite region between $f(x) = 1/x^2$ and the x–axis for $x \geq 1$ (Fig. 1)

(b) The area between g(x) and the x–axis for $0 \leq x \leq 4$ (Fig. 2)

Solution: (a) $\int_1^\infty \frac{1}{x^2}\,dx$ (b) $\int_0^4 g(x)\,dx$

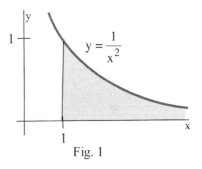

Fig. 1

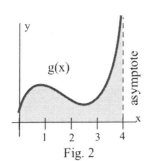

Fig. 2

Practice 1: Represent each quantity as an improper definite integral.

(a) The volume swept out when the infinite region between $f(x) = 1/x$ and the x–axis is revolved about the x–axis for $x \geq 4$ (Fig. 3)

(b) The area between the curves in Fig. 4 for $1 \leq x \leq 3$.

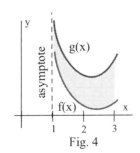

Fig. 4

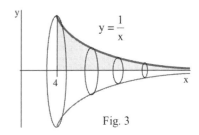

Fig. 3

General Strategy For Improper Integrals

Our general strategy for evaluating improper integrals is to shrink the interval of integration so we have a definite integral we can evaluate. Then as we let the interval grow to approach the interval of integration we want, the value of the integral on the growing intervals approaches the value of the improper integral. The value of the improper integral is the limiting value of the definite integrals as the interval grows to the interval we want, provided that the limit exists.

Infinitely Long Intervals of Integration

We evaluate an improper integral on an infinitely long interval by
- replacing the inifinitely long interval with a finite interval,
- evaluating the integral on the finite interval, and, finally,
- letting the finite interval grow longer and longer, approaching the interval we want.

Example 2: Evaluate $\int_1^\infty \frac{1}{x^2} \, dx$.

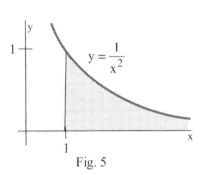

Fig. 5

Solution: The interval $[1, \infty)$ is infinitely long, but we can evaluate the integral on the finite intervals $[1, 2], [1, 10], [1, 1000]$, and in general $[1, C]$ (Fig. 5).

$$\int_1^2 \frac{1}{x^2} \, dx = -\frac{1}{x} \Big|_1^2 = \left(-\frac{1}{2}\right) - \left(-\frac{1}{1}\right) = 1 - \frac{1}{2} = \frac{1}{2}.$$

Similarly, $\int_1^{10} \frac{1}{x^2} \, dx = 1 - \frac{1}{10} = .9$, $\int_1^{1000} \frac{1}{x^2} \, dx = 1 - \frac{1}{1000} = .999$, and, in general,

$\int_1^C \frac{1}{x^2} \, dx = -\frac{1}{x} \Big|_1^C = \left(-\frac{1}{C}\right) - \left(-\frac{1}{1}\right) = 1 - \frac{1}{C}$. As the value of C gets larger, the length of the interval $[1, C]$ increases, and the value of $\int_1^C \frac{1}{x^2} \, dx$ approaches the value of $\int_1^\infty \frac{1}{x^2} \, dx$.

The value of $\int_1^\infty \frac{1}{x^2} \, dx$ is the limit of the values of $\int_1^C \frac{1}{x^2} \, dx$ as C approaches infinity:

$$\int_1^\infty \frac{1}{x^2} \, dx = \lim_{C \to \infty} \left\{ \int_1^C \frac{1}{x^2} \, dx \right\} = \lim_{C \to \infty} \left\{ 1 - \frac{1}{C} \right\} = 1.$$

We say that the improper integral $\int_1^\infty \frac{1}{x^2} \, dx$ is **convergent** and **converges to 1**.

If the following limits exist,

the value of $\int_a^{\infty} f(x)\,dx$ is defined to be the value of $\lim_{C \to \infty} \left\{ \int_a^C f(x)\,dx \right\}$, and

the value of $\int_{-\infty}^b f(x)\,dx$ is defined to be the value of $\lim_{C \to -\infty} \left\{ \int_C^b f(x)\,dx \right\}$.

In each case, first evaluate the proper integral and then take the limit.

If the limit is a finite number, we say the improper integral is **convergent**.

If the limit does not exist or if it is infinite, we say the improper integral is **divergent**.

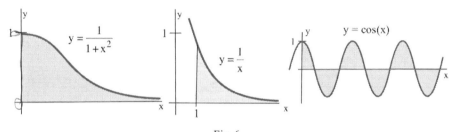

Fig. 6

Example 3: Evaluate (a) $\int_0^{\infty} \frac{1}{1+x^2}\,dx$, (b) $\int_1^{\infty} \frac{1}{x}\,dx$ and (c) $\int_0^{\infty} \cos(x)\,dx$. (Fig. 6)

Solution: (a) $\int_0^{\infty} \frac{1}{1+x^2}\,dx = \lim_{C \to \infty} \left\{ \int_0^C \frac{1}{1+x^2}\,dx \right\} = \lim_{C \to \infty} \left\{ \arctan(x) \Big|_0^C \right\}$

$= \lim_{C \to \infty} \left\{ \arctan(C) - \arctan(0) \right\} = \frac{\pi}{2} - 0 = \frac{\pi}{2}$.

so we say that $\int_0^{\infty} \frac{1}{1+x^2}\,dx$ is **convergent**.

(b) $\int_1^{\infty} \frac{1}{x}\,dx = \lim_{C \to \infty} \left\{ \int_1^C \frac{1}{x}\,dx \right\} = \lim_{C \to \infty} \left\{ \ln(x) \Big|_1^C \right\} = \lim_{C \to \infty} \left\{ \ln(C) - \ln(1) \right\} = \infty$

so we say the improper integral $\int_1^{\infty} \frac{1}{x}\,dx$ is **divergent**.

(c) $\displaystyle\int_0^\infty \cos(x)\, dx = \lim_{C\to\infty}\left\{\int_0^C \cos(x)\, dx\right\} = \lim_{C\to\infty}\left\{\sin(x)\Big|_0^C\right\}$

$= \lim_{C\to\infty}\{\sin(C) - \sin(0)\} = \lim_{C\to\infty}\sin(C)$.

As C increases the values of sin(C) oscilate between –1 and 1 and do not approach a single value so the last limit does not exist and the improper integral $\displaystyle\int_1^\infty \cos(x)\, dx$ is **divergent**.

Practice 2: Evaluate (a) $\displaystyle\int_1^\infty \frac{1}{x^3}\, dx$ and (b) $\displaystyle\int_0^\infty \sin(x)\, dx$.

Functions Undefined At An Endpoint Of The Interval Of Integration

If the function we want to integrate is unbounded at one of the endpoints of an interval of finite length, we can shrink the interval so the function is bounded at both endpoints of the new, smaller interval, evaluate the integral over the smaller interval, and finally, let the smaller interval grow to approach the original interval.

Example 4: Evaluate $\displaystyle\int_0^1 \frac{1}{\sqrt{x}}\, dx$. (Fig. 7)

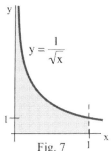

Fig. 7

Solution: The function $1/\sqrt{x}$ is not bounded at $x = 0$, the lower endpoint of integration, but the function is bounded on the intervals [.36, 1], [0.09, 1], and, in general, on the interval [C, 1] for any C > 0.

$\displaystyle\int_{.36}^1 \frac{1}{\sqrt{x}}\, dx = 2\sqrt{x}\,\Big|_{.36}^1 = 2\sqrt{1} - 2\sqrt{.36} = 2 - 1.2 = 0.8$. Similarly,

$\displaystyle\int_{0.09}^1 \frac{1}{\sqrt{x}}\, dx = 2\sqrt{x}\,\Big|_{0.09}^1 = 2\sqrt{1} - 2\sqrt{0.09} = 1.4$ and $\displaystyle\int_C^1 \frac{1}{\sqrt{x}}\, dx = 2\sqrt{x}\,\Big|_C^1 = 2 - 2\sqrt{C}$.

As C approaches 0 from the right, the interval [C, 1] approaches the interval [0, 1] and the value of $2 - 2\sqrt{C}$ approaches 2. We say that $\displaystyle\int_0^1 \frac{1}{\sqrt{x}}\, dx$ converges to 2 and write $\displaystyle\int_0^1 \frac{1}{\sqrt{x}}\, dx = 2$.

> If $f(x)$ is not bounded (not defined) at $x = b$,
>
> the value of $\int_a^b f(x)\, dx$ is defined to be the value of $\lim_{C \to b^-} \left\{ \int_a^C f(x)\, dx \right\}$
>
> if this limit exists.
>
> If $f(x)$ is not bounded (not defined) at $x = a$,
>
> the value of $\int_a^b f(x)\, dx$ is defined to be the value of $\lim_{C \to a^+} \left\{ \int_C^b f(x)\, dx \right\}$.
>
> if this limit exists.
>
> In each case, first evaluate the proper integral and then take the limit.
>
> If the limit is a finite number, we say the improper integral is **convergent**.
>
> If the limit does not exist or if it is infinite, we say the improper integral is **divergent**.

Practice 3: Show that (a) $\int_1^{10} \frac{1}{\sqrt{10 - x}}\, dx = 6$ and (b) $\int_0^1 \frac{1}{x}\, dx$ is divergent.

If the function is unbounded at one or more points inside the interval of integration, we can split the original improper integral into several improper integrals on subintervals so the function is unbounded at one endpoint of each subinterval.

Testing For Convergence: The Comparison Test and P–Test

Sometimes the only thing that matters about an improper integral is whether or not it converges to a finite number. There are ways to determine its convergence even though we may not be able to or may not want to determine the exact value of the integral. In the remainder of this section we consider two methods for testing the convergence of an improper integral. Neither method gives us the value of the improper integral, but each enables us to determine whether some improper integrals are convergent. The Comparison Test For Integrals enables us to determine the convergence (or divergence) of some integrals by comparing them with some integrals we already know converge or diverge. The Comparison Test, however, requires that we know the convergence or divergence of the integrals we compare against, and the P–Test provides examples of known convergent and divergent integrals to use for this comparison.

We start with the P–Test in order to have some examples to use when we consider the Comparison Test.

> **P–Test for Integrals**
>
> For any $a > 0$, the improper integral $\displaystyle\int_a^\infty \frac{1}{x^p}\, dx$ $\begin{cases} \text{converges} & \text{if } p > 1 \\ \text{diverges} & \text{if } p \leq 1 \end{cases}$

Proof: It is easiest to consider three cases: $p = 1$, $p > 1$, and $p < 1$.

Case $p = 1$: Then $\displaystyle\int_a^\infty \frac{1}{x^p}\, dx = \int_a^\infty \frac{1}{x}\, dx = \lim_{C \to \infty}\left\{ \int_a^C \frac{1}{x}\, dx \right\}$

$= \displaystyle\lim_{C \to \infty}\left\{ \ln(x)\Big|_a^C \right\} = \lim_{C \to \infty}\left\{ \ln(C) - \ln(a) \right\} = \infty$ so $\displaystyle\int_a^\infty \frac{1}{x^p}\, dx$ diverges.

For the other two cases, $p \neq 1$, so $\displaystyle\int_a^\infty \frac{1}{x^p}\, dx = \lim_{C \to \infty}\left\{ \int_a^C \frac{1}{x^p}\, dx \right\} = \lim_{C \to \infty}\left\{ \int_a^C x^{-p}\, dx \right\}$

$= \displaystyle\lim_{C \to \infty}\left\{ \frac{1}{1-p} \cdot x^{1-p}\Big|_a^C \right\} = \lim_{C \to \infty}\left\{ C^{1-p} - a^{1-p} \right\}.$

Case $p > 1$: Then $1 - p < 0$ so $\displaystyle\lim_{C \to \infty} C^{1-p} = 0$ and

$\displaystyle\int_a^\infty \frac{1}{x^p}\, dx = \lim_{C \to \infty} \frac{1}{1-p}\left\{ C^{1-p} - a^{1-p} \right\} = -\frac{a^{1-p}}{1-p} = \frac{a^{1-p}}{p-1}$, a finite number.

Case $p < 1$: Then $1 - p > 0$ so $\displaystyle\lim_{C \to \infty} C^{1-p} = \infty$ and $\displaystyle\int_a^\infty \frac{1}{x^p}\, dx$ diverges.

Example 5: Determine the convergence or divergence of (a) $\displaystyle\int_5^\infty \frac{1}{x^2}\, dx$, (b) $\displaystyle\int_1^\infty \frac{1}{\sqrt{x}}\, dx$ and (c) $\displaystyle\int_1^8 \frac{1}{x^{1/3}}\, dx$.

Solution: (a) $\displaystyle\int_5^\infty \frac{1}{x^2}\, dx$ matches the form for the P–Test with $p = 2 > 1$, so the integral is convergent.

(b) $\displaystyle\int_1^\infty \frac{1}{\sqrt{x}}\, dx = \int_1^\infty \frac{1}{x^{1/2}}\, dx$ matches the form for the P–Test with $p = 1/2 < 1$ so the integral is divergent.

(c) $\displaystyle\int_1^8 \frac{1}{x^{1/3}}\, dx$ does not match the form for the P–Test because the interval $[1, 8]$ is finite.

$\displaystyle\int_1^8 \frac{1}{x^{1/3}}\, dx = \int_1^8 x^{-1/3}\, dx = \frac{3}{2} x^{2/3}\Big|_1^8 = \frac{3}{2}\left\{ 8^{2/3} - 1^{2/3} \right\} = \frac{3}{2}\{ 4 - 1 \} = \frac{9}{2}.$

The following Comparison Test enables us to determine the convergence or divergence of an improper integral of a new positive function by comparing the new function with functions whose improper integrals we already know converge or diverge.

Comparison Test for Integrals of Positive Functions

(a) If the new integral is smaller than one we know converges, then the new integral converges (Fig. 8):

if $0 \leq f(x) \leq g(x)$ and $\int_a^\infty g(x)\, dx$ converges, then $\int_a^\infty f(x)\, dx$ converges.

(b) If the new integral is larger than one which diverges, then the new integral diverges (Fig. 9):

if $f(x) \geq g(x) \geq 0$ and $\int_a^\infty g(x)\, dx$ diverges, then $\int_a^\infty f(x)\, dx$ diverges.

(c) If the new integral is larger than a convergent integral or smaller than a divergent integral, then we can draw no immediate conclusion about the new integral — the new integral may converge or diverge (Fig. 10).

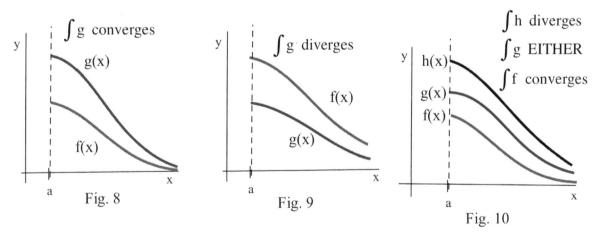

Fig. 8   Fig. 9   Fig. 10

The proof is a straightforward application of the definition of the value of an improper integral and facts about limits.

Example 6: Determine whether each of these integrals is convergent by comparing it with an appropriate integral which you know converges or diverges.

(a) $\int_1^\infty \dfrac{7}{x^3 + 5}\, dx$ (b) $\int_1^\infty \dfrac{3 + \sin(x)}{x^2}\, dx$ (c) $\int_6^\infty \dfrac{9}{\sqrt{x-5}}\, dx$

Solution: (a) The dominant power of $\frac{7}{x^3+5}$ is $\frac{1}{x^3}$ so we should compare with $\frac{1}{x^3}$.

$$\int_1^\infty \frac{7}{x^3+5}\,dx < \int_1^\infty \frac{7}{x^3}\,dx = 7\int_1^\infty \frac{1}{x^3}\,dx.$$ We know $\int_1^\infty \frac{1}{x^3}\,dx$ is convergent

by the P–Test (p = 3 > 1), so we can conclude that $\int_1^\infty \frac{7}{x^3+5}\,dx$ is convergent.

(b) We know $0 < 3 + \sin(x) \le 4$ and the dominant term is $\frac{1}{x^2}$ so we should compare with $\frac{1}{x^2}$.

$$\int_1^\infty \frac{3+\sin(x)}{x^2}\,dx \le \int_1^\infty \frac{4}{x^2}\,dx = 4\int_1^\infty \frac{1}{x^2}\,dx.$$ We know $\int_1^\infty \frac{1}{x^2}\,dx$ is convergent

by the P–Test (p = 2 > 1), so we can conclude that $\int_1^\infty \frac{3+\sin(x)}{x^2}\,dx$ is convergent.

(c) The dominant power of $\frac{9}{\sqrt{x-5}}$ is $\frac{1}{\sqrt{x}}$ so we should compare with $\frac{1}{x^{1/2}}$.

$$\int_6^\infty \frac{9}{\sqrt{x-5}}\,dx > \int_1^\infty \frac{9}{\sqrt{x}}\,dx = 9\int_1^\infty \frac{1}{x^{1/2}}\,dx.$$ We know $\int_1^\infty \frac{1}{x^{1/2}}\,dx$ is divergent

by the P–Test (p = 1/2 < 1), so we can conclude that $\int_6^\infty \frac{9}{\sqrt{x-5}}\,dx$ is divergent.

PROBLEMS

In 1–21, use the definition of an improper integral to evaluate the given integral.

1. $\int_{10}^\infty \frac{1}{x^3}\,dx$

2. $\int_e^\infty \frac{5}{x \cdot \ln(x)^2}\,dx$

3. $\int_3^\infty \frac{2}{1+x^2}\,dx$

4. $\int_1^\infty \frac{2}{e^x}\,dx$

5. $\int_e^\infty \frac{5}{x \cdot \ln(x)}\,dx$

6. $\int_0^\infty \frac{x}{1+x^2}\,dx$

7. $\int_3^\infty \frac{1}{x-2}\,dx$

8. $\int_3^\infty \frac{1}{(x-2)^2}\,dx$

9. $\int_3^\infty \frac{1}{(x-2)^3}\,dx$

10. $\int_3^\infty \frac{1}{x+2}\,dx$

11. $\int_3^\infty \frac{1}{(x+2)^2}\,dx$

12. $\int_3^\infty \frac{1}{(x+2)^3}\,dx$

13. $\int_0^4 \frac{1}{\sqrt{x}}\,dx$

14. $\int_0^8 \frac{1}{\sqrt[3]{x}}\,dx$

15. $\int_0^{16} \frac{1}{\sqrt[4]{x}}\,dx$

16. $\int_0^2 \frac{1}{\sqrt{2-x}}\,dx$

17. $\int_0^2 \frac{1}{\sqrt{4-x^2}}\,dx$

18. $\int_0^2 \frac{3x^2}{\sqrt{8-x^3}}\,dx$

19. $\int_{-2}^{\infty} \sin(x)\,dx$

20. $\int_{\pi}^{\infty} \sin(x)\,dx$

21. $\int_0^{\pi/2} \tan(x)\,dx$

22. Example 3(b) showed that $\int_1^C \frac{1}{x}\,dx$ grew arbitrarily large as C grew arbitrarily large, so no finite amount of paint would cover the area bounded between the x–axis and the graph of f(x) = 1/x for x > 1 (Fig. 11a). Show that the volume obtained when the area in Fig. 11a is revolved about the x–axis (Fig. 11b) is finite so the 3–dimensional trumpet–shaped region can be filled with a finite amount of paint. Is there a contradiction here?

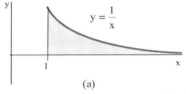
(a)

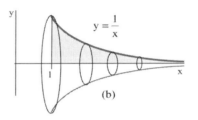
(b)

Fig. 11

23. In the **Lifting a Payload** discussion in Section 5.3, we determined that the amount of work needed to lift a payload from the surface of a moon to an altitude of A above the moon's surface was

work = $\int_R^{R+A} \frac{R^2 P}{x^2}\,dx$.

(a) Calculate the amount of work required to lift the payload to an altitude of R miles and 2R miles.

(b) Calculate the amount of work needed to lift the payload arbitrarily high. (Calculate the limit of the work integral as "A → ∞.")

In problems 24–32, determine whether the improper integral is convergent or divergent. Do not evaluate the integral.

24. $\int_3^{\infty} \frac{7}{x^2+5}\,dx$

25. $\int_3^{\infty} \frac{1}{x^3+x}\,dx$

26. $\int_7^{\infty} \frac{1}{x-2}\,dx$

27. $\int_3^{\infty} \frac{7}{x+\ln(x)}\,dx$

28. $\int_3^{\infty} \frac{1}{x^2-1}\,dx$

29. $\int_7^{\infty} \frac{1+\cos(x)}{x^2}\,dx$

30. The volume obtained when the area between the x–axis for $x \geq 1$ and the graph of $f(x) = \frac{\sin(x)}{x}$ (Fig. 12) is revolved about the x–axis.

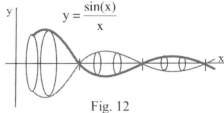

Fig. 12

31. (a) The volume obtained when the area between the positive x–axis ($x \geq 0$) and the graph of $f(x) = \frac{1}{x^2 + 1}$ (Fig. 13a) is revolved about the x–axis.

   (b) The volume obtained when the area between the positive x–axis ($x \geq 0$) and the graph of $f(x) = \frac{1}{x^2 + 1}$ (Fig. 13b) is revolved about the **y–axis**. (Use the method of "tubes" from section 5.5.)

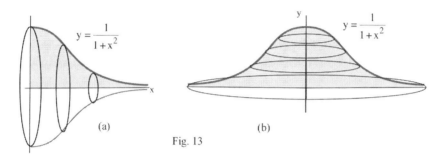

Fig. 13

32. (a) The volume obtained when the area between the positive x–axis ($x \geq 0$) and the graph of $f(x) = \frac{1}{e^x}$ is revolved about the x–axis.

   (b) The volume obtained when the area between the positive x–axis ($x \geq 0$) and the graph of $f(x) = \frac{1}{e^x}$ is revolved about the **y–axis**. (Use the method of "tubes" from section 5.5.

33. (a) Use Fig. 14a to help determine which is larger: $\int_1^A \frac{1}{x} dx$ or $\sum_{k=1}^{A-1} \frac{1}{k}$.

   (b) Use Fig. 14b to help determine which is larger: $\int_1^A \frac{1}{x} dx$ or $\sum_{k=2}^{A} \frac{1}{k}$.

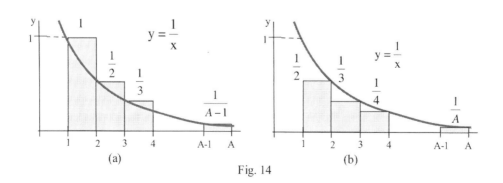

Fig. 14

34. (a) Use Fig. 15a to help determine which is larger: $\int_{1}^{A} \frac{1}{x^2} dx$ or $\sum_{k=1}^{A-1} \frac{1}{k^2}$.

(b) Use Fig. 15b to help determine which is larger: $\int_{1}^{A} \frac{1}{x^2} dx$ or $\sum_{k=2}^{A} \frac{1}{k^2}$.

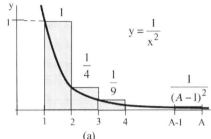

Fig. 15

Section 8.1 Practice Answers

Practice 1: (a) $V = \int_{4}^{\infty} \pi \left(\frac{1}{x} \right)^2 dx$ (b) $A = \int_{1}^{3} g(x) - f(x) \, dx$

Practice 2: (a) $\int_{1}^{\infty} \frac{1}{x^3} dx = \lim_{A \to \infty} \int_{1}^{A} \frac{1}{x^3} dx = \lim_{A \to \infty} \left\{ -\frac{1}{2x^2} \right\} \Big|_{1}^{A} = \lim_{A \to \infty} \frac{1}{2} - \frac{1}{2A^2} = \frac{1}{2}$.

(b) $\int_{0}^{\infty} \sin(x) \, dx = \lim_{A \to \infty} \int_{1}^{A} \sin(x) \, dx = \lim_{A \to \infty} \{ -\cos(x) \} \Big|_{0}^{A} = \lim_{A \to \infty} \{ 1 - \cos(A) \}$ DNE.

$\int_{0}^{\infty} \sin(x) \, dx$ is DIVERGENT (or DIVERGES).

Practice 3: (a) $\int_{1}^{10} \frac{1}{\sqrt{10 - x}} dx = \lim_{A \to 10^-} \int_{1}^{A} \frac{1}{\sqrt{10 - x}} dx = \lim_{A \to 10^-} \left\{ -2(10 - x)^{1/2} \right\} \Big|_{1}^{A}$

$= \lim_{A \to 10^-} \left\{ -2(10 - A)^{1/2} + 2(10 - 1)^{1/2} \right\} = \lim_{A \to 10^-} \left\{ 6 + 2(10 - A)^{1/2} \right\} = 6.$

(b) $\int_{0}^{1} \frac{1}{x} dx = \lim_{B \to 0^+} \int_{B}^{1} \frac{1}{x} dx = \lim_{B \to 0^+} \{ \ln(x) \} \Big|_{B}^{1} = \lim_{B \to 0^+} \{ \ln(1) - \ln(B) \} = -(-\infty) = \infty.$

$\int_{0}^{1} \frac{1}{x} dx$ is DIVERGENT (or DIVERGES).

8.2 FINDING ANTIDERIVATIVES: A REVIEW

Success at integration is primarily a matter of recognizing standard patterns and being able to manipulate functions into the form of these recognizable patterns. Integral tables such as the short one given below and the longer one inside the covers of this book give antiderivatives for patterns of functions and their derivatives, but often it is necessary to change the variable in order to see the pattern. For most people, skill with these patterns comes with practice, and this section provides a variety of problems to review and hone your skills.

Table of Common Integral Formulas

Constant $\quad \int k \, du = ku + C$

Powers $\quad \int u^n \, du = \dfrac{u^{n+1}}{n+1} + C \quad (n \neq -1) \qquad \int u^{-1} \, du = \int \dfrac{1}{u} \, du = \ln|u| + C$

Trigonometric $\quad \int \sin(u) \, du = -\cos(u) + C \qquad \int \cos(u) \, du = \sin(u) + C$

$\int \tan(u) \, du = -\ln|\cos(u)| + C \qquad \int \cot(u) \, du = \ln|\sin(u)| + C$

$\int \sec(u) \, du = \ln|\sec(u) + \tan(u)| + C \qquad \int \csc(u) \, du = -\ln|\csc(u) + \cot(u)| + C$

Exponential $\quad \int e^u \, du = e^u + C \qquad \int a^u \, du = \dfrac{a^u}{\ln(a)} + C$

Inverse Trigonometric $\quad \int \dfrac{1}{a^2 + u^2} \, du = \dfrac{1}{a} \arctan\left(\dfrac{u}{a}\right) + C$

$\int \dfrac{1}{\sqrt{a^2 - u^2}} \, du = \arcsin\left(\dfrac{u}{a}\right) + C$

$\int \dfrac{1}{|u|\sqrt{u^2 - a^2}} \, du = \dfrac{1}{a} \operatorname{arcsec}\left(\dfrac{u}{a}\right) + C$

The most generally useful and powerful integration technique, Changing the Variable, was introduced in Chapter 4, and has been used extensively since then. The first problems at the end of this section provide additional practice changing variables to calculate integrals. As we develop more complicated and more specialized techniques for finding antiderivatives, your first thought should still be whether the integral can be simplified by changing the variable.

A Particular Situation: An Irreducible Quadratic Denominator

Sometimes the appropriate change of variable is not obvious, and we may need to manipulate the function first.

Example 1: Evaluate $\int \frac{18}{x^2 - 6x + 10} \, dx$.

Solution: When the denominator of a rational function is an irreducible quadratic ($ax^2 + bx + c$ with $b^2 - 4ac < 0$), we can write it as the sum of two squares by using the algebraic technique of "completing the square:" $x^2 - 6x + 10 = (x - 3)^2 + 1$. Then

$$\int \frac{18}{x^2 - 6x + 10} \, dx = \int \frac{18}{(x - 3)^2 + 1} \, dx$$ and the change of variable $u = x - 3$ makes the

arctangent pattern more evident. If $u = x - 3$, then $du = dx$ and

$$\int \frac{18}{x^2 - 6x + 10} \, dx = \int \frac{18}{u^2 + 1} \, du = 18 \arctan(u) + C = 18 \arctan(x - 3) + C.$$

Completing The Square: $x^2 + bx + c = (x + \frac{b}{2})^2 + (c - \frac{b^2}{4})$.

Practice 1: Evaluate $\int \frac{5}{x^2 + 8x + 25} \, dx$ by completing the square and changing the variable.

Example 2: Evaluate $\int \frac{6x}{x^2 - 6x + 10} \, dx$.

Solution: This would be an easy problem if the numerator was $6x - 18$. Then the numerator would be three times the derivative of the denominator and the pattern of the integral would be $\int \frac{3}{u} \, du$ with $u = x^2 - 6x + 10$. Fortunately, we can rewrite the numerator as $6x - 18 + 18$. Then

$$\int \frac{6x}{x^2 - 6x + 10} \, dx = \int \frac{6x - 18 + 18}{x^2 - 6x + 10} \, dx$$

$$= \int \frac{6x - 18}{x^2 - 6x + 10} \, dx + \int \frac{18}{x^2 - 6x + 10} \, dx$$

$$= \int \frac{3}{u} \, du + \int \frac{18}{w^2 + 1} \, dw \quad \text{where } u = x^2 - 6x + 10 \text{ and } w = x - 3$$

$$= 3 \cdot \ln|u| + 18 \cdot \arctan(w) + C = 3 \cdot \ln(x^2 - 6x + 10) + 18 \cdot \arctan(x - 3) + C.$$

8.2 Integration Review

Practice 2: Evaluate $\int \frac{4x+21}{x^2+8x+25} dx$. (Hint: if $u = x^2 + 8x + 25$, then $2 du = (4x + 16) dx$.)

The "logarithm plus an arctangent" pattern of these answers is quite typical for integrals of linear functions divided by irreducible quadratics. If the quadratic denominator can be factored into a product of two linear factors, we use the technique discussed in Section 8.3, Partial Fraction Decomposition.

Problems

In problems 1 – 42, find the indefinite integrals and evaluate the definite integrals. In some of the problems a particular change of variable is suggested.

1. $\int 6x(x^2+7)^2 dx \qquad u = x^2 + 7$

2. $\int 6x(x^2-1)^3 dx \qquad u = x^2 - 1$

3. $\int_2^4 \frac{6x}{\sqrt{x^2-3}} dx \qquad u = x^2 - 3$

4. $\int_0^\pi 12 \cos(x) \cdot \{\sin(x) + 2\}^2 dx \quad u = \sin(x) + 2$

5. $\int \frac{12x}{x^2+3} dx$

6. $\int \frac{\cos(x)}{2+\sin(x)} dx$

7. $\int \sin(3x+2) dx$

8. $\int \cos(x/5) dx$

9. $\int_0^1 e^x \sec^2(e^x + 3) dx \qquad u = e^x + 3$

10. $\int_0^{\pi/2} \cos(x) \sqrt{1+\sin(x)} \, dx \quad u = 1 + \sin(x)$

11. $\int \frac{\ln(x)}{x} dx \qquad u = \ln(x)$

12. $\int \frac{\cos(\sqrt{x})}{\sqrt{x}} dx \qquad u = \sqrt{x}$

13. $\int \cos(x) \cdot e^{\sin(x)} dx$

14. $\int e^x \sin(e^x) dx$

15. $\int_1^3 \frac{5}{1+9x^2} dx \qquad u = 3x$

16. $\int_0^1 \frac{7}{1+(x+3)^2} dx \qquad u = x+3$

17. $\int_1^2 \frac{1}{x^2} \cos(\frac{1}{x}) dx \qquad u = \frac{1}{x}$

18. $\int_1^e \frac{\sec(2+\ln(x))}{x} dx \qquad u = 2 + \ln(x)$

19. $\int \frac{6 \sin(x) \cos(x)}{5 + \sin^2(x)} dx \qquad u = 5 + \sin^2(x)$

20. $\int \frac{6 \cos(x)}{5 + \sin^2(x)} dx \qquad u = \sin(x)$

21. $\int \frac{10}{2x+5} dx$

22. $\int \frac{3}{8x+1} dx$

23. $\int_{1}^{3} \dfrac{20x}{5x^2+3}\, dx$

24. $\int_{1}^{5} \dfrac{4x}{x^2+9}\, dx$

25. $\int_{0}^{1} \dfrac{7}{(x+3)^2+4}\, dx \qquad u = x+3$

26. $\int_{-2.1}^{-2.3} \dfrac{1}{\sqrt{1-(x+2)^2}}\, dx \qquad u = x+2$

27. $\int \dfrac{e^x}{1+e^{2x}}\, dx$

28. $\int \dfrac{4x+10}{x^2+5x+9}\, dx \qquad u = x^2+5x+9$

29. $\int_{1}^{e} \dfrac{3}{x \cdot \{1+\ln(x)\}}\, dx$

30. $\int_{0}^{1} \dfrac{e^x}{1+e^x}\, dx$

31. $\int_{0}^{1} 2x\sqrt{1-x^2}\, dx$

32. $\int_{0}^{3} \dfrac{2x}{\sqrt{5+x^2}}\, dx$

33. $\int \cos(x)(1+\sin(x))^3\, dx$

34. $\int \cos(x)\sin^4(x)\, dx$

35. $\int_{1}^{e} \dfrac{\sqrt{\ln(x)}}{x}\, dx$

36. $\int_{1}^{2} e^x \sqrt{2+e^x}\, dx$

37. $\int \dfrac{\sec^2(x)}{5+\tan(x)}\, dx$

38. $\int \dfrac{6x}{(x^2-1)^3}\, dx$

39. $\int \tan(x-5)\, dx$

40. $\int (x^3+3)^2\, dx$

41. $\int_{0}^{1} e^{5x}\, dx$

42. $\int \sec(2+3x)\, dx$

In problems 43 – 48, complete the square in the denominator, make the appropriate substitution, and integrate.

43. $\int \dfrac{7}{x^2+4x+5}\, dx$

44. $\int \dfrac{3}{x^2+4x+29}\, dx$

45. $\int \dfrac{2}{x^2-6x+58}\, dx$

46. $\int \dfrac{11}{x^2-2x+10}\, dx$

47. $\int \dfrac{3}{x^2+10x+29}\, dx$

48. $\int \dfrac{5}{x^2+2x+5}\, dx$

In problems 49 – 54, evaluate the first integral as a sum of two integrals.

49. $\int \dfrac{2x+11}{x^2+4x+5}\, dx = \int \dfrac{2x+4}{x^2+4x+5}\, dx + \int \dfrac{7}{x^2+4x+5}\, dx$

50. $\int \dfrac{4x+11}{x^2+4x+5}\, dx = \int \dfrac{4x+8}{x^2+4x+5}\, dx + \int \dfrac{3}{x^2+4x+5}\, dx$

51. $\int \dfrac{4x+7}{x^2-6x+10}\, dx = \int \dfrac{4x-12}{x^2-6x+10}\, dx + \int \dfrac{19}{x^2-6x+10}\, dx$

52. $\int \dfrac{6x + 28}{x^2 + 10x + 34} \, dx$

53. $\int \dfrac{6x + 5}{x^2 - 4x + 13} \, dx$

54. $\int \dfrac{4x + 9}{x^2 + 6x + 13} \, dx$

Section 8.2   **PRACTICE Answers**

Practice 1: $\int \dfrac{5}{x^2 + 8x + 25} \, dx = \int \dfrac{5}{(x+4)^2 + 9} \, dx$ (Put $u = x + 4$, $du = dx$)

$= \int \dfrac{5}{(u)^2 + 9} \, du = \dfrac{5}{3} \arctan\left(\dfrac{u}{3}\right) + C = \dfrac{5}{3} \arctan\left(\dfrac{x+4}{3}\right) + C$.

Practice 2: $\int \dfrac{4x + 21}{x^2 + 8x + 25} \, dx = \int \dfrac{4x + 16}{x^2 + 8x + 25} \, dx + \int \dfrac{5}{x^2 + 8x + 25} \, dx = (a) + (b)$

(a) Put $u = x^2 + 8x + 25$. Then $du = (2x + 8) \, dx$ so $2 \, du = (4x + 16) \, dx$.

$\int \dfrac{4x + 16}{x^2 + 8x + 25} \, dx = \int \dfrac{2}{u} \, du = 2 \cdot \ln|u| + C = 2 \cdot \ln(x^2 + 8x + 25) + C$

(b) $\int \dfrac{5}{x^2 + 8x + 25} \, dx = \dfrac{5}{3} \arctan\left(\dfrac{x+4}{3}\right) + C$. (see Practice 1)

Therefore, $\int \dfrac{4x + 21}{x^2 + 8x + 25} \, dx = \mathbf{2 \cdot \ln(x^2 + 8x + 25) + \dfrac{5}{3} \arctan\left(\dfrac{x+4}{3}\right) + C}$.

8.3 INTEGRATION BY PARTS

Integration by parts is an integration method which enables us to find antiderivatives of some new functions such as $\ln(x)$ and $\arctan(x)$ as well as antiderivatives of products of functions such as $x^2 \cdot \ln(x)$ and $e^x \cdot \sin(x)$. It is the method used to derive many of the general integral formulas in the Table of Integrals. The Integration By Parts Formula for integrals comes from the Product Rule for derivatives.

For functions $u = u(x)$ and $v = v(x)$, the Product Rule for derivatives is

$$\frac{d(uv)}{dx} = u\frac{dv}{dx} + v\frac{du}{dx} \quad \text{or, in the form using differentials,} \quad d(uv) = u\,dv + v\,du.$$

Algebraically solving for $u\,dv$, we have $u\,dv = d(uv) - v\,du$ which can then be integrated to give

$$\int u\,dv = \int d(uv) - \int v\,du = uv - \int v\,du.$$

This last formula is called the Integration By Parts Formula, and it enables us to find antiderivatives for many functions which we have not been able to integrate using the substitution method. In practice, the Integration By Parts Formula allows us to exchange the problem of finding one integral, $\int u\,dv$, for the problem of finding a different integral, $\int v\,du$. This trade of one integral for another may not look very useful, but we can often arrange the exchange so we trade a difficult integral for a much easier one.

INTEGRATION BY PARTS FORMULA

If u, v, u', and v' are continuous functions,

then $\int u\,dv = u \cdot v - \int v\,du$.

For definite integrals, the Integration By Parts Formula is

$$\int_a^b u\,dv = u \cdot v \Big|_a^b - \int_a^b v\,du = \{u(b) \cdot v(b) - u(a) \cdot v(a)\} - \int_a^b v\,du.$$

Example 1: Use Integration By Parts to evaluate $\int x \cdot \cos(x)\,dx$

and $\int_0^\pi x \cdot \cos(x)\,dx$. (Fig. 1)

Solution: Our first step is to write this integral in the form of the

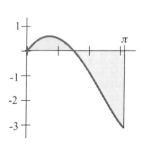

Fig. 1: $y = x \cdot \cos(x)$

Integration By Parts Formula, $\int u \, dv$. If we put $u = x$, then we **must** have $dv = \cos(x) \, dx$ so that $u \, dv$ completely represents the integrand $x \cdot \cos(x)$. In order to use integration by parts, we also need to calculate du and v:

Since $u = x$ and $dv = \cos(x) \, dx$
then $du = dx$ and $v = \sin(x)$.

Putting these pieces into the Integration By Parts Formula, we have

$$\int x \cdot \cos(x) \, dx = x \cdot \sin(x) - \int \sin(x) \, dx = x \cdot \sin(x) + \cos(x) + C.$$

(To check this result, differentiate $x \cdot \sin(x) + \cos(x)$ to verify that its derivative is $x \cdot \cos(x)$.)

$$\int_0^\pi x \cdot \cos(x) \, dx = x \cdot \sin(x) + \cos(x) \Big|_0^\pi = \{\pi \cdot \sin(\pi) + \cos(\pi)\} - \{0 \cdot \sin(0) + \cos(0)\} = -1 - 1 = -2.$$

The Integration By Parts formula allowed us to exchange the problem of evaluating $\int x \cdot \cos(x) \, dx$ for the much easier problem of evaluating $\int \sin(x) \, dx$.

Practice 1: Use the Integration By Parts Formula on $\int x \cdot \cos(x) \, dx$ with the choice $u = \cos(x)$ and $dv = x \, dx$. Why does this lead to a poor exchange?

Example 2: Use integration by parts to evaluate $\int x \cdot e^{3x} \, dx$ and $\int_0^1 x \cdot e^{3x} \, dx$. (Let $u = x$.)

Solution: Let $u = x$. Then $dv = e^{3x} \, dx$

so $du = dx$ and $v = \frac{1}{3} e^{3x}$.

Using the Integration By Parts Formula, we get

$$\int x \cdot e^{3x} \, dx = x \cdot \frac{1}{3} \cdot e^{3x} - \int \frac{1}{3} \cdot e^{3x} \, dx = \frac{x}{3} \cdot e^{3x} - \frac{1}{9} \cdot e^{3x} + C.$$

$$\int_0^1 x \cdot e^{3x} \, dx = \frac{x}{3} \cdot e^{3x} - \frac{1}{9} \cdot e^{3x} \Big|_0^1 = \left\{ \frac{1}{3} \cdot e^3 - \frac{1}{9} \cdot e^3 \right\} - \left\{ 0 - \frac{1}{9} \right\} = \frac{2}{9} \cdot e^3 + \frac{1}{9}.$$

In this Example, it is valid to choose $u = e^{3x}$ and $dv = x \, dx$, but that choice results in an integral that is more difficult than the original one. If we put $u = e^{3x}$ and $dv = x \, dx$, then $du = 3e^{3x} \, dx$ and $v = \frac{x^2}{2}$, and the Integration By Parts Formula gives

$$\int x \cdot e^{3x}\, dx = e^{3x} \cdot \frac{x^2}{2} - \int \frac{x^2}{2}\, 3e^{3x}\, dx\ .$$

We end up exchanging the integral $\int x \cdot e^{3x}\, dx$ for the more difficult integral $\int \frac{x^2}{2}\, 3e^{3x}\, dx$.

Practice 2: Evaluate $\int x \cdot \sin(x)\, dx$ and $\int x \cdot e^{5x}\, dx$. (In each integral, let $u = x$.)

Once we have chosen u and dv to represent the integrand as $u\, dv$, we need to calculate du and v. The "du" calculation is usually easy, but finding v from dv can be difficult for some choices of dv. In practice, you need to select u so dv is a simple enough part of the integrand so you can find v, the antiderivative of dv.

Example 3: Evaluate $\int 2x \cdot \ln(x)\, dx$.

Solution: The choice $u = 2x$ seems fine until we go a little further with the process. If $u = 2x$, then $dv = \ln(x)\, dx$ and we need to find du and v. Finding $du = 2\, dx$ is simple, but then we have the difficult problem of finding an antiderivative v for our choice $dv = \ln(x)\, dx$.

In this Example, the choice $u = \ln(x)$ results in easier calculations.

Let $u = \ln(x)$. Then $dv = 2x\, dx$
so $du = \frac{1}{x}\, dx$ and $v = x^2$.

Then the Integration By Parts Formula gives

$$\int 2x \cdot \ln(x)\, dx = \ln(x)\, x^2 - \int x^2 \cdot \frac{1}{x}\, dx$$

$$= x^2 \cdot \ln(x) - \int x\, dx = x^2 \cdot \ln(x) - \frac{x^2}{2} + C\ .$$

If you can not find a v for your original choice of dv, try a different choice for u and dv.

Integration by parts also enables us to evaluate the integrals of the inverse trigonometric functions and of the logarithm.

Example 4: Evaluate $\int \arctan(x)\, dx$.

Solution: Let $u = \arctan(x)$. Then $dv = dx$

so $du = \frac{1}{1 + x^2}\, dx$ and $v = x$.

Then $\int \arctan(x)\, dx = x\arctan(x) - \int x \cdot \frac{1}{1+x^2}\, dx$. We can evaluate the new integral

$\int x \cdot \frac{1}{1+x^2}\, dx$ by changing the variable using $w = 1 + x^2$. Then $dw = 2x\, dx$, so

$\int x \cdot \frac{1}{1+x^2}\, dx = \int \frac{1}{2} \frac{1}{w}\, dw = \frac{1}{2} \ln|w| = \frac{1}{2} \ln|1 + x^2|$. Putting this all together,

$$\int \arctan(x)\, dx = x\arctan(x) - \frac{1}{2}\ln|1 + x^2| + C.$$

Practice 3: Evaluate $\int \ln(x)\, dx$ and $\int_1^e \ln(x)\, dx$.

Notes:
1. Once u is chosen, then dv is completely determined: dv = rest of the integrand.
2. Since we need to find an antiderivative of dv to get v, pick u and dv so an antiderivative v can be found for the chosen dv.
3. The Integration By Parts Formula allows us to trade one integral for another one.
   (a) If the new integral is more difficult than the original integral, then we have made a poor choice of u and dv. Try a different choice for u and dv or try a different technique.
   (b) To evaluate the new integral $\int v\, du$ we may need to use substitution, integration by parts again, or some other technique such as the ones discussed later in this chapter.

More General Uses of Integration By Parts

The Integration By Parts Formula is also used to derive many of the entries in the Table of Integrals. For some integrands such as $x^n \cdot \ln(x)$, the result is simply a function, an antiderivative of the integrand. For some integrands such as $\sin^n(x)$, the result is a **reduction formula**, a formula which still contains an integral, but the new integrand is the sine function raised to a smaller power, $\sin^{n-2}(x)$. By repeatedly applying the reduction formula, we can evaluate the integral of sine raised to any positive integer power.

General Patterns

Example 5: Evaluate $\int x^n \cdot \ln(x)\, dx$ for $n \neq -1$.

Solution: Let $u = \ln(x)$. Then $dv = x^n\, dx$

so $du = \frac{1}{x}\, dx$ and $v = \frac{1}{n+1} x^{n+1}$.

Then $\int x^n \ln(x)\, dx = \frac{1}{n+1} \cdot x^{n+1} \cdot \ln(x) - \int \frac{1}{n+1} \cdot x^{n+1} \cdot \frac{1}{x}\, dx$.

But $\int \frac{1}{n+1} \cdot x^{n+1} \cdot \frac{1}{x}\, dx = \frac{1}{n+1} \int x^n\, dx = \frac{1}{n+1} \cdot \frac{1}{n+1} \cdot x^{n+1} = \frac{x^{n+1}}{(n+1)^2}$, so

$\int x^n \cdot \ln(x)\, dx = \frac{x^{n+1}}{n+1} \cdot \ln(x) - \frac{x^{n+1}}{(n+1)^2} + C = \frac{x^{n+1}}{n+1} \left\{ \ln(x) - \frac{1}{n+1} \right\} + C$ for $n \neq -1$.

Practice 4: Use the **result** of Example 5 to evaluate $\int x^2 \cdot \ln(x)\, dx$ and $\int \ln(x)\, dx$.

Reduction Formulas

Sometimes the general pattern still contains an integral, but a simpler one with a smaller exponent. In that case we can reuse the reduction pattern until the resulting integral is simple enough to integrate completely.

Example 6: Evaluate $\int x^n e^x\, dx$ and use the result to evaluate $\int x^2 e^x\, dx$.

Solution: Put $u = x^n$. Then $dv = e^x\, dx$,
so $du = n x^{n-1}\, dx$ and $v = e^x$.

The Integration By Parts Formula gives $\int x^n e^x\, dx = x^n e^x - n \int x^{n-1} e^x\, dx$, a reduction formula since we have reduced the power of x by 1 and have succeeded in trading the integral $\int x^n e^x\, dx$ for the "reduced" integral $\int x^{n-1} e^x\, dx$.

$\int x^2 e^x\, dx$ has the form of the general pattern $\int x^n e^x\, dx$ with $n = 2$, so

$$\int x^2 e^x\, dx = x^2 e^x - 2 \int x^1 e^x\, dx.$$

Using the pattern on $\int x^1 e^x\, dx$ with $n = 1$, we have

$$\int x^2 e^x\, dx = x^2 \cdot e^x - 2 \int x^1 e^x\, dx$$

$$= x^2 \cdot e^x - 2\left\{ x \cdot e^x - \int e^x\, dx \right\}$$

$$= x^2 \cdot e^x - 2x \cdot e^x + 2e^x + C \text{ or } e^x \left\{ x^2 - 2x + 2 \right\} + C.$$

Practice 5: Derive the reduction formula $\int x^n \cdot \sin(x)\, dx = -x^n \cdot \cos(x) + n \int x^{n-1} \cdot \cos(x)\, dx$.

The Reappearing Integral

Sometimes the integral we are trying to evaluate shows up on both sides of the equation during our calculations in such a way that we can solve for the integral algebraically.

Example 7: Evaluate $\int e^x \cos(x)\, dx$.

Solution: Let $u = e^x$. Then $dv = \cos(x)\, dx$
so $du = e^x\, dx$ and $v = \sin(x)$.

Then $\int e^x \cos(x)\, dx = e^x \sin(x) - \int e^x \sin(x)\, dx$. The new integral does not look any easier than the original one, but lets try to evaluate the new integral using integration by parts again.

To evaluate $\int e^x \sin(x)\, dx$., let $u = e^x$ and $dv = \sin(x)\, dx$. Then $du = e^x\, dx$ and $v = -\cos(x)$ so

$$\int e^x \sin(x)\, dx = -e^x \cos(x) + \int e^x \cos(x)\, dx.$$

Putting this result back into the original problem, we get

$$\int e^x \cos(x)\, dx = e^x \sin(x) - \int e^x \sin(x)\, dx = e^x \sin(x) - \left\{ -e^x \cos(x) + \int e^x \cos(x)\, dx \right\}$$

$$= e^x \sin(x) + e^x \cos(x) - \int e^x \cos(x)\, dx.$$

The integral of $e^x \cos(x)$ appears on each side of this last equation, and we can algebraically solve for it to get

$$2 \int e^x \cos(x)\, dx = e^x \sin(x) + e^x \cos(x) \text{, and finally,}$$

$$\int e^x \cos(x)\, dx = \frac{1}{2} \left\{ e^x \sin(x) + e^x \cos(x) \right\} + C.$$

Practice 6: Derive the formula $\int e^x \sin(x)\, dx = \frac{1}{2} \left\{ e^x \sin(x) - e^x \cos(x) \right\} + C$.

PROBLEMS

In problems 1–6, a function u or dv is given. Find the piece u or dv which is not given, calculate du and v, and apply the Integration by Parts Formula.

1. $\int 12x \cdot \ln(x)\, dx$ $u = \ln(x)$

2. $\int x \cdot e^{-x}\, dx$ $u = x$

3. $\int x^4 \ln(x)\, dx$ $dv = x^4\, dx$

4. $\int x \cdot \sec^2(3x)\, dx$ $dv = \sec^2(3x)\, dx$

5. $\int x \cdot \arctan(x)\, dx$ $dv = x\, dx$

6. $\int x \cdot (5x+1)^{19}\, dx$ $u = x$

In problems 7–24, evaluate the integrals.

7. $\int_0^1 \dfrac{x}{e^{3x}}\, dx$

8. $\int_0^1 10x \cdot e^{3x}\, dx$

9. $\int x \cdot \sec(x) \cdot \tan(x)\, dx$

10. $\int_0^\pi 5x \cdot \sin(2x)\, dx$

11. $\int_{\pi/3}^{\pi/2} 7x \cdot \cos(3x)\, dx$

12. $\int 6x \cdot \sin(x^2 + 1)\, dx$

13. $\int 12x \cdot \cos(3x^2)\, dx$

14. $\int x^2 \cos(x)\, dx$

15. $\int_1^3 \ln(2x + 5)\, dx$

16. $\int x^3 \ln(5x)\, dx$

17. $\int_1^e (\ln(x))^2\, dx$

18. $\int_1^e \sqrt{x} \cdot \ln(x)\, dx$

19. $\int \arcsin(x)\, dx$

20. $\int x^2 e^{5x}\, dx$

21. $\int x \cdot \arctan(3x)\, dx$

22. $\int x \ln(x + 1)\, dx$

23. $\int_1^2 \dfrac{\ln(x)}{x}\, dx$

24. $\int_1^2 \dfrac{\ln(x)}{x^2}\, dx$

These reduction formulas can all be derived using integration by parts. In problems 25–30, use them to help evaluate the integrals. (These are entries 19, 20 and 23 in the Table of Integrals with $a = 1$.)

$$\int \sin^n(x)\, dx = \dfrac{1}{n}\left\{ -\sin^{n-1}(x) \cdot \cos(x) + (n-1) \int \sin^{n-2}(x)\, dx \right\} + C$$

$$\int \cos^n(x)\, dx = \dfrac{1}{n}\left\{ \cos^{n-1}(x) \cdot \sin(x) + (n-1) \int \cos^{n-2}(x)\, dx \right\} + C$$

$$\int \sec^n(x)\, dx = \dfrac{1}{n-1}\left\{ \sec^{n-2}(x) \cdot \tan(x) + (n-2) \int \sec^{n-2}(x)\, dx \right\} + C$$

25. (a) $\int \sin^3(x)\,dx$ (b) $\int \sin^4(x)\,dx$ (c) $\int \sin^5(x)\,dx$

26. (a) $\int \cos^3(x)\,dx$ (b) $\int \cos^4(x)\,dx$ (c) $\int \cos^5(x)\,dx$

27. (a) $\int \sec^3(x)\,dx$ (b) $\int \sec^4(x)\,dx$ (c) $\int \sec^5(x)\,dx$

28. $\int \sin^3(5x-2)\,dx$ 29. $\int \cos^3(2x+3)\,dx$ 30. $\int \sec^3(7x-1)\,dx$

31. $\int x\cdot(2x+5)^{19}\,dx$ can be evaluated using integration by parts or a change of variable. (a) Evaluate the integral using integration by parts with $u = x$ and $dv = (2x+5)^{19}\,dx$. (b) Evaluate the integral using change of variable with $u = 2x+5$. (c) Which method is easier?

32. $\int \dfrac{x}{\sqrt{1+x}}\,dx$ can be evaluated using integration by parts or using a change of variable. (a) Evaluate the integral using integration by parts with $u = x$ and $dv = \dfrac{1}{\sqrt{1+x}}\,dx$. (b) Evaluate the integral using change of variable with $u = 1+x$.

33. (a) Before evaluating the integrals, which do you think is larger, $\int_0^1 x\cdot\sin(x)\,dx$ or $\int_0^1 \sin(x)\,dx$? Why?

    (b) Evaluate $\int_0^1 x\cdot\sin(x)\,dx$ and $\int_0^1 \sin(x)\,dx$. Was your prediction in part (a) correct?

34. (a) Before evaluating the integrals, which do you think is larger, $\int_0^\pi x\cdot\sin(x)\,dx$ or $\int_0^\pi \sin(x)\,dx$? Why?

    (b) Evaluate $\int_0^\pi x\cdot\sin(x)\,dx$ and $\int_0^\pi \sin(x)\,dx$. Was your prediction in part (a) correct?

35. In Fig. 2, the volume swept out when region A is revolved about the x–axis is $\int_{x=1}^{e} \pi(\ln(x))^2\,dx$ (using the disk method), and the volume swept out when region B is revolved about the x–axis is

    $\int_{y=0}^{1} 2\pi y\cdot e^y\,dy$ (using the tube method).

    (a) Before evaluating the integrals, which volume do you think is larger? Why?

    (b) Evaluate the integrals. Was your prediction in part (a) correct?

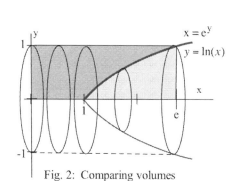

Fig. 2: Comparing volumes

36. Use the tube method to calculate the volume when the region between the x–axis and the graph of $y = \sin(x)$ for $0 \leq x \leq \pi$ is rotated about the y–axis.

37. We derived the Integration by Parts Formula analytically, but the formula also has a geometric interpretation. In Fig. 3, let D be the large rectangle formed by the regions A, B, and C so we have the area equation

    (area of C) = (area of D) – (area of A) – (area of B).

    (a) Represent the area of the large rectangle D as a function of u_2 and v_2.

    (b) Represent the area of the small rectangle (region A) as a function of u_1 and v_1.

    (c) Represent the area of region C as an integral with respect to the variable u.

    (d) Represent the area of region B as an integral with respect to the variable v.

    (e) Rewrite the area equation using the representations in parts (a) – (d). This result should look very familiar.

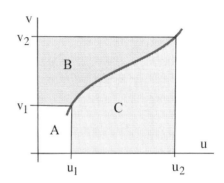

D is the region obtained by putting together regions A, B and C

Fig. 3: Graphic version of Integration by Parts

38. $\int x \cdot (\ln(x))^2 \, dx$

39. $\int x^2 \cdot \arctan(x) \, dx$

40. $\int_0^1 e^{-x} \sin(x) \, dx$

41. $\int_0^1 \frac{\cos(x)}{e^x} \, dx$

42. $\int \sin(\ln(x)) \, dx$

43. $\int \cos(\ln(x)) \, dx$

44. $\int e^{3x} \sin(x) \, dx$

45. $\int e^x \cos(3x) \, dx$

46. Use integration by parts to evaluate $\int \sec^3(x) \, dx$.

47. Derive a reduction formula for $\int x^n e^{ax} \, dx$.

48. Derive a reduction formula for $\int x^n \sin(ax) \, dx$.

49. Derive a reduction formula for $\int x \cdot (\ln(x))^n \, dx$.

50. Suppose f and f' are continuous and bounded on the interval $[0, 2\pi]$ ($|f(x)| < M$ and $|f'(x)| < M$ for all $0 \le x \le 2\pi$). The nth Fourier Sine Coefficient of f is defined as the value of

$$S_n = \int_0^{2\pi} f(x) \cdot \sin(nx)\, dx .$$

(a) Use the Integration by Parts Formula with $u = f(x)$ and $dv = \sin(nx)\, dx$ to represent the formula for S_n in a different way.

(b) Use the new representation of S_n in part (a) to determine what happens to the values of S_n when n is very large ($n \to \infty$). (Hint: $|f'(x) \cdot \cos(nx)| \le |f'(x)| \cdot |\cos(nx)| < M \cdot 1 = M$.)

(c) What happens to the values of the nth Fourier Cosine Coefficient $C_n = \int_0^{2\pi} f(x) \cdot \cos(nx)\, dx$

when n is very large.

Section 8.3   PRACTICE Answers

Practice 1: $\int_a^b x \cdot \cos(x)\, dx$. Put $u = \cos(x)$ and $dv = x\, dx$. Then $du = -\sin(x)\, dx$ and $v = \frac{1}{2} x^2$.

$$uv - \int v\, du = (\cos(x))(\tfrac{1}{2} x^2) - \int (\tfrac{1}{2} x^2)(-\sin(x))\, dx$$

$$= \tfrac{1}{2} x^2 \cos(x) + \tfrac{1}{2} \int x^2 \sin(x)\, dx .$$

The last integral is worse than the original integral.

Practice 2: (a) $\int_a^b x \cdot \sin(x)\, dx$. Put $u = x$ and $dv = \sin(x)\, dx$. Then $du = dx$ and $v = -\cos(x)\, dx$.

$$uv - \int v\, du = (x)(-\cos(x)) - \int -\cos(x)\, dx = -\mathbf{x \cdot \cos(x) + \sin(x) + C} .$$

(b) $\int_a^b x \cdot e^{5x}\, dx$. Put $u = x$ and $dv = e^{5x}\, dx$. Then $du = dx$ and $v = \tfrac{1}{5} e^{5x}$.

$$uv - \int v\, du = (x)(\tfrac{1}{5} e^{5x}) - \int \tfrac{1}{5} e^{5x}\, dx = \mathbf{\tfrac{1}{5} x \cdot e^{5x} - \tfrac{1}{25} e^{5x} + C} .$$

Practice 3: $\int \ln(x)\, dx$. Put $u = \ln(x)$ and $dv = dx$. Then $du = \tfrac{1}{x}\, dx$ and $v = x$.

$$uv - \int v\, du = \ln(x) \cdot x - \int x \cdot \tfrac{1}{x}\, dx = x \cdot \ln(x) - \int 1\, dx = \mathbf{x \cdot \ln(x) - x + C} .$$

$$\int_1^e \ln(x)\, dx = x \cdot \ln(x) - x \Big|_1^e = \{e \cdot \ln(e) - e\} - \{1 \cdot \ln(1) - 1\} = \{e - e\} - \{0 - 1\} = 1 .$$

Practice 4: Example 5: $\int x^n \ln(x)\, dx = \dfrac{x^{n+1}}{n+1} \left\{ \ln(x) - \dfrac{1}{n+1} \right\} + C$.

(a) $n = 2$: $\int x^2 \ln(x)\, dx = \dfrac{x^3}{3} \left\{ \ln(x) - \dfrac{1}{3} \right\} + C$.

(b) $n = 0$: $\int \ln(x)\, dx = \dfrac{x^1}{1} \left\{ \ln(x) - \dfrac{1}{1} \right\} + C = x \cdot \ln(x) - x + C$.

Practice 5: $\int x^n \sin(x)\, dx$. Put $u = x^n$ and $dv = \sin(x)\, dx$. Then $du = nx^{n-1}\, dx$ and $v = -\cos(x)$.

$uv - \int v\, du = (x^n)(-\cos(x)) - \int -\cos(x)\, nx^{n-1}\, dx$

$= -x^n \cos(x) + n \int x^{n-1} \cos(x)\, dx$.

Practice 6: (Similar to Example 7)

$\int e^x \sin(x)\, dx$. Put $u = e^x$ and $dv = \sin(x)\, dx$. Then $du = e^x\, dx$ and $v = -\cos(x)\, dx$.

$uv - \int v\, du = (e^x)(-\cos(x)) - \int -\cos(x)\, e^x\, dx$

$= -e^x \cos(x) + \int e^x \cos(x)\, dx$.

(For the last integral, put $u = e^x$, $dv = \cos(x)\, dx$, $du = e^x\, dx$, $v = \sin(x)\, dx$)

$= -e^x \cos(x) + \int e^x \cos(x)\, dx = -e^x \cos(x) + \left\{ e^x \sin(x) - \int e^x \sin(x)\, dx \right\}$

So $2 \int e^x \sin(x)\, dx = -e^x \cos(x) + e^x \sin(x)$

and $\int e^x \sin(x)\, dx = \dfrac{1}{2} \left\{ -e^x \cos(x) + e^x \sin(x) \right\} + C$.

8.4 PARTIAL FRACTION DECOMPOSITION

Rational functions (polynomials divided by polynomials) and their integrals are important in mathematics and applications, but if you look through a table of integral formulas, you will find very few formulas for their integrals. Partly that is because the general formulas are rather complicated and have many special cases, and partly it is because they can all be reduced to just a few cases using the algebraic technique discussed in this section, Partial Fraction Decomposition.

In algebra you learned to add rational functions to get a single rational function. Partial Fraction Decomposition is a technique for reversing that procedure to "decompose" a single rational function into a **sum** of simpler rational functions. Then the integral of the single rational function can be evaluated as the sum of the integrals of the simpler functions.

Example 1: Use the algebraic decomposition $\dfrac{17x - 35}{2x^2 - 5x} = \dfrac{7}{x} + \dfrac{3}{2x - 5}$ to evaluate $\displaystyle\int \dfrac{17x - 35}{2x^2 - x}\, dx$.

Solution: The decomposition allows us to exchange the original integral for two much easier ones:

$$\int \frac{17x-35}{2x^2-5x}\, dx = \int \frac{7}{x} + \frac{3}{2x-5}\, dx = \int \frac{7}{x}\, dx + \int \frac{3}{2x-5}\, dx$$

$$= 7\ln|x| + \frac{3}{2}\ln|2x-5| + C.$$

Practice 1: Use the algebraic decomposition $\dfrac{7x - 11}{3x^2 - 8x - 3} = \dfrac{4}{3x + 1} + \dfrac{1}{x - 3}$ to evaluate $\displaystyle\int \dfrac{7x - 11}{3x^2 - 8x - 3}\, dx$.

The Example illustrates how to use a "decomposed" fraction with integrals, but it does not show how to achieve the decomposition. The algebraic basis for the Partial Fraction Decomposition technique is that every polynomial can be factored into a product of linear factors $ax + b$ and irreducible quadratic factors $ax^2 + bx + c$ (with $b^2 - 4ac < 0$). These factors may not be easy to find, and they will typically be more complicated than the examples in this section, but every polynomial has such factors. Before we apply the Partial Fraction Decomposition technique, the fraction must have the following form:

> (i) (the degree of the numerator) < (degree of the denominator)
>
> (ii) The denominator has been factored into a product of linear factors and irreducible quadratic factors.

If assumption (i) is not true, we can use polynomial division until we get a remainder which has a smaller degree than the denominator. If assumption (ii) is not true, we simply cannot use the Partial Fraction Decomposition technique.

Example 2: Put each fraction into a form for Partial Fraction Decomposition:

(a) $\dfrac{2x^2 + 4x - 6}{x^2 - 2x}$ 　　(b) $\dfrac{3x^3 - 3x^2 - 9x + 8}{x^2 - x - 6}$ 　　(c) $\dfrac{7x^2 + 12x - 12}{x^3 - 4x}$

Solution: (a) $\dfrac{2x^2 + 4x - 6}{x^2 - 2x} = 2 + \dfrac{8x - 6}{x^2 - 2x} = 2 + \dfrac{8x - 6}{x(x - 2)}$

(b) $\dfrac{3x^3 - 3x^2 - 9x + 8}{x^2 - x - 6} = 3x + \dfrac{9x + 8}{x^2 - x - 6} = 3x + \dfrac{9x + 8}{(x + 2)(x - 3)}$

(c) $\dfrac{7x^2 + 12x - 12}{x^3 - 4x} = \dfrac{7x^2 + 12x - 12}{x(x + 2)(x - 2)}$

Distinct Linear Factors

If the denominator can be factored into a product of distinct linear factors, then the original fraction can be written as the **sum** of fractions of the form $\dfrac{\text{number}}{\text{linear factor}}$. Our job is to find the values of the numbers in the numerators, and that typically requires solving a system of equations.

Example 3: Find values for A and B so $\dfrac{17x - 35}{x(2x - 5)} = \dfrac{A}{x} + \dfrac{B}{2x - 5}$.

Solution: We can combine the two terms on the right by putting them over the common denominator $x(2x - 5)$. Multiplying the $\dfrac{A}{x}$ term by $\dfrac{2x - 5}{2x - 5}$ and multiplying the $\dfrac{B}{2x - 5}$ term by $\dfrac{x}{x}$, we have

$$\dfrac{A}{x} \cdot \dfrac{2x - 5}{2x - 5} + \dfrac{B}{2x - 5} \cdot \dfrac{x}{x} = \dfrac{A2x - 5A + Bx}{x(2x - 5)} = \dfrac{(2A + B)x - 5A}{x(2x - 5)}.$$

Since $\dfrac{(2A + B)x - 5A}{x(2x - 5)} = \dfrac{17x - 35}{x(2x - 5)}$, the coefficients of like terms in the numerators must be equal:

　　coefficients of x:　　$2A + B = 17$
　　constant terms:　　$-5A = -35$

Solving this system of two equations with two unknowns, we get $A = 7$ and $B = 3$ so

$$\dfrac{17x - 35}{x(2x - 5)} = \dfrac{7}{x} + \dfrac{3}{2x - 5}.$$

As a check, add $\frac{7}{x}$ and $\frac{3}{2x-5}$ and verify that the sum is $\frac{17x-35}{x(2x-5)}$.

Practice 2: Find values of A and B so $\frac{6x-7}{(x+3)(x-2)} = \frac{A}{x+3} + \frac{B}{x-2}$.

In general, there is one unknown coefficient for each distinct linear factor of the denominator. However, if the number of distinct linear factors is large, we would need to solve a large system of equations for the unknowns.

Example 4: Find values for A, B, and C so $\frac{2x^2+7x+9}{x(x+1)(x+3)} = \frac{A}{x} + \frac{B}{x+1} + \frac{C}{x+3}$.

Solution:
$$\frac{A}{x} + \frac{B}{x+1} + \frac{C}{x+3} = \frac{A}{x}\frac{(x+1)(x+3)}{(x+1)(x+3)} + \frac{B}{(x+1)}\frac{x(x+3)}{x(x+3)} + \frac{C}{(x+3)}\frac{x(x+1)}{x(x+1)}$$

$$= \frac{A(x+1)(x+3) + Bx(x+3) + Cx(x+1)}{x(x+1)(x+3)}$$

$$= \frac{(A+B+C)x^2 + (4A+3B+C)x + (3A)}{x(x+1)(x+3)} = \frac{2x^2+7x+9}{x(x+1)(x+3)}.$$

The coefficients of the like terms in the numerators must be equal:

coefficients of x^2: $A + B + C = 2$
coefficients of x: $4A + 3B + C = 7$
constant terms: $3A = 9$ so $A = 3, B = -2$, and $C = 1$.

Finally, $\frac{2x^2+7x+9}{x(x+1)(x+3)} = \frac{3}{x} + \frac{-2}{x+1} + \frac{1}{x+3}$.

Practice 3: Use the result of Example 4 to evaluate $\int \frac{2x^2+7x+9}{x(x+1)(x+3)} \, dx$.

Practice 4: How large would the system be for a Partial Fraction Decomposition of $\frac{\text{something}}{5^{\text{th}} \text{ degree polynomial}}$?

The next two subsections describe how to decompose fractions whose denominators contain irreducible quadratic factors and repeated factors. We will not discuss why the suggestions work except to note that they provide enough, but not too many, unknown coefficients for the decomposition.

Distinct Irreducible Quadratic Factors

If the factored denominator includes a distinct irreducible quadratic factor, then the Partial Fraction Decomposition **sum** contains a fraction of the form of a linear polynomial with unknown coefficients divided by the irreducible quadratic factor:

$$\frac{\text{linear polynomial}}{\text{irreducible quadratic factor}} \quad \text{or} \quad \frac{Ax + B}{\text{irreducible quadratic factor}}.$$

Once again we will solve a system of equations to find the values of the unknown coefficients A and B.

Example 5: Find values for A, B, and C so $\dfrac{x^2 + 3x - 15}{(x^2 + 2x + 5)x} = \dfrac{Ax + B}{x^2 + 2x + 5} + \dfrac{C}{x}$.

Solution:
$$\frac{Ax + B}{x^2 + 2x + 5} + \frac{C}{x} = \frac{Ax + B}{x^2 + 2x + 5}\left(\frac{x}{x}\right) + \frac{C}{x}\left(\frac{x^2 + 2x + 5}{x^2 + 2x + 5}\right)$$

$$= \frac{Ax^2 + Bx + Cx^2 + 2Cx + 5C}{x(x^2 + 2x + 5)}$$

$$= \frac{(A + C)x^2 + (B + 2C)x + 5C}{x(x^2 + 2x + 5)} = \frac{x^2 + 3x - 15}{(x^2 + 2x + 5)x}.$$

Then $A + C = 1$, $B + 2C = 3$, and $5C = -15$ so $C = -3$, $B = 9$, and $A = 4$.

In general, there are 2 unknown coefficients for each distinct irreducible quadratic factor of the denominator. We would start the decomposition of

$$\frac{6x^3 + 36x^2 + 50x + 53}{(x^2 + 4)(x^2 + 4x + 5)} \quad \text{by writing it as the sum} \quad \frac{Ax + B}{x^2 + 4} + \frac{Cx + D}{x^2 + 4x + 5}.$$

We would finish this decomposition by solving the system of 4 equations with 4 unknowns, $A + C = 6$, $4A + B + D = 36$, $5A + 4B + 4C = 50$, and $5B + 4D = 53$ to get $A = 6$, $B = 5$, $C = 0$, and $D = 7$.

Repeated Factors

If the factored denominator contains a linear factor raised to a power (greater than one), then we need to start the decomposition with several terms. There should be one term with one unknown coefficient for each power of the linear factor. For example,

$$\frac{\text{something}}{(x + 1)(x - 2)^3} = \frac{A}{x + 1} + \frac{B}{x - 2} + \frac{C}{(x - 2)^2} + \frac{D}{(x - 2)^3}.$$

Similarly, if the factored denominator contains an irreducible quadratic factor raised to a power greater than one), then we need to start the decomposition with several terms. There should be one term with two unknown coefficients for each power of the irreducible quadratic. For example,

$$\frac{\text{something}}{x^2(x^2+9)^3} = \frac{A}{x} + \frac{B}{x^2} + \frac{Cx+D}{x^2+9} + \frac{Ex+F}{(x^2+9)^2} + \frac{Gx+H}{(x^2+9)^3}.$$

This leads to a system of 8 equations with 8 unknowns.

Example 6: Decompose $\frac{-4x^2+5x+3}{x(x-1)^2}$ and evaluate $\int \frac{-4x^2+5x+3}{x(x-1)^2} \, dx$.

Solution: $\frac{-4x^2+5x+3}{x(x-1)^2} = \frac{A}{x} + \frac{B}{x-1} + \frac{C}{(x-1)^2}$

$$= \frac{A}{x}\left(\frac{(x-1)^2}{(x-1)^2}\right) + \frac{B}{x-1}\left(\frac{x(x-1)}{x(x-1)}\right) + \frac{C}{(x-1)^2}\left(\frac{x}{x}\right)$$

$$= \frac{A(x-1)^2 + Bx(x-1) + Cx}{x(x-1)^2}$$

$$= \frac{(A+B)x^2 + (-2A-B+C)x + A}{x(x-1)^2} = \frac{-4x^2+5x+3}{x(x-1)^2}.$$

Then $A+B=-4$, $-2A-B+C=5$, and $A=3$ so $A=3$, $B=-7$, and $C=4$. Finally,

$$\int \frac{-4x^2+5x+3}{x(x-1)^2} \, dx = \int \frac{3}{x} + \frac{-7}{x-1} + \frac{4}{(x-1)^2} \, dx = 3\ln|x| - 7\ln|x-1| + \frac{-4}{x-1} + C.$$

Practice 5: Decompose $\frac{2x^2+27x+85}{(x+5)^2}$ and evaluate $\int \frac{2x^2+27x+85}{(x+5)^2} \, dx$.

The primary use of the partial fraction technique in this course is to put rational functions in a form that is easier to integrate, but this algebraic technique can also be used to simplify the differentiation of some rational functions. The next example illustrates the use of partial fractions to make a differentiation problem easier.

Example 7: For $f(x) = \frac{2x+13}{x^2+x-2}$, calculate $f'(x)$, $f''(x)$, and $f'''(x)$.

Solution: You already know how to calculate these derivatives using the quotient rule, but that process is rather tedious for the second and third derivatives here. Instead, we can use the partial fraction technique to rewrite f as $f(x) = \frac{5}{x-1} - \frac{3}{x+2} = 5(x-1)^{-1} - 3(x+2)^{-1}$. Then the derivatives are very straightforward:

$$f'(x) = -5(x-1)^{-2} + 3(x+2)^{-2},$$
$$f''(x) = 10(x-1)^{-3} - 6(x+2)^{-3}, \text{ and}$$
$$f'''(x) = -30(x-1)^{-4} + 18(x+2)^{-4}.$$

Practice 6: Use the partial fraction decomposition of $g(x) = \frac{9x+1}{x^2 - 2x - 3}$ to calculate $g'(x)$, $g''(x)$, and $g^{(4)}(x)$.

PROBLEMS

In problems 1 – 12, decompose the fractions.

1. $\dfrac{7x+2}{x(x+1)}$
2. $\dfrac{7x+9}{(x+3)(x-1)}$
3. $\dfrac{11x+25}{x^2+9x+8}$
4. $\dfrac{3x+7}{x^2-1}$

5. $\dfrac{2x^2+15x+25}{x^2+5x}$
6. $\dfrac{3x^3+3x^2}{x^2+x-2}$
7. $\dfrac{6x^2+9x-15}{x(x+5)(x-1)}$
8. $\dfrac{6x^2-x-1}{x^3-x}$

9. $\dfrac{8x^2-x+3}{x^3+x}$
10. $\dfrac{9x^2+13x+15}{x^3+2x^2-3x}$
11. $\dfrac{11x^2+23x+6}{x^2(x+2)}$
12. $\dfrac{6x^2+14x-9}{x(x+3)^2}$

In problems 13 – 30, evaluate the integrals.

13. $\int \dfrac{3x+13}{(x+2)(x-5)}\, dx$
14. $\int \dfrac{2x+11}{(x-7)(x-2)}\, dx$
15. $\int_2^5 \dfrac{2}{x^2-1}\, dx$

16. $\int_1^3 \dfrac{5x^2+5x+3}{x^3+x}\, dx$
17. Integrate the functions in problems 1 – 4.

18. Integrate the functions in problems 5 – 8.

19. $\int \dfrac{2x^2+5x+3}{x^2-1}\, dx$
20. $\int \dfrac{2x^2+19x+22}{x^2+x-12}\, dx$
21. $\int \dfrac{3x^2+19x+24}{x^2+6x+5}\, dx$

22. $\int \dfrac{7x^2+8x-2}{x^2+2x}\, dx$
23. $\int \dfrac{3x^2-1}{x^3-x}\, dx$
24. $\int \dfrac{x^4+5x^3+x-15}{x^2+5x}\, dx$

25. $\int \dfrac{x^3 + 3x^2 - 4x + 30}{x^2 + 3x - 10} \, dx$ 26. $\int \dfrac{2x + 5}{(x + 1)^2} \, dx$ 27. $\int \dfrac{12x^2 + 19x - 6}{x^3 + 3x^2} \, dx$

28. $\int \dfrac{7x^3 + x^2 + 7x + 10}{x^4 + 2x^3} \, dx$ 29. $\int \dfrac{7x^2 + 3x + 7}{x^3 + x} \, dx$ 30. $\int \dfrac{7x^2 - 4x + 4}{x^3 + 1} \, dx$

31. Integrals are very sensitive to small changes in the integrand. Evaluate

(a) $\int \dfrac{1}{x^2 + 2x + \mathbf{2}} \, dx$ (b) $\int \dfrac{1}{x^2 + 2x + \mathbf{1}} \, dx$ (c) $\int \dfrac{1}{x^2 + 2x + \mathbf{0}} \, dx$.

32. Evaluate (a) $\int \dfrac{1}{x^2 - 6x + \mathbf{8}} \, dx$ (b) $\int \dfrac{1}{x^2 - 6x + \mathbf{9}} \, dx$ (c) $\int \dfrac{1}{x^2 - 6x + \mathbf{10}} \, dx$.

33. Use the partial fraction decomposition of the functions in problems 1 and 2 to calculate their first and second derivatives.

34. Use the partial fraction decomposition of the functions in problems 3 and 4 to calculate their first and second derivatives.

35. Use the partial fraction decomposition of the functions in problems 5 and 6 to calculate their first and second derivatives.

The following two applications involve a type of differential equation which can be solved by separating the variables and using a partial fraction decomposition to help calculate the antiderivatives. The same type of differential equation is also used to model the spread of rumors and diseases as well as some populations and chemical reactions.

Logistic Growth: The growth rate of many different populations depends not only on the number of individuals (leading to exponential growth) but also on a "carrying capacity" of the environment. If x is the population at time t and the growth rate of x is proportional to the **product** of the population and the carrying capacity M minus the population, then the growth rate is described by the differential equation

$$\dfrac{dx}{dt} = k \cdot x \cdot (M - x)$$

where k and M are constants for a given species in a given environment.

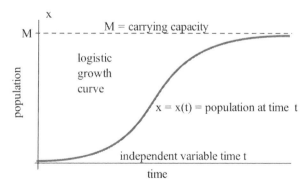

Fig. 1: Logistic growth curve

36. Let $k = 1$ and $M = 100$, and assume the initial population is $x(0) = 5$.

   (a) Solve the differential equation $\frac{dx}{dt} = x(100 - x)$ for x.

   (b) Graph the population $x(t)$ for $0 \le t \le 20$.

   (c) When will the population be 20? 50? 90? 100?

   (d) What is the population after a "long" time? (Find the limit, as t becomes arbitrarily large, of x.)

   (e) Explain the shape of the graph in (a) in terms of a population of bacteria.

   (f) When is the growth rate largest? (Maximize dx/dt.)

   (g) What is the population when the growth rate is largest?

37. Let $k = 1$ and $M = 100$, and assume the initial population is $x(0) = 150$.

   (a) Solve the differential equation $\frac{dx}{dt} = x(100 - x)$ for x and graph $x(t)$ for $0 \le t \le 20$.

   (b) When will the population be 120? 110? 100?

   (c) What is the population after a "long" time? (Find the limit, as t becomes arbitrarily large, of x.)

   (d) Explain the shape of the graph in (a) in terms of a population of bacteria.

38. Let k and M be positive constants, and assume the initial population is $x(0) = x_0$.

   (a) Solve the differential equation $\frac{dx}{dt} = k \cdot x \cdot (M - x)$ for x.

   (b) What is the population after a "long" time? (Find the limit, as t becomes arbitrarily large, of x.)

   (c) When is the growth rate largest? (Maximize dx/dt.)

   (d) What is the population when the growth rate is largest?

Chemical Reaction: In some chemical reactions, a new material X is formed from materials A and B, and the rate at which X forms is proportional to the **product** of the amount of A and the amount of B remaining in the solution. Let x represent the amount of material X present at time t, and assume that the reaction begins with a grams of A, b grams of B, and no material X ($x(0) = 0$). Then the rate of formation of material X can be described by the differential equation

$$\frac{dx}{dt} = k(a - x)(b - x).$$

39. Solve the differential equation $\frac{dx}{dt} = k(a - x)(b - x)$ for x if $k = 1$ and the reaction begins with

   (i) 7 grams of A and 5 grams of B, and (ii) 6 grams of A and 6 grams of B.

40. Solve the differential equation $\frac{dx}{dt} = k(a - x)(b - x)$ for x if $k = 1$ and the reaction begins with

   (i) a grams of A and b grams of B with $a \ne b$, and (ii) c grams of A and c grams of B ($c \ne 0$).

Section 8.4   PRACTICE Answers

Practice 1: $\int \dfrac{7x-11}{3x^2-8x-3}\,dx = \int \dfrac{4}{3x+1}\,dx + \int \dfrac{1}{x-3}\,dx$

$$= \dfrac{4}{3}\cdot\ln|3x+1| + \ln|x-3| + C.$$

Practice 2: $\dfrac{6x-7}{(x+3)(x-2)} = \dfrac{A}{x+3} + \dfrac{B}{x-2} = \dfrac{A(x-2)+B(x+3)}{(x+3)(x-2)} = \dfrac{(A+B)x+(-2A+3B)}{(x+3)(x-2)}.$

This gives us the system: $A+B=6$ and $-2A+3B=-7$ so (solving) **A = 5** and **B = 1**.

$$\dfrac{6x-7}{(x+3)(x-2)} = \dfrac{5}{x+3} + \dfrac{1}{x-2}.$$

Practice 3: From Example 4,

$$\int \dfrac{2x^2+7x+7}{x(x+1)(x+3)}\,dx = \int \dfrac{3}{x} + \dfrac{-2}{x+1} + \dfrac{1}{x+3}\,dx$$

$$= 3\cdot\ln|x| - 2\cdot\ln|x+1| + \ln|x+3| + C.$$

Practice 4: If the 5th degree polynomial can be factored into a product of 5 distinct linear terms, then we would have

$$\dfrac{\text{something}}{5^{\text{th}}\text{ degree polynomial}} = \dfrac{A}{1^{\text{st}}\text{ term}} + \dfrac{B}{2^{\text{nd}}\text{ term}} + \ldots + \dfrac{E}{5^{\text{th}}\text{ term}}.$$

Practice 5: $\dfrac{2x^2+27x+85}{(x+5)^2} = \dfrac{2x^2+27x+85}{x^2+10x+25} = 2 + \dfrac{7x+35}{(x+5)^2} = 2 + \dfrac{7}{x+5}.$

Then $\int \dfrac{2x^2+27x+85}{(x+5)^2}\,dx = \int 2 + \dfrac{7}{x+5}\,dx = \mathbf{2x + 7\cdot\ln|x+5| + C}.$

Practice 6: $g(x) = \dfrac{9x+1}{(x-3)(x+1)} = \dfrac{A}{x-3} + \dfrac{B}{x+1} = \dfrac{A(x+1)+B(x-3)}{(x-3)(x+1)} = \dfrac{(A+B)x+(A-3B)}{(x-3)(x+1)}.$

This gives us the system $A+B=9$ and $A-3B=1$ so (solving) $A=7$ and $B=2$.

$g(x) = \dfrac{7}{x-3} + \dfrac{2}{x+1} = 7(x-3)^{-1} + 2(x+1)^{-1}$. Then

$g'(x) = -7(x-3)^{-2} - 2(x+1)^{-2}$,

$g''(x) = 14(x-3)^{-3} + 4(x+1)^{-3}$,

$g'''(x) = -42(x-3)^{-4} - 12(x+1)^{-4}$, and

$g''''(x) = 168(x-3)^{-5} + 48(x+1)^{-5}.$

8.5 Trigonometric Substitution — Another Change of Variable

Changing the variable is a very powerful technique for finding antiderivatives, and by now you have probably found a lot of integrals by setting u = something. This section also involves a change of variable, but for more specialized patterns, and the change is more complicated. Another difference from previous work is that instead of setting u equal to a function of x we will be replacing x with a function of θ.

The next three examples illustrate the typical steps involved making trigonometric substitutions. After these examples, we examine each step in more detail and consider how to make the appropriate decisions.

Example 1: In the expression $\sqrt{9-x^2}$ replace x with $3\sin(\theta)$ and simplify the result.

Solution: Replacing x with $3\sin(\theta)$, $\sqrt{9-x^2}$ becomes

$$\sqrt{9-(3\sin(\theta))^2} = \sqrt{9-9\sin^2(\theta)} = \sqrt{9\cdot(1-\sin^2(\theta))} = 3\cos(\theta).$$

Example 2: Evaluate $\int \sqrt{9-x^2}\, dx$ using the change of variable $x = 3\sin(\theta)$ and then use the antiderivative to evaluate $\int_0^3 \sqrt{9-x^2}\, dx$.

Solution: If $x = 3\sin(\theta)$, then $dx = 3\cos(\theta)\, d\theta$ and $\sqrt{9-x^2} = 3\cos(\theta)$. With this change of variable, the integral becomes

$$\int \sqrt{9-x^2}\, dx = \int 3\cos(\theta)\, 3\cos(\theta)\, d\theta = 9\int \cos^2(\theta)\, d\theta = 9\left\{\frac{\theta}{2} - \frac{\sin(2\theta)}{4}\right\} + C$$

$$= 9\left\{\frac{\theta}{2} - \frac{2\sin(\theta)\cos(\theta)}{4}\right\} + C = \frac{9}{2}\left\{\theta - \sin(\theta)\cos(\theta)\right\} + C$$

This antiderivative, a function of the variable θ, can be converted back to a function of the variable x. Since $x = 3\sin(\theta)$ we can solve for θ to get $\theta = \arcsin(x/3)$. Replacing θ with $\arcsin(x/3)$ in the antiderivative, we get

$$\frac{9}{2}\left\{\theta - \sin(\theta)\cos(\theta)\right\} + C = \frac{9}{2}\left\{\mathbf{arcsin(x/3)} - \sin(\mathbf{arcsin(x/3)})\cos(\mathbf{arcsin(x/3)})\right\} + C$$

$$= \frac{9}{2}\left\{\arcsin\left(\frac{x}{3}\right) + \frac{x}{3}\frac{\sqrt{9-x^2}}{3}\right\} + C = \frac{9}{2}\arcsin\left(\frac{x}{3}\right) + \frac{1}{2}x\sqrt{9-x^2} + C.$$

Using this antiderivative, we can evaluate the definite integral:

$$\int_0^3 \sqrt{9-x^2}\, dx = \frac{9}{2} \arcsin\left(\frac{x}{3}\right) + \frac{x}{2}\sqrt{9-x^2}\,\Big|_0^3$$

$$= \left\{\frac{9}{2}\arcsin\left(\frac{3}{3}\right) + \frac{3}{2}\sqrt{9-3^2}\right\} - \left\{\frac{9}{2}\arcsin\left(\frac{0}{3}\right) + \frac{0}{2}\sqrt{9-0^2}\right\} = \frac{9\pi}{4}$$

Example 3: The definite integral $\int_0^3 \sqrt{9-x^2}\, dx$ represents the area of what region?

Solution: The area of one fourth of the circle of radius 3 which lies in the first quadrant (Fig. 1). The area of this quarter circle is

$$\frac{\text{area of whole circle}}{4} = \frac{1}{4}\pi r^2 = \frac{1}{4}\pi 3^2 = \frac{9}{4}\pi$$

which agrees with the value found in the previous example.

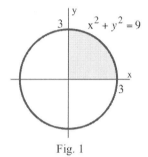

Fig. 1

Each Trigonometric Substitution involves four major steps:

1. Choose which substitution to make, x = a trigonomteric function of θ.
2. Rewrite the original integral in terms of θ and dθ.
3. Find an antiderivative of the new integral.
4. Write the antiderivative in step 3 in terms of the original variable x.

The rest of this section discusses each of these steps. The first step requires you to make a decision. Then the other three steps follow from that decision. For most students, the key to success with the Trigonometric Substitution technique is to THINK TRIANGLES.

Step 1: Choosing the substitution

The first step requires that you make a decision, and the pattern of the familiar Pythagorean Theorem can help you make the correct choice.

Pythagorean Theorem: $(\text{side})^2 + (\text{side})^2 = (\text{hypotenuse})^2$ or $(\text{side})^2 = (\text{hypotenuse})^2 - (\text{side})^2$.

The pattern $3^2 + x^2$ matches the Pythagorean pattern if 3 and x are sides of a right triangle. For a right triangle with sides 3 and x (Fig. 2), we know $\tan(\theta) = \text{opposite/adjacent} = x/3$ so $x = 3\tan(\theta)$.

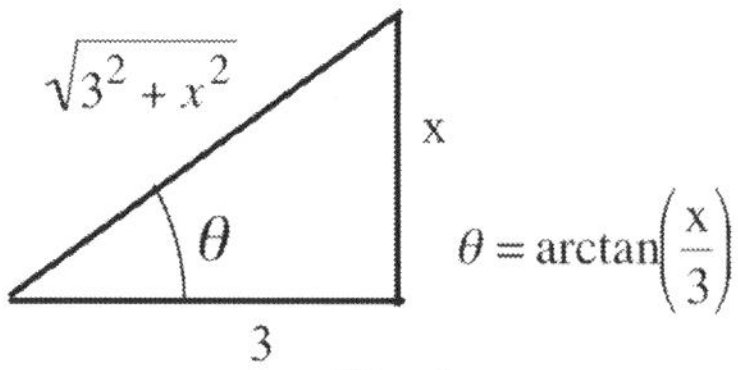

Fig. 2

The pattern $3^2 - x^2$ matches the Pythagorean pattern if 3 is the hypotenuse and x is a side of a right triangle (Fig. 3). Then $\sin(\theta) = $ opposite/hypotenuse $= x/3$ so $x = 3 \sin(\theta)$.

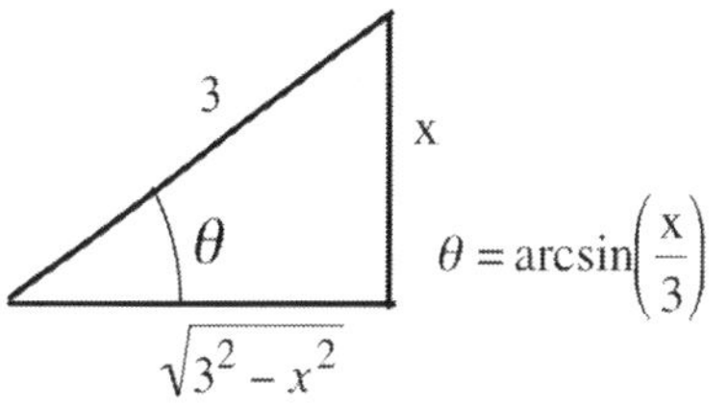

Fig. 3

The pattern $x^2 - 3^2$ matches the Pythagorean pattern if x is the hypotenuse and 3 is a side of a right triangle (Fig. 4). Then $\sec(\theta) = $ hypotenuse/adjacent $= x/3$ so $x = 3 \sec(\theta)$.

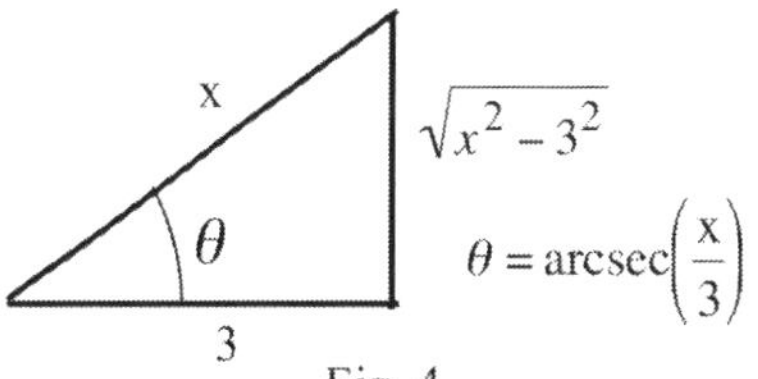

Fig. 4

Once the choice has been made for the substitution, then several things follow automatically:

- dx can be calculated by differentiating x with respect to θ,
- θ can be found by solving the substitution equation for θ,
  (if $x = 3 \tan(\theta)$ then $\tan(\theta) = x/3$ so $\theta = \arctan(x/3)$), and
- the patterns $3^2 + x^2$, $3^2 - x^2$, and $x^2 - 3^2$ can be simplified using algebra and the trigonometric identities $1 + \tan^2(\theta) = \sec^2(\theta)$, $1 - \sin^2(\theta) = \cos^2(\theta)$, and $\sec^2(\theta) - 1 = \tan^2(\theta)$.

These results are collected in the table below.

$3^2 + x^2$ (Fig. 2)	$3^2 - x^2$ (Fig. 3)	$x^2 - 3^2$ (Fig. 4)
Put $x = 3 \tan(\theta)$.	Put $x = 3 \sin(\theta)$.	Put $x = 3 \sec(\theta)$.
Then $dx = 3 \sec^2(\theta) \, d\theta$	Then $dx = 3 \cos(\theta) \, d\theta$	Then $dx = 3 \sec(\theta) \tan(\theta) \, d\theta$
$\theta = \arctan(\frac{x}{3})$	$\theta = \arcsin(\frac{x}{3})$	$\theta = \text{arcsec}(\frac{x}{3})$
$3^2 + x^2$ $= 3^2 + 3^2 \tan^2(\theta)$ $= 3^2(1 + \tan^2(\theta))$ $= 3^2 \sec^2(\theta)$	$3^2 - x^2$ $= 3^2 - 3^2 \sin^2(\theta)$ $= 3^2(1 - \sin^2(\theta))$ $= 3^2 \cos^2(\theta)$	$x^2 - 3^2$ $= 3^2 \sec^2(\theta) - 3^2$ $= 3^2(\sec^2(\theta) - 1)$ $= 3^2 \tan^2(\theta)$

Example 4: For the patterns $16 - x^2$ and $5 + x^2$, (a) decide on the appropriate substitution for x, (b) calculate dx and θ, and (c) use the substitution to simplify the pattern.

Solution: $16 - x^2$: This matches the Pythagorean pattern if 4 is a hypotenuse and x is the side of a right triangle. Then $\sin(\theta) = $ opposite/hypotenuse $= x/4$ so $\mathbf{x = 4\sin(\theta)}$. For $x = 4\sin(\theta)$, $dx = 4\cos(\theta)\, d\theta$ and $\theta = \arcsin(x/4)$. Finally,

$$16 - x^2 = 16 - (4\sin(\theta))^2 = 16 - 16\sin^2(\theta) = 16(1 - \sin^2(\theta)) = 16\cos^2(\theta).$$

$5 + x^2$: This matches the Pythagorean pattern if x and $\sqrt{5}$ are the sides of a right triangle. Then $\tan(\theta) = $ opposite/adjacent $= x/\sqrt{5}$ so $\mathbf{x = \sqrt{5}\tan(\theta)}$. For $x = \sqrt{5}\tan(\theta)$, $dx = \sqrt{5}\sec^2(\theta)\, d\theta$ and $\theta = \arctan(x/\sqrt{5})$. Finally,

$$5 + x^2 = 5 + (\sqrt{5}\tan(\theta))^2 = 5 + 5\tan^2(\theta) = 5(1 + \tan^2(\theta)) = 5\sec^2(\theta).$$

Practice 1: For the patterns $25 + x^2$ and $x^2 - 13$, (a) decide on the appropriate substitution for x, (b) calculate dx and θ, and (c) use the substitution to simplify the pattern.

Step 2: Rewriting the integral in terms of θ and $d\theta$

Once we decide on the appropriate substitution, calculate dx, and simplify the the pattern, then the second step is very straightforward.

Example 5: Use the substitution $x = 5\tan(\theta)$ to rewrite the integral $\int \dfrac{1}{\sqrt{25 + x^2}}\, dx$ in terms of θ and $d\theta$.

Solution: Since $x = 5\tan(\theta)$, then $dx = 5\sec^2(\theta)\, d\theta$ and

$$25 + x^2 = 25 + (5\tan(\theta))^2 = 25 + 25\tan^2(\theta) = 25\{1 + \tan^2(\theta)\} = 25\sec^2(\theta). \text{ Finally,}$$

$$\int \frac{1}{\sqrt{25 + x^2}}\, dx = \int \frac{1}{\sqrt{25\sec^2(\theta)}} \, 5\sec^2(\theta)\, d\theta = \int \frac{5\sec^2(\theta)}{5\sec(\theta)}\, d\theta = \int \sec(\theta)\, d\theta.$$

Practice 2: Use the substitution $x = 5\sin(\theta)$ to rewrite the integral $\int \dfrac{1}{\sqrt{25 - x^2}}\, dx$ in terms of θ and $d\theta$.

Steps 3 & 4: Finding an antiderivative of the new integral & writing the answer in terms of x

After changing the variable, the new integral typically involves trigonometric functions and we can use any of our previous methods (a change of variable, integration by parts, a trigonometric identity, or the integral tables) to find an antiderivative.

Once we have an antiderivative, usually a trigonometric function of θ, we can replace θ with the appropriate inverse trigonometric function of x and simplify. Since the antiderivatives commonly contain trigonometric functions, we frequently need to simplify a trigonometric function of an inverse trigonometric function, and it is **very** helpful to refer back to the right triangle we used at the beginning of the substitution process.

Example 6: By replacing x with $5\tan(\theta)$, $\int \frac{1}{\sqrt{25+x^2}}\,dx$ becomes $\int \sec(\theta)\,d\theta$. Evaluate $\int \sec(\theta)\,d\theta$ and write the resulting antiderivative in terms of the variable x.

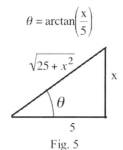

Fig. 5

Solution: $x = 5\tan(\theta)$ so $\theta = \arctan(x/5)$ (Fig. 5). Then

$$\int \sec(\theta)\,d\theta = \ln|\sec(\theta) + \tan(\theta)| + C = \ln|\sec(\arctan(x/5)) + \tan(\arctan(x/5))| + C$$

By referring to the right triangle in Fig. 5, we see that

$$\sec(\arctan(x/5)) = \frac{\sqrt{25+x^2}}{5} \quad \text{and} \quad \tan(\arctan(x/5)) = \frac{x}{5} \quad \text{so}$$

$$\ln|\sec(\arctan(x/5)) + \tan(\arctan(x/5))| + C = \ln\left|\frac{\sqrt{25+x^2}}{5} + \frac{x}{5}\right| + C .$$

Putting these pieces together, we have $\int \frac{1}{x\sqrt{25+x^2}}\,dx = \ln\left|\frac{\sqrt{25+x^2}}{5} + \frac{x}{5}\right| + C .$

Practice 3: Show that by replacing x with $3\sin(\theta)$, $\int \frac{1}{x^2\sqrt{9-x^2}}\,dx$ becomes $\frac{1}{9}\int \csc^2(\theta)\,d\theta$. Evaluate $\frac{1}{9}\int \csc^2(\theta)\,d\theta$ and write the resulting antiderivative in terms of the variable x.

Sometimes it is useful to "complete the square" in an irreducible quadratic to make the pattern more obvious.

Example 7: Rewrite $x^2 + 2x + 26$ by completing the square and evaluate $\int \frac{1}{\sqrt{x^2+2x+26}}\,dx$.

Solution: $x^2 + 2x + 26 = (x+1)^2 + 25$ so $\int \frac{1}{\sqrt{x^2+2x+26}}\,dx = \int \frac{1}{\sqrt{(x+1)^2+25}}\,dx .$

Put $u = x + 1$. Then $du = dx$, and

$$\int \frac{1}{\sqrt{x^2 + 2x + 26}} \, dx = \int \frac{1}{\sqrt{(x+1)^2 + 25}} \, dx$$

$$= \int \frac{1}{\sqrt{u^2 + 25}} \, du$$

$$= \ln\left| \frac{\sqrt{25 + u^2}}{5} + \frac{u}{5} \right| + C \quad \text{(using the result of Example 6)}$$

$$= \ln\left| \frac{\sqrt{25 + (x+1)^2}}{5} + \frac{x+1}{5} \right| + C = \ln\left| \frac{\sqrt{x^2 + 2x + 26}}{5} + \frac{x+1}{5} \right| + C$$

THINK TRIANGLES. The first and last steps of the method (choosing the substitution and writing the answer interms of x) are easier if you understand the triangles (Figures 2, 3, and 4) and have drawn the appropriate triangle for the problem. Of course, you also need to practice the method.

PROBLEMS

In problems 1–6, (a) make the given substitution and simplify the result, and (b) calculate dx.

1. $x = 3 \cdot \sin(\theta)$ in $\dfrac{1}{\sqrt{9 - x^2}}$
2. $x = 3 \cdot \tan(\theta)$ in $\dfrac{1}{\sqrt{x^2 + 9}}$

3. $x = 3 \cdot \sec(\theta)$ in $\dfrac{1}{\sqrt{x^2 - 9}}$
4. $x = 6 \cdot \sin(\theta)$ in $\dfrac{1}{36 - x^2}$

5. $x = \sqrt{2} \cdot \tan(\theta)$ in $\dfrac{1}{\sqrt{2 + x^2}}$
6. $x = \sec(\theta)$ in $\dfrac{1}{x^2 - 1}$

In problems 7–12, (a) solve for θ as a function of x,
(b) replace θ in $f(\theta)$ with you result in part (a), and (c) simplify.

7. $x = 3 \cdot \sin(\theta)$, $f(\theta) = \cos(\theta) \cdot \tan(\theta)$
8. $x = 3 \cdot \tan(\theta)$, $f(\theta) = \sin(\theta) \cdot \tan(\theta)$

9. $x = 3 \cdot \sec(\theta)$, $f(\theta) = \sqrt{1 + \sin^2(\theta)}$
10. $x = 5 \cdot \sin(\theta)$, $f(\theta) = \dfrac{\cos(\theta)}{1 + \sec(\theta)}$

11. $x = 5 \cdot \tan(\theta)$, $f(\theta) = \dfrac{\cos^2(\theta)}{1 + \cot(\theta)}$
12. $x = 5 \cdot \sec(\theta)$, $f(\theta) = \cos(\theta) + 7 \cdot \tan^2(\theta)$

8.5 Trigonometric Substitution   Contemporary Calculus

In problems 13–36, evaluate the integrals. (More than one method works for some of the integrals.)

13. $\int \dfrac{1}{x^2\sqrt{9-x^2}}\, dx$

14. $\int \dfrac{x^2}{\sqrt{9-x^2}}\, dx$

15. $\int \dfrac{1}{\sqrt{x^2+49}}\, dx$

16. $\int \dfrac{1}{\sqrt{x^2+1}}\, dx$

17. $\int \sqrt{36-x^2}\, dx$

18. $\int \sqrt{1-36x^2}\, dx$

19. $\int \dfrac{1}{\sqrt{36+x^2}}\, dx$

20. $\int \dfrac{1}{x\sqrt{25-x^2}}\, dx$

21. $\int \dfrac{1}{\sqrt{49+x^2}}\, dx$

22. $\int \dfrac{\sqrt{25-x^2}}{x^2}\, dx$

23. $\int \dfrac{x}{\sqrt{25-x^2}}\, dx$

24. $\int \dfrac{1}{x^2+49}\, dx$

25. $\int \dfrac{x}{x^2+49}\, dx$

26. $\int \dfrac{1}{49x^2+25}\, dx$

27. $\int \dfrac{1}{(x^2-9)^{3/2}}\, dx$

28. $\int \dfrac{1}{(4x^2-1)^{3/2}}\, dx$

29. $\int \dfrac{5}{2x\sqrt{x^2-25}}\, dx$

30. $\int \dfrac{1}{x\sqrt{3-x^2}}\, dx$

31. $\int \dfrac{1}{25-x^2}\, dx$

32. $\int \dfrac{1}{a^2+x^2}\, dx$

33. $\int \dfrac{1}{\sqrt{a^2+x^2}}\, dx$

34. $\int \dfrac{1}{x\cdot\sqrt{a^2+x^2}}\, dx$

35. $\int \dfrac{1}{x^2\sqrt{a^2+x^2}}\, dx$

36. $\int \dfrac{1}{(a^2+x^2)^{3/2}}\, dx$

In problems 37–42, first complete the square, make the appropriate substitutions, and evaluate the integral.

37. $\int \dfrac{1}{\sqrt{(x+1)^2+9}}\, dx$

38. $\int \dfrac{1}{\sqrt{(x+3)^2+1}}\, dx$

39. $\int \dfrac{1}{x^2+10x+29}\, dx$

40. $\int \dfrac{1}{x^2-4x+13}\, dx$

41. $\int \dfrac{1}{\sqrt{x^2+4x+3}}\, dx$

42. $\int \dfrac{1}{\sqrt{x^2-6x-16}}\, dx$

Section 8.5   Practice Answers

Practice 1: $25 + x^2$:   (a) Put $x = 5\cdot\tan(\theta)$

(b) Then $dx = 5\cdot\sec^2(\theta)\,d\theta$ and $\theta = \arctan(x/5)$

(c) $25 + x^2 = 25 + 25\cdot\tan^2(\theta) = 25(1 + \tan^2(\theta)) = 25\cdot\sec^2(\theta)$

$x^2 - 13$:   (a) Put $x = \sqrt{13}\cdot\sec(\theta)$

(b) Then $dx = \sqrt{13}\cdot\sec(\theta)\cdot\tan(\theta)\,d\theta$ and $\theta = \text{arcsec}(x/\sqrt{13})$

(c) $x^2 - 13 = 13\cdot\sec^2(\theta) - 13 = 13(\sec^2(\theta) - 1) = 13\cdot\tan^2(\theta)$

Practice 2: $x = 5\cdot\sin(\theta)$ so $dx = 5\cdot\cos(\theta)\,d\theta$ and $25 - x^2 = 25(1 - \sin^2(\theta)) = 25\cdot\cos^2(\theta)$.

Then $\displaystyle\int \frac{1}{\sqrt{25 - x^2}}\,dx = \int \frac{1}{\sqrt{25 - \sin^2(\theta)}}\,5\cdot\cos(\theta)\,d\theta$

$\displaystyle = \int \frac{1}{\sqrt{25\cdot\cos^2(\theta)}}\,5\cdot\cos(\theta)\,d\theta = \int 1\,d\theta = \theta + C = \arcsin(x/5) + C$

Practice 3: $x = 3\cdot\sin(\theta)$ so $dx = 3\cdot\cos(\theta)\,d\theta$ and $9 - x^2 = 9(1 - \sin^2(\theta)) = 9\cdot\cos^2(\theta)$.

Then $\displaystyle\int \frac{1}{x^2\sqrt{9 - x^2}}\,dx = \int \frac{1}{9\cdot\sin^2(\theta)\sqrt{9 - \sin^2(\theta)}}\,3\cdot\cos(\theta)\,d\theta = \int \frac{1}{9\cdot\sin^2(\theta)\sqrt{9\cdot\cos^2(\theta)}}\,3\cdot\cos(\theta)\,d\theta$

$\displaystyle = \frac{1}{9}\int \csc^2(\theta)\,d\theta = -\frac{1}{9}\cot(\theta) + C = -\frac{1}{9}\cot(\arcsin(x/3)) + C = -\frac{1}{9}\frac{\sqrt{9 - x^2}}{x} + C.$

8.6 Integrals of Trigonometric Functions

There are an overwhelming number of combinations of trigonometric functions which appear in integrals, but fortunately they fall into a few patterns and most of their integrals can be found using reduction formulas and tables of integrals. This section examines some of the patterns of these combinations and illustrates how some of their integrals can be derived.

Products of Sine and Cosine: $\int \sin(ax)\cdot\sin(bx)\, dx$, $\int \cos(ax)\cdot\cos(bx)\, dx$, $\int \sin(ax)\cdot\cos(bx)\, dx$

All of these integrals are handled by referring to the trigonometric identities for sine and cosine of sums and differences:

$$\sin(A + B) = \sin(A)\cos(B) + \cos(A)\sin(B)$$
$$\sin(A - B) = \sin(A)\cos(B) - \cos(A)\sin(B)$$
$$\cos(A + B) = \cos(A)\cos(B) - \sin(A)\sin(B)$$
$$\cos(A - B) = \cos(A)\cos(B) + \sin(A)\sin(B)$$

By adding or subtracting the appropriate pairs of identities, we can write the various products such as $\sin(ax)\cos(bx)$ as a sum or difference of single sines or cosines. For example, by adding the first two identities we get $2\sin(A)\cos(B) = \sin(A+B) + \sin(A-B)$ so $\sin(A)\cos(B) = \frac{1}{2}\{\sin(A+B) + \sin(A-B)\}$. Using this last identity, the integral of $\sin(ax)\cos(bx)$ for $a \neq b$ is relatively easy:

$$\int \sin(ax)\cos(bx)\, dx = \int \frac{1}{2}\{\sin((a+b)x) + \sin((a-b)x)\}\, dx = \frac{1}{2}\left\{\frac{-\cos((a-b)x)}{a-b} + \frac{-\cos((a+b)x)}{a+b}\right\} + C.$$

The other integrals of products of sine and cosine follow in a similar manner.

If $a \neq b$, then

$$\int \sin(ax)\cdot\sin(bx)\, dx = \frac{1}{2}\left\{\frac{\sin((a-b)x)}{a-b} - \frac{\sin((a+b)x)}{a+b}\right\} + C$$

$$\int \cos(ax)\cdot\cos(bx)\, dx = \frac{1}{2}\left\{\frac{\sin((a-b)x)}{a-b} + \frac{\sin((a+b)x)}{a+b}\right\} + C$$

$$\int \sin(ax)\cdot\cos(bx)\, dx = \frac{-1}{2}\left\{\frac{\cos((a-b)x)}{a-b} + \frac{\cos((a+b)x)}{a+b}\right\} + C$$

If a = b, we have patterns we have already used.

$$\int \sin^2(ax)\,dx = \frac{x}{2} - \frac{\sin(2ax)}{4a} + C = \frac{x}{2} - \frac{\sin(ax)\cdot\cos(ax)}{2a} + C$$

$$\int \cos^2(ax)\,dx = \frac{x}{2} + \frac{\sin(2ax)}{4a} + C = \frac{x}{2} + \frac{\sin(ax)\cdot\cos(ax)}{2a} + C$$

$$\int \sin(ax)\cdot\cos(ax)\,dx = \frac{\sin^2(ax)}{2a} + C = \frac{1-\cos(2ax)}{4a} + C$$

The first and second of these integral formulas follow from the identities $\sin^2(ax) = \frac{1-\cos(2ax)}{2}$ and $\cos^2(ax) = \frac{1+\cos(2ax)}{2}$, and the third can be derived by changing the variable to $u = \sin(ax)$.

Powers of Sine and Cosine Alone: $\int \sin^n(x)\,dx$, $\int \cos^n(x)\,dx$

All of these antiderivatives can be found using integration by parts or the reduction formulas (formulas 19 and 20 in the integral tables) which were derived using integration by parts. For small values of m and n it is just as easy to find the antiderivatives directly.

Even Powers of Sine or Cosine Alone

For **even** powers of sine or cosine, we can successfully reduce the size of the exponent by repeatedly applying the identities $\sin^2(x) = \frac{1-\cos(2x)}{2}$ and $\cos^2(x) = \frac{1+\cos(2x)}{2}$.

Example 1: Evaluate $\int \sin^4(x)\,dx$.

Solution: $\sin^4(x) = \{\sin^2(x)\}^2 = \{\frac{1}{2}[1-\cos(2x)]\}^2 = \frac{1}{4}\{1 - 2\cos(2x) + \cos^2(2x)\}$ so

$$\int \sin^4(x)\,dx = \int \frac{1}{4}\{1 - 2\cos(2x) + \cos^2(2x)\}\,dx$$

$$= \frac{1}{4}\{x + \sin(2x) + \frac{x}{2} + \frac{\sin(2x)\cos(2x)}{2}\} + C.$$

Practice 1: Evaluate $\int \cos^4(x)\,dx$.

Odd Powers of Sine or Cosine Alone

For odd powers of sine or cosine we can split off one factor of sine or cosine, reduce the remaining even exponent using the identities $\sin^2(x) = 1 - \cos^2(x)$ or $\cos^2(x) = 1 - \sin^2(x)$, and finally integrate by changing the variable.

Example 2: Evaluate $\int \sin^5(x)\, dx$.

Solution: $\sin^5(x) = \sin^4(x)\sin(x) = \{\sin^2(x)\}^2 \sin(x) = \{1 - \cos^2(x)\}^2 \sin(x)$
$$= \{1 - 2\cos^2(x) + \cos^4(x)\} \sin(x).$$

Then $\int \sin^5(x)\, dx = \int \sin(x)\, dx - 2\int \cos^2(x)\sin(x)\, dx + \int \cos^4(x)\sin(x)\, dx$.

The first integral is easy, and the last two can be evaluated by changing the variable to $u = \cos(x)$:

$$\int \sin^5(x)\, dx = -\cos(x) - 2\left\{ -\frac{\cos^3(x)}{3} \right\} + \left\{ -\frac{\cos^5(x)}{5} \right\} + C.$$

Practice 2: Evaluate $\int \cos^5(x)\, dx$.

Patterns for $\int \sin^m(x)\cos^n(x)\, dx$

If the exponent of sine is odd, we can split off one factor $\sin(x)$ and use the identity $\sin^2(x) = 1 - \cos^2(x)$ to rewrite the remaining even power of sine in terms of cosine. Then the change of variable $u = \cos(x)$ makes all of the integrals straightforward.

Example 3: Evaluate $\int \sin^3(x)\cos^6(x)\, dx$.

Solution: $\sin^3(x)\cos^6(x) = \sin(x)\,\mathbf{\sin^2(x)}\,\cos^6(x) = \sin(x)\{\mathbf{1 - \cos^2(x)}\}\cos^6(x)$
$$= \sin(x)\cos^6(x) - \sin(x)\cos^8(x).$$

Then $\int \sin^3(x)\cos^6(x)\, dx = \int \sin(x)\cos^6(x) - \sin(x)\cos^8(x)\, dx$ (put $u = \cos(x)$)

$$= -\frac{\cos^7(x)}{7} + \frac{\cos^9(x)}{9} + C.$$

Practice 3: Evaluate $\int \sin^3(x)\cos^4(x)\, dx$.

If the **exponent of cosine is odd**, we can split off one factor $\cos(x)$ and use the identity $\cos^2(x) = 1 - \sin^2(x)$ to rewrite the remaining even power of cosine in terms of sine. Then the change of variable $u = \sin(x)$ makes all of the integrals straightforward.

If **both exponents are even**, we can use the identities $\sin^2(x) = \frac{1}{2}(1 - \cos(2x))$ and $\cos^2(x) = \frac{1}{2}(1 + \cos(2x))$ to rewrite the integral in terms of powers of $\cos(2x)$ and then proceed with integrating even powers of cosine.

Powers of Secant and Tangent Alone: $\int \sec^n(x)\,dx$, $\int \tan^n(x)\,dx$

All of the integrals of powers of secant and tangent can be evaluated by knowing

$$\int \sec(x)\,dx = \ln|\sec(x) + \tan(x)| + C \text{ and}$$

$$\int \tan(x)\,dx = -\ln|\cos(x)| + C = \ln|\sec(x)| + C$$

and then using the reduction formulas

$$\int \sec^n(x)\,dx = \frac{\sec^{n-2}(x)\cdot\tan(x)}{n-1} + \frac{n-2}{n-1}\int \sec^{n-2}(x)\,dx \text{ and}$$

$$\int \tan^n(x)\,dx = \frac{\tan^{n-1}(x)}{n-1} - \int \tan^{n-2}(x)\,dx.$$

Example 4: Evaluate $\int \sec^3(x)\,dx$.

Solution: Using the reduction formula with $n = 3$,

$$\int \sec^3(x)\,dx = \frac{\sec(x)\cdot\tan(x)}{2} + \frac{1}{2}\int \sec(x)\,dx = \frac{\sec(x)\cdot\tan(x)}{2} + \frac{1}{2}\ln|\sec(x) + \tan(x)| + C.$$

Practice 4: Evaluate $\int \tan^3(x)\,dx$ and $\int \sec^5(x)\,dx$.

Patterns for $\int \sec^m(x)\cdot\tan^n(x)\,dx$

The patterns for evaluating $\int \sec^m(x)\cdot\tan^n(x)\,dx$ are similar to those for $\int \sin^m(x)\cdot\cos^n(x)\,dx$ because we treat the even and odd powers differently and we use the identities $\tan^2(x) = \sec^2(x) - 1$ and $\sec^2(x) = \tan^2(x) + 1$.

If the **exponent of secant is even**, factor off $\sec^2(x)$, replace the other even powers (if any) of secant using $\sec^2(x) = \tan^2(x) + 1$, and make the change of variable $u = \tan(x)$ (then $du = \sec^2(x)\,dx$).

If the **exponent of tangent is odd**, factor off $\sec(x)\tan(x)$, replace the remaining even powers (if any) of tangent using $\tan^2(x) = \sec^2(x) - 1$, and make the change of variable $u = \sec(x)$ (then $du = \sec(x)\tan(x)\,dx$).

If the **exponent of secant is odd and the exponent of tangent is even**, replace the even powers of tangent using $\tan^2(x) = \sec^2(x) - 1$. Then the integral contains only powers of secant, and we can use the patterns for integrating powers of secant alone.

Example 5: Evaluate $\int \sec(x)\cdot\tan^2(x)\,dx$.

Solution: Since the exponent of secant is odd and and the exponent of tangent is even, we can use the last method mentions: replace the even powers of tangent using $\tan^2(x) = \sec^2(x) - 1$. Then

$$\int \sec(x)\cdot\tan^2(x)\,dx = \int \sec(x)\cdot\{\sec^2(x) - 1\}\,dx$$

$$= \int \sec^3(x) - \sec(x)\,dx = \int \sec^3(x)\,dx - \int \sec(x)\,dx$$

$$= \left\{\frac{\sec(x)\cdot\tan(x)}{2} + \frac{1}{2}\ln|\sec(x) + \tan(x)|\right\} - \ln|\sec(x) + \tan(x)| + C$$

$$= \frac{\sec(x)\cdot\tan(x)}{2} - \frac{1}{2}\ln|\sec(x) + \tan(x)| + C.$$

Practice 5: Evaluate $\int \sec^4(x)\cdot\tan^2(x)\,dx$.

Wrap Up

Even if you use tables of integrals (or computers) for most of your future work, it is important to realize that most of the integral formulas can be derived from some basic facts using the techniques we have discussed in this and earlier sections.

PROBLEMS

Evaluate the integrals. (More than one method works for some of the integrals.)

1. $\int \sin^2(3x)\,dx$

2. $\int \cos^2(5x)\,dx$

3. $\int e^x\cdot\sin(e^x)\cdot\cos(e^x)\,dx$

4. $\int \frac{1}{x}\cdot\sin^2(\ln(x))\,dx$

5. $\int_0^\pi \sin^4(3x)\,dx$

6. $\int_0^\pi \cos^4(5x)\,dx$

7. $\displaystyle\int_0^\pi \sin^3(7x)\,dx$ 　　　 8. $\displaystyle\int_0^\pi \cos^3(5x)\,dx$ 　　　 9. $\displaystyle\int \sin(7x)\cdot\cos(7x)\,dx$

10. $\displaystyle\int \sin(7x)\cdot\cos^2(7x)\,dx$ 　　　 11. $\displaystyle\int \sin(7x)\cdot\cos^3(7x)\,dx$ 　　　 12. $\displaystyle\int \sin^2(3x)\cdot\cos(3x)\,dx$

13. $\displaystyle\int \sin^2(3x)\cdot\cos^2(3x)\,dx$ 　　　 14. $\displaystyle\int \sin^2(3x)\cdot\cos^3(3x)\,dx$ 　　　 15. $\displaystyle\int \sec^2(5x)\cdot\tan(5x)\,dx$

16. $\displaystyle\int \sec^2(3x)\cdot\tan^2(3x)\,dx$ 　　　 17. $\displaystyle\int \sec^3(3x)\cdot\tan(3x)\,dx$ 　　　 18. $\displaystyle\int \sec^3(5x)\cdot\tan^2(5x)\,dx$

The definite integrals of various combinations of sine and cosine on the interval $[0, 2\pi]$ exhibit a number of interesting patterrns. For now these patterns are simply curiousities and a source of additional problems for practice, but the patterns are very important as the foundation for an applied topic, Fourier Series, that you may encounter in more advanced courses.

The next three problems ask you to show that the definite integral on $[0, 2\pi]$ of $\sin(mx)$ multiplied by almost any other combination of $\sin(nx)$ or $\cos(nx)$ is 0. The only nonzero value comes when $\sin(mx)$ is multiplied by itself.

19. Show that if m and n are integers with $m \neq n$, then $\displaystyle\int_0^{2\pi} \sin(mx)\cdot\sin(nx)\,dx = 0$.

20. Show that if m and n are integers, then $\displaystyle\int_0^{2\pi} \sin(mx)\cdot\cos(nx)\,dx = 0$. (Consider $m = n$ and $m \neq n$.)

21. Show that if $m \neq 0$ is an integer, then $\displaystyle\int_0^{2\pi} \sin(mx)\cdot\sin(mx)\,dx = \pi$.

22. Suppose $P(x) = 5\cdot\sin(x) + 7\cdot\cos(x) - 4\cdot\sin(2x) + 8\cdot\cos(2x) - 2\cdot\sin(3x)$. (This is called a trigonometric polynomial.) Use the **results** of problems 19–21 to quickly evaluate

   (a) $a_1 = \dfrac{1}{\pi}\displaystyle\int_0^{2\pi} \sin(1x)\cdot P(x)\,dx$ 　　　 (b) $a_2 = \dfrac{1}{\pi}\displaystyle\int_0^{2\pi} \sin(2x)\cdot P(x)\,dx$

   (c) $a_3 = \dfrac{1}{\pi}\displaystyle\int_0^{2\pi} \sin(3x)\cdot P(x)\,dx$ 　　　 (d) $a_4 = \dfrac{1}{\pi}\displaystyle\int_0^{2\pi} \sin(4x)\cdot P(x)\,dx$

   (e) Describe how the values of a_i are related to the coeffiecients of $P(x)$.

   (f) Make up your own trigonometric polynomial $P(x)$ and see if your description in part (e) holds for the a_i values calculated from the new $P(x)$.

   (g) Just by knowing the a_i values we can "rebuild" part of $P(x)$. Find a similar method for getting the coefficients of the cosine terms of $P(x)$: $b_i = $??

23. Show that if n is a positive, **odd** integer, then $\int_0^{2\pi} \sin^n(x)\, dx = 0$.

24. It is straightforward (using formula 19 in the integral table) to show that $\int_0^{2\pi} \sin^2(x)\, dx = \pi$, $\int_0^{2\pi} \sin^4(x)\, dx = \frac{3}{4}\pi$, and $\int_0^{2\pi} \sin^6(x)\, dx = \frac{5\cdot 3}{6\cdot 4}\pi$. (a) Evaluate $\int_0^{2\pi} \sin^8(x)\, dx$.

   (b) Predict the value of $\int_0^{2\pi} \sin^{10}(x)\, dx$ and then evalaute the integral.

Section 8.6   Practice Answers

Practice 1: $\int \cos^4(x)\, dx \quad \{\text{Use } \cos^2(x) = \frac{1}{2}(1+\cos(2x))\}$

$= \int \cos^2(x)\cdot\cos^2(x)\, dx = \int \frac{1}{2}(1+\cos(2x))\frac{1}{2}(1+\cos(2x))\, dx$

$= \frac{1}{4}\int 1 + 2\cos(2x) + \cos^2(2x)\, dx = \frac{1}{4}\int 1 + 2\cos(2x) + \frac{1}{2}\{1+\cos(4x)\}\, dx$

$= \frac{1}{4}\int \frac{3}{2} + 2\cos(2x) + \frac{1}{2}\cos(4x)\, dx = \frac{3}{8}x + \frac{1}{4}\sin(2x) + \frac{1}{32}\sin(4x) + C$.

Practice 2: $\int \cos^5(x)\, dx = \int \cos^2(x)\cdot\cos^2(x)\cdot\cos(x)\, dx = \int (1-\sin^2(x))(1-\sin^2(x))\cos(x)\, dx$

$= \int \{1 - 2\sin^2(x) + \sin^4(x)\}\cos(x)\, dx$

$= \int \cos(x)\, dx - 2\int \sin^2(x)\cdot\cos(x)\, dx + \int \sin^4(x)\cdot\cos(x)\, dx \quad (\text{Use } u = \sin(x),\ du = \cos(x)\, dx)$

$= \sin(x) - \frac{2}{3}\sin^3(x) + \frac{1}{5}\sin^5(x) + C$.

Practice 3: $\int \sin^3(x)\cdot\cos^4(x)\, dx = \int \sin(x)\cdot\sin^2(x)\cdot\cos^4(x)\, dx = \int \sin(x)\cdot(1-\cos^2(x))\cdot\cos^4(x)\, dx$

$= \int \sin(x)\cdot\cos^4(x)\, dx - \int \sin(x)\cdot\cos^6(x)\, dx \quad (\text{Use } u = \cos(x),\ du = -\sin(x)\, dx)$

$= -\frac{1}{5}\cos^5(x) + \frac{1}{7}\cos^7(x) + C$

Practice 4: $\int \tan^3(x)\, dx = \frac{1}{2}\tan^2(x) - \int \tan(x)\, dx = \frac{1}{2}\tan^2(x) - \ln|\sec(x)| + C$.

$$\int \sec^5(x)\, dx = \frac{1}{2}\sec^3(x)\cdot\tan(x) + \frac{3}{4}\int \sec^3(x)\, dx$$

$$= \frac{1}{2}\sec^3(x)\cdot\tan(x) + \frac{3}{4}\left\{\frac{1}{2}\sec(x)\cdot\tan(x) + \frac{1}{2}\int \sec(x)\, dx\right\}$$

$$= \frac{1}{2}\sec^3(x)\cdot\tan(x) + \frac{3}{8}\sec(x)\cdot\tan(x) + \frac{3}{8}\ln|\sec(x) + \tan(x)| + C.$$

Practice 5: $\int \sec^4(x)\cdot\tan^2(x)\, dx = \int \sec^2(x)\cdot\sec^2(x)\cdot\tan^2(x)\, dx$

$$= \int \sec^2(x)\cdot(\tan^2(x) + 1)\cdot\tan^2(x)\, dx$$

$$= \int \sec^2(x)\cdot\tan^4(x)\, dx + \int \sec^2(x)\cdot\tan^2(x)\, dx \quad (\text{Use } u = \tan(x),\ du = \sec^2(x)\, dx)$$

$$= \frac{1}{5}\tan^5(x) + \frac{1}{3}\tan^3(x) + C.$$

PROBLEM ANSWERS Chapter Eight

Section 8.1

1. $\int = \lim_{A \to \infty} \{ -\frac{1}{2x^2} \Big|_{10}^{A} \} = \lim_{A \to \infty} \{ (-\frac{1}{2A^2}) - (-\frac{1}{200}) \} = \lim_{A \to \infty} \{ \frac{1}{200} - \frac{1}{2A^2} \} = \frac{1}{200}$.

3. $\int = \lim_{A \to \infty} \{ 2 \cdot \arctan(x) \Big|_{3}^{A} \} = \lim_{A \to \infty} \{ 2 \cdot \arctan(A) - 2 \cdot \arctan(3) \} = 2(\frac{\pi}{2}) - 2 \cdot \arctan(3) \approx 0.644$.

5. Use $u = \ln(x)$.

    $\int = \lim_{A \to \infty} \{ 5 \cdot \ln(\ln(x)) \Big|_{e}^{A} \} = \lim_{A \to \infty} \{ 5 \cdot \ln(\ln(A)) - 5 \cdot \ln(\ln(e)) \} = \lim_{A \to \infty} \{ 5 \cdot \ln(\ln(A)) - 0 \} = \infty$.
    $\int$ DIVERGES.

7. $\int = \lim_{A \to \infty} \{ \ln(x-2) \Big|_{3}^{A} \} = \lim_{A \to \infty} \{ \ln(A-2) - \ln(3-2) \} = \lim_{A \to \infty} \{ \ln(A-2) - 0 \} = \infty$. $\int$ DIVERGES.

9. $\int = \lim_{A \to \infty} \{ \frac{-1}{2(x-2)^2} \Big|_{3}^{A} \} = \lim_{A \to \infty} \{ \frac{-1}{2(A-2)^2} - \frac{-1}{2(3-2)^2} \} = \lim_{A \to \infty} \{ \frac{-1}{2(A-2)^2} + \frac{1}{2} \} = \frac{1}{2}$.

11. $\int = \lim_{A \to \infty} \{ \frac{-1}{x+2} \Big|_{3}^{A} \} = \lim_{A \to \infty} \{ \frac{-1}{A+2} - \frac{-1}{3+2} \} = \lim_{A \to \infty} \{ \frac{-1}{A+2} + \frac{1}{5} \} = \frac{1}{5}$.

13. $\int = \lim_{A \to \infty} \{ 2\sqrt{x} \Big|_{A}^{4} \} = \lim_{A \to \infty} \{ 2\sqrt{4} - 2\sqrt{A} \} = 4$.

15. $\int = \lim_{A \to 0^+} \{ \frac{4}{3} x^{3/4} \Big|_{A}^{16} \} = \lim_{A \to 0^+} \{ \frac{4}{3}(16)^{3/4} - \frac{4}{3} A^{3/4} \} = \lim_{A \to 0^+} \{ \frac{32}{3} - \frac{4}{3} A^{3/4} \} = \frac{32}{3}$.

17. $\int = \lim_{A \to 2^-} \{ \arcsin(\frac{x}{2}) \Big|_{0}^{A} \} = \lim_{A \to 2^-} \{ \arcsin(\frac{A}{2}) - \arcsin(\frac{0}{2}) \} = \frac{\pi}{2} - 0 = \frac{\pi}{2}$.

19. $\int = \lim_{A \to \infty} \{ -\cos(x) \Big|_{-2}^{A} \} = \lim_{A \to \infty} \{ -\cos(A) - -\cos(-2) \} = \lim_{A \to \infty} -0.416 - \cos(A)$ DNE. $\int$ DIVERGES.

21. $\int = \lim_{A \to \pi/2} \{ -\ln|\cos(x)| \Big|_{0}^{A} \} = \lim_{A \to \pi/2} \{ -\ln|\cos(A)| - -\ln|\cos(0)| \} = \lim_{A \to \pi/2} 0 - \ln|\cos(A)|$ DNE.
    $\int$ DIVERGES.

23. (a) $A = R$: $\displaystyle\int_R^{2R} \frac{R^2P}{x^2}\,dx = R^2P(\frac{-1}{x})\Big|_R^{2R} = R^2P(\frac{-1}{2R} - \frac{-1}{R}) = \frac{1}{2}RP$.

$A = 2R$: $\displaystyle\int_R^{3R} \frac{R^2P}{x^2}\,dx = R^2P(\frac{-1}{x})\Big|_R^{3R} = R^2P(\frac{-1}{3R} - \frac{-1}{R}) = \frac{2}{3}RP$.

(b) $\displaystyle\lim_{A\to\infty} \int_R^{R+A} \frac{R^2P}{x^2}\,dx = \lim_{A\to\infty} R^2P(\frac{-1}{x})\Big|_R^{R+A} = \lim_{A\to\infty} R^2P(\frac{-1}{R+A} - \frac{-1}{R}) = R^2P(\frac{1}{R}) = RP$.

25. $\displaystyle\int_3^\infty \frac{1}{x(x^2+1)}\,dx = \int_3^\infty \frac{1}{x^3+x}\,dx < \int_3^\infty \frac{1}{x^3}\,dx$ which converges by the p–test.

Therefore, $\displaystyle\int_3^\infty \frac{1}{x(x^2+1)}\,dx$ converges.

27. For $x > 0$, $\ln(x) < x$ so $x + \ln(x) < 2x$ and $\frac{7}{x+\ln(x)} > \frac{7}{2x}$.

Then $\displaystyle\int_3^\infty \frac{7}{2x}\,dx = \frac{7}{2}\int_3^\infty \frac{1}{x}\,dx$ which diverges by the p–test. Therefore, $\displaystyle\int_3^\infty \frac{7}{x+\ln(x)}\,dx$ diverges.

29. $-1 \le \cos(x) \le 1$ so $0 \le 1+\cos(x) \le 2$ and $0 \le \frac{1+\cos(x)}{x^2} \le \frac{2}{x^2}$. Then

$\displaystyle\int_7^\infty \frac{1+\cos(x)}{x^2}\,dx \le \int_7^\infty \frac{2}{x^2}\,dx$ which converges by the p–test. Therefore, $\displaystyle\int_7^\infty \frac{1+\cos(x)}{x^2}\,dx$ converges.

31. $V = \displaystyle\int_0^\infty \pi(\frac{1}{x^2+1})^2\,dx < \pi\int_0^\infty \frac{1}{x^4}\,dx$ which converges by the p–test.

Therefore, $V = \displaystyle\int_0^\infty \pi(\frac{1}{x^2+1})^2\,dx$ converges.

33. (a) $\displaystyle\int_1^A \frac{1}{x}\,dx < \sum_{k=1}^{A-1} \frac{1}{k}$ (b) $\displaystyle\int_1^A \frac{1}{x}\,dx > \sum_{k=2}^{A} \frac{1}{k}$

Section 8.2

1. $u = x^2 + 7$, $du = 2x\,dx$, $3\,du = 6x\,dx$: $\int = \int u^2\,3\,du = u^3 + C = (x^2+7)^3 + C$.

3. $u = x^2 - 3$, $du = 2x\,dx$, $3\,du = 6x\,dx$: $\int = \int \frac{1}{\sqrt{u}}\,3\,du = 3\cdot 2\sqrt{u} = 6\sqrt{x^2-3}\Big|_2^4 = 6\sqrt{13} - 6\sqrt{1} \approx 15.6$

5. $u = x^2 + 3$: $\int = 6\ln(x^2+3) + C$.   7. $u = 3x+2$: $\int = -\frac{1}{3}\cos(3x+2) + C$.

9. $u = e^x + 3, du = e^x\, dx$: $\int = \int \sec^2(u)\, du = \tan(u) = \tan(e^x + 3)\Big|_0^1 = \tan(e + 3) - \tan(1+3) \approx -1.79$

11. $u = \ln(x), du = \frac{1}{x}\, dx$: $\int = \int u\, du = \frac{1}{2} u^2 + C = \frac{1}{2}(\ln(x))^2 + C$.

13. $u = \sin(x), du = \cos(x)\, dx$: $\int = \int e^u\, du = e^u + C = e^{\sin(x)} + C$.

15. $u = 3x, du = 3\, dx$: $\int = \int \frac{5}{1+u^2}\, du = \frac{5}{3} \arctan(u) = \frac{5}{3} \arctan(3x)\Big|_1^3 = \frac{5}{3} \arctan(9) - \frac{5}{3} \arctan(3) \approx 0.35$

17. $u = \frac{1}{x}, du = \frac{-1}{x^2}\, dx$: $\int = \int -\cos(u)\, du = -\sin(u) = -\sin(\frac{1}{x})\Big|_1^2 = (-\sin(\frac{1}{2})) - (-\sin(1)) \approx 0.36$

19. $u = 5 + \sin^2(x), du = 2 \cdot \sin(x) \cdot \cos(x)\, dx$: $\int = \int \frac{1}{u}\, 3\, du = 3 \ln|u| + C = 3 \ln|5 + \sin^2(x)| + C$.

21. $\int = 5 \ln|2x + 5| + C$.

23. $\int = 2 \ln|5x^2 + 3|\Big|_1^3 = 2 \ln|48| - 2 \ln|8| = 2 \ln|\frac{48}{8}| = \ln(36) \approx 3.58$

25. $\int = \frac{7}{2} \arctan(\frac{x+3}{2})\Big|_0^1 = \frac{7}{2} \arctan(2) - \frac{7}{2} \arctan(1.5) \approx 0.44$

27. $u = e^x, du = e^x\, dx$: $\int = \int \frac{1}{1+u^2}\, du = \arctan(u) + C = \arctan(e^x) + C$.

29. $u = 1 + \ln(x), du = \frac{1}{x}\, dx$: $\int = \int \frac{3}{u}\, du = 3 \ln|u| = 3 \ln|1 + \ln(x)|\Big|_1^e = 3 \ln|2| \approx 2.08$

31. $u = 1 - x^2, du = -2x\, dx$: $\int = \int -\sqrt{u}\, du = \frac{-2}{3} u^{3/2} = \frac{-2}{3}(1 - x^2)^{3/2}\Big|_0^1 = \frac{-2}{3}(0)^{3/2} - \frac{-2}{3}(1)^{3/2} = \frac{2}{3}$

33. $u = 1 + \sin(x), du = \cos(x)\, dx$: $\int = \int u^3\, du = \frac{1}{4} u^4 + C = \frac{1}{4}(1 + \sin(x))^4 + C$.

35. $u = \ln(x), du = \frac{1}{x}\, dx$: $\int = \int \sqrt{u}\, du = \frac{2}{3} u^{3/2} = \frac{2}{3}(\ln(x))^{3/2}\Big|_1^e = \frac{2}{3}(\ln(e))^{3/2} - \frac{2}{3}(\ln(1))^{3/2} = \frac{2}{3}$.

37. $u = 5 + \tan(x), du = \sec^2(x)\, dx$: $\int = \int \frac{1}{u}\, du = \ln|u| + C = \ln|5 + \tan(x)| + C$.

39. $u = x - 5$, $du = dx$: $\int = \int \tan(u)\, du = \ln|\sec(u)| + C = \ln|\sec(x-5)| + C$.

41. $u = 5x$, $du = 5\, dx$: $\int = \int \frac{1}{5} e^u\, du = \frac{1}{5} e^u = \frac{1}{5} e^{5x} \Big|_0^1 = \frac{1}{5} e^5 - \frac{1}{5} e^0 \approx 29.48$

43. $\int = \int \frac{7}{(x+2)^2 + 1}\, dx = 7\cdot\arctan(x+2) + C$.   45. $\int = \int \frac{2}{(x-3)^2 + 49}\, dx = \frac{2}{7}\cdot\arctan\left(\frac{x-3}{7}\right) + C$.

47. $\int = \int \frac{3}{(x+5)^2 + 4}\, dx = \frac{3}{2}\cdot\arctan\left(\frac{x+5}{2}\right) + C$.

49. $\int = \ln|x^2 + 4x + 5| + 7\cdot\arctan(x+2) + C$.   51. $\int = 2\cdot\ln|x^2 - 6x + 10| + 19\cdot\arctan(x-3) + C$.

53. $\int = \int \frac{6x - 12}{x^2 - 4x + 13}\, dx + \int \frac{17}{(x-2)^2 + 9}\, dx = 3\cdot\ln|x^2 - 4x + 13| + \frac{17}{3}\cdot\arctan\left(\frac{x-2}{3}\right) + C$.

Section 8.3

1. $\int 12x\cdot\ln(x)\, dx \qquad u = \ln(x)$. Then $dv = 12x\, dx$, $du = \frac{1}{x}\, dx$, and $v = 6x^2$.

   $= uv - \int v\, du = \ln(x)\cdot 6x^2 - \int 6x^2 \cdot \frac{1}{x}\, dx = 6x^2 \ln(x) - \int 6x\, dx = \mathbf{6x^2 \ln(x) - 3x^2 + C}$.

3. $\int x^4 \ln(x)\, dx \qquad dv = x^4\, dx$. Then $u = \ln(x)$, $du = \frac{1}{x}\, dx$, and $v = \frac{1}{5} x^5$.

   $= uv - \int v\, du = \ln(x)\cdot \frac{1}{5} x^5 - \int \frac{1}{5} x^5 \cdot \frac{1}{x}\, dx = \frac{1}{5} x^5 \cdot \ln(x) - \int \frac{1}{5} x^4\, dx = \mathbf{\frac{1}{5} x^5 \cdot \ln(x) - \frac{1}{25} x^5 + C}$.

5. $\int x\cdot\arctan(x)\, dx \qquad dv = x\, dx$. Then $u = \arctan(x)$, $du = \frac{1}{1+x^2}\, dx$, and $v = \frac{1}{2} x^2$.

   $= uv - \int v\, du = \arctan(x)\cdot \frac{1}{2} x^2 - \int \frac{1}{2} x^2 \cdot \frac{1}{1+x^2}\, dx = \frac{1}{2} x^2 \cdot \arctan(x) - \frac{1}{2} \int 1 - \frac{1}{1+x^2}\, dx$

   $= \frac{1}{2} x^2 \cdot \arctan(x) - \frac{1}{2} \{x - \arctan(x)\} + C = \mathbf{\frac{1}{2} x^2 \cdot \arctan(x) - \frac{1}{2} \cdot x + \frac{1}{2} \cdot \arctan(x) + C}$.

7. $\int_0^1 \frac{x}{e^{3x}}\, dx = \int_0^1 x\cdot e^{-3x}\, dx$. Put $u = x$. Then $dv = e^{-3x}\, dx$, $du = dx$, and $v = \frac{-1}{3} e^{-3x}$.

   $= uv - \int v\, du = x\cdot \frac{-1}{3} e^{-3x} - \int \frac{-1}{3} e^{-3x}\, dx = x\cdot \frac{-1}{3} e^{-3x} - \frac{1}{9} e^{-3x} \Big|_0^1$

   $= \left\{ 1\cdot \frac{-1}{3} e^{-3(1)} - \frac{1}{9} e^{-3(1)} \right\} - \left\{ 0\cdot \frac{-1}{3} e^{-3(0)} - \frac{1}{9} e^{-3(0)} \right\} = \mathbf{\frac{1}{9} - \frac{4}{9} e^{-3}}$.

9. $\int x\cdot\sec(x)\cdot\tan(x)\, dx$. \qquad Put $u = x$. Then $dv = \sec(x)\cdot\tan(x)\, dx$, $du = dx$, and $v = \sec(x)$.

   $= uv - \int v\, du = x\cdot\sec(x) - \int \sec(x)\, dx = \mathbf{x\cdot\sec(x) - \ln|\sec(x) + \tan(x)| + C}$.

11. $\int_{\pi/3}^{\pi/2} 7x \cdot \cos(3x)\, dx$ Put $u = 7x$. Then $dv = \cos(3x)\, dx$, $du = 7\, dx$, and $v = \frac{1}{3}\sin(3x)$.

$= uv - \int v\, du = 7x \cdot \frac{1}{3}\sin(3x) - \int \frac{1}{3}\sin(3x) \cdot 7\, dx = \frac{7}{3} x \cdot \sin(3x) + \frac{7}{9}\cos(3x) \Big|_{\pi/3}^{\pi/2}$

$= \{ \frac{7}{3} \cdot \frac{\pi}{2} \cdot \sin(3 \cdot \frac{\pi}{2}) + \frac{7}{9}\cos(3 \cdot \frac{\pi}{2}) \} - \{ \frac{7}{3} \cdot \frac{\pi}{3} \cdot \sin(3 \cdot \frac{\pi}{3}) + \frac{7}{9}\cos(3 \cdot \frac{\pi}{3}) \} \approx -2.887$.

13. $\int 12x \cdot \cos(3x^2)\, dx$. **Use u–substitution!** Put $u = 3x^2$. Then $du = 6x\, dx$ and $2\, du = 12x\, dx$.

$\int = \int \cos(u) \cdot 2\, du = 2 \cdot \sin(u) + C = \mathbf{2 \cdot \sin(3x^2) + C}$.

15. $\int_1^3 \ln(2x+5)\, dx$.   Put $u = \ln(2x+5)$. Then $dv = dx$, $du = \frac{2}{2x+5}\, dx$, and $v = x$.

$= uv - \int v\, du = \ln(2x+5) \cdot x - \int x \cdot \frac{2}{2x+5}\, dx = x \cdot \ln(2x+5) - \int 1 - \frac{5}{2x+5}\, dx$

$= x \cdot \ln(2x+5) - \{ x - \frac{5}{2} \cdot \ln|2x+5| \} \Big|_1^3 = \{ 3 \cdot \ln(11) - 3 + \frac{5}{2} \cdot \ln|11| \} - \{ 1 \cdot \ln(7) - 1 + \frac{5}{2} \cdot \ln|7| \}$

$= \frac{11}{2} \cdot \ln(11) - \frac{7}{2} \cdot \ln(7) - 2 \approx \mathbf{4.38}$.

17. $\int_1^e (\ln(x))^2\, dx$.   Put $u = (\ln(x))^2$. Then $dv = dx$, $du = 2 \cdot \ln(x) \cdot \frac{1}{x}\, dx$, and $v = x$.

$= uv - \int v\, du = (\ln(x))^2 x - \int x \cdot 2 \cdot \ln(x) \cdot \frac{1}{x}\, dx$

$= x \cdot (\ln(x))^2 - \int 2 \cdot \ln(x)\, dx = x \cdot (\ln(x))^2 - 2\{ x \cdot \ln(x) - x \} \Big|_1^e$

$= \{ e(\ln(e))^2 - 2e \cdot \ln(e) + 2e \} - \{ 1(\ln(1))^2 - 2 \cdot \ln(1) + 2 \} = e - 2 \approx \mathbf{0.718}$.

19. $\int \arcsin(x)\, dx$.   Put $u = \arcsin(x)$. Then $dv = dx$, $du = \frac{1}{\sqrt{1-x^2}}\, dx$, and $v = x$.

$= uv - \int v\, du = \arcsin(x) \cdot x - \int x \cdot \frac{1}{\sqrt{1-x^2}}\, dx = x \cdot \arcsin(x) - \int \frac{x}{\sqrt{1-x^2}}\, dx$ (use u–sub with $u = 1 - x^2$)

$= x \cdot \arcsin(x) - \int \frac{1}{\sqrt{u}} \left(\frac{-1}{2} \right) du = x \cdot \arcsin(x) + \sqrt{u} + C = \mathbf{x \cdot \arcsin(x) + \sqrt{1-x^2} + C}$.

21. $\int x \cdot \arctan(3x)\, dx$.   Put $u = \arctan(3x)$. Then $dv = x\, dx$, $du = \dfrac{3}{1+9x^2}\, dx$, and $v = \dfrac{1}{2}x^2$.

$= uv - \int v\, du = \arctan(x) \cdot \dfrac{1}{2}x^2 - \int \dfrac{1}{2}x^2 \cdot \dfrac{3}{1+9x^2}\, dx = \dfrac{1}{2}x^2 \cdot \arctan(3x) - \dfrac{1}{2}\int \dfrac{1}{3} - \dfrac{1/3}{1+9x^2}\, dx$

$= \dfrac{1}{2}x^2 \cdot \arctan(3x) - \dfrac{1}{2}\{\dfrac{x}{3} - \dfrac{1}{9} \cdot \arctan(3x)\} + C = \dfrac{1}{2}x^2 \cdot \arctan(3x) - \dfrac{1}{6}x + \dfrac{1}{18}\arctan(3x)\} + C$.

23. $\int_1^2 \dfrac{\ln(x)}{x}\, dx$.   **Use u–substitution!** Put $u = \ln(x)$. Then $du = \dfrac{1}{x}\, dx$.

$\int = \int u\, du = \dfrac{1}{2}u^2 = \dfrac{1}{2}(\ln(x))^2 \Big|_1^2 = \dfrac{1}{2}(\ln(2))^2 - \dfrac{1}{2}(\ln(1))^2 = \dfrac{1}{2}(\ln(2))^2 \approx \mathbf{0.240}$.

25. (a) $\int \sin^3(x)\, dx = \dfrac{1}{3}\{-S^2 \cdot C + 2\int S\, dx\} = \dfrac{1}{3}\{-S^2 \cdot C - 2C\} + K$

$= \dfrac{1}{3}\{-\sin^2(x) \cdot \cos(x) - 2 \cdot \cos(x)\} + K$.

(b) $\int \sin^4(x)\, dx = \dfrac{1}{4}\{-S^3 \cdot C + 3\int S^2\, dx\} = \dfrac{1}{4}\{-S^4 \cdot C + 3[\dfrac{1}{2}(-SC + x)]\} + K$

$= \dfrac{1}{4}\{-S^3 \cdot C - \dfrac{3}{2}SC + \dfrac{3}{2}x\} + K$

$= \dfrac{1}{4}\{-\sin^3(x) \cdot \cos(x) - \dfrac{3}{2}\sin(x) \cdot \cos(x) + \dfrac{3}{2}x\} + K$ (c) on your own.

27. (a) $\int \sec^3(x)\, dx = \dfrac{1}{2}\{\sec(x)\tan(x) + \int \sec(x)\, dx\} = \dfrac{1}{2}\{\sec(x) \cdot \tan(x) + \ln|\sec(x) + \tan(x)|\} + K$

(b) and (c)   on your own.

29. $\int \cos^3(2x+3)\, dx$. First do a substitution: $u = 2x + 3$. Then $du = 2\, dx$ and $dx = \dfrac{1}{2}\, du$.

$= \int \dfrac{1}{2}\cos^3(u)\, du = \dfrac{1}{2}\{\dfrac{1}{3}(C^2 \cdot S + 2\int C\, du)\} = \dfrac{1}{6}\{C^2 \cdot S + 2S\} + K$

$= \dfrac{1}{6}\{\cos^2(u) \cdot \sin(u) + 2\sin(u)\} + K = \dfrac{1}{6}\{\cos^2(2x+3) \cdot \sin(2x+3) + 2\sin(2x+3)\} + K$

31. $\int x \cdot (2x+5)^{19}\, dx$.

(a) By parts: put $u = x$. Then $dv = (2x+5)^{19}\, dx$, $du = dx$, and $v = \dfrac{1}{40}(2x+5)^{20}$.

$\int = uv - \int v\, du = x \cdot \dfrac{1}{40}(2x+5)^{20} - \int \dfrac{1}{40}(2x+5)^{20}\, dx$

$= x \cdot \dfrac{1}{40}(2x+5)^{20} - \dfrac{1}{40}\dfrac{1}{42}(2x+5)^{21} + C$.

(b) Substitution: put $u = 2x + 5$. Then $du = 2\, dx$ and $dx = \dfrac{1}{2}\, du$. Also, $x = \dfrac{1}{2}(u-5)$.

$\int = \int \dfrac{1}{2}(u-5) \cdot u^{19} \dfrac{1}{2}\, du = \dfrac{1}{4}\int u^{20} - 5u^{19}\, du = \dfrac{1}{4}\{\dfrac{1}{21}u^{21} - \dfrac{5}{20}u^{20}\} + C$

$= \dfrac{1}{84}(2x+5)^{21} - \dfrac{5}{80}(2x+5)^{20} + C$.

The answers (antiderivatives) in parts (a) and (b) look different, but you can check that the derivative of each answer is $x \cdot (2x+5)^{19}$.

Odd Answers: Chapter Eight   Contemporary Calculus

33. (a) Make an informed prediction.

(b) $\int_0^1 \sin(x)\,dx = -\cos(x)\Big|_0^1 = (-\cos(1)) - (-\cos(0)) = \cos(0) - \cos(1) \approx 1 - 0.54 = \mathbf{0.46}$.

$\int_0^1 x\cdot\sin(x)\,dx = -x\cdot\cos(x) + \sin(x)\Big|_0^1 = \{-1\cdot\cos(1) + \sin(1)\} - \{-0\cdot\cos(0) + \sin(0)\} = \sin(1) - \cos(1) \approx \mathbf{0.30}$.

35. (a) Make an informed prediction.

(b) See problem 17: $V_{x-axis} = \int_{x=1}^{e} \pi(\ln(x))^2\,dx = \pi(e-2) \approx \mathbf{2.257}$.

$V_{y-axis} = \int_{y=0}^{1} 2\pi\cdot y\cdot e^y\,dy = 2\pi\int_{y=0}^{1} y\cdot e^y\,dy$ (use integration by parts with $u = y$, $dv = e^y\,dy$: see Example 2)

$= 2\pi(y\cdot e^y - e^y)\Big|_0^1 = 2\pi(1\cdot e^1 - e^1) - 2\pi(0\cdot e^0 - e^0) = 2\pi(0) - 2\pi(-1) = 2\pi \approx \mathbf{6.283}$.

37. On your own.

39. $\int x^2\cdot\arctan(x)\,dx$.   Put $u = \arctan(x)$. Then $dv = x^2\,dx$, $du = \dfrac{1}{x^2+1}\,dx$, and $v = \dfrac{1}{3}x^3$.

$= uv - \int v\,du = \arctan(x)\cdot\dfrac{1}{3}x^3 - \int \dfrac{1}{3}x^3\dfrac{1}{x^2+1}\,dx$

$= \dfrac{1}{3}x^3\cdot\arctan(x) - \dfrac{1}{3}\int x - \dfrac{x}{x^2+1}\,dx$ (dividing x^3 by x^2+1)

$= \dfrac{1}{3}x^3\cdot\arctan(x) - \dfrac{1}{3}\{\dfrac{1}{2}x^2 - \dfrac{1}{2}\ln(x^2+1)\} + C$

$= \dfrac{1}{3}x^3\cdot\mathbf{arctan(x)} - \dfrac{1}{6}x^2 + \dfrac{1}{6}\ln(x^2+1)\} + C$.

40 – 50. On your own.

Section 8.4

1. $= \dfrac{A}{x} + \dfrac{B}{x+1} = \dfrac{2}{x} + \dfrac{5}{x+1}$

3. $= \dfrac{A}{x+1} + \dfrac{B}{x+8} = \dfrac{2}{x+1} + \dfrac{9}{x+8}$

5. Divide first: $\dfrac{2x^2 + 15x + 25}{x^2 + 5x} = 2 + \dfrac{5x + 25}{x^2 + 5x} = 2 + \dfrac{5(x+5)}{x(x+5)} = 2 + \dfrac{5}{x}$.

7. $\dfrac{6x^2 + 9x - 15}{x(x+5)(x-1)} = \dfrac{A}{x} + \dfrac{B}{x+5} + \dfrac{C}{x-1} = \dfrac{A(x^2 + 4x - 5) + B(x^2 - x) + C(x^2 + 5x)}{x(x+5)(x-1)}$.

Solving x^2: $A + B + C = 6$
          x: $4A - B + 5C = 9$
          k: $-5A = -15$ we get $A = 3$, $B = 3$, and $C = 0$ so

$\dfrac{6x^2 + 9x - 15}{x(x+5)(x-1)} = \dfrac{3}{x} + \dfrac{3}{x+5} + \dfrac{0}{x-1} = \dfrac{3}{x} + \dfrac{3}{x+5}$.

Odd Answers: Chapter Eight   Contemporary Calculus

9. $\dfrac{8x^2 - x + 3}{x(x^2 + 1)} = \dfrac{A}{x} + \dfrac{Bx + C}{x^2 + 1} = \dfrac{A(x^2 + 1) + x(Bx + C)}{x(x^2 + 1)}$.

Solving x^2: $A + B = 8$
         x: $C = -1$
         k: $A = 3$ we get $A = 3, B = 5,$ and $C = -1$ so

$\dfrac{8x^2 - x + 3}{x(x^2 + 1)} = \dfrac{3}{x} + \dfrac{5x - 1}{x^2 + 1}$.

11. $\dfrac{11x^2 + 23x + 6}{x^2(x + 2)} = \dfrac{A}{x} + \dfrac{B}{x^2} + \dfrac{C}{x + 2} = \dfrac{A(x)(x + 2) + B(x + 2) + C(x^2)}{x^2(x + 2)}$.

Solving x^2: $A + C = 11$
         x: $2A + B = 23$
         k: $2B = 6$ we get $A = 10, B = 3,$ and $C = 1$ so

$\dfrac{11x^2 + 23x + 6}{x^2(x + 2)} = \dfrac{10}{x} + \dfrac{3}{x^2} + \dfrac{1}{x + 2}$.

13. $\int \dfrac{3x + 13}{(x + 2)(x - 5)}\, dx = \int \dfrac{-1}{x + 2} + \dfrac{4}{x - 5}\, dx = -\ln|x + 2| + 4\cdot\ln|x - 5| + C$.

15. $\int_2^5 \dfrac{2}{x^2 - 1}\, dx = \int_2^5 \dfrac{-1}{x + 1} + \dfrac{1}{x - 1}\, dx = -\ln|x + 1| + \ln|x - 1|$

$= \ln\left|\dfrac{x - 1}{x + 1}\right|\Big|_2^5 = \ln(\tfrac{4}{6}) - \ln(\tfrac{1}{3}) \approx \mathbf{0.693}$.

17. (1) $\int \dfrac{2}{x} + \dfrac{5}{x + 1}\, dx = \mathbf{2\cdot\ln|x| + 5\cdot\ln|x + 1| + C}$

    (2) $\int \dfrac{3}{x + 3} + \dfrac{4}{x - 1}\, dx = \mathbf{3\cdot\ln|x + 3| + 4\cdot\ln|x - 1| + C}$

19. $\int \dfrac{2x^2 + 5x + 3}{x^2 - 1}\, dx = \int 2 + \dfrac{5}{x - 1}\, dx = \mathbf{2x + 5\cdot\ln|x - 1| + C}$.

21. $\int \dfrac{3x^2 + 19x + 24}{x^2 + 6x + 5}\, dx = \int 3 + \dfrac{2}{x + 1} + \dfrac{-1}{x + 5}\, dx = \mathbf{3x + 2\cdot\ln|x + 1| - \ln|x + 5| + C}$.

23. $\int \dfrac{3x^2 - 1}{x^3 - x}\, dx$. Use u–substitution with $u = x^3 - x$. Then $du = 3x^2 - 1$ so

$\int = \int \dfrac{1}{u}\, du = \ln|u| + C = \ln|x^3 - x| + C$. A partial fraction decomposition also works but takes longer.

25. $\int \dfrac{x^3 + 3x^2 - 4x + 30}{x^2 + 3x - 10}\, dx = \int x + \dfrac{6}{x - 2}\, dx = \mathbf{\tfrac{1}{2} x^2 + 6\ln|x - 2| + C}$.

Odd Answers: Chapter Eight   Contemporary Calculus

27. $\int \dfrac{12x^2 + 19x - 6}{x^3 + 3x^2} \, dx = \int \dfrac{7}{x} + \dfrac{-2}{x^2} + \dfrac{5}{x+3} \, dx = 7\cdot\ln|x| + \dfrac{2}{x} + 5\cdot\ln|x+3| + C$.

29. $\int \dfrac{7x^2 + 3x + 7}{x^3 + x} \, dx = \int \dfrac{7}{x} + \dfrac{3}{x^2 + 1} \, dx = 7\cdot\ln|x| + 3\cdot\arctan(x) + C$.

31. (a) $\int \dfrac{1}{x^2 + 2x + 2} \, dx = \int \dfrac{1}{(x+1)^2 + 1} \, dx = \arctan(x+1) + C$.

  (b) $\int \dfrac{1}{x^2 + 2x + 1} \, dx = \int \dfrac{1}{(x+1)^2} \, dx = -(x+1)^{-1} + C = \dfrac{-1}{x+1} + C$.

  (c) $\int \dfrac{1}{x^2 + 2x + 0} \, dx = \int \dfrac{1}{x(x+2)} \, dx = \int \dfrac{1/2}{x} + \dfrac{-1/2}{x+2} \, dx = \dfrac{1}{2}\ln|x| - \dfrac{1}{2}\ln|x+2| + C$.

33. Prob. 1: $f(x) = \dfrac{2}{x} + \dfrac{5}{x+1} = 2\cdot(x)^{-1} + 5\cdot(x+1)^{-1}$. Then

$f'(x) = -2\cdot(x)^{-2} - 5\cdot(x+1)^{-2}$ and $f''(x) = 4\cdot(x)^{-3} + 10\cdot(x+1)^{-3}$.

Prob. 3: $g(x) = \dfrac{3}{x+3} + \dfrac{4}{x-1} = 3\cdot(x+3)^{-1} + 4\cdot(x-1)^{-1}$. Then

$g'(x) = -3\cdot(x+3)^{-2} - 4\cdot(x-1)^{-2}$ and $g''(x) = 6\cdot(x+3)^{-3} + 8\cdot(x-1)^{-3}$.

35. Prob. 5: $f(x) = \dfrac{2x^2 + 15x + 15}{x^2 + 5x} = 2 + \dfrac{5}{x}$ so $f'(x) = \dfrac{-5}{x^2}$ and $f''(x) = \dfrac{10}{x^3}$.

Prob. 6:   On your own.

37. (a) Solve $\dfrac{dx}{dt} = x(100 - x)$.

Separate the variables: $\dfrac{1}{x(100-x)} \, dx = dt$.

Use partial fractions: $\dfrac{1}{x(100-x)} = \dfrac{A}{x} + \dfrac{B}{100-x} = \dfrac{0.01}{x} + \dfrac{0.01}{100-x}$ (solving $-A+B = 0$ and $100A = 1$)

so $\left\{ \dfrac{0.01}{x} + \dfrac{0.01}{100-x} \right\} dx = dt$ and, integrating, $\int \dfrac{0.01}{x} + \dfrac{0.01}{100-x} \, dx = \int 1 \, dt$. Then

$0.01\cdot\ln|x| - 0.01\cdot\ln|100-x| = t + C$ so $\ln\left|\dfrac{x}{100-x}\right| = 100t + K$. ($K = 100C$ is a constant)

Using the initial condition $x(0) = 150$: $\ln\left|\dfrac{150}{100-150}\right| = \ln(3) = 100\cdot(0) + K$ so $K = \ln(3)$.

Finally, $\ln\left|\dfrac{x}{100-x}\right| = 100\cdot t + \ln(3)$ so $\left|\dfrac{x}{100-x}\right| = e^{100t}\cdot e^{\ln(3)} = 3\cdot e^{100t}$ and

$x = \dfrac{-300\cdot e^{100t}}{1 - 3\cdot e^{100t}}$. Graph this on your own.

  (b) In this part it is easier to use the form $\left|\dfrac{x}{100-x}\right| = 3\cdot e^{100t}$.

Put $x = 120$ and solve $\left|\dfrac{120}{100-120}\right| = 3\cdot e^{100t}$: $6 = 3\cdot e^{100t}$ so $t = \dfrac{1}{100}\ln(2) \approx \mathbf{0.0069}$.

Put $x = 110$ and solve $\left|\frac{110}{100-110}\right| = 3 \cdot e^{100t}$: $11 = 3 \cdot e^{100t}$ so $t = \frac{1}{100} \ln(\frac{11}{3}) \approx \mathbf{0.013}$.

Put $x = 100$. Then $\left|\frac{x}{100-x}\right|$ is undefined (division by 0) so $x(t)$ is never equal to 100.

(c) limit (as t becomes arbitrarily large) of $x = \frac{-300 \cdot e^{100t}}{1 - 3 \cdot e^{100t}}$ is $\frac{-300}{-3} = 100$.

(d) The population $x(t)$ is decling to the "carrying capacity" $M = 100$ of the enviroment.

39. (a) Solve $\frac{dx}{dt} = (7-x)(5-x)$ by separating the variables and using partial fractions to rewrite the fraction.

$\frac{7-x}{5-x} = \frac{7}{5} e^{2kt}$ and $x(t) = \frac{7 \cdot e^{2kt} - 7}{\frac{7}{5} \cdot e^{2kt} - 1}$. (As t gets big, x approaches 5.)

(b) Solve $\frac{dx}{dt} = (6-x)(6-x) = (6-x)^2$ by separating the variables: $\frac{1}{(6-x)^2} dx = dt$ and integrating.

$\frac{1}{6-x} = t + C$ $(C = \frac{1}{6})$ so $x = \frac{6t+1}{t+\frac{1}{6}} = \frac{36t+6}{6t+1}$. (As t gets big, x approaches 6.)

Section 8.5

1. $x = 3 \cdot \sin(\theta)$ (a) $9 - x^2 = 9 - 9\sin^2(\theta) = 9(1 - \sin^2(\theta)) = 9\cos^2(\theta)$ so $\frac{1}{\sqrt{9-x^2}} = \frac{1}{3\cos(\theta)}$.

   (b) $dx = 3\cos(\theta) \, d\theta$

3. $x = 3 \cdot \sec(\theta)$ (a) $x^2 - 9 = 9\sec^2(\theta) - 9 = 9(\sec^2(\theta) - 1) = 9\tan^2(\theta)$ so $\frac{1}{\sqrt{x^2-9}} = \frac{1}{3\tan(\theta)}$.

   (b) $dx = 3\sec(\theta)\tan(\theta) \, d\theta$

5. $x = \sqrt{2} \tan(\theta)$ (a) $2 + x^2 = 2 + 2\tan^2(\theta) = 2(1 + \tan^2(\theta)) = 2\sec^2(\theta)$ so $\frac{1}{\sqrt{2+x^2}} = \frac{1}{\sqrt{2}\sec(\theta)}$.

   (b) $dx = \sqrt{2} \sec^2(\theta) \, d\theta$

7. $x = 3 \cdot \sin(\theta)$ (a) $\theta = \arcsin(x/3)$

   (b) & (c) $f(\theta) = \cos(\theta) \cdot \tan(\theta) = \cos(\arcsin(x/3)) \cdot \tan(\arcsin(x/3)) = \frac{\sqrt{9-x^2}}{3} \cdot \frac{x}{\sqrt{9-x^2}} = \frac{x}{3}$.

9. $x = 3 \cdot \sec(\theta)$ (a) $\theta = \text{arcsec}(x/3)$

   (b) & (c) $f(\theta) = \sqrt{1 + \sin^2(\theta)} = \sqrt{1 + \sin^2(\text{arcsec}(x/3))} = \sqrt{1 + \left(\frac{\sqrt{x^2-9}}{x}\right)^2}$

11. $x = 5 \cdot \tan(\theta)$ (a) $\theta = \arctan(x/5)$

(b) & (c) $f(\theta) = \dfrac{\cos^2(\theta)}{1 + \cot(\theta)} = \dfrac{\left(\dfrac{5}{\sqrt{25+x^2}}\right)^2}{1 + \left(\dfrac{5}{x}\right)} = \dfrac{\dfrac{25}{25+x^2}}{1 + \dfrac{5}{x}}$

13. Same as Practice 3. Sorry.

15. $x = 7 \cdot \tan(\theta)$. $dx = 7 \sec^2(\theta)\, d\theta$. $x^2 + 49 = 49\tan^2(\theta) + 49 = 49(\tan^2(\theta) + 1) = 49\sec^2(\theta)$.

$$\int \dfrac{1}{\sqrt{x^2+49}}\, dx = \int \dfrac{1}{\sqrt{49\sec^2(\theta)}}\, 7\sec^2(\theta)\, d\theta$$

$$= \int \sec(\theta)\, d\theta = \ln|\sec(\theta) + \tan(\theta)| + C = \ln|\sec(\arctan(x/7)) + \tan(\arctan(x/7))| + C$$

$$= \ln\left| \dfrac{\sqrt{x^2+49}}{7} + \dfrac{x}{7} \right| + C.$$

17. $x = 6 \cdot \sin(\theta)$. $dx = 6\cos(\theta)\, d\theta$. $36 - x^2 = 36\cos^2(\theta)$.

$$\int \sqrt{36-x^2}\, dx = \int \sqrt{36 - \cos^2(\theta)}\, 6\cos(\theta)\, d\theta = 36 \int \cos^2(\theta)\, d\theta = \text{(use Table \#14)}$$

$$= 36\left\{\dfrac{1}{2}\theta + \dfrac{1}{2}\sin(\theta)\cdot\cos(\theta)\right\} + C = 36\left\{\dfrac{1}{2}\arcsin(x/6) + \dfrac{1}{2}\left(\dfrac{x}{6}\right)\cdot\left(\dfrac{\sqrt{36-x^2}}{6}\right)\right\} + C.$$

19. $x = 6 \cdot \tan(\theta)$. $dx = 6 \cdot \sec^2(\theta)\, d\theta$. $36 + x^2 = 36\sec^2(\theta)$.

$$\int \sqrt{36+x^2}\, dx = \int \dfrac{1}{\sqrt{36+x^2}}\, 6\cdot\sec^2(\theta)\, d\theta = \int \sec(\theta)\, d\theta = \text{(use Table \#11)}$$

$$= \ln|\sec(\theta) + \tan(\theta)| + C = \ln\left|\dfrac{\sqrt{36+x^2}}{6} + \dfrac{x}{6}\right| + C \text{ or } \ln|\sqrt{36+x^2} + x| + K.$$

21. Similar to 19: $x = 7\cdot\tan(\theta)$. $\displaystyle\int \dfrac{1}{\sqrt{49+x^2}}\, dx = \ln|\sqrt{49+x^2} + x| + C.$

23. $x = 5\cdot\sin(\theta)$. $\displaystyle\int = -5\cos(\arcsin(x/5)) + C = -5\dfrac{\sqrt{25-x^2}}{5} + C = -\sqrt{25-x^2} + C.$

25. $x = 7\cdot\tan(\theta)$. $\displaystyle\int = -\ln|\cos(\arctan(x/7))| + C = -\ln\left|\dfrac{7}{\sqrt{49+x^2}}\right| + C$ (now some algebra)

$$= -\ln|7| + \ln|\sqrt{49+x^2}| + C = \dfrac{1}{2}\ln|49+x^2| + K.$$

(A u–substitution with $u = 49 + x^2$ is **much** easier.)

Odd Answers: Chapter Eight   Contemporary Calculus

27. $x = 3 \cdot \sec(\theta)$. $\int = \frac{1}{9} \int \frac{\cos(\theta)}{\sin^2(\theta)} d\theta = $ (put $u = \sin(\theta)$) $\frac{-1}{9} \frac{1}{\sin(\theta)} + C = -\csc(\theta) + C = \frac{-1}{9} \frac{x}{\sqrt{x^2+9}} + C$.

29. $x = 5 \cdot \sec(\theta)$. $\int = \frac{1}{2} \theta + C = \frac{1}{2} \operatorname{arcsec}(\frac{x}{5}) + C$.

31. $x = 5 \cdot \sin(\theta)$. $\int = \frac{1}{5} \ln| \sec(\theta) + \tan(\theta) | + C = \frac{1}{5} \ln\left| \frac{5}{\sqrt{25-x^2}} + \frac{x}{\sqrt{25-x^2}} \right| + C$

$= $ (after lots of algebra) $\frac{1}{10} \ln\left| \frac{5+x}{5-x} \right| + C$.

33. Similar to 19. $\int = \ln\left| \frac{\sqrt{a^2+x^2}}{a} + \frac{x}{a} \right| + C = \ln| \sqrt{a^2+x^2} + x | + K$.

35. $x = a \cdot \tan(\theta)$. $\int = \frac{-1}{a^2} \frac{\sqrt{a^2+x^2}}{x} + C$.

37. $x + 1 = u$. Then $u = 3 \cdot \tan(\theta)$. $\int = \ln| \sec(\theta) + \tan(\theta) | + C = \ln\left| \frac{\sqrt{u^2+9}}{3} + \frac{u}{3} \right| + C$

$= \ln| \sqrt{u^2+9} + u | - \ln(3) + C = \ln| \sqrt{u^2+9} + u | + K = \ln| \sqrt{(x+1)^2+9} + (x+1) | + K$.

39. $\frac{1}{2} \arctan(\frac{x+5}{2}) + C$ 41. $\ln| (x+2) + \sqrt{x^2+4x+3} | + C$.

Section 8.6

1. $\int \sin^2(3x) dx = \frac{x}{2} - \frac{\sin(6x)}{12} + C = \frac{x}{2} - \frac{\sin(3x) \cdot \cos(3x)}{6} + C$.

3. Put $u = \sin(e^x)$. $\int = \frac{1}{2} \sin^2(e^x) + C$. (If you put $w = \cos(e^x)$ then $\int = -\frac{1}{2} \cos^2(e^x) + C$.)

5. $\frac{3}{8} \pi$ 7. $\frac{4}{21}$

9. $\frac{1}{14} \sin^2(7x) + C$ or $-\frac{1}{14} \cos^2(7x) + C$

11. Put $u = \cos(7x)$. $\int = -\frac{1}{28} \cos^4(7x) + C$.

13. $\int \sin^2(3x)\cos^2(3x)\,dx = \int \frac{1}{2}(1-\cos(6x))\frac{1}{2}(1+\cos(6x))\,dx$

$= \frac{1}{4}\int 1-\cos^2(6x)\,dx = \frac{x}{4} - \frac{1}{4}\int \cos^2(6x)\,dx$

$= \frac{x}{4} - \frac{1}{4}\left\{\frac{x}{2} + \frac{\sin(12x)}{24}\right\} + C$.

15. $\int = \frac{1}{10}\tan^2(5x) + C$.   17. $\int = \frac{1}{9}\sec^3(3x) + C$.

19. $m \neq n$. $\int_0^{2\pi} \sin(mx)\sin(nx)\,dx = \frac{1}{2}\left\{\frac{\sin((m-n)x)}{m-n} - \frac{\sin((m+n)x)}{m+n}\right\}\Big|_0^{2\pi}$

$= \frac{1}{2}\left\{\frac{\sin((m-n)2\pi)}{m-n} - \frac{\sin((m+n)2\pi)}{m+n}\right\} - \frac{1}{2}\left\{\frac{\sin((m-n)0)}{m-n} - \frac{\sin((m+n)0)}{m+n}\right\}$

$= \frac{1}{2}\{0-0\} - \frac{1}{2}\{0-0\} = 0$.

21. $\int_0^{2\pi} \sin(mx)\sin(mx)\,dx = \int_0^{2\pi} \sin^2(mx)\,dx = \frac{x}{2} - \frac{\sin(mx)\cos(mx)}{2m}\Big|_0^{2\pi}$

$= \left\{\frac{2\pi}{2} - \frac{\sin(m2\pi)\cos(m2\pi)}{2m}\right\} - \left\{\frac{0}{2} - \frac{\sin(0)\cos(0)}{2m}\right\} = \pi$.

23. On your own.

9.1 POLAR COORDINATES

The rectangular coordinate system is immensely useful, but it is not the only way to assign an address to a point in the plane and sometimes it is not the most useful. In many experimental situations, our location is fixed and we or our instruments, such as radar, take readings in different directions (Fig. 1); this information can be graphed using rectangular coordinates (e.g., with the angle on the horizontal axis and the measurement on the vertical axis). Sometimes, however, it is more useful to plot the information in a way similar to the way in which it was collected, as magnitudes along radial lines (Fig. 2). This system is called the Polar Coordinate System.

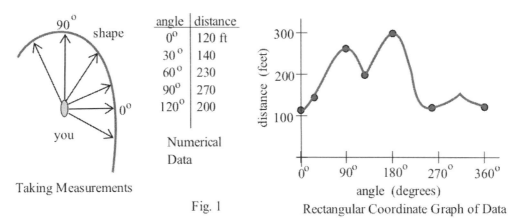

Fig. 1

In this section we introduce polar coordinates and examine some of their uses. We start with graphing points and functions in polar coordinates, consider how to change back and forth between the rectangular and polar coordinate systems, and see how to find the slopes of lines tangent to polar graphs. Our primary reasons for considering polar coordinates, however, are that they appear in applications, and that they provide a "natural" and easy way to represent some kinds of information.

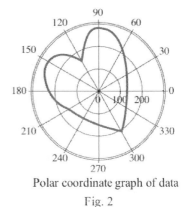

Polar coordinate graph of data
Fig. 2

Example 1: SOS! You've just received a distress signal from a ship located at A on your radar screen (Fig. 3). Describe its location to your captain so your vessel can speed to the rescue.

Solution: You could convert the relative location of the other ship to rectangular coordinates and then tell your captain to go due east for 7.5 miles and north for 13 miles,

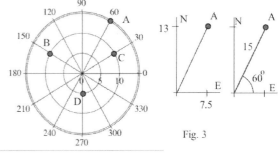

Fig. 3

but that certainly is not the quickest way to reach the other ship. It is better to tell the captain to sail for 15 miles in the direction of 60°. If the distressed ship was at B on the radar screen, your vessel should sail for 10 miles in the direction 150°. (Real radar screens have 0° at the top of the screen, but the convention in mathematics is to put 0° in the direction of the positive x–axis and to measure positive angles counterclockwise from there. And a real sailor speaks of "bearing" and "range" instead of direction and magnitude.)

Practice 1: Describe the locations of the ships at C and D in Fig. 3 by giving a distance and a direction to those ships from your current position at the center of the radar screen.

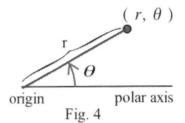

Fig. 4

Points in Polar Coordinates

To construct a polar coordinate system we need a starting point (called the **origin** or **pole**) for the magnitude measurements and a starting direction (called the **polar axis**) for the angle measurements (Fig. 4).

> A polar coordinate pair for a point P in the plane is an ordered pair (r, θ) where r is the directed distance along a radial line from O to P, and θ is the angle formed by the polar axis and the segment OP (Fig. 4).

The angle θ is positive when the angle of the radial line OP is measured counterclockwise from the polar axis, and θ is negative when measured clockwise.

Degree or Radian Measure for θ? Either degree or radian measure can be used for the angle in the polar coordinate system, but when we differentiate and integrate trigonometric functions of θ we will want all of the angles to be given in radian measure. From now on, we will primarily use radian measure. You should assume that all angles are given in radian measure unless the units " ° " ("degrees") are shown.

Example 2: Plot the points with the given polar coordinates: $A(2, 30°)$, $B(3, \pi/2)$, $C(-2, \pi/6)$, and $D(-3, 270°)$.

Solution: To find the location of A, we look along the ray that makes an angle of 30° with the polar axis, and then take two steps in that direction (assuming 1 step = 1 unit). The locations of A and B are shown in Fig. 5.

To find the location of C, we look along the ray which makes an angle of $\pi/6$ with the polar axis, and then we take two steps backwards since $r = -2$ is negative. Fig. 6 shows the locations of C and D.

Notice that the points B and D have different addresses, $(3, \pi/2)$ and $(-3, 270°)$, but the same location.

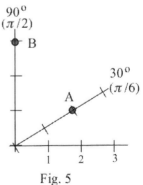

Fig. 5

Practice 2: Plot the points with the given polar coordinates: A(2, π/2), B(2, –120°), C(–2, π/3), D(–2, –135°), and E(2, 135°). Which two points coincide?

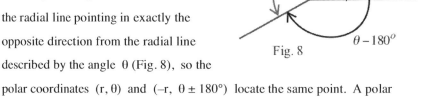

Fig. 6

Each polar coordinate pair (r, θ) gives the location of one point, but each location has lots of different addresses in the polar coordinate system: the polar coordinates of a point are not unique. This nonuniqueness of addresses comes about in two ways. First, the angles θ, θ ± 360°, θ ± 2·360°, . . . all describe the same radial line (Fig. 7), so the polar coordinates (r, θ), (r, θ ± 360°), (r, θ ± 2·360°) , . . . all locate the same point.

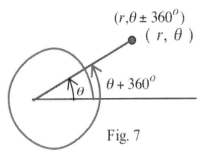

Secondly, the angle θ ± 180° describes the radial line pointing in exactly the opposite direction from the radial line described by the angle θ (Fig. 8), so the polar coordinates (r, θ) and (–r, θ ± 180°) locate the same point. A polar coordinate pair gives the location of exactly one point, but the location of one point is described by many (an infinite number) different polar coordinate pairs.

Fig. 7

Fig. 8

Note: In the rectangular coordinate system we use (x, y) and y = f(x): first variable independent and second variable dependent. In the polar coordinate system we use (r, θ) and r = f(θ): first variable dependent and second variable independent, a reversal from the rectangular coordinate usage.

Practice 3: Table 1 contains measurements to the edge of a plateau taken by a remote sensor which crashed on the plateau. Fig. 9 shows the data plotted in rectangular coordinates. Plot the data in polar coordinates and determine the shape of the top of the plateau.

angle	distance	angle	distance
0°	28 feet	150°	22 feet
20°	30	180°	18
40°	36	210°	21
60°	27	230°	13
80°	24	270°	10
100°	24	330°	18
130°	30	340°	30

Table 1

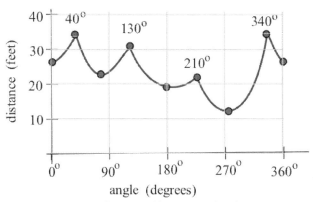

Rectangular Coordinate Graph of Data

Fig. 9

Graphing Functions in the Polar Coordinate System

In the rectangular coordinate system, we have worked with functions given by tables of data, by graphs, and by formulas. Functions can be represented in the same ways in polar coordinates.

- If a function is given by a table of data, we can graph the function in polar coordinates by plotting individual points in a polar coordinate system and connecting the plotted points to see the shape of the graph. By hand, this is a tedious process; by calculator or computer, it is quick and easy.

- If the function is given by a rectangular coordinate graph of magnitude as a function of angle, we can read coordinates of points on the rectangular graph and replot them in polar coodinates. In essence, as we go from the rectangular coordinate graph to the polar coordinate graph we "wrap" the rectangular graph around the "pole" at the origin of the polar coordinate system. (Fig. 10)

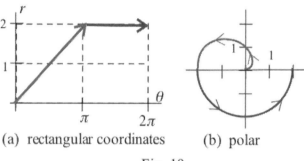

(a) rectangular coordinates   (b) polar

Fig. 10

- If the function is given by a formula, we (or our calculator) can graph the function to help us obtain information about its behavior. Typically, a graph is created by evaluating the function at a lot of points and then plotting the points in the polar coordinate system. Some of the following examples illustrate that functions given by simple formulas may have rather exotic graphs in the polar coordinate system.

If a function is already given by a polar coordinate graph, we can use the graph to answer questions about the behavior of the function. It is usually easy to locate the maximum value(s) of r on a polar coordinate graph, and, by moving counterclockwise around the graph, we can observe where r is increasing, constant, or decreasing.

Example 3: Graph $r = 2$ and $r = \pi - \theta$ in the polar coordinate system for $0 \leq \theta \leq 2\pi$.

Solution: $r = 2$: In every direction θ, we simply move 2 units along the radial line and plot a point. The resulting polar graph (Fig. 11b) is a circle centered at the origin with a radius of 2. In the rectangular coordinate system, the graph of a constant $y = k$ is a horizontal line. In the polar coordinate system, the graph of a constant $r = k$ is a circle with radius $|k|$.

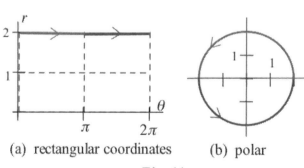

(a) rectangular coordinates   (b) polar

Fig. 11

$r = \pi - \theta$: The rectangular coordinate graph of $r = \pi - \theta$ is shown in Fig. 12a. If we read the values of r and θ from the rectangular coordinate graph and plot them in polar coordinates, the result is the shape in Fig. 12b. The different line thicknesses are used in the figures to help you see which values from the rectangular graph became which parts of the loop in the polar graph.

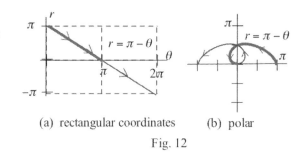
(a) rectangular coordinates   (b) polar
Fig. 12

Practice 4:   Graph $r = -2$ and $r = \cos(\theta)$ in the polar coordinate system.

Example 4:   Graph $r = \theta$ and $r = 1 + \sin(\theta)$ in the polar coordinate system.

Solution: $r = \theta$: The rectangular coordinate graph of $r = \theta$ is a straight line (Fig. 13a). If we read the values of r and θ from the rectangular coordinate graph and plot them in polar coordinates, the result is the spiral, called an Archimedean spiral, in Fig. 13b.

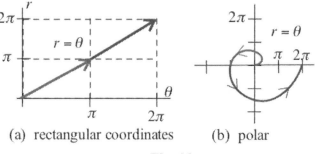
(a) rectangular coordinates   (b) polar
Fig. 13

$r = 1 + \sin(\theta)$: The rectangular coordinate graph of $r = 1 + \sin(\theta)$ is shown in Fig. 14a, and it is the graph of the sine curve shifted up 1 unit. In polar coordinates, the result of adding 1 to sine is much less obvious and is shown in Fig. 14b.

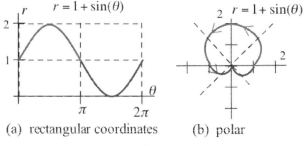
(a) rectangular coordinates   (b) polar
Fig. 14

Practice 5:   Plot the points in Table 2 in the polar coordinate system and connect them with a smooth curve. Describe the shape of the graph in words.

angle (radians)	distance (meters)
0	3.0
π/6	1.9
π/4	1.7
π/3	1.6
π/2	2.0

Table 2

Fig. 15 shows the effects of adding various constants to the rectangular and polar graphs of r = sin(θ). In rectangular coordinates the result is a graph shifted up or down by k units. In polar coordinates, the result **may** be a graph with an entirely different shape (Fig. 16).

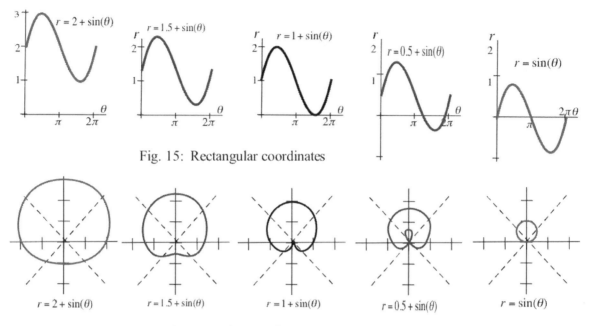

Fig. 15: Rectangular coordinates

Fig. 16: Polar coordinates

Fig. 17 shows the effects of adding a constant to the independent variable in rectangular coordinates, and the result is a horizontal shift of the original graph. In polar coordinates, Fig. 18, the result is a rotation of the original graph. Generally it is difficult to find formulas for rotated figures in rectangular coordinates, but rotations are easy in polar coordinates.

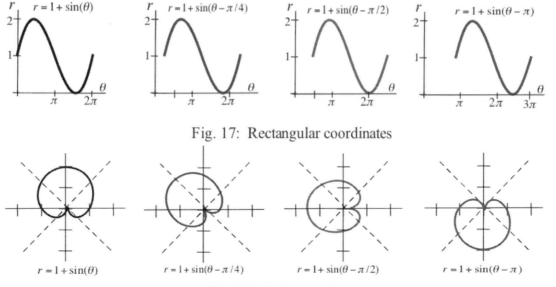

Fig. 17: Rectangular coordinates

Fig. 18: Polar coordinates

The formulas and names of several functions with exotic shapes in polar coordinates are given in the problems. Many of them are difficult to graph "by hand," but by using a graphing calculator or computer you can enjoy the shapes and easily examine the effects of changing some of the constants in their formulas.

Converting Between Coordinate Systems

Sometimes both rectangular and polar coordinates are needed in the same application, and it is necessary to change back and forth between the systems. In such a case we typically place the two origins together and align the polar axis with the positive x–axis. Then the conversions are straightforward exercises using trigonometry and right triangles (Fig. 19).

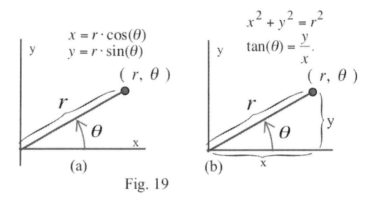

Fig. 19

Polar to Rectangular (Fig. 19a)

$x = r \cdot \cos(\theta)$

$y = r \cdot \sin(\theta)$

Rectangular to Polar (Fig. 19b)

$r^2 = x^2 + y^2$

$\tan(\theta) = \frac{y}{x}$ (if $x \neq 0$)

Example 5: Convert (a) the polar coordinate point $P(7, 0.4)$ to rectangular coordinates, and (b) the rectangular coordinate point $R(12, 5)$ to polar coordinates.

Solution: (a) $r = 7$ and $\theta = 0.4$ (Fig. 20) so $x = r \cdot \cos(\theta) = 7 \cdot \cos(0.4) \approx 7(0.921) = 6.447$ and $y = 7 \cdot \sin(0.4) \approx 7(0.389) = 2.723$.

(b) $x = 12$ and $y = 5$ so $r^2 = x^2 + y^2 = 144 + 25 = 169$ and $\tan(\theta) = y/x = 5/12$ so we can take $r = 13$ and $\theta = \arctan(5/12) \approx 0.395$. The polar coordinate addresses $(13, 0.395 \pm n \cdot 2\pi)$ and $(-13, 0.395 \pm (2n+1) \cdot \pi)$ give the location of the same point.

The conversion formulas can also be used to convert function equations from one system to the other.

Example 6: Convert the rectangular coordinate linear equation $y = 3x + 5$ (Fig. 21) to a polar coordinate equation.

Solution: This simply requires that we replace x with r·cos(θ) and y with r·sin(θ). Then

y = 3x + 5 becomes r·sin(θ) = 3r·cos(θ) + 5

so r·(sin(θ) – 3cos(θ)) = 5 and r = 5/(sin(θ) – 3cos(θ)). This final representation is valid only for θ such that sin(θ) – 3cos(θ) ≠ 0.

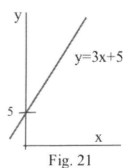

Fig. 21

Practice 6: Convert the polar coordinate equation $r^2 = 4r·\sin(θ)$ to a rectangular coordinate equation.

Example 7: **Robotic Arm**: A robotic arm has a hand at the end of a 12 inch long forearm which is connected to an 18 inch long upper arm (Fig. 22). Determine the position of the hand, relative to the shoulder, when θ = 45° (π/4) and φ = 30° (π/6).

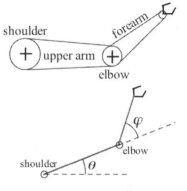

Fig. 22: Robot arm

Solution: The hand is 12·cos(π/4 + π/6) ≈ 3.1 inches to the right of the elbow (Fig. 23) and 12sin(π/4 + π/6) ≈ 11.6 inches above the elbow. Similarly, the elbow is 18·cos(π/4) ≈ 12.7 inches to the right of the shoulder and 18·sin(π/4) ≈ 12.7 inches above the shoulder. Finally, the hand is approximately 3.1 + 12.7 = 15.8 inches to the right of the shoulder and approximately 11.6 + 12.7 = 24.3 inches above the shoulder. In polar coordinates, the hand is approximately 29 inches from the shoulder, at an angle of about 57° (about 0.994 radians) above the horizontal.

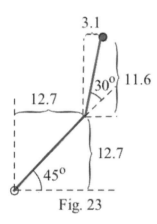

Fig. 23

Practice 7: Determine the position of the hand, relative to the shoulder, when θ = 30° and φ = 45°.

Graphing Functions in Polar Coordinates on a Calculator or Computer

Some calculators and computers are programmed to graph polar functions simply by keying in the formula for r, either as a function of θ or of t, but others are only designed to display rectangular coordinate graphs. However, we can graph polar functions on most of them as well by using the rectangular to polar conversion formulas, selecting the parametric mode (and the radian mode) on the calculator, and then graphing the resulting parametric equations in the rectangular coordinate system:

To graph r = r(θ) for θ between 0 and 3π,

define x(t) = r(t)·cos(t) and y(t) = r(t)·sin(t)

and graph the parametric equations x(t), y(t) for t taking values from 0 to 9.43.

Which Coordinate System Should You Use?

There are no rigid rules. Use whichever coordinate system is easier or more "natural" for the problem or data you have. Sometimes it is not clear which system to use until you have graphed the data both ways, and some problems are easier if you switch back and forth between the systems.

Generally, the polar coordinate system is easier if

- the data consists of measurements in various directions (radar)
- your problem involves locations in relatively featureless locations (deserts, oceans, sky)
- rotations are involved.

Typically, the rectangular coordinate system is easier if

- the data consists of measurements given as functions of time or location (temperature, height)
- your problem involves locations in situations with an established grid (a city, a chess board)
- translations are involved.

PROBLEMS

1. Give the locations in polar coordinates (using radian measure) of the points labeled A, B, and C in Fig. 24.

2. Give the locations in polar coordinates (using radian measure) of the points labeled D, E, and F in Fig. 24.

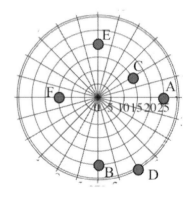

3. Give the locations in polar coordinates (using radian measure) of the points labeled A, B, and C in Fig. 25.

Fig. 24

4. Give the locations in polar coordinates (using radian measure) of the points labeled D, E, and F in Fig. 25.

In problems 5–8, plot the points A – D in polar coordinates, connect the dots by line segments in order (A to B to C to D to A), and name the approximate shape of the resulting figure.

5. A(3, 0°), B(2, 120°), C(2, 200°), and D(2.8, 315°).

6. A(3, 30°), B(2, 130°), C(3, 150°), and D(2, 280°).

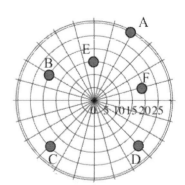

7. A(2, 0.175), B(3, 2.269), C(2, 2.618), and D(3, 4.887).

Fig. 25

8. A(3, 0.524), B(2, 2.269), C(3, 2.618), and D(2, 4.887).

In problems 9–14, the rectangular coordinate graph of a function $r = r(\theta)$ is shown. Sketch the polar coordinate graph of $r = r(\theta)$.

9. The graph in Fig. 26.   10. The graph in Fig. 27.   11. The graph in Fig. 28.

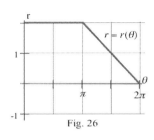
Fig. 26

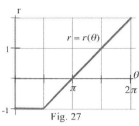
Fig. 27

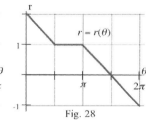
Fig. 28

12. The graph in Fig. 29.   13. The graph in Fig. 30.   14. The graph in Fig. 31.

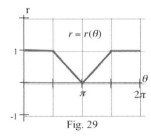
Fig. 29

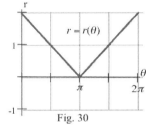
Fig. 30

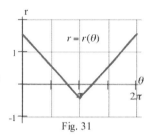
Fig. 31

15. The rectangular coordinate graph of $r = f(\theta)$ is shown in Fig. 32.
    (a) Sketch the rectangular coordinate graphs of $r = 1 + f(\theta)$, $r = 2 + f(\theta)$, and $r = -1 + f(\theta)$.
    (b) Sketch the polar coordinate graphs of $r = f(\theta)$, $r = 1 + f(\theta)$, $r = 2 + f(\theta)$, and $r = -1 + f(\theta)$.

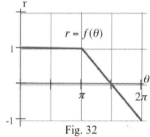
Fig. 32

16. The rectangular coordinate graph of $r = g(\theta)$ is shown in Fig. 33.
    (a) Sketch the rectangular coordinate graphs of $r = 1 + g(\theta)$, $r = 2 + g(\theta)$, and $r = -1 + g(\theta)$.
    (b) Sketch the polar coordinate graphs of $r = g(\theta)$, $r = 1 + g(\theta)$, $r = 2 + g(\theta)$, and $r = -1 + g(\theta)$.

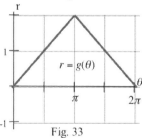
Fig. 33

17. The rectangular coordinate graph of $r = f(\theta)$ is shown in Fig. 34.
    (a) Sketch the rectangular coordinate graphs of $r = 1 + f(\theta)$, $r = 2 + f(\theta)$, and $r = -1 + f(\theta)$.
    (b) Sketch the polar coordinate graphs of $r = f(\theta)$, $r = 1 + f(\theta)$, $r = 2 + f(\theta)$, and $r = -1 + f(\theta)$.

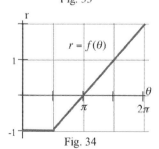
Fig. 34

18. The rectangular coordinate graph of $r = g(\theta)$ is shown in Fig. 35.

    (a) Sketch the rectangular coordinate graphs of $r = 1 + g(\theta)$, $r = 2 + g(\theta)$, and $r = -1 + g(\theta)$.

    (b) Sketch the polar coordinate graphs of $r = g(\theta)$, $r = 1 + g(\theta)$, $r = 2 + g(\theta)$, and $r = -1 + g(\theta)$.

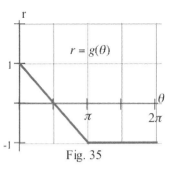

Fig. 35

19. Suppose the rectangular coordinate graph of $r = f(\theta)$ has the horizontal asymptote $r = 3$ as θ grows arbitrarily large. What does that tell us about the polar coordinate graph of $r = f(\theta)$ for large values of θ?

20. Suppose the rectangular coordinate graph of $r = f(\theta)$ has the vertical asymptote $\theta = \pi/6$: $\lim\limits_{\theta \to \pi/6} f(\theta) = +\infty$. What does that tell us about the polar coordinate graph of $r = f(\theta)$ for values of θ near $\pi/6$?

A computer or graphing calculator is recommended for the problems marked with a *.

In problems 21–40, graph the functions in polar coordinates for $0 \leq \theta \leq 2\pi$.

21. $r = -3$
22. $r = 5$
23. $\theta = \pi/6$
24. $\theta = 5\pi/3$

25. $r = 4 \cdot \sin(\theta)$
26. $r = -2 \cdot \cos(\theta)$
27. $r = 2 + \sin(\theta)$
28. $r = -2 + \sin(\theta)$

29. $r = 2 + 3 \cdot \sin(\theta)$
30. $r = \sin(2\theta)$
*31. $r = \tan(\theta)$
*32. $r = 1 + \tan(\theta)$

*33. $r = \dfrac{3}{\cos(\theta)}$
*34. $r = \dfrac{2}{\sin(\theta)}$
*35. $r = \dfrac{1}{\sin(\theta) + \cos(\theta)}$
36. $r = \dfrac{\theta}{2}$

37. $r = 2 \cdot \theta$
38. $r = \theta^2$
39. $r = \dfrac{1}{\theta}$
40. $r = \sin(2\theta) \cdot \cos(3\theta)$

*41. $r = \sin(m\theta) \cdot \cos(n\theta)$ produces lovely graphs for various small integer values of m and n. Go exploring with a graphic calculator to find values of m and n which result in shapes you like.

*42. Graph $r = \dfrac{1}{1 + 0.5 \cdot \cos(\theta + a)}$, $0 \leq \theta \leq 2\pi$, for $a = 0, \pi/6, \pi/4$, and $\pi/2$. How are the graphs related?

*43. Graph $r = \dfrac{1}{1 + 0.5 \cdot \cos(\theta - a)}$, $0 \leq \theta \leq 2\pi$, for $a = 0, \pi/6, \pi/4$, and $\pi/2$. How are the graphs related?

*44. Graph $r = \sin(n\theta)$, $0 \leq \theta \leq 2\pi$, for $n = 1, 2, 3$, and 4 and count the number of "petals" on each graph. Predict the number of "petals" for the graphs of $r = \sin(n\theta)$ for $n = 5, 6$, and 7, and then test your prediction by creating those graphs.

*45. Repeat the steps in problem 44 but using $r = \cos(n\theta)$.

In problems 46–49, convert the rectangular coordinate locations to polar coordinates.

46. $(0, 3), (5, 0)$, and $(1, 2)$

47. $(-2, 3), (2, -3)$, and $(0, -4)$.

48. $(0, -2), (4, 4)$, and $(3, -3)$

49. $(3, 4), (-1, -3)$, and $(-7, 12)$.

In problems 50–53, convert the polar coordinate locations to rectangular coordinates.

50. $(3, 0), (5, 90°)$, and $(1, \pi)$

51. $(-2, 3), (2, -3)$, and $(0, -4)$.

52. $(0, 3), (5, 0)$, and $(1, 2)$

53. $(2, 3), (-2, -3)$, and $(0, 4)$.

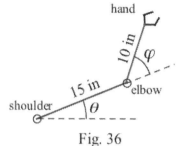

Fig. 36

Problems 54–60 refer to the robotic arm in Fig. 36.

54. Determine the position of the hand, realtive to the shoulder, when $\theta = 60°$ and $\phi = -45°$.

55. Determine the position of the hand, relative to the shoulder, when $\theta = -30°$ and $\phi = 30°$.

56. Determine the position of the hand, relative to the shoulder, when $\theta = 0.6$ and $\phi = 1.2$.

57. Determine the position of the hand, relative to the shoulder, when $\theta = -0.9$ and $\phi = 0.4$.

58. Suppose the robot's shoulder can pivot so that $-\pi/2 \leq \theta \leq \pi/2$, but the elbow is broken and ϕ is always $0°$. Sketch the points the hand can reach.

59. Suppose the robot's shoulder can pivot so that $-\pi/2 \leq \theta \leq \pi/2$, and the elbow can pivot so that $-\pi/2 \leq \phi \leq \pi/2$. Sketch the points the hand can reach.

60. Suppose the robot's shoulder can pivot so that $-\pi/2 \leq \theta \leq \pi/2$, and the elbow can pivot completely so $-\pi \leq \phi \leq \pi$. Sketch the points the hand can reach.

*61. Graph $r = \dfrac{1}{1 + a \cdot \cos(\theta)}$ for $0 \leq \theta \leq 2\pi$ and $a = 0.5, 0.8, 1, 1.5,$ and 2. What shapes do the various values of a produce?

*62. Repeat problem 61 with $r = \dfrac{1}{1 + a \cdot \sin(\theta)}$.

Some Exotic Curves (and Names)

Many of the following curves were discovered and named even before polar coordinates were invented. In most cases the path of a point moving on or around some object is described. You may enjoy using your calculator to graph some of these curves or you can invent your own exotic shapes. (An inexpensive source for these shapes and names is A Catalog Of Special Plane Curves by J. Dennis Lawrence, Dover Publications, 1972, and the page references below are to that book)

Some Classics:

Cissoid ("like ivy") of Diocles (about 200 B.C.): $r = a \sin(\theta) \cdot \tan(\theta)$ p. 98

Right Strophoid ("twisting") of Barrow (1670): $r = a(\sec(\theta) - 2\cos(\theta))$ p. 101

Trisectrix of Maclaurin (1742): $r = a \sec(\theta) - 4a \cos(\theta)$ p. 105

Lemniscate ("ribbon") of Bernoulli (1694): $r^2 = a^2 \cos(2\theta)$ p. 122

Conchoid ("shell") of Nicomedes (225 B.C.): $r = a + b \cdot \sec(\theta)$ p. 137

Hippopede ("horse fetter") of Proclus (about 75 B.C.): $r^2 = 4b(a - b \sin^2(\theta))$ p. 144   b = 3, a = 1, 2, 3, 4

Devil's Curve of Cramer (1750): $r^2(\sin^2(\theta) - \cos^2(\theta)) = a^2 \sin^2(\theta) - b^2 \cos^2(\theta)$ p. 151   a= 2, b=3

Nephroid ("kidney") of Freeth: $r = a \cdot (1 + 2 \sin(\frac{\theta}{2}))$ p. 175   a = 3

Some of our own: (Based on their names, what shapes do you expect for the following curves?)

Piscatoid of Pat (1992): $r = \frac{1}{\cos(\theta)} - 3\cos(\theta)$ for $-1.1 \leq \theta \leq 1.1$ Window x: (–2, 1) and y: (–1, 1)

Kermitoid of Kelcey (1992) :

$r = 2.5 \cdot \sin(2\theta) \cdot (\theta - 4.71) \cdot \text{INT}(\theta/\pi) + \{ 5 \cdot \sin^3(\theta) - 3 \cdot \sin^9(\theta) \} \cdot \{ 1 - \text{INT}(\theta/\pi) \}$ for $0 \leq \theta \leq 2\pi$
Window x: (–3, 3) and y: (–1, 4)

Bovine Oculoid: $r = 1 + \text{INT}(\theta/(2\pi))$ for $0 \leq \theta \leq 6\pi$ (≈ 18.85)   Window x: (–5, 5) and y: (–4, 4)

A Few Reference Facts

The polar form of the linear equation $Ax + By + C = 0$ is $r \cdot (A \cdot \cos(\theta) + B \cdot \sin(\theta)) + C = 0$

The equation of the line through the polar coordinate points (r_1, θ_1) and (r_2, θ_2) is

$r \cdot \{ r_1 \cdot \sin(\theta - \theta_1) + r_2 \cdot \sin(\theta_2 - \theta) \} = r_1 \cdot r_2 \cdot \sin(\theta_2 - \theta_1)$

The graph of $r = a \cdot \sin(\theta) + b \cdot \cos(\theta)$ is a circle through the origin with center (b/2, a/2) and radius $\frac{1}{2}\sqrt{a^2 + b^2}$. (Hint: multiply each side by r, and then convert to rectangular coordinates.)

The equations $r = \frac{1}{1 \pm a \cdot \cos(\theta)}$ and $r = \frac{1}{1 \pm a \cdot \sin(\theta)}$ are conic sections with one focus at the origin.

If $a < 1$, the denominator is **never** 0 for $0 \leq \theta < 2\pi$ and the graph is an **ellipse**.

If $a = 1$, the denominator is 0 for **one** value of $\theta, 0 \leq \theta < 2\pi$, and the graph is a **parabola**.

If $a > 1$, the denominator is 0 for **two** values of $\theta, 0 \leq \theta < 2\pi$, and the graph is a **hyperbola**.

Section 9.1 PRACTICE Answers

Practice 1: Point C is at a distance of 10 miles in the direction 30°. D is 5 miles at 270°.

Practice 2: The points are plotted in Fig. 37.

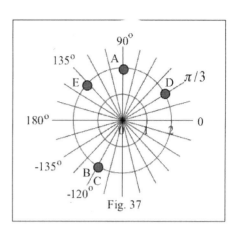
Fig. 37

Practice 3: See Fig. 38.

The top of the plateau is roughly rectangular.

Practice 4: The graphs are shown in Figs. 39 and 40.

Note that the graph of $r = \cos(\theta)$ traces out a circle **twice**; once as θ goes from 0 to π, and a second time as θ goes from π to 2π.

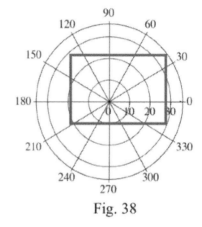
Fig. 38

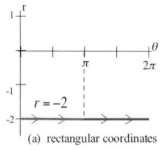
(a) rectangular coordinates

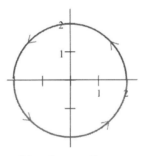
(b) polar coordinates

Fig. 39

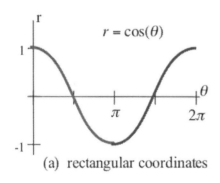
(a) rectangular coordinates

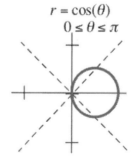
(b) polar coordinates

Fig. 40

Practice 5: The points are plotted in Fig. 41.

The points (almost) lie on a straight line.

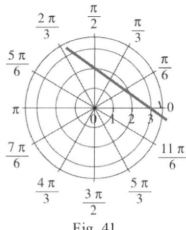

Fig. 41

Practice 6: $r^2 = x^2 + y^2$ and $r \cdot \sin(\theta) = y$ so $r^2 = 4r \cdot \sin(\theta)$ becomes $x^2 + y^2 = 4y$.

Putting this last equation into the standard form for a circle (by completing the square) we have $x^2 + (y - 2)^2 = 4$, the equation of a circle with center at $(0, 2)$ and radius 2.

Practice 7: See Fig. 42.

For point A, the "elbow," relative to O, the "shoulder:"

$x = 18 \cdot \cos(30^o) \approx 15.6$ inches and $y = 18 \cdot \sin(30^o) = 9$ inches.

For point B, the "hand," relative to A:

$x = 12 \cdot \cos(75^o) \approx 3.1$ inches and $y = 12 \cdot \sin(75^o) \approx 11.6$ inches.

Then the retangular coordinate location of the B relative to O is

$x \approx 15.6 + 3.1 = 18.7$ inches and $y \approx 9 + 11.6 = 20.6$ inches.

The polar coordinate location of B relative to O is

$r = \sqrt{x^2 + y^2} \approx 27.8$ inches and $\theta \approx 47.7^o$ (or 0.83 radians)

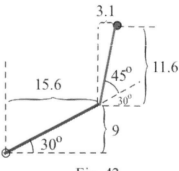

Fig. 42

9.2 CALCULUS IN THE POLAR COORDINATE SYSTEM

The previous section introduced the polar coordinate system and discussed how to plot points, how to create graphs of functions (from data, a rectangular graph, or a formula), and how to convert back and forth between the polar and rectangular coordinate systems. This section examines calculus in polar coordinates: rates of changes, slopes of tangent lines, areas, and lengths of curves. The results we obtain may look different, but they all follow from the approaches used in the rectangular coordinate system.

Polar Coordinates and Derivatives

In the rectangular coordinate system, the derivative dy/dx measured both the rate of change of y with respect to x and the slope of the tangent line. In the polar coordinate system two different derivatives commonly appear, and it is important to distinguish between them.

> $\dfrac{dr}{d\theta}$ measures the **rate of change** of r with respect to θ.
>
> The sign of $\dfrac{dr}{d\theta}$ tells us whether r is increasing or decreasing as θ increases.
>
> $\dfrac{dy}{dx}$ measures the **slope** $\dfrac{\Delta y}{\Delta x}$ **of the tangent line** to the polar graph of r.

We can use our usual rules for derivatives to calculate the derivative of a polar coordinate equation r with respect to θ, and dr/dθ tells us how r is changing with respect to (increasing) θ. For example, if dr/dθ > 0 then the directed distance r is increasing as θ increases (Fig. 1). However, dr/dθ is **NOT** the slope of the line tangent to the polar graph of r. For the simple spiral r = θ (Fig. 2), $\dfrac{dr}{d\theta} = 1 > 0$ for all values of θ; but the slope of the tangent line, $\dfrac{dy}{dx}$, may be positive (at A and C) or negative (at B and D).

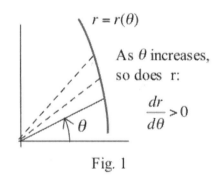

Fig. 1

Similarly, $\dfrac{dx}{d\theta}$ is the rate of change of the x-coordinate of the graph with respect to (increasing) θ, and $\dfrac{dy}{d\theta}$ is the rate of change of the y-coordinate of the graph with respect to (increasing) θ. The values of the derivatives dy/dθ and dx/dθ depend on the location on the graph. They will also be used to calculate the slope $\dfrac{dy}{dx}$ of the tangent line, and also to express the formula for arc length in polar coordinates.

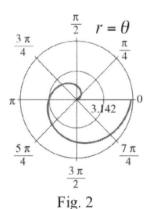

Fig. 2

Example 1: State whether the values of $dr/d\theta$, $dx/d\theta$, $dy/d\theta$, and dy/dx are + (positive), – (negative), 0 (zero), or U (undefined) at the points A and B on the graph in Fig. 3.

Solution: The values of the derivatives at A and B are given in Table 1.

Practice 1: Fill in the rest of Table 1 for points labeled C and D.

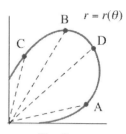

Fig. 3

When r is given by a formula we can calculate dy/dx, the slope of the tangent line, by using the polar–rectangular conversion formulas and the Chain Rule. By the Chain Rule $\frac{dy}{d\theta} = \frac{dy}{dx} \frac{dx}{d\theta}$, so we can solve for $\frac{dy}{dx}$ by dividing each side of the equation by $\frac{dx}{d\theta}$.

Point	$\frac{dx}{d\theta}$	$\frac{dy}{d\theta}$	$\frac{dr}{d\theta}$	$\frac{dy}{dx}$
A	+	+	+	+
B	–	0	–	0
C				
D				

Table 1

Then the slope $\frac{dy}{dx}$ of the line tangent to the polar coordinate graph of $r(\theta)$ is

(1) $\quad \frac{dy}{dx} = \frac{\frac{dy}{d\theta}}{\frac{dx}{d\theta}} = \frac{\frac{d(r \cdot \sin(\theta))}{d\theta}}{\frac{d(r \cdot \cos(\theta))}{d\theta}}$.

Since r is a function of θ, $r = r(\theta)$, we may use the product rule and the Chain Rule for derivatives to calculate each derivative and to obtain

(2) $\quad \frac{dy}{dx} = \frac{r \cdot \cos(\theta) + r' \cdot \sin(\theta)}{-r \cdot \sin(\theta) + r' \cdot \cos(\theta)}$ (with $r' = dr/d\theta$) .

The result in (2) is difficult to remember, but the starting point (1) and derivation are straightforward.

Example 2: Find the slopes of the lines tangent to the spiral $r = \theta$ (shown in Fig. 2) at the points $P(\pi/2, \pi/2)$ and $Q(\pi, \pi)$.

Solution: $y = r \cdot \sin(\theta) = \theta \cdot \sin(\theta)$ and $x = r \cdot \cos(\theta) = \theta \cdot \cos(\theta)$ so

$$\frac{dy}{dx} = \frac{\frac{d(r \cdot \sin(\theta))}{d\theta}}{\frac{d(r \cdot \cos(\theta))}{d\theta}} = \frac{\frac{d(\theta \cdot \sin(\theta))}{d\theta}}{\frac{d(\theta \cdot \cos(\theta))}{d\theta}} = \frac{\theta \cdot \cos(\theta) + 1 \cdot \sin(\theta)}{-\theta \cdot \sin(\theta) + 1 \cdot \cos(\theta)} .$$

At the point P, $\theta = \pi/2$ and $r = \pi/2$ so $\frac{dy}{dx} = \frac{\frac{\pi}{2} \cdot 0 + 1 \cdot (1)}{-\frac{\pi}{2} \cdot (1) + 1 \cdot (0)} = -\frac{2}{\pi} \approx -0.637$.

At the point Q, $\theta = \pi$ and $r = \pi$ so $\frac{dy}{dx} = \frac{\pi \cdot (-1) + 1 \cdot (0)}{-\pi \cdot (0) + 1 \cdot (-1)} = \frac{-\pi}{-1} = \pi \approx 3.142$.

The function $r = \theta$ is steadily increasing, but the slope of the line tangent to the polar graph can negative or positive or zero or even undefined (where?).

Practice 2: Find the slopes of the lines tangent to the cardioid $r = 1 - \sin(\theta)$ (Fig. 4) when $\theta = 0, \pi/4,$ and $\pi/2$.

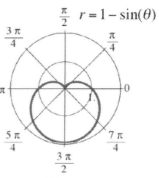

Fig. 4

Areas in Polar Coordinates

The patterns for calculating areas in rectangular and polar coordinates look different, but they are derived in the same way: partition the area into pieces, calculate areas of the pieces, add the small areas together to get a Riemann sum, and take the limit of the Riemann sum to get a definite integral. The major difference is the shape of the pieces: we use thin rectangular pieces in the rectangular system and thin sectors (pieces of pie) in the polar system. The formula we need for the area of a sector can be found by using proportions (Fig. 5):

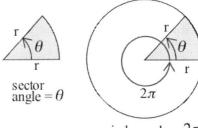

$$\frac{\text{area of sector}}{\text{area of whole circle}} = \frac{\text{sector angle}}{\text{angle of whole circle}} = \frac{\theta}{2\pi}$$

so (area of sector) $= \dfrac{\theta}{2\pi}$ (area of whole circle) $= \dfrac{\theta}{2\pi}(\pi r^2) = \dfrac{1}{2} r^2 \theta$.

Figures 6 and 7 refer to the area discussion after the figures.

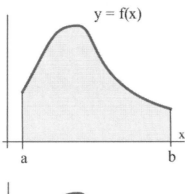

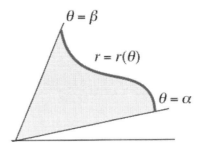

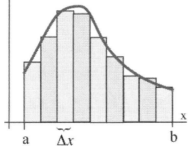

Fig. 6 Area with rectangular coordinates

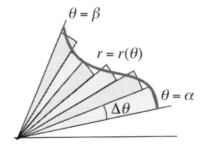

Fig. 7: Area with polar coordinates

Area in Rectangular Coordinates (Fig. 6)	**Area in Polar Coordinates** (Fig. 7)
Partition the domain x of the rectangular coordinate function into small pieces of width Δx.	Partition the domain θ of the polar coordinate function into small pieces of angular width $\Delta \theta$.
Build rectangles on each piece of the domain.	Build "nice" shapes (pieces of pie shaped sectors) along each piece of the domain.
Calculate the area of each piece (rectangle): $\text{area}_i = (\text{base}_i) \cdot (\text{height}_i) = f(x_i) \cdot \Delta x_i$.	Calculate the area of each piece (sector): $\text{area}_i = \frac{1}{2}(\text{radius}_i)^2 (\text{angle}_i) = \frac{1}{2} r_i^2 \Delta \theta_i$.
Approximate the total area by adding the small areas together, a Riemann sum: $\text{total area} \approx \sum \text{area}_i = \sum f(x_i) \cdot \Delta x_i$.	Approximate the total area by adding the small areas together, a Riemann sum: $\text{total area} \approx \sum \text{area}_i = \sum \frac{1}{2} r_i^2 \Delta \theta_i$.
The limit of the Riemann sum is a definite integral: $\text{Area} = \int_{x=a}^{b} f(x)\, dx$.	The limit of the Riemann sum is a definite integral: $\text{Area} = \int_{\theta=\alpha}^{\beta} \frac{1}{2} r^2(\theta)\, d\theta$.

If r is a continuous function of θ, then the limit of the Riemann sums is a finite number, and we have a formula for the area of a region in polar coordinates.

Area In Polar Coordinates

The area of the region bounded by a continuous function $r(\theta)$ and radial lines at angles $\theta = \alpha$ and $\theta = \beta$ is

$$\text{area} = \int_{\theta=\alpha}^{\beta} \frac{1}{2} r^2(\theta)\, d\theta.$$

Example 3: Find the area inside the cardioid $r = 1 + \cos(\theta)$. (Fig. 8)

Solution: This is a straightforward application of the area formula.

$$\text{Area} = \int_{\theta=0}^{2\pi} \frac{1}{2}(1+\cos(\theta))^2 \, d\theta = \frac{1}{2}\int_{\theta=0}^{2\pi} \{1 + 2\cos(\theta) + \cos^2(\theta)\} \, d\theta$$

$$= \frac{1}{2}\left\{\theta + 2\sin(\theta) + \frac{1}{2}\left[\theta + \frac{1}{2}\sin(2\theta)\right]\right\}\Big|_0^{2\pi}$$

$$= \frac{1}{2}\left\{\left[2\pi + 0 + \frac{1}{2}(2\pi+0)\right] - [0+0+0]\right\} = \frac{3}{2}\pi.$$

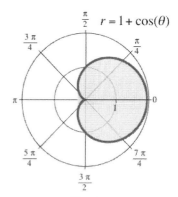

Fig. 8

We could also have used the symmetry of the region and determined this area by integrating from 0 to π (Fig. 9) and multiplying the result by 2.

Practice 3: Find the area inside one "petal" of the rose $r = \sin(3\theta)$. (Fig. 10)

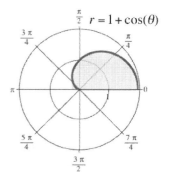

Fig. 9

We can also calculate the area between curves in polar coordinates.

The area of the region (Fig. 11) between the continuous curves $r_1(\theta) \leq r_2(\theta)$ for $\alpha \leq \theta \leq \beta$ is

$$\int_{\theta=\alpha}^{\beta} \frac{1}{2}r_2^2(\theta) \, d\theta - \int_{\theta=\alpha}^{\beta} \frac{1}{2}r_1^2(\theta) \, d\theta$$

$$= \int_{\theta=\alpha}^{\beta} \frac{1}{2}\{r_2^2(\theta) - r_1^2(\theta)\} \, d\theta.$$

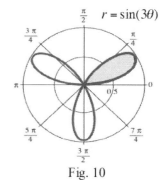

Fig. 10

It is a good idea to sketch the graphs of the curves to help determine the endpoints of integration.

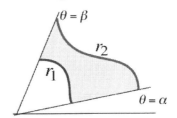

Fig. 11

Example 4: Find the area of the shaded region in Fig. 12.

Solution: A_1 = area between the circle and the origin = $\int_{\theta=0}^{\pi/2} \frac{1}{2} 1^2 \, d\theta$

$= \frac{1}{2} \theta \Big|_0^{\pi/2} = \frac{\pi}{4} \approx 0.785$.

A_2 = area between the cardioid and the origin = $\int_{\theta=0}^{\pi/2} \frac{1}{2}(1 + \cos(\theta))^2 \, d\theta$

$= \frac{3}{4} \theta + \sin(\theta) + \frac{1}{8} \sin(2\theta) \Big|_0^{\pi/2} = \{ \frac{3\pi}{8} + 1 + 0 \} - \{ 0 + 0 + 0 \} \approx 2.178$.

The area we want is $A_2 - A_1 = 1 + \frac{3\pi}{8} - \frac{\pi}{4} = 1 + \frac{\pi}{8} \approx 1.393$.

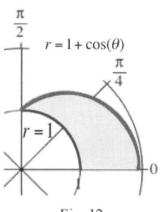

Fig. 12

Practice 4: Find the area of the region outside the cardioid $1 + \cos(\theta)$ and inside the circle $r = 2$. (Fig. 13)

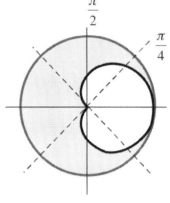

Fig. 13

Arc Length in Polar Coordinates

The patterns for calculating the lengths of curves in rectangular and polar coordinates look different, but they are derived from the Pythagorean Theorem and the same sum we used in Section 5.2 (Fig. 14):

$$\text{length} \approx \sum \sqrt{(\Delta x)^2 + (\Delta y)^2} = \sum \sqrt{\left(\frac{\Delta x}{\Delta \theta}\right)^2 + \left(\frac{\Delta y}{\Delta \theta}\right)^2} \, \Delta\theta .$$

If x and y are differentiable functions of θ, then as $\Delta\theta$ approaches 0, $\Delta x/\Delta\theta$ approaches $dx/d\theta$, $\Delta y/\Delta\theta$ approaches $dy/d\theta$, and the Riemann sum approaches the definite integral

$$\text{length} = \int_{\theta=\alpha}^{\beta} \sqrt{\left(\frac{dx}{d\theta}\right)^2 + \left(\frac{dy}{d\theta}\right)^2} \, d\theta .$$

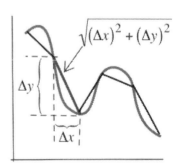

Fig. 14

Replacing x with $r \cdot \cos(\theta)$ and y with $r \cdot \sin(\theta)$, we have $dx/d\theta = -r \cdot \sin(\theta) + r' \cdot \cos(\theta)$ and $dy/d\theta = r \cdot \cos(\theta) + r' \cdot \sin(\theta)$. Then $(dx/d\theta)^2 + (dy/d\theta)^2$ inside the square root simplifies to $r^2 + (r')^2$ and we have a more useful form of the integral for arc length in polar coordinates.

Arc Length

If r is a differentiable function of θ for $\alpha \leq \theta \leq \beta$, then the length of the graph of r is

$$\text{Length} = \int_{\theta=\alpha}^{\beta} \sqrt{(r)^2 + \left(\frac{dr}{d\theta}\right)^2}\, d\theta.$$

Problems

Derivatives

In problems 1–4, fill in the table for each graph with + (positive), – (negative), 0 (zero), or U (undefined) for each derivative at each labeled point.

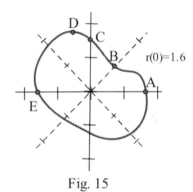

Fig. 15

1. Use Fig. 15.

Point	$\frac{dr}{d\theta}$	$\frac{dx}{d\theta}$	$\frac{dy}{d\theta}$	$\frac{dy}{dx}$
A				
B				
C				
D				
E				

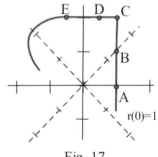

Fig. 16

2. Use Fig. 16.

Point	$\frac{dr}{d\theta}$	$\frac{dx}{d\theta}$	$\frac{dy}{d\theta}$	$\frac{dy}{dx}$
A				
B				
C				
D				
E				

3. Use Fig. 17.

Point	$\frac{dr}{d\theta}$	$\frac{dx}{d\theta}$	$\frac{dy}{d\theta}$	$\frac{dy}{dx}$
A				
B				
C				
D				
E				

Fig. 17

4. Use Fig. 18.

Point	$\dfrac{dr}{d\theta}$	$\dfrac{dx}{d\theta}$	$\dfrac{dy}{d\theta}$	$\dfrac{dy}{dx}$
A				
B				
C				
D				
E				

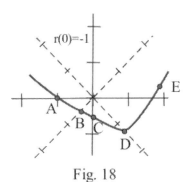

Fig. 18

In problems 5–8, sketch the graph of the polar coordinate function $r = r(\theta)$ for $0 \leq \theta \leq 2\pi$, label the points with the given polar coordinates on the graph, and calculate the values of $\dfrac{dr}{d\theta}$ and $\dfrac{dy}{dx}$ at the points with the given polar coordinates.

5. $r = 5$ at $A(5, \pi/4)$, $B(5, \pi/2)$, and $C(5, \pi)$.

6. $r = 2 + \cos(\theta)$ at $A(2 + \dfrac{\sqrt{2}}{2}, \pi/4)$, $B(2, \pi/2)$, and $C(1, \pi)$.

7. $r = 1 + \cos^2(\theta)$ at $A(2, 0)$, $B(3/2, \pi/4)$, and $C(1, \pi/2)$.

8. $r = \dfrac{6}{2 + \cos(\theta)}$ at $A(2, 0)$, $B(3, \pi/2)$, and $C(\dfrac{24 - 6\sqrt{2}}{7}, \pi/4) \approx (2.216, \pi/4)$.

9. Graph $r = 1 + 2 \cdot \cos(\theta)$ for $0 \leq \theta \leq 2\pi$, and show that the graph goes through the origin when $\theta = 2\pi/3$ and $\theta = 4\pi/3$. Calculate dy/dx when $\theta = 2\pi/3$ and $\theta = 4\pi/3$. How can a curve have two different tangent lines (and slopes) when it goes through the origin?

10. Graph the cardiod $r = 1 + \sin(\theta)$ for $0 \leq \theta \leq 2\pi$.

    (a) At what points on the cardioid does $dx/d\theta = 0$? (b) At what points on the cardiod does $dy/d\theta = 0$?

    (c) At what points on the cardioid does $dr/d\theta = 0$? (d) At what points on the cardiod does $dy/dx = 0$?

11. Show that if a polar coordinate graph goes through the origin when the angle is θ_0 (and if $dr/d\theta$ exists and does not equal 0 there), then the slope of the tangent line at the origin is $\tan(\theta_0)$. (Suggestion: Evaluate formula (2) for dy/dx at the point $(0, \theta_0)$.)

Areas

In problems 12–20, represent each area as a definite integral. Then evaluate the integral exactly or using Simpson's rule (with n = 100).

12. The area of the shaded region in Fig. 19.

13. The area of the shaded region in Fig. 20.

14. The area of the shaded region in Fig. 21.

15. The area in the first quadrant outside the circle $r = 1$ and inside the cardiod $r = 1 + \cos(\theta)$.

16. The region in the second quadrant bounded by $r = \theta$ and $r = \theta^2$.

17. The area inside one "petal" of the graph of (a) $r = \sin(3\theta)$ and (b) $r = \sin(5\theta)$.

18. The area (a) inside the "peanut" $r = 1.5 + \cos(2\theta)$ and (b) inside $r = a + \cos(2\theta)$ (a > 1).

19. The area inside the circle $r = 4 \cdot \sin(\theta)$.

20. The area of the shaded region in Fig. 22.

21. Goat and Square Silo: (This problem does not require calculus.)
    One end of a 40 foot long rope is attached to the middle of a wall of a 20 foot square silo, and the other end is tied to a goat (Fig. 23).
    (a) Sketch the region that the goat can reach.
    (b) Find the area of the region that the goat can reach.
    (c) Can the goat reach more area if the rope is tied to the corner of the silo?

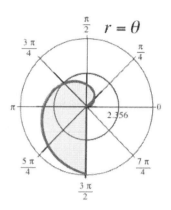

Fig. 19

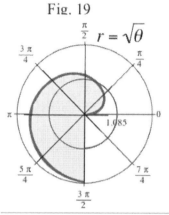

Fig. 20

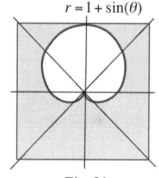

Fig. 21

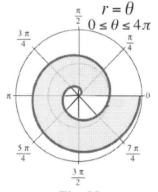

Fig. 22

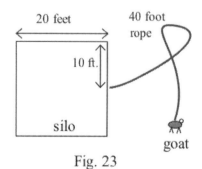

Fig. 23

22. **Goat and Round Silo:** One end of a 10π foot long rope is attached to the wall of a round silo that has a radius of 10 feet, and the other end is tied to a goat (Fig. 24).

   (a) Sketch the region the goat can reach.

   (b) Justify that the area of the region in Fig. 25 as the goat goes around the silo from having θ feet of rope taut against the silo to having $\theta + \Delta\theta$ feet taut against the silo is approximately
   $$\frac{1}{2}(10\pi - 10\cdot\theta)^2 \Delta\theta.$$

   (c) Use the result from part (b) to help calculate the area of the region that the goat can reach.

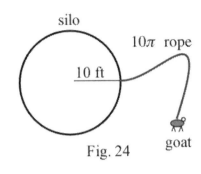

Fig. 24

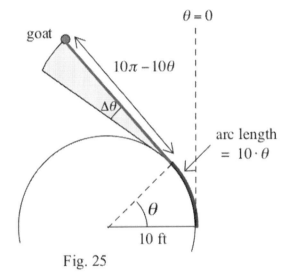

Fig. 25

Arc Lengths

In problems 23–29, represent the length of each curve as a definite integral. Then evaluate the integral exactly or using your calculator.

23. The length of the spiral $r = \theta$ from $\theta = 0$ to $\theta = 2\pi$.

24. The length of the spiral $r = \theta$ from $\theta = 2\pi$ to $\theta = 4\pi$.

25. The length of the cardiod $r = 1 + \cos(\theta)$.

26. The length of $r = 4\cdot\sin(\theta)$ from $\theta = 0$ to $\theta = \pi$.

27. The length of the circle $r = 5$ from $\theta = 0$ to $\theta = 2\pi$.

28. The length of the "peanut" $r = 1.2 + \cos(2\theta)$.

29. The length (a) of one "petal" of the graph of $r = \sin(3\theta)$ and (b) of one "petal" of $r = \sin(5\theta)$.

30. Assume that r is a differentiable function of θ. Verify that $\{\frac{dx}{d\theta}\}^2 + \{\frac{dy}{d\theta}\}^2 = \{r\}^2 + \{\frac{dr}{d\theta}\}^2$ by replacing x with $r\cdot\cos(\theta)$ and y with $r\cdot\sin(\theta)$ in the left side of the equation, differentiating, and then simplifying the result to obtain the right side of the equation.

Section 9.2　　　　　　　　　　PRACTICE Answers

Practice 1:　　The values are shown in Fig. 26.

Point	$\frac{dx}{d\theta}$	$\frac{dy}{d\theta}$	$\frac{dr}{d\theta}$	$\frac{dy}{dx}$
A	+	+	+	+
B	−	0	−	0
C	−	−	−	+
D	−	+	+	−

Fig. 26

Practice 2:　　$r = 1 - \sin(\theta)$ and $r' = -\cos(\theta)$.

$$\frac{dy}{dx} = \frac{\frac{dy}{d\theta}}{\frac{dx}{d\theta}} = \frac{r \cdot \cos(\theta) + r' \cdot \sin(\theta)}{-r \cdot \sin(\theta) + r' \cdot \cos(\theta)}$$

$$= \frac{(1 - \sin(\theta)) \cdot \cos(\theta) + (-\cos(\theta)) \cdot \sin(\theta)}{-(1 - \sin(\theta)) \cdot \sin(\theta) + (-\cos(\theta)) \cdot \cos(\theta)} = \frac{\cos(\theta) - 2 \cdot \sin(\theta) \cdot \cos(\theta)}{-\sin(\theta) + \sin^2(\theta) - \cos^2(\theta)}.$$

When $\theta = 0$, $\frac{dy}{dx} = \frac{1 - 0}{-0 + 0 - 1} = -1$.

When $\theta = \frac{\pi}{4}$, $\frac{dy}{dx} = \frac{(1/\sqrt{2}) - 2(1/\sqrt{2})(1/\sqrt{2})}{-(1/\sqrt{2}) + (1/\sqrt{2})^2 - (1/\sqrt{2})^2} = \frac{1/\sqrt{2} - 1}{-1/\sqrt{2} + \frac{1}{2} - \frac{1}{2}} = \sqrt{2} - 1 \approx 0.414$.

When $\theta = \frac{\pi}{2}$, $\frac{dy}{dx} = \frac{0 - 0}{-1 + 1 - 0}$ which is undefined. Why does this result make sense in terms of the graph of the cardioid $r = 1 - \sin(\theta)$?

Practice 3:　　One "petal" of the rose $r = \sin(3\theta)$ is swept out as θ goes from 0 to $\pi/3$ (see Fig. 10) so the endpoints of the area integral are 0 and $\pi/3$.

$$\text{area} = \int_{\theta=\alpha}^{\beta} \frac{1}{2} r^2(\theta) \, d\theta = \int_{\theta=0}^{\pi/3} \frac{1}{2} \{\sin(3\theta)\}^2 \, d\theta \quad \text{(then using integral table entry \#13)}$$

$$= \frac{1}{2} \left\{ \frac{1}{2}\theta - \frac{1}{4(3)} \sin(2 \cdot 3\theta) \right\} \Big|_0^{\pi/3} = \frac{1}{2} \left\{ \left[\frac{\pi}{6} - \frac{1}{12}(0) \right] - [0 - 0] \right\} = \frac{\pi}{12} \approx 0.262.$$

Practice 4:　　The area we want in Fig. 13 is

{area of circle} − {area of cardioid from Example 3} = $\pi(2)^2 - \frac{3}{2}\pi = \frac{5}{2}\pi \approx 7.85$.

9.3 PARAMETRIC EQUATIONS

Some motions and paths are inconvenient, difficult or impossible for us to describe by a single function or formula of the form y = f(x).

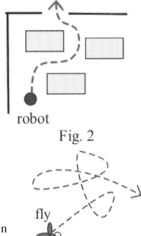

Fig. 1

- A rider on the "whirligig" (Fig. 1) at the carnival goes in circles at the end of a rotating bar.
- A robot delivering supplies in a factory (Fig. 2) needs to avoid obstacles.
- A fly buzzing around the room (Fig. 3) and a molecule in a solution follow erratic paths.
- A stone caught in the tread of a rolling wheel has a smooth path with some sharp corners (Fig. 4).

Parametric equations provide a way to describe all of these motions and paths. And parametric equations generalize easily to describe paths and motions in 3 dimensions.

Parametric equations were used briefly in earlier sections (2.5: Applications of the Chain Rule and 5.2: Arc Length). In those sections the equations were always given. In this section we look at functions given parametrically by data, graphs, and formulas and examine how to build formulas to describe some motions parametrically. The last curve in this section is the cycloid, one of the most famous curves in mathematics. The next section considers calculus with parametric equations: slopes of tangent lines, arc lengths, and areas.

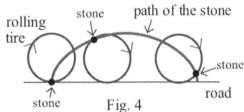

Fig. 4

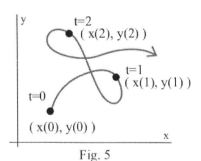

Fig. 5

Parametric equations describe the location of a point (x,y) on a graph or path as a function of a single independent variable t, a "parameter" often representing time. In 2 dimensions, the coordinates x and y are functions of the variable t: $x = x(t)$ and $y = y(t)$ (Fig. 5). In 3 dimensions, the z coordinate is also a function of t: $z = z(t)$. With parametric equations we can also analyze the forces acting on an object separately in each coordinate direction and then combine the results to see the overall behavior of the object. Parametric equations often provide an easier way to understand and build equations for complicated motions.

Graphing Parametric Equations

The data for creating a parametric equation graph can be given as a table of values, as graphs of (t, x(t)) and (t, y(t)), or as formulas for x and y as functions of t.

Example 1: Table 1 is a record of the location of a roller coaster car relative to its starting location. Use the data to sketch a graph of the car's path for the first 7 seconds.

t	x(t)	y(t)	t	x(t)	y(t)
0	0	70	7	90	55
1	30	20	8	105	85
2	70	50	9	125	100
3	60	75	10	130	80
4	30	70	11	150	65
5	32	35	12	180	75
6	60	15	13	200	30

Table 1

Solution: Figure 6 is a plot of the (x, y) locations of the car for t = 0 to 7 seconds. The points are connected by a smooth curve to show a possible path of the car.

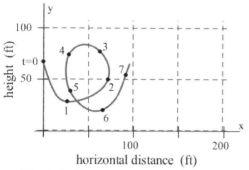

Fig. 6: Roller coaster for 0≤t≤7 seconds

Practice 1: Use the data in Table 1 to sketch the path of the roller coaster for the next 6 seconds.

Note: Clearly the graph in Fig. 6 is not the graph of a function y = f(x). But every y = f(x) function has an easy parametric representation by setting x(t) = t and y(t) = f(t).

Sometimes a parametric graph can show patterns that are not clearly visible in individual graphs.

Example 2: Figures 7a and 7b are graphs of the populations of rabbits and foxes on an island. Use these graphs to sketch a parametric graph of rabbits (x–axis) versus foxes (y–axis) for 0 ≤ t ≤ 10 years.

Solution: The separate rabbit and fox population graphs give us information about each population separately, but the parametric graph helps us see the effects of the interaction between the rabbits and the foxes more clearly.

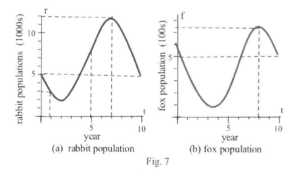

Fig. 7
(a) rabbit population
(b) fox population

For each time t we can read the rabbit and fox populations from the separate graphs (e.g., when t = 1, there are approximately 3000 rabbits and 400 foxes so x ≈ 3000 and y ≈ 400) and then combine this information to plot a single point on the parametric graph. If we repeat this process for a large number of values of t, we get a graph (Fig. 8) of the "motion" of the rabbit and fox populations over a period of time, and we can ask questions about why the populations might show this behavior.

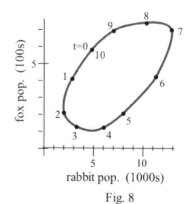

Fig. 8

The type of graph in Fig. 8 is very common for "predator–prey" interactions. Some two–species populations tend to approach a "steady state" or "fixed point" (Fig. 9). However, many two–species population graphs tend to cycle over a period of time as in Fig. 9.

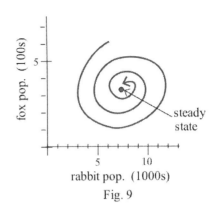

Fig. 9

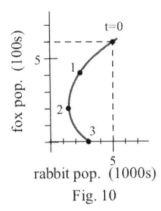

Fig. 10

Practice 2: What would it mean if the rabbit–fox parametric equation graph hit the horizontal axis as in Fig. 10?

Example 3: Graph the pair of parametric equations $x(t) = 2t - 2$ and $y(t) = 3t + 1$.

t	x(t)	y(t)
0	–2	1
1	0	4
2	2	7
–1	–4	–2

Table 2

Solution: Table 2 shows the values of x and y for several values of t. These points are plotted in Fig. 11, and the graph appears to be a straight line.

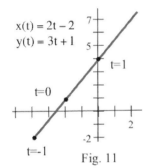

Fig. 11

Usually it is not possible to write y as a simple function of x, but in this case we can do so. By solving $x = 2t - 2$ for $t = \frac{1}{2}x + 1$ and then replacing the t in the equation $y = 3t + 1$, we get $y = 3t + 1 = 3\{\frac{1}{2}x + 1\} + 1 = \frac{3}{2}x + 4$, a linear function of x.

Practice 3: Graph the pair of parametric equations $x(t) = 3 - t$ and $y(t) = t^2 + 1$. Write y as a function of x alone and identify the shape of the graph.

Example 4: Graph the pair of parametric equations $x(t) = 3 \cdot \cos(t)$ and $y(t) = 2 \cdot \sin(t)$ for $0 \leq t \leq 2\pi$, and show that these equations satisfy the relation $\frac{x^2}{9} + \frac{y^2}{4} = 1$ for all values of t.

Solution: The graph, an ellipse, is shown in Fig. 12.

$$\frac{x^2}{9} + \frac{y^2}{4} = \frac{3^2 \cdot \cos^2(t)}{9} + \frac{2^2 \cdot \sin^2(t)}{4} = \cos^2(t) + \sin^2(t) = 1.$$

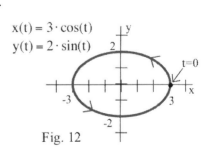

Fig. 12

Practice 4: Graph the pair of parametric equations $x(t) = \sin(t)$ and $y(t) = 5 \cdot \cos(t)$ for $0 \leq t \leq 2\pi$, and show that these equations satisfy the relation $\frac{x^2}{1} + \frac{y^2}{25} = 1$ for all values of t.

Example 5: Describe the motion of a point whose position is

$x(t) = -R \cdot \sin(t)$ and $y(t) = -R \cdot \cos(t)$.

Solution: The point starts at $x(0) = -R \cdot \sin(0) = 0$ and $y(0) = -R \cdot \cos(0) = -R$. By plotting $x(t)$ and $y(t)$ for several other values of t (Fig. 13), we can see that the point is rotating clockwise around the origin. Since $x^2(t) + y^2(t) = R^2 \sin^2(t) + R^2 \cos^2(t) = R^2$, we know the point is always on the circle of radius R which is centered at the origin.

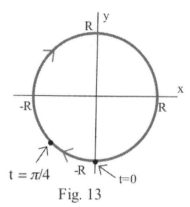

Fig. 13

Practice 5: The path of each parametric equation given below is a circle with radius 1 and center at the origin. If an object is located at the point (x, y) at time t seconds:
(a) Where is the object at $t = 0$? (b) Is the object traveling clockwise or counterclockwise around the circle? (c) How long does it take the object to make 1 revolution?

A: $x = \cos(2t), y = \sin(2t)$ B: $x = -\cos(3t), y = \sin(3t)$ C: $x = \sin(4t), y = -\cos(4t)$

Putting Motions Together

If we know how an object moves horizontally and how it moves vertically, then we can put these motions together to see how it moves in the plane.

If an object is thrown straight upward with an initial velocity of A feet per second, then its height after t seconds is $y(t) = A \cdot t - \frac{1}{2} g \cdot t^2$ feet where $g = 32$ feet/second^2 is the downward acceleration of gravity (Fig. 14a). If an object is thrown horizontally with an initial velocity of B feet per second, then its horizontal distance from the starting place after t seconds is $x(t) = B \cdot t$ feet (Fig. 14b).

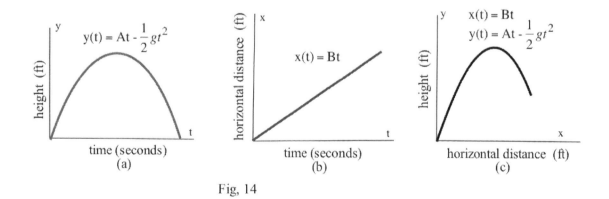

Fig. 14

Example 6: Write an equation for the position at time t (Fig. 14c) of an object thrown at an angle of 30° with the ground (horizontal) with an initial velocity 100 feet per second.

Solution: If the object travels 100 feet along a line at an angle of 30° to the horizontal ground (Fig. 15), then it travels $100 \cdot \sin(30°) = 50$ feet upward and $100 \cdot \cos(30°) \approx 86.6$ feet sideways, so $A = 50$ and $B = 86.6$. The position of the object at time t is

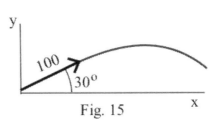

Fig. 15

$$y(t) = 50 \cdot t - \tfrac{1}{2} g t^2 \text{ and } x(t) = 86.6 \cdot t.$$

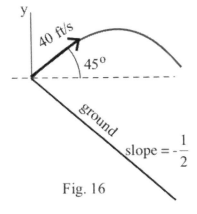

Fig. 16

Practice 6: A ball is thrown upward at an angle of 45° (Fig. 16) with an initial velocity of 40 ft/sec.

(a) Write the parametric equations for the position of the ball as a function of time.

(b) Use the parametric equations to find when and then where the ball will hit the sloped ground. (Suggestion: set $y(t) = -0.5 x(t)$ from part (a) and solve for t. Then use that value of t to evaluate $x(t)$ and $y(t)$.)

Sometimes the location or motion of an object is measured by an instrument which is in motion itself (e.g., tracking a pod of migrating whales from a moving ship), and we want to determine the path of the object independent of the location of the instrument. In that case, the "absolute" location of the object with respect to the origin is the sum of the relative location of the object (pod of whales) with respect to the instrument (ship) and the location of the instrument (ship) with respect to the origin. The same approach works for describing the motion of linked objects such as connected gears.

Example 7: **Carnival Ride** The car (Fig. 17) makes one counterclockwise revolution ($r = 8$ feet) about the pivot point A every 2 seconds and the long arm ($R = 20$ feet) makes one counterclockwise revolution about its pivot point (the origin) every 5 seconds. Assume that the ride begins with the two arms along the positive x–axis and sketch the path you think the car will follow. Find a pair of parametric equations to describe the position of the car at time t.

Solution: The position of the car relative to its pivot point A is $x_c(t) = 8 \cdot \cos(\tfrac{2\pi}{2} t)$ and $y_c(t) = 8 \cdot \sin(\tfrac{2\pi}{2} t)$.

The position of the pivot point A relative to the origin is $x_p(t) = 20 \cdot \cos(\tfrac{2\pi}{5} t)$ and $y_p(t) = 20 \cdot \sin(\tfrac{2\pi}{5} t)$, so the location of the car, relative to the origin, is

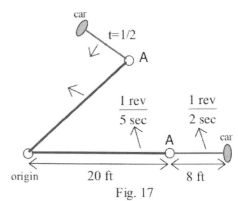

Fig. 17

$$x(t) = x_p(t) + x_c(t) = 20 \cdot \cos(\tfrac{2\pi}{5} t) + 8 \cdot \cos(\tfrac{2\pi}{2} t) \text{ and}$$

$$y(t) = y_p(t) + y_c(t) = 20 \cdot \sin(\tfrac{2\pi}{5} t) + 8 \cdot \sin(\tfrac{2\pi}{2} t) .$$

Use a graphing calculator to graph the path of the car for 5 seconds.

Example 8: **Cycloid**   A light is attached to the edge of a wheel of radius R which is rolling along a level road (Fig. 18). Find parametric equations to describe the location of the light.

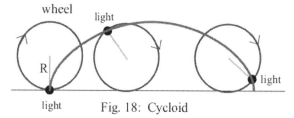

Fig. 18: Cycloid

Solution: We can describe the location of the axle of the wheel, the location of the light relative to the axle, and then put the results together to get the location of the light.

The axle of the wheel is always R inches off the ground, so the y coordinate of the axle is $y_a(t) = R$ (Fig. 19). When the wheel has rotated t radians about its axle, the wheel has rolled a distance of R·t along the road, and the x coordinate of the axle is $x_a(t) = R \cdot t$.

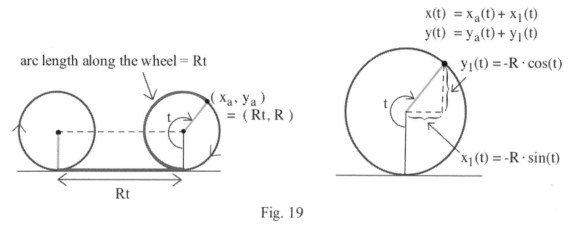

Fig. 19

The position of the light relative to the axle is $x_l(t) = -R \cdot \sin(t)$ and $y_l(t) = -R \cdot \cos(t)$ (see Example 3) so the position of the light is

$$x(t) = x_a(t) + x_l(t) = R \cdot t - R \cdot \sin(t) = R \cdot \{ t - \sin(t) \} \text{ and}$$

$$y(t) = y_a(t) + y_l(t) = R - R \cdot \cos(t) = R \cdot \{ 1 - \cos(t) \} .$$

This curve is called a **cycloid**, and it is one of the most famous and interesting curves in mathematics. Many great mathematicians and physicists (Mersenne, Galileo, Newton, Bernoulli, Huygens, and others) examined the cycloid, determined its properties, and used it in physical applications.

Practice 7: A light is attached r units from the axle of an R inch radius wheel (r < R) that is rolling along a level road (Fig. 20). Use the approach of the solution to Example 8 to find parametric equations to describe the location of the light. The resulting curve is called an curate cycloid.

The cycloid, the path of a point on a rolling circle, was studied in the early 1600's by Mersenne (1588–1648) who thought the path might be part of an ellipse (it isn't). In 1634 Roberval determined the parametric form of the cycloid and found the area under the cycloid as did Descartes and Fermat. This was done before Newton (1642–1727) was even born; they used various specialized geometric approaches to solve the area problem. About the same time Galileo determined the area experimentally by cutting a cycloid region from a sheet of lead and balancing it against a number of circular regions (with the same radius as the circle which generated the cycloid) cut from the same material. How many circles do you think balanced the cycloid region's area (Fig. 21)?

However, the most amazing properties of the cycloid involve motion along a cycloid–shaped path, and their discovery had to wait for Newton and the calculus. These calculus–based properties are discussed at the end of the next section.

PROBLEMS

For problems 1–4, use the data in each table to create three graphs: (a) (t, x(t)), (b) (t, y(t)), and (c) the parametric graph (x(t), y(t)). (Connect the points with straight line segments to create the graph.)

1. Use Table 3.
2. Use Table 4.
3. Use Table 5.
4. Use Table 6.

t	x(t)	y(t)
0	2	1
1	2	0
2	−1	0
3	1	−1

Table 3

t	x(t)	y(t)
0	0	1
1	1	1
2	1	−1
3	2	0

Table 4

t	x(t)	y(t)
0	1	2
1	−1	−1
2	1	2
3	0	2

Table 5

t	x(t)	y(t)
0	0	1
1	−1	0
2	0	−2
3	3	1

Table 6

For problems 5–8, use the data in the given graphs of $(t, x(t))$ and $(t, y(t))$ to sketch the parametric graph $(x(t), y(t))$.

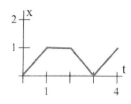

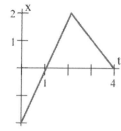

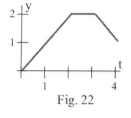

5. Use x and y from Fig. 22.

6. Use x and y from Fig. 23.

7. Use x and y from Fig. 24.

8. Use x and y from Fig. 25.

9. Graph $x(t) = 3t - 2$, $y(t) = 1 - 2t$. What shape is this graph?

10. Graph $x(t) = 2 - 3t$, $y(t) = 3 + 2t$. What shape is this graph?

Fig. 22

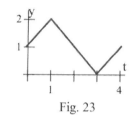

Fig. 23

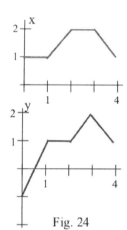

Fig. 24

11. Calculate the slope of the line through the points $P = (x(0), y(0))$ and $Q = (x(1), y(1))$ for $x(t) = at + b$ and $y(t) = ct + d$.

12. Graph $x(t) = 3 + 2 \cdot \cos(t)$, $y(t) = -1 + 3 \cdot \sin(t)$ for $0 \le t \le 2\pi$. Describe the shape of the graph.

13. $x(t) = -2 + 3 \cdot \cos(t)$, $y(t) = 1 - 4 \cdot \sin(t)$ for $0 \le t \le 2\pi$. Describe the shape of the graph.

14. Graph (a) $x(t) = t^2$, $y(t) = t$, (b) $x(t) = \sin^2(t)$, $y(t) = \sin(t)$, and (c) $x(t) = t$, $y(t) = \sqrt{t}$. Describe the similarities and the differences among these graphs.

Fig. 25

15. Graph (a) $x(t) = t$, $y(t) = t$, (b) $x(t) = \sin(t)$, $y(t) = \sin(t)$, and (c) $x(t) = t^2$, $y(t) = t^2$. Describe the similarities and the differences among these graphs.

16. Graph $x(t) = (4 - \frac{1}{t})\cos(t)$, $y(t) = (4 - \frac{1}{t})\sin(t)$ for $t \ge 1$. Describe the behavior of the graph.

17. Graph $x(t) = \frac{1}{t} \cdot \cos(t)$, $y(t) = \frac{1}{t} \cdot \sin(t)$ for $t \ge \pi/4$. Describe the behavior of the graph.

18. Graph $x(t) = t + \sin(t)$, $y(t) = t^2 + \cos(t)$ for $0 \le t \le 2\pi$. Describe the behavior of the graph.

Problems 19–22 refer to the rabbit–fox population graph shown in Fig. 26 which shows several different population cycles depending on the various numbers of rabbits and foxes. Wildlife biologists sometimes try to control animal populations by "harvesting" some of the animals, but it needs to be done with care. The thick dot on the graph is the fixed point for this two-species population.

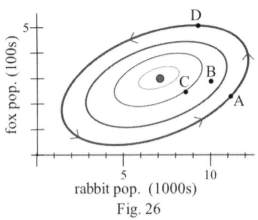

Fig. 26

19. Suppose there are currently 11,000 rabbits and 200 foxes (point A on the graph), and 1,000 rabbits are "harvested" (removed from the population). Does the harvest shift the populations onto a cycle closer to or farther from the fixed point?

20. Suppose there are currently 10,000 rabbits and 300 foxes (point B on the graph), and 100 foxes are "harvested." Does the harvest shift the populations onto a cycle closer to or farther from the fixed point?

21. Suppose there are currently 8,000 rabbits and 250 foxes (point C on the graph), and 1,000 rabbits die during a hard winter. Does the wildlife biologist need to take action to maintain the population balance? Justify your response.

22. Suppose there are currently 9,000 rabbits and 500 foxes (point D on the graph), and 2,000 rabbits die during a hard winter. Does the wildlife biologists need to take action to maintain the population balance? Justify your response.

23. Suppose x and y are functions of the form $x(t) = a \cdot t + b$ and $y(t) = c \cdot t + d$ with $a \neq 0$ and $c \neq 0$. Write y as a function of x alone and show that the parametric graph (x, y) is a straight line. What is the slope of the resulting line?

24. The parametric equations given in (a) – (e) all satisfy $x^2 + y^2 = 1$, and, for $0 \leq t \leq 2\pi$, the path of each object is a circle with radius 1 and center at the origin. Explain how the motions of the objects **differ**.
    (a) $x(t) = \cos(t), y(t) = \sin(t)$,   (b) $x(t) = \cos(-t), y(t) = \sin(-t)$,   (c) $x(t) = \cos(2t), y(t) = \sin(2t)$,
    (d) $x(t) = \sin(t), y(t) = \cos(t)$, and   (e) $x(t) = \cos(t + \pi/2), y(t) = \sin(t + \pi/2)$

25. From a tall building you observe a person is walking along a straight path while twirling a light (parallel to the ground) at the end of a string. (a) If the person is walking slowly, sketch the path of the light. (b) How would the graph change if the person was running? (c) Sketch the path for a person walking (running) along a parabolic path.

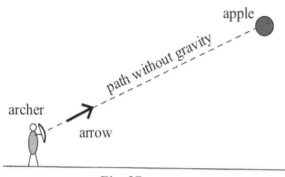

Fig. 27

26. **William Tell and the Falling Apple:** William Tell is aiming directly at an apple, and releases the arrow at exactly the same instant that the apple stem breaks. In a world without gravity (or air resistance), the apple remains in place after the stem breaks, and the arrow flies in a straight line to hit the apple (Fig. 27). Sketch the path of the apple and the arrow in a world with gravity (but still no air). Does the arrow still hit the apple? Why or why not?

27. Find the radius R of a circle which generates a cycloid starting at the point (0,0) and

  (a) passing through the point $(10\pi, 0)$ on its first complete revolution ($0 \le t \le 2\pi$)..

  (b) passing through the point (5, 2) on its first complete revolution. (A calculator is helpful here.)

  (c) passing through the point (2, 3) on its first complete revolution. (A calculator is helpful here.)

  (d) passing through the point $(4\pi, 8)$ on its first complete revolution.

The Ferris Wheel and the Apple (problems 28 – 30).

28. Your friends are on the Ferris wheel illustrated in Fig. 28, and at time t seconds, their location is given parametrically as
$(-20 \sin(\frac{2\pi}{15} t) , 30 - 20 \cos(\frac{2\pi}{15} t))$.

  (a) Is the Ferris wheel turning clockwise of counterclockwise?

  (b) How many seconds does it take the Ferris wheel to make a revolution?

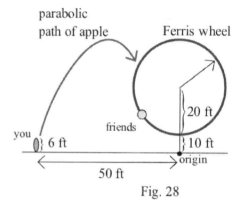

Fig. 28

29. You are 50 feet to the left of the Ferris wheel in problem 28, and you toss an apple from a height of 6 feet above the ground at an angle of 45°. Write parametric equations for the location of the apple (relative to the origin in Fig. 29) at time t if

  (a) its initial velocity is 30 feet per second, and (b) its initial velocity is V feet per second.

30. Help — the Ferris wheel won't stop! To keep your friends on the Ferris wheel in problem 28 from getting too hungry, you toss an apple to them (at time t = 0). Write an equation for the distance between the apple and your friends at time t. Somehow, find a value for the initial velocity V of the apple so that it comes close enough for your friend to catch it, within 2 feet. (Note: A calculator or computer is probably required for this problem.)

Section 9.3 PRACTICE Answers

Practice 1: A possible path for the car is shown in Fig. 29.

Practice 2: If the (rabbit, fox) parametric graph touches the horizontal axis, then there are 0 foxes: the foxes are extinct.

Practice 3: $x = 3 - t$ and $y = t^2 + 1$.

Then $t = 3 - x$ and $y = (3 - x)^2 + 1 = x^2 - 6x + 10$.

The graph in Fig. 30 is parabola, opening upward, with vertex at $(3,1)$.

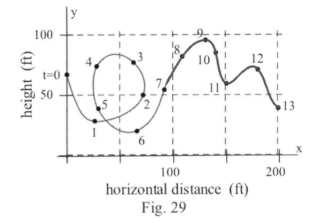

Fig. 29

Practice 4: The parametric graph of $x(t) = \sin(t)$ and $y(t) = 5\cos(t)$ is shown in Fig. 31. For all t,

$$\frac{x^2}{1} + \frac{y^2}{25} = \frac{\sin^2(t)}{1} + \frac{25\cos^2(t)}{25} = \sin^2(t) + \cos^2(t) = 1.$$

Practice 5:
- A: Starts at $(1,0)$, travels counterclockwise, and takes $2\pi/2 = \pi$ seconds to make one revolution.
- B: Starts at $(-1,0)$, travels clockwise, and takes $2\pi/3$ seconds to make one revolution.
- C: Starts at $(0,-1)$, travels counterclockwise, and takes $2\pi/4 = \pi/2$ seconds to make one revolution.

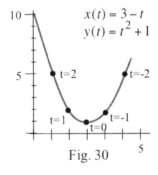

Fig. 30

Fig. 31

Practice 6: (a) $x(t) = 40 \cdot \cos(45°) \cdot t$, $y(t) = 40 \cdot \sin(45°) \cdot t - 16t^2$.

(b) Let $A = 40 \cdot \sin(45°) = 40 \cdot \cos(45°) \approx 28.284$.

Then the ball is at $x(t) = At$ and $y(t) = At - 16t^2$.

Along the ground line, $y = -\frac{1}{2}x$ so the ball intersects the ground when $y(t) = -\frac{1}{2}x(t)$: $At - 16t^2 = -\frac{1}{2}At$.

When $t \neq 0$, we can solve $At - 16t^2 = -\frac{1}{2}At$ for $t = \frac{3}{32}A$.

Putting $t = \frac{3}{32}A$ into the equations for the location of the ball, we have

$x(\frac{3}{32}A) = A \cdot (\frac{3}{32}A) = \frac{3}{32}A^2$ and $y(\frac{3}{32}A) = A \cdot (\frac{3}{32}A) - 16(\frac{3}{32}A)^2 = -\frac{3}{64}A^2$.

The ball hits the ground after $t = \frac{3}{32}A = \frac{3}{32} \cdot 40 \cdot \sin(45°) \approx \textbf{2.652 seconds}$.

The ball hits the ground at the location $x = \frac{3}{32}A^2 = \textbf{75 feet}$ and $y = -\frac{3}{64}A^2 = \textbf{-37.5 feet}$.

Practice 7: Axle: $x_a = R \cdot t$ and $y_a = R$. Light relative to the axle: $x_l = -r \cdot \sin(t)$ and $y_l = -r \cdot \cos(t)$.

Then $x(t) = x_a + x_l = R \cdot t - r \cdot \sin(t)$ and $y(t) = y_a + y_l = R - r \cdot \cos(t)$.

9.4 CALCULUS AND PARAMETRIC EQUATIONS

The previous section discussed parametric equations, their graphs, and some of their uses for visualizing and analyzing information. This section examines some of the ideas and techniques of calculus as they apply to parametric equations: slope of a tangent line, speed, arc length, and area. Slope, speed, and arc length were considered earlier (in optional parts of sections 2.5 and 5.2), and the presentation here is brief. The material on area is new and is a variation on the Riemann sum development of the integral. This section ends with a presentation of some of the properties of the cycloid.

Slope (also see section 2.5)

If $x(t)$ and $y(t)$ are differentiable functions of t, then the derivatives dx/dt and dy/dt measure the rates of change of x and y with respect to t: dx/dt and dy/dt tell how fast each variable is changing. The derivative dy/dx measures the slope of the line tangent to the parametric graph $(x(t), y(t))$. To calculate dy/dx we need to use the Chain Rule:

$$\frac{dy}{dt} = \frac{dy}{dx} \cdot \frac{dx}{dt} .$$

Dividing each side of the Chain Rule by $\frac{dx}{dt}$, we have $\frac{dy}{dx} = \frac{dy/dt}{dx/dt}$.

Slope with Parametric Equations

If $x(t)$ and $y(t)$ are differentiable functions of t and $\frac{dx}{dt} \neq 0$,

then the **slope** of the line tangent to the parametric graph is $\frac{dy}{dx} = \frac{dy/dt}{dx/dt}$.

Example 1: The location of an object is given by the parametric equations $x(t) = t^3 + 1$ feet and $y(t) = t^2 + t$ feet at time t seconds.

(a) Evaluate $x(t)$ and $y(t)$ at $t = -2, -1, 0, 1,$ and 2, and then graph the path of the object for $-2 \leq t \leq 2$.

(b) Evaluate dy/dx for $t = -2, -1, 0, 1,$ and 2. Do your calculated values for dy/dx agree with the shape of your graph in part (a)?

t	x	y	dy/dx
–2	–7	2	–3/12 = –1/4
–1	0	0	–1/3
0	1	0	undefined
1	2	2	3/3 = 1
2	9	6	5/12

Table 1

Solution: (a) When $t = -2$,
$x = (-2)^3 + 1 = -7$ and
$y = (-2)^2 + (-2) = 2$. The other values for x and y are given in Table 1.
The graph of (x, y) is shown in Fig. 1.

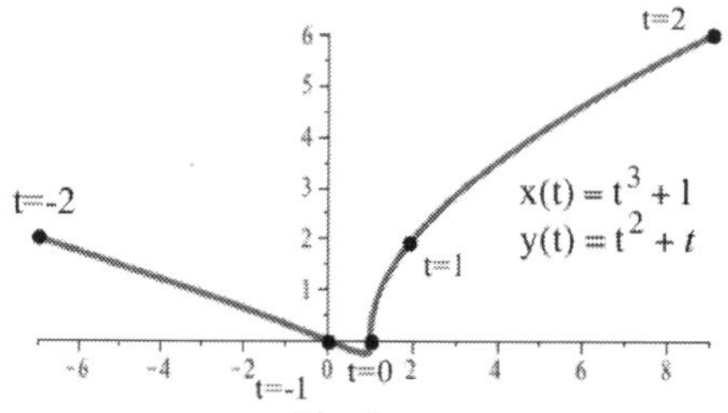

Fig. 1

(b) $dy/dt = 2t + 1$ and $dx/dt = 3t^2$ so $\dfrac{dy}{dx} = \dfrac{2t+1}{3t^2}$. When $t = -2$, $\dfrac{dy}{dx} = \dfrac{-3}{12}$. The other values for dy/dx are given in Table 1.

Practice 1: Find the equation of the line tangent to the graph of the parametric equations in Example 1 when $t = 3$.

An object can "visit" the same location more than once, and a parametric graph can go through the same point more than once.

Example 2: Fig. 2 shows the x and y coordinates of an object at time t.

Fig. 2

(a) Sketch the parametric graph $(x(t), y(t))$, the position of the object at time t.

(b) Give the coordinates of the object when $t = 1$ and $t = 3$.

(c) Find the slopes of the tangent lines to the parametric graph when $t = 1$ and $t = 3$.

Solution: (a) By reading the x and y values on the graphs in Fig. 2, we can plot points on the parametric graph. The parametric graph is shown in Fig. 3.

(b) When $t = 1$, $x = 2$ and $y = 2$ so the parametric graph goes through the point $(2,2)$. When $t = 3$, the parametric graph goes through the same point $(2,2)$.

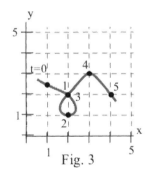

Fig. 3

(c) When $t = 1$, $dy/dt \approx -1$ and $dx/dt \approx +1$ so $\dfrac{dy}{dx} = \dfrac{dy/dt}{dx/dt} \approx \dfrac{-1}{+1} = -1$.

When $t = 3$, $dy/dt \approx +1$ and $dx/dt \approx +1$ so $\dfrac{dy}{dx} \approx \dfrac{+1}{+1} = +1$.

These values agree with the appearance of the parametric graph in Fig. 3.

The object goes through the point $(2,2)$ twice (when $t=1$ and $t=3$), but it is traveling in a different direction each time.

Practice 2: (a) Estimate the slopes of the lines tangent to the parametric graph when $t = 2$ and $t = 5$.

(b) When does $y'(t) = 0$ in Fig. 2?

(c) When does the parametric graph in Fig. 3 have a maximum? A minimum?

(d) How are the maximum and minimum points on a parametric graph related to the derivatives of $x(t)$ and $y(t)$?

Speed

If we know how fast an object is moving in the x direction (dx/dt) and how fast in the y direction (dy/dt), it is straightforward to determine the speed of the object, how fast it is moving in the xy–plane.

If, during a short interval of time Δt, the object's position changes Δx in the x direction and Δy in the y direction (Fig. 4), then the object has moved $\sqrt{(\Delta x)^2 + (\Delta y)^2}$ in Δt time. Then

$$\text{average speed} = \frac{\text{distance moved}}{\text{time change}} = \frac{\sqrt{(\Delta x)^2 + (\Delta y)^2}}{\Delta t}$$

$$= \sqrt{\left(\frac{\Delta x}{\Delta t}\right)^2 + \left(\frac{\Delta y}{\Delta t}\right)^2} \ .$$

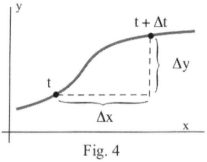

Fig. 4

If x(t) and y(t) are differentiable functions of t, and if we take the limit of the average speed as Δt approaches 0, then

$$\text{speed} = \lim_{\Delta t \to 0} \{\text{average speed}\} = \lim_{\Delta t \to 0} \sqrt{\left(\frac{\Delta x}{\Delta t}\right)^2 + \left(\frac{\Delta y}{\Delta t}\right)^2} = \sqrt{\left(\frac{dx}{dt}\right)^2 + \left(\frac{dy}{dt}\right)^2} \ .$$

Speed with Parametric Equations

If   an object is located at (x(t), y(t)) at time t, and x(t) and y(t) are differentiable functions of t,

then   the **speed** of the object is $\sqrt{\left(\frac{dx}{dt}\right)^2 + \left(\frac{dy}{dt}\right)^2}$.

Example 3: At time t seconds an object is located at (cos(t) feet, sin(t) feet) in the plane. Sketch the path of the object and show that it is travelling at a constant speed.

Solution: The object is moving in a circular path (Fig. 5). dx/dt = –sin(t) feet/second and dy/dt = cos(t) feet/second so at all times the speed of the object is

$$\sqrt{\left(\frac{dx}{dt}\right)^2 + \left(\frac{dy}{dt}\right)^2} = \sqrt{(-\sin(t))^2 + (\cos(t))^2}$$

$$= \sqrt{\sin^2(t) + \cos^2(t)} = \sqrt{1} = 1 \text{ foot per second.}$$

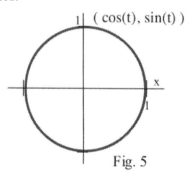

Fig. 5

Practice 3: Is the object in Example 2 traveling faster when t = 1 or when t = 3? When t = 1 or when t = 2?

Arc Length (also see section 5.2)

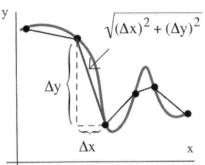

Fig. 6

In section 5.2 we approximated the total length L of a curve by partitioning the curve into small pieces (Fig. 6), approximating the length of each piece using the distance formula, and then adding the lengths of the pieces together to get

$$L \approx \sum \sqrt{(\Delta x)^2 + (\Delta y)^2}$$

$$= \sum \sqrt{\left(\frac{\Delta x}{\Delta x}\right)^2 + \left(\frac{\Delta y}{\Delta x}\right)^2} \, \Delta x \text{ , a Riemann sum.}$$

As Δx approaches 0, the Riemann sum approaches the definite integral

$$L = \int_{x=a}^{b} \sqrt{1 + \left(\frac{dy}{dx}\right)^2} \, dx.$$

A similar approach also works for parametric equations, but in this case we factor out a Δt from the original summation:

$$L \approx \sum \sqrt{(\Delta x)^2 + (\Delta y)^2} = \sum \sqrt{\left(\frac{\Delta x}{\Delta t}\right)^2 + \left(\frac{\Delta y}{\Delta t}\right)^2} \, \Delta t \quad \text{(a Riemann sum)}$$

$$\longrightarrow \int_{t=a}^{b} \sqrt{\left(\frac{dx}{dt}\right)^2 + \left(\frac{dy}{dt}\right)^2} \, dt \text{ as } \Delta t \to 0.$$

Arc Length with Parametric Equations

If   x(t) and y(t) are differentiable functions of t

then   the length of the parametric graph from (x(a), y(a)) to (x(b), y(b)) is

$$L = \int_{t=a}^{t=b} \sqrt{\left(\frac{dx}{dt}\right)^2 + \left(\frac{dy}{dt}\right)^2} \, dt$$

Example 4: Find the length of the cycloid

$x = R(t - \sin(t))$ $y = R(1 - \cos(t))$

for $0 \le t \le 2\pi$. (Fig. 7)

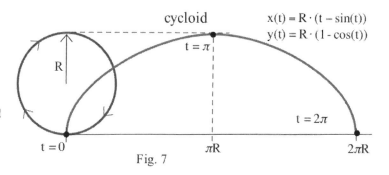

Fig. 7

Solution: Since $dx/dt = R(1 - \cos(t))$ and $dy/dt = R \cdot \sin(t)$,

$$L = \int_{t=a}^{b} \sqrt{\left(\frac{dx}{dt}\right)^2 + \left(\frac{dy}{dt}\right)^2}\, dt =$$

$$\int_{t=0}^{2\pi} \sqrt{(R(1-\cos(t)))^2 + (R\cdot\sin(t))^2}\, dt$$

$$= R\int_{t=0}^{2\pi} \sqrt{1 - 2\cos(t) + \cos^2(t) + \sin^2(t)}\, dt = R\int_{t=0}^{2\pi} \sqrt{2 - 2\cos(t)}\, dt.$$

By replacing θ with $t/2$ in the formula $\sin^2(\theta) = \dfrac{1 - \cos(2\theta)}{2}$ we have $\sin^2(t/2) = \dfrac{1 - \cos(t)}{2}$

so $2 - 2\cos(t) = 4\sin^2(t/2)$, and the integral becomes

$$L = R\int_{t=0}^{2\pi} 2\sin(t/2)\, dt = 2R\{-2\cos(t/2)\}\Big|_{0}^{2\pi} = 2R\{-2\cos(\pi) + 2\cos(0)\} = \mathbf{8R}.$$

The length of a cycloid arch is 8 times the radius of the rolling circle that generated the cycloid.

Practice 4: Represent the length of the ellipse $x = 3\cdot\cos(t)$ $y = 2\cdot\sin(t)$ for for $0 \le t \le 2\pi$ (Fig. 8). as a definite integral. Use Simpson's rule with n= 20 to approximate the value of the integral.

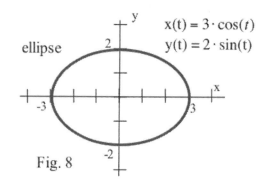

Fig. 8

Area

When we first discussed area and developed the definite integral, we approximated the area of a positive function y (Fig. 9) by partitioning the domain $a \leq x \leq b$ into pieces of length Δx, finding the areas of the thin rectangles, and approximating the total area by adding the little areas together:

$$A \approx \sum y \, \Delta x \quad \text{(a Riemann sum)}.$$

As Δx approached 0, the Riemann sum approached the definite integral $\int_{x=a}^{x=b} y \, dx$.

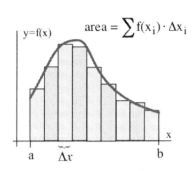

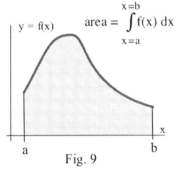

Fig. 9

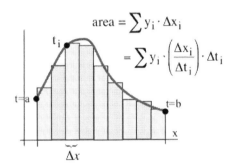

For parametric equations, the independent variable is t and the domain is an interval $[a, b]$.

If x is an increasing function of t, then a partition of the t–interval $[a, b]$ into pieces of length Δt induces a partition along the x–axis (Fig. 10), and we can use the induced partition of the x–axis to approximate the total area by

$$A \approx \sum y \, \Delta x = \sum y \, \frac{\Delta x}{\Delta t} \, \Delta t \quad \text{which approaches the definite}$$

integral $A = \int_{t=a}^{t=b} y \cdot \left(\frac{dx}{dt}\right) dt$ as Δt approaches 0.

Fig. 10

Area with Parametric Equations

If y and dx/dt do not change sign for $a \leq t \leq b$,

then the **area** between the graph (x , y) and the x–axis is $A = \left| \int_{t=a}^{t=b} y \cdot \left(\frac{dx}{dt}\right) dt \right|$.

The requirement that y not change sign for $a \leq t \leq b$ is to prevent the parametric graph from being above the x–axis sometimes and below the x–axis sometimes. The requirement that dx/dt not change sign for $a \leq t \leq b$ is to prevent the graph from "turning around" (Fig. 11). If either of those situations occurs, some of the area is evaluated as positive and some of the area is evaluated as negative.

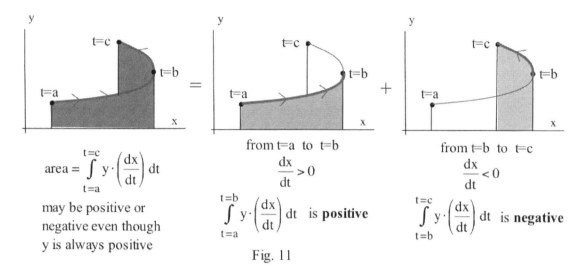

$$\text{area} = \int_{t=a}^{t=c} y \cdot \left(\frac{dx}{dt}\right) dt$$

may be positive or negative even though y is always positive

from t=a to t=b
$$\frac{dx}{dt} > 0$$
$$\int_{t=a}^{t=b} y \cdot \left(\frac{dx}{dt}\right) dt \text{ is \textbf{positive}}$$

from t=b to t=c
$$\frac{dx}{dt} < 0$$
$$\int_{t=b}^{t=c} y \cdot \left(\frac{dx}{dt}\right) dt \text{ is \textbf{negative}}$$

Fig. 11

Example 5: Find the area of the ellipse $x = a \cdot \cos(t)$, $y = b \cdot \sin(t)$ ($a, b > 0$) in the first quadrant (Fig. 12).

Solution: The derivative $dx/dt = -a \cdot \sin(t)$, and in the first quadrant $0 \leq t \leq \pi/2$. Then the

$$\text{area of the ellipse in first quadrant} = \left| \int_{t=a}^{b} y \cdot \left(\frac{dx}{dt}\right) dt \right|$$

$$= \left| \int_{t=0}^{\pi/2} \{ b \cdot \sin(t) \} \cdot (-a \cdot \sin(t)) \, dt \right|$$

$$= \left| -ab \int_{t=0}^{\pi/2} \sin^2(t) \, dt \right| = ab \int_{t=0}^{\pi/2} \sin^2(t) \, dt \quad \text{(replace } \sin^2(t) \text{ with } \frac{1-\cos(2t)}{2} \text{)}$$

$$= \frac{1}{2} ab \int_{t=0}^{\pi/2} 1 - \cos(2t) \, dt = \frac{1}{2} ab \left\{ t - \frac{1}{2} \cdot \sin(2t) \right\} \Big|_{0}^{\pi/2} = \frac{1}{4} ab\pi .$$

The area of the whole ellipse is $4\{ \frac{1}{4} ab\pi \} = \mathbf{\pi ab}$.

If $a = b$, the ellipse is a circle with radius $r = a = b$, and its area is πr^2 as expected.

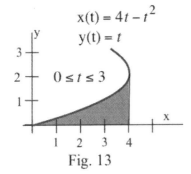

Fig. 13

Practice 5: Let $x(t) = 4t - t^2$ and $y(t) = t$ (Fig. 13).

(a) Represent the shaded area in Fig. 13 as an integral and evaluate the integral.

(b) Evaluate $\int_{t=0}^{t=3} t \cdot (4 - 2t) \, dt$. Does this value represent an area?

Area under a Cycloid: (Fig. 7) For all $t \geq 0$, $x = R(t - \sin(t)) \geq 0$, $y = R(1 - \cos(t)) \geq 0$, and $dx/dt = R(1 - \cos(t)) \geq 0$ so we can use the area formula. Then

$$\text{area} = \left| \int_{t=a}^{b} y \cdot \left(\frac{dx}{dt}\right) dt \right| = \left| \int_{t=0}^{2\pi} \{R(1 - \cos(t))\} \cdot (R(1 - \cos(t))) \, dt \right|$$

$$= R^2 \int_{t=0}^{2\pi} 1 - 2\cos(t) + \cos^2(t) \, dt \quad (\text{replace } \cos^2(t) \text{ with } \frac{1 + \cos(2t)}{2} \text{ and integrate})$$

$$= R^2 \{ t - 2\sin(t) + \frac{1}{2} t + \frac{1}{4} \sin(2t) \} \Big|_0^{2\pi} = R^2 \{ 2\pi + \pi \} = 3\pi R^2.$$

The area under one arch of a cycloid is 3 times the area of the circle that generates the cycloid.

Properties of the Cycloid

Suppose you and a friend decide to have a contest to see who can build a slide that gets a person from point A to point B (Fig. 14) in the shortest time. What shape should you make your slide — a straight line, part of a circle, or something else? Assuming that the slide is frictionless and that the only acceleration is due to gravity, John Bernoulli showed that the **shortest time** ("brachistochrone" for "brachi" = short and "chrone" = time) path is a cycloid that starts at A that also goes through the point B. Fig. 15 shows the cycloid paths for A and B as well as the cycloid paths for two other "finish" points, C and D.

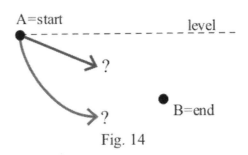

Fig. 14

Even before Bernoulli solved the brachistochrone problem, the astronomer (physicist, mathematician) Huygens was trying to design an accurate pendulum clock. On a standard pendulum clock (Fig. 16), the path of the bob is part of a circle, and the period of the swing depends on the displacement angle of the bob. As friction slows the bob, the displacement angle gets smaller and the clock slows

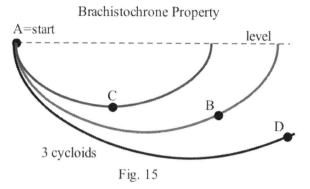

Fig. 15

down. Huygens designed a clock (Fig. 17) whose bob swung in a curve so that the period of the swing did not depend on the displacement angle. The curve Huygens found to solve the **same time** ("tautochrone" for "tauto" = same and "chrone" = time) problem was the cycloid. Beads strung on a wire in the shape of a cycloid (Fig. 18) reach the bottom in the same amount of time, no matter where along the wire (except the bottom point) they are released.

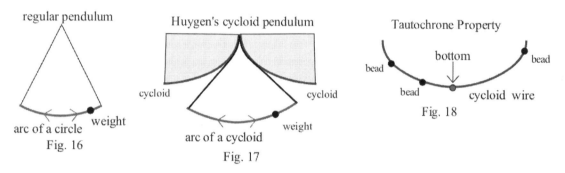

The brachistochane and tautochrone problems are examples from a field of mathematics called the Calculus of Variations. Typical optimization problems in calculus involve finding a point or number that maximizes or minimizes some quantity. Typical optimization problems in the Calculus of Variations involve finding the curve or function that maximizes or minimizes some quantity. For example, what curve or shape with a given length encloses the greatest area? (Answer: a circle) Modern applications of Calculus of Variations include finding routes for airliners and ships to minimize travel time or fuel consumption depending on prevailing winds or currents.

PROBLEMS

Slope

For problems 1–8, (a) sketch the parametric graph (x,y),

(b) find the slope of the line tangent to the parametric graph at the given values of t, and

(c) find the points (x,y) at which dy/dx is either 0 or undefined.

1. $x(t) = t - t^2$, $y(t) = 2t + 1$ at $t = 0, 1$, and 2.

2. $x(t) = t^3 + t$, $y(t) = t^2$ at $t = 0, 1$, and 2.

3. $x(t) = 1 + \cos(t)$, $y(t) = 2 + \sin(t)$ at $t = 0, \pi/4$, and $\pi/2$.

4. $x(t) = 1 + 3 \cdot \cos(t)$, $y(t) = 2 + 2 \cdot \sin(t)$ at $t = 0, \pi/4, \pi/2$, and π.

5. $x(t) = \sin(t)$, $y(t) = \cos(t)$ at $t = 0, \pi/4, \pi/2$, and 17.3.

6. $x(t) = 3 + \sin(t)$, $y(t) = 2 + \sin(t)$ at $t = 0, \pi/4, \pi/2$, and 17.3.

7. $x(t) = \ln(t)$, $y(t) = 1 - t^2$ at $t = 1, 2$, and e.

8. $x(t) = \arctan(t)$, $y(t) = e^t$ at $t = 0, 1$, and 2.

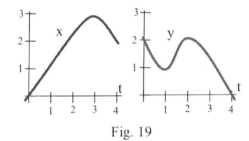

Fig. 19

In problems 9–12, the graphs of x(t) and y(t) are given.

Use this graphical information to estimate

(a) the slope of the line tangent to the parametric graph

at $t = 0, 1, 2$, and 3, and

(b) the points (x,y) at which dy/dx is either 0 or undefined.

9. x(t) and y(t) in Fig. 19.

10. x(t) and y(t) in Fig. 20.

11. x(t) and y(t) in Fig. 21.

12. x(t) and y(t) in Fig. 22.

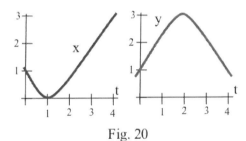

Fig. 20

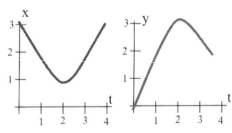

Fig. 21

Speed

For problems 13–20, the locations x(t) and y(t) (in feet) of an object are given at time t seconds. Find the speed of the object at the given times.

13. $x(t) = t - t^2$, $y(t) = 2t + 1$ at $t = 0, 1$, and 2.

14. $x(t) = t^3 + t$, $y(t) = t^2$ at $t = 0, 1$, and 2.

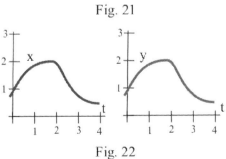

Fig. 22

15. $x(t) = 1 + \cos(t)$, $y(t) = 2 + \sin(t)$ at $t = 0, \pi/4, \pi/2$, and π.

16. $x(t) = 1 + 3\cdot\cos(t)$, $y(t) = 2 + 2\cdot\sin(t)$ at $t = 0, \pi/4, \pi/2$, and π.

17. x and y in Fig. 19 at $t = 0, 1, 2, 3$, and 4.

18. x and y in Fig. 20 at $t = 0, 1, 2$, and 3.

19. x and y in Fig. 21 at $t = 0, 1, 2$, and 3.

20. x and y in Fig. 22 at $t = 0, 1, 2$, and 3.

21. At time t seconds an object is located at the point $x(t) = R\cdot(t - \sin(t))$, $y(t) = R\cdot(1 - \cos(t))$ (in feet).
    (a) Find the speed of the object at time t. (b) At what time is the object traveling fastest?
    (c) Where is the object on the cycloid when it is traveling fastest?

22. At time t seconds an object is located at the point $x(t) = 5\cdot\cos(t)$, $y(t) = 2\cdot\sin(t)$ (in feet).
    (a) Find the speed of the object at time t. (b) At what time is the object traveling fastest?
    (c) Where is the object on the ellipse when it is traveling fastest?

Arc Length

For problems 23–28, (a) represent the arc length of each parametric function as a definite integral, and
(b) evaluate the integral (if necessary, use your calculator's **fnInt()** feature to evaluate the integral).

23. $x(t) = t - t^2$, $y(t) = 2t + 1$ for $t = 0$ to 2.

24. $x(t) = t^3 + t$, $y(t) = t^2$ for $t = 0$ to 2.

25. $x(t) = 1 + \cos(t)$, $y(t) = 2 + \sin(t)$ for $t = 0$ to π.

26. $x(t) = 1 + 3\cdot\cos(t)$, $y(t) = 2 + 2\cdot\sin(t)$ for $t = 0$ to π.

27. x and y in Fig. 23 for $t = 1$ to 3.

28. x and y in Fig. 22 for $t = 0$ to 2.

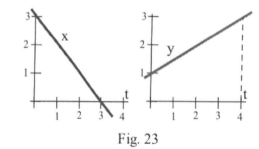

Fig. 23

Area

For problems 29–34, (a) represent the area of each region as a definite integral, and
(b) evaluate the integral (if necessary, use your calculator's **fnInt()** feature to evaluate the integral).

29. $x(t) = t^2$, $y(t) = 4t^2 - t^4$ for $0 \le t \le 2$.

30. $x(t) = 1 + \sin(t)$, $y(t) = 2 + \sin(t)$ for $0 \le t \le \pi$.

31. $x(t) = t^2$, $y(t) = 1 + \cos(t)$ for $0 \le t \le 2$.

32. $x(t) = \cos(t)$, $y(t) = 2 - \sin(t)$ for $0 \le t \le \pi/2$.

33. "Cycloid" with a square wheel: Find the area under one "arch" of the path of a point on the corner of a "rolling" square that has sides of length R. (This problem does not require calculus.)

34. The region bounded between the x–axis and the curate cycloid $x(t) = R \cdot t - r \cdot \sin(t)$, $y(t) = R - r \cdot \cos(t)$ for $0 \le t \le 2\pi$.

Section 9.4   **PRACTICE Answers**

Practice 1: $\frac{dy}{dt} = 2t + 1$, so when $t = 3$, $\frac{dy}{dt} = 7$. $\frac{dx}{dt} = 3t^2$, so when $t = 3$, $\frac{dx}{dt} = 27$.

Finally, $\frac{dy}{dx} = \frac{dy/dt}{dx/dt}$ so when $t = 3$, $\frac{dy}{dx} = \frac{7}{27}$.

When $t = 3$, $x = 28$ and $y = 12$ so the equation of the tangent line is $y - 12 = \frac{7}{27}(x - 28)$.

Practice 2: (a) When $t = 2$, $dy/dx \approx 0$. When $t = 5$, $dy/dx \approx -1$.

(b) In Fig. 2, $\frac{dy}{dt} = 0$ when $t \approx 2$ and $t \approx 4$.

(c) In Fig. 3, a minimum occurs when $t \approx 2$ and a maximum when $t \approx 4$.

(d) If the parametric graph has a maximum or minimum at $t = t^*$, then dy/dt is either 0 or undefined when $t = t^*$.

Practice 3: When $t = 1$, speed $= \sqrt{(dx/dt)^2 + (dy/dt)^2} \approx \sqrt{(1)^2 + (-1)^2} = \sqrt{2} \approx 1.4$ ft/sec.
When $t = 2$, speed $= \sqrt{(dx/dt)^2 + (dy/dt)^2} \approx \sqrt{(-1)^2 + (0)^2} = \sqrt{1} = 1$ ft/sec.
When $t = 3$, speed $= \sqrt{(dx/dt)^2 + (dy/dt)^2} \approx \sqrt{(1)^2 + (1)^2} = \sqrt{2} \approx 1.4$ ft/sec.

Practice 4: Length $= \int_{t=0}^{2\pi} \sqrt{(-3\sin(t))^2 + (2\cos(t))^2} \, dt$

≈ 15.87 (using my calculator's **fnInt()** feature)

Practice 5: (a) $A = \int_{t=0}^{2} t \cdot (4 - 2t) \, dt = 2 \cdot t^2 - \frac{2}{3} t^3 \Big|_0^2 = \{ 8 - \frac{16}{3} \} - \{ 0 - 0 \} = \frac{8}{3}$.

(b) $\int_{t=0}^{3} t \cdot (4 - 2t) \, dt = 2 \cdot t^2 - \frac{2}{3} t^3 \Big|_0^3 = \{ 18 - 18 \} - \{ 0 - 0 \} = 0$.

This integral represents {shaded area in Fig. 13} – {area from $t = 2$ to $t = 3$}.

9.4½ BEZIER CURVES — Getting the shape you want

Historically, parametric equations were often used to model the motion of objects, and that is the approach we have seen so far. But more recently, as computers became more common in design work and manufacturing, a need arose to efficiently find formulas for shapes such as airplane wings and automobile bodies and even letters of the alphabet that designers or artists had created.

One simple but inefficient method for describing and storing the shape of a curve is to measure the location and save the coordinates of hundreds or thousands of points along the curve. This result is called a "bitmap" of the shape. However, bitmaps typically require a large amount of computer memory, and when the bitmap is reconverted from stored coordinates back into a graphic image, originally smooth curves often appear jagged (Fig. 1). Also, when these bitmapped shapes are stretched or rotated, the new location of every one of the points must be calculated, a relatively slow process.

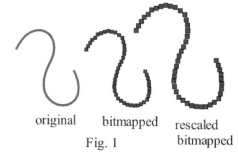
original   bitmapped   rescaled
          Fig. 1   bitmapped

A second method, still simple but more efficient than bitmaps, is to store fewer points along the curve, but to automatically connect consecutive points with line segments (Fig. 2). Less computer memory is required since fewer coordinates are stored, and stretches and rotations are calculated more quickly since the new locations of fewer points are needed. This method is commonly used in computer graphics to store and redraw surfaces (Fig. 3). Sometimes instead of saving the coordinates of each point, a "vector" is used to describe how to get to the next point from the previous point, and the result is a "vector map" of the curve. The major drawback of this method is that the stored and redrawn curve consists of straight segments and corners even though the original curve may have been smooth.

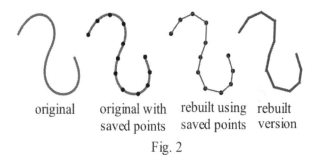
original   original with   rebuilt using   rebuilt
            saved points   saved points   version
                       Fig. 2

original surface

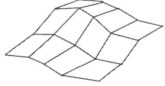
surface rebuilt using
points and line segments

Fig. 3

The primary building block for curves and surfaces represented as line segments is the line segment given by parametric equations.

Example 1: Show that the parametric equations

$x(t) = (1-t) \cdot x_0 + t \cdot x_1$ and

$y(t) = (1-t) \cdot y_0 + t \cdot y_1$

for $0 \le t \le 1$ go through the points $P_0 = (x_0, y_0)$ and $P_1 = (x_1, y_1)$ (Fig. 4) and that the slope is

$\dfrac{dy}{dx} = \dfrac{y_1 - y_0}{x_1 - x_0}$

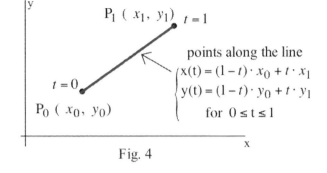

Fig. 4

for all values of t between 0 and 1.

We typically abbreviate the pair of parametric equations as $P(t) = (1-t) \cdot P_0 + t \cdot P_1$.

Solution: $x(0) = (1-0) \cdot x_0 + 0 \cdot x_1 = x_0$ and $y(0) = (1-0) \cdot y_0 + 0 \cdot y_1 = y_0$ so $P(0) = P_0$.

$x(1) = (1-1) \cdot x_0 + 1 \cdot x_1 = x_1$ and $y(1) = (1-1) \cdot y_0 + 1 \cdot y_1 = y_1$ so $P(1) = P_1$.

$\dfrac{dy(t)}{dt} = -y_0 + y_1$ and $\dfrac{dx(t)}{dt} = -x_0 + x_1$

so $\dfrac{dy}{dx} = \dfrac{dy/dt}{dx/dt} = \dfrac{y_1 - y_0}{x_1 - x_0}$, the slope of the line segment from P_0 to P_1.

Practice 1: Use the pattern of Example 1 to write parametric equations for the line segments that

(a) connect (1, 2) to (5, 4), (b) connect (5, 4) to (1, 2), and (c) connect (6, –2) to (3, 1).

Bezier Curves

One solution to the problem of efficiently saving and redrawing a smooth curve was independently developed in the 1960s by two French automobile engineers, Pierre Bezier (pronounced "bez–ee–ay") who worked for Renault automobile company and P. de Casteljau who worked for Citroen. Originally, the solutions were considered industrial secrets, but Bezier's work was eventually published first. The curves that result using Bezier's method are called Bezier curves. The method of Bezier curves allow us to efficiently store information about smooth (and not–so–smooth) shapes and to quickly stretch, rotate and distort these shapes. Bezier curves are now commonly used in computer–aided design work and in most computer drawing programs. They are also used to specify the shapes of letters of the alphabet in different fonts. By using this method, a computer and a laser printer can have many different fonts in many different sizes available without using a large amount of memory. (Bezier curves were used to produce most of the graphs in this book.)

Here we define Bezier curves and examine some of their properties. At the end of this section some optional material describes the mathematical construction of Bezier curves.

Definition: Bezier Curve

The Bezier curve $B(t)$ defined for the four points $P_0, P_1, P_2,$ and P_3 (Fig. 5) is

$$B(t) = (1-t)^3 \cdot P_0 + 3(1-t)^2 \cdot t \cdot P_1 + 3(1-t) \cdot t^2 \cdot P_2 + t^3 \cdot P_3$$

for $0 \le t \le 1$:

$$x(t) = (1-t)^3 \cdot x_0 + 3(1-t)^2 \cdot t \cdot x_1 + 3(1-t) \cdot t^2 \cdot x_2 + t^3 \cdot x_3 \text{ and}$$
$$y(t) = (1-t)^3 \cdot y_0 + 3(1-t)^2 \cdot t \cdot y_1 + 3(1-t) \cdot t^2 \cdot y_2 + t^3 \cdot y_3 \ .$$

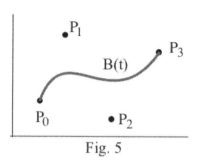

Fig. 5

The four points $P_0, P_1, P_2,$ and P_3 are called **control points** for the Bezier curve. Fig. 6 shows Bezier curves for several sets of control points. The dotted lines connecting the control points in Fig. 6 are shown to help illustrate the relationship between the graph of $B(t)$ and the control points.

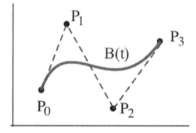

Example 2: Plot the points $P_0 = (0,3), P_1 = (1,5), P_2 = (3,-1),$ and $P_3 = (4,0)$, and determine the equation of the Bezier curve for these control points. Then graph the Bezier curve.

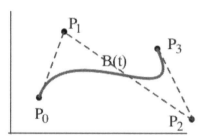

Solution:

$$x(t) = (1-t)^3 \cdot 0 + 3(1-t)^2 \cdot t \cdot 1 + 3(1-t) \cdot t^2 \cdot 3 + t^3 \cdot 4 = -2t^3 + 3t^2 + 3t \ .$$
$$y(t) = (1-t)^3 \cdot 3 + 3(1-t)^2 \cdot t \cdot 5 + 3(1-t) \cdot t^2 \cdot (-1) + t^3 \cdot 0$$
$$= 15t^3 - 24t^2 + 6t + 3 \ .$$

The control points and the graph of $B(t) = (x(t), y(t))$ are shown in Fig. 7.

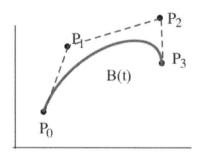

Fig. 6: Several Bezier curves as P_2 is moved

Practice 2: Plot the points $P_0 = (0, 4)$, $P_1 = (1, 2)$, $P_2 = (4, 2)$, and $P_3 = (4, 4)$, and determine the equation of the Bezier curve for these control points. Then graph the Bezier curve.

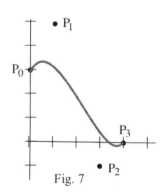

Fig. 7

Properties of Bezier Curves

Bezier curves have a number of properties that make them particularly useful for design work, and some of them are stated below. These properties are verified at the end of this section.

(1) $B(0) = P_0$ and $B(1) = P_3$ so the Bezier curve goes through the points P_0 and P_3.

> This property guarantees that $B(t)$ goes through specified points. If we want two Bezier curves to fit together, it is important that the value at the end of one curve matches the starting value of the next curve. This property guarantees that we can control the values of the Bezier curves at their endpoints by choosing appropriate values for the control points P_0 and P_3.

(2) $B(t)$ is a cubic polynomial.

> This is an important property because it guarantees that $B(t)$ is continuous and differentiable at each point so its graph is connected and smooth at each point. It also guarantees that the graph of $B(t)$ does not "wiggle" too much between control points.

(3) $B'(0) =$ slope of the line segment from P_0 to P_1; $B'(1) =$ slope of the line segment from P_2 to P_3.

> This is an important property because it means we can match the ending slope of one curve with the starting slope of the next curve to result in a smooth connection. We can see in Fig. 6 that the dotted line from P_0 to P_1 is tangent to the graph of $B(t)$ at the point P_0.

(4) For $0 \le t \le 1$, the graph of $B(t)$ is in the region whose corners are the control points.

> Visually, property (4) means that if we put a rubber band around the four control points P_0, P_1, P_2, and P_3 (Fig. 8), then the graph of $B(t)$ will be inside the rubber banded region. This is an important property of Bezier curves because it guarantees that the graph of $B(t)$ does not get too far from the four control points.

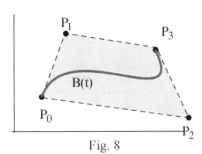

Fig. 8

Example 3: Find a formula for a Bezier curve that goes through the points $(1, 1)$ and $(0, 0)$ and is shaped like an "S."

Solution: Since we want the curve to begin at the point $(1,1)$ and end at $(0,0)$ we can put $P_0 = (1,1)$ and $P_3 = (0,0)$. A little experimentation with values for P_1 and P_2 indicates that $P_1 = (-1, 2)$ and $P_2 = (2, -1)$ gives a mediocre "S" shape (Fig. 9). Then the formula for $B(t)$ is

$$B(t) = (1-t)^3 \cdot P_0 + 3(1-t)^2 \cdot t \cdot P_1 + 3(1-t) \cdot t^2 \cdot P_2 + t^3 \cdot P_3$$

for $0 \le t \le 1$, and

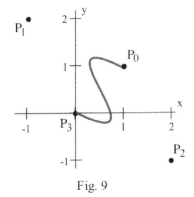

Fig. 9

$$x(t) = (1-t)^3 \cdot x_0 + 3(1-t)^2 \cdot t \cdot x_1 + 3(1-t) \cdot t^2 \cdot x_2 + t^3 \cdot x_3 = (1-t)^3 \cdot 1 + 3(1-t)^2 \cdot t \cdot (-1) + 3(1-t) \cdot t^2 \cdot 2 + t^3 \cdot 0$$

and

$$y(t) = (1-t)^3 \cdot y_0 + 3(1-t)^2 \cdot t \cdot y_1 + 3(1-t) \cdot t^2 \cdot y_2 + t^3 \cdot y_3 = (1-t)^3 \cdot 1 + 3(1-t)^2 \cdot t \cdot 2 + 3(1-t) \cdot t^2 \cdot (-1) + t^3 \cdot 0$$

Certainly other values of P_1 and P_2 can give similar shapes.

Practice 3: Find a formula for a Bezier curve that goes through the points $(0,0)$ and $(0,1)$ and is shaped like a "C."

Example 4: Find a formula for a Bezier curve that goes through the point $(0,5)$ with a slope of 2 and through the point $(6,1)$ with a slope of 3.

Solution: Since we want to curve to begin at $(0,5)$ and end at $(6,1)$, we put $P_0 = (0,5)$ and $P_3 = (6,1)$. To get the slopes we need to pick P_1 so the slope of the line segment from P_0 to P_1 is 2: going "over 1 and up 2" to get $P_1 = (1,7)$ works fine as do several other points. Similarly, to get the right slope at P_3 we can go "back 1 and down 3" to get $P_2 = (5,-2)$. The Bezier curve for $P_0 = (0,5), P_1 = (1,7), P_2 = (5,-2)$, and $P_3 = (6,1)$ is

$$x(t) = (1-t)^3 \cdot 0 + 3(1-t)^2 \cdot t \cdot 1 + 3(1-t) \cdot t^2 \cdot 5 + t^3 \cdot 6 = -6t^3 + 9t^2 + 3t \text{ and}$$
$$y(t) = (1-t)^3 \cdot 5 + 3(1-t)^2 \cdot t \cdot 7 + 3(1-t) \cdot t^2 \cdot (-2) + t^3 \cdot 1 = 23t^3 - 33t^2 + 6t + 5.$$

The graph of this B(t) is shown in Fig. 10.

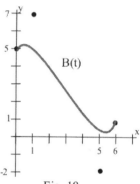
Fig. 10

Keeping the previous values of P_0, P_1 and P_3, we could pick P_2 by going "over 1 and up 3" to $P_2 = (7,4)$, and the graph of the B(t) for this choice of P_2 is shown in Fig. 11. The graph for B(t) when $P_1 = (2,9)$ and $P_2 = (5.5, -0.5)$ is shown in Fig. 12. These, and other choices of P_1 and P_2 satisfy the conditions specified in the problem: the choice of which one you use depends on the other properties of the shape that you want the curve to have.

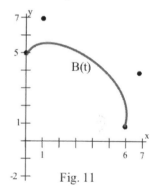
Fig. 11

Practice 4: Find a formula for a Bezier curve that goes through the point $(1,3)$ with a slope of -2 and through the point $(5,3)$ with a slope of -1.

Example 5: Find formulas for a pair of Bezier curves so that the first starts at the point $A = (0,3)$ with slope 2, the second ends at the point $C = (7,2)$ with slope 1, and the curves connect at the point $B = (4,6)$ with slope 0.

Solution: For the first Bezier curve B(t) take $P_0 = A = (0,3), P_1 = (1,5)$ (to get the slope 2),

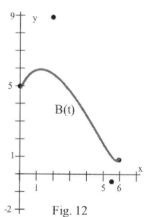
Fig. 12

$P_3 = B = (4,6)$, and $P_2 = (3,6)$ (to get the slope 0 at the connecting point). Then,

for $0 \le t \le 1$, $x(t) = (1-t)^3 \cdot 0 + 3(1-t)^2 \cdot t \cdot 1 + 3(1-t) \cdot t^2 \cdot 3 + t^3 \cdot 4$ and

$y(t) = (1-t)^3 \cdot 3 + 3(1-t)^2 \cdot t \cdot 5 + 3(1-t) \cdot t^2 \cdot 6 + t^3 \cdot 6$.

For the second Bezier curve $C(t)$ take $P_0 = B = (4,6)$, $P_1 = (5,6)$ (to get the slope 0 at the connecting point), $P_3 = C = (7,2)$, and $P_2 = (6,1)$ (to get the slope 1). Then,

for $0 \le t \le 1$, $x(t) = (1-t)^3 \cdot 4 + 3(1-t)^2 \cdot t \cdot 5 + 3(1-t) \cdot t^2 \cdot 6 + t^3 \cdot 7$ and

$y(t) = (1-t)^3 \cdot 6 + 3(1-t)^2 \cdot t \cdot 6 + 3(1-t) \cdot t^2 \cdot 1 + t^3 \cdot 2$.

Fig. 13 shows the graphs of $B(t)$ and $C(t)$ and illustrates how they connect, continuously and smoothly, at the common point $(4,6)$.

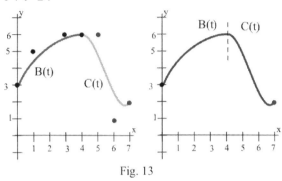

Fig. 13

Using Bezier Curves

In practice, Bezier curves are usually used in computer design or manufacturing programs, and the user of Bezier curves does not have to know the mathematics behind them. But the program creator does!

Typically a designer sketches a crude shape for an object and then moves certain points to locations specified by the plans. Sometimes the designer adds additional points along the curve to "fix" the location of the curve. These "fixed" points along the curve become the endpoints P_0 and P_3 for each of the sections of the curve that will be described by a Bezier formula. Then, for each section of the curve, the designer visually experiments with different locations of the interior control points P_1 and P_2 to get the shape "just right."

Meanwhile, the computer program adjusts the formulas for the Bezier curves based on the current locations of the control points for each section, and, when the design is complete, saves the locations of the control points.

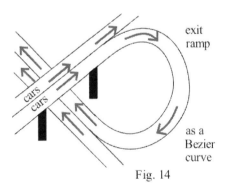

Fig. 14

You may never need to calculate the formulas for Bezier curves (outside of a mathematics class), but if you do any computer–aided design work you will certainly be using these curves. And the ideas and formulas for Bezier curves in two dimensions extend very easily and naturally to describe paths in three dimensions such as the route of a highway exit ramp (Fig. 14) or the path of a hydraulic hose for the landing gear of an airplane.

Mathematical Construction of Bezier Curves & Verifications of Their Properties

In order to use and program Bezier curves we don't need to know where the formulas came from, but their construction is a beautiful piecing–together of simple geometric ideas.

The first idea is the parametric representation of a line segment from point
P_A to P_B as $L(t) = (1-t) \cdot P_A + t \cdot P_B$ for $0 \leq t \leq 1$.

This parametric pattern for a line is used in the construction of a Bezier curve.

When $t=0$, the point is $L(0) = P_A$. When $t=1$, $L(1) = P_B$. When $t=0.5$, $L(0.5)$ is the midpoint of the line from P_A to P_B (Fig. 15). When $t=0.2$, $L(0.2)$ is 20% of the way along the line L from P_A to P_B.

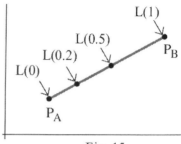

Fig. 15

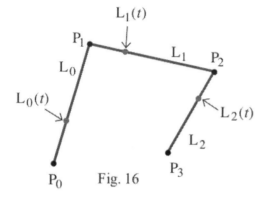

Fig. 16

To construct the Bezier curve for the four control points $P_0, P_1, P_2,$ and P_3 we start by fixing a value of t between 0 and 1. Then we find the point $L_0(t)$ along the parametric line from P_0 to P_1, the point $L_1(t)$ along the parametric line from P_1 to P_2, and the point $L_2(t)$ along the parametric line from P_2 to P_3 (Fig. 16):

$$L_0(t) = (1-t) \cdot P_0 + t \cdot P_1,$$
$$L_1(t) = (1-t) \cdot P_1 + t \cdot P_2,$$
$$L_2(t) = (1-t) \cdot P_2 + t \cdot P_3.$$

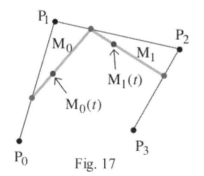

Fig. 17

For the fixed value of t, we then find the point $M_0(t)$ along the parametric line from the point $L_0(t)$ to the point $L_1(t)$, and the point $M_1(t)$ along the parametric line from the point $L_1(t)$ to the point $L_2(t)$ (Fig. 17):

$$M_0(t) = (1-t) \cdot L_0(t) + t \cdot L_1(t)$$
$$M_1(t) = (1-t) \cdot L_1(t) + t \cdot L_2(t).$$

For the same fixed value of t, we finally find the point B(t) along the parametric line from the point $M_0(t)$ to the point $M_1(t)$ (Fig. 18):

$$B(t) = (1-t) \cdot M_0(t) + t \cdot M_1(t).$$

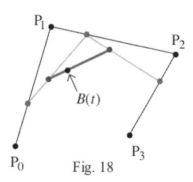

Fig. 18

As the variable t takes on different values between 0 and 1, the points $L_0(t), L_1(t),$ and $L_2(t)$ move along the lines connecting $P_0, P_1, P_2,$ and P_3. Similarly, the points $M_0(t)$ and $M_1(t)$ move along the lines connecting $L_0(t), L_1(t),$ and $L_2(t)$, and the point B(t) moves along the line connecting $M_0(t)$ and $M_1(t)$. It is all quite dynamic.

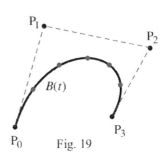

Fig. 19

Fig. 19 shows these points and lines for several values of t between 0 and 1.

We can use the previous geometric construction to obtain the formula for B(t) given in the definition of Bezier curves by working backwards from $B(t) = (1-t) \cdot M_0(t) + t \cdot M_1(t)$:

$$
\begin{aligned}
B(t) &= (1-t) \cdot M_0 + t \cdot M_1 \\
&= (1-t) \cdot \{ (1-t) \cdot L_0 + t \cdot L_1 \} + t \cdot \{ (1-t) \cdot L_1 + t \cdot L_2 \} \quad \text{replacing } M_0 \text{ and } M_1 \text{ in terms of } L_0, L_1 \text{ and } L_2 \\
&= (1-t)^2 \cdot L_0 + 2(1-t) \cdot t \cdot L_1 + t^2 \cdot L_2 \quad \text{simplifying} \\
&= (1-t)^2 \cdot \{ (1-t) \cdot P_0 + t \cdot P_1 \} + 2(1-t) \cdot t \cdot \{ (1-t) \cdot P_1 + t \cdot P_2 \} + t^2 \cdot \{ (1-t) \cdot P_2 + t \cdot P_3 \} \\
&\quad \text{replacing } L_0, L_1 \text{ and } L_2 \text{ in terms of } P_0, P_1, P_2 \text{ and } P_3 \\
&= (1-t)^3 \cdot P_0 + 3(1-t)^2 \cdot t \cdot P_1 + 3(1-t) \cdot t^2 \cdot P_2 + t^3 \cdot P_3 \quad \text{simplifying}
\end{aligned}
$$

Verifications of Properties (1) – (4)

Property (1) is easy to verify by evaluating B(0) and B(1):

$B(0) = (1-0)^3 \cdot P_0 + 3(1-0)^2 \cdot 0 \cdot P_1 + 3(1-0) \cdot t^2 \cdot P_2 + 0^3 \cdot P_3 = P_0$. Similarly,

$B(1) = (1-1)^3 \cdot P_0 + 3(1-1)^2 \cdot 1 \cdot P_1 + 3(1-1) \cdot 1^2 \cdot P_2 + 1^3 \cdot P_3 = P_3$.

Property (2) is clear from the defining formula for B(t), or we can expand the powers of $1-t$ and t and collect the similar terms to rewrite B(t) as

$B(t) = (-P_0 + 3P_1 - 3P_2 + P_3) \cdot t^3 + (3P_0 - 6P_1 + 3P_2) \cdot t^2 + (-3P_0 + 3P_1) \cdot t + (P_0)$.

Property (3) can be verified using the rewritten form from Property 2,

$x(t) = (-x_0 + 3x_1 - 3x_2 + x_3) \cdot t^3 + (3x_0 - 6x_1 + 3x_2) \cdot t^2 + (-3x_0 + 3x_1) \cdot t + (x_0)$ and

$y(t) = (-y_0 + 3y_1 - 3y_2 + y_3) \cdot t^3 + (3y_0 - 6y_1 + 3y_2) \cdot t^2 + (-3y_0 + 3y_1) \cdot t + (y_0)$.

Then $\dfrac{d\, x(t)}{d\, t} = 3(-x_0 + 3x_1 - 3x_2 + x_3) \cdot t^2 + 2(3x_0 - 6x_1 + 3x_2) \cdot t + (-3x_0 + 3x_1)$ and

$\dfrac{d\, y(t)}{d\, t} = 3(-y_0 + 3y_1 - 3y_2 + y_3) \cdot t^2 + 2(3y_0 - 6y_1 + 3y_2) \cdot t + (-3y_0 + 3y_1)$.

When $t=0$, $\dfrac{d\, x(t)}{d\, t} = -3x_0 + 3x_1 = -3(x_1 - x_0)$ and $\dfrac{d\, x(t)}{d\, t} = -3(y_1 - y_0)$ so

$\dfrac{d\, B(t)}{d\, t} = \dfrac{d\, y(t)/dt}{d\, x(t)/dt} = \dfrac{-3(y_1 - y_0)}{-3(x_1 - x_0)} = \dfrac{y_1 - y_0}{x_1 - x_0}$ = slope of the line from P_0 to P_1.

The verification that $B'(1)$ equals the slope of the line segment from P_2 to P_3 is similar: evaluate $x'(1), y'(1)$ and $B'(1) = \dfrac{y'(1)}{x'(1)}$.

We will not verify Property (4) here, but it follows from the fact that each point on the Bezier curve is a "weighted average" of the four control points. For $0 \leq t \leq 1$, each of the coefficients $(1-t)^3$, $3(1-t)^2 \cdot t$, $3(1-t) \cdot t^2$ and t^3 is between (or equal to) 0 and 1, and they always add up to 1 (just expand the powers and add them to check this statement).

Problems

In problems 1 – 6, pairs of points, P_A and P_B, are given. In each problem (a) sketch the line segment L from P_A and P_B, and (b) plot the locations of the points $L(0.2)$, $L(0.5)$, and $L(0.9)$ on the line segment in part (a). Finally, (c) determine the equation of the line segment $L(t)$ from P_A and P_B and graph it.

1. P_A and P_B are given in Fig. 20.   2. P_A and P_B are given in Fig. 21.

3. P_A and P_B are given in Fig. 22.   4. P_A and P_B are given in Fig. 23.

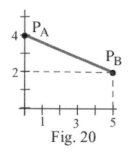
Fig. 20

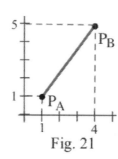
Fig. 21

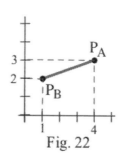
Fig. 22

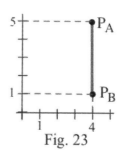
Fig. 23

5. $P_A = (1, 4)$ and $P_B = (5, 1)$.   6. $P_A = (8, 5)$ and $P_B = (4, 3)$.

7. Show that the parametric equations for the line segment given in Example 1,
    $x(t) = (1 - t) \cdot x_0 + t \cdot x_1$ and $y(t) = (1 - t) \cdot y_0 + t \cdot y_1$ for $0 \leq t \leq 1$,
    is equivalent to the parametric equations
    $x(t) = x_0 + t \cdot \Delta x$ and $y(t) = y_0 + t \cdot \Delta y$ for $0 \leq t \leq 1$ where $\Delta x = x_1 - x_0$ and $\Delta y = y_1 - y_0$.

In problems 8 – 13, find the parametric equations for a Bezier curve with the given control points or the given properties.

8. $P_0 = (1, 0)$, $P_1 = (2, 3)$, $P_2 = (5, 2)$, $P_3 = (6, 3)$ 9. $P_0 = (0, 5)$, $P_1 = (2, 3)$, $P_2 = (1, 4)$, $P_3 = (4, 2)$

10. $P_0 = (5, 1)$, $P_1 = (3, 3)$, $P_2 = (3, 5)$, $P_3 = (2, 1)$ 11. $P_0 = (6, 5)$, $P_1 = (6, 3)$, $P_2 = (2, 5)$, $P_3 = (2, 0)$

12. $P_0 = (0, 1)$, $P_3 = (4, 1)$ and $B'(0) = 1$, $B'(1) = -3$ 13. $P_0 = (5, 1)$, $P_3 = (1, 3)$ and $B'(0) = 2$, $B'(1) = 3$

In problems 14 – 17, sets of control points P_0, P_1, P_2, and P_3 are shown. Sketch a reasonable Bezier curve for the given control points.

14. P_0, P_1, P_2, and P_3 are given in Fig. 24.

15. P_0, P_1, P_2, and P_3 are given in Fig. 25.

16. P_0, P_1, P_2, and P_3 are given in Fig. 26.

17. P_0, P_1, P_2, and P_3 are given in Fig. 27.

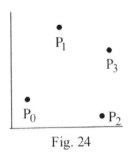
Fig. 24

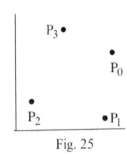
Fig. 25

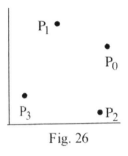
Fig. 26

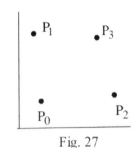
Fig. 27

In problems 18 – 21, sets of control points P_0, P_1, P_2, and P_3 are shown as well as a curve $C(t)$. For each problem explain why we can be certain that $C(t)$ is NOT the Bezier curve for the given control points (state which property or properties of Bezier curves $C(t)$ does not have).

18. P_0, P_1, P_2, and P_3 are given in Fig. 28.

19. P_0, P_1, P_2, and P_3 are given in Fig. 29.

20. P_0, P_1, P_2, and P_3 are given in Fig. 30.

21. P_0, P_1, P_2, and P_3 are given in Fig. 31.

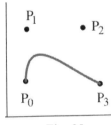
Fig. 28

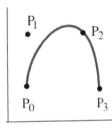
Fig. 29

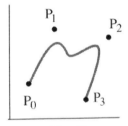
Fig. 30

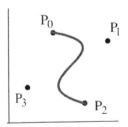
Fig. 31

In problems 22 – 23, find a pair of Bezier curves that satisfy the given conditions.

22. $B(0) = (0, 5)$, $B'(0) = 2$, $B(1) = C(0) = (3, 1)$, $B'(1) = C'(0) = 2$, $C(1) = (6, 2)$, and $C'(1) = 4$.

23. $B(0) = (0, 5)$, $B'(0) = 2$, $B(1) = C(0) = (3, 1)$, $B'(1) = C'(0) = 2$, $C(1) = (6, 2)$, and $C'(1) = 4$

Some Applications of Bezier Curves

The following Applications illustrate just a few of the wide variety of design applications of Bezier curves. This combination of differentiation and algebra is very powerful.

Applications

For each of the following applications, write the equation of a Bezier curve $B(t)$ that satisfies the requirements of the application, and then use your calculator/computer to graph $B(t)$.
(Typically in these applications, the starting and ending points, P_0 and P_3, are specified, but several choices of the control points P_1 and P_2 meet the requirements of the application. Select P_1 and P_2 so the resulting graph of $B(t)$ satisfies the requirements of the application and is also "visually pleasing.")

1. You have been hired to design an escalator for a shopping mall, and the design requirements are that the entrance and exit of the escalator must be horizontal (Fig. 32), the total rise is 20 feet, and the total run is 30 feet.

    (a) Find the equation of a Bezier curve $B(t)$ that meets these design requirements and graph it.
    (Suggestion: Place the origin at the lower left end of the escalator.)

    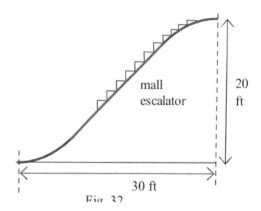
    Fig. 32

    (b) In practice, the middle section of an escalator is straight, and each end consists of a curved section (a Bezier curve) that smoothly converts our horizontal motion to motion along the straight section and then to horizontal motion again for our exit. Find the equation of a Bezier curve that models the curve at the entrance to the up escalator (Fig. 33), and the equation of the straight line section.

    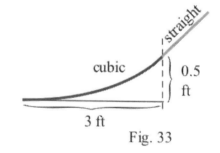
    Fig. 33

2. To design the exit ramp on a highway (Fig. 34), you need to find the equation of a Bezier curve so that at the beginning of the exit the elevation is 20 feet with a slope of –0.05 (about 3°), and 600 feet later, measured horizontally, the elevation is 0 feet and the ramp is horizontal.

    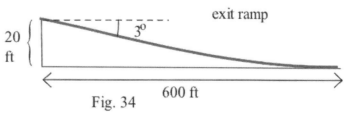
    Fig. 34

3. Find a Bezier curve that describes the final 60 feet of the ski jump shown in Fig. 35.

4. Find a Bezier curve that describes the left half of the arch shown in Fig. 36.

5. Find two Bezier curves B(t) and C(t) that describe the pieces of the curve for the top half of the hull of the Concordia Yawl that is 40 feet long, has a beam of 10 feet, and a transom width of 2 feet (Fig. 37).

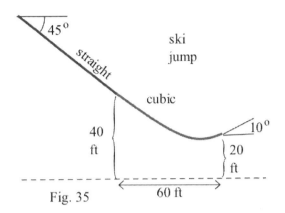

Fig. 35

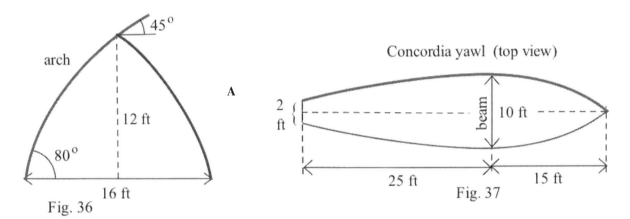

Fig. 36

Fig. 37

Final Note: The examples and applications given here have consisted of only one or two Bezier curves, and they are intended only as an introduction to the ideas and techniques of fitting Bezier curves to particular situations. But these ideas extend very nicely to curves that require pieces of several different Bezier curves for a good fit (Fig. 38) and even to curves and surfaces in three dimensions. Unfortunately, the systems of equations tend to grow very large for these extended applications.

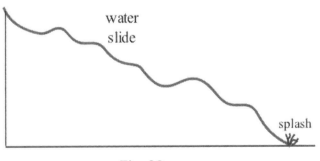

Fig. 38

Practice Answers

Practice 1: (a) $x(t) = (1-t) \cdot x_0 + t \cdot x_1 = (1-t) \cdot 1 + t \cdot 5 = 1 + 4t$,

$y(t) = (1-t) \cdot y_0 + t \cdot y_1 = (1-t) \cdot 2 + t \cdot 4 = 2 + 2t$.

(b) $x(t) = 5 - 4t$, $y(t) = 4 - 2t$

(c) $x(t) = 6 - 3t$, $y(t) = -2 + 3t$

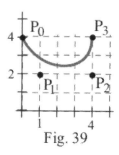

Fig. 39

Practice 2: $P_0 = (0, 4)$, $P_1 = (1, 2)$, $P_2 = (4, 2)$, and $P_3 = (4, 4)$. Then

$x(t) = (1-t)^3 \cdot 0 + 3(1-t)^2 \cdot t \cdot 1 + 3(1-t) \cdot t^2 \cdot 4 + t^3 \cdot 4 = -5t^3 + 6t^2 + 3t$.

$y(t) = (1-t)^3 \cdot 4 + 3(1-t)^2 \cdot t \cdot 2 + 3(1-t) \cdot t^2 \cdot 2 + t^3 \cdot 4 = 6t^2 - 6t + 4$.

The control points and the graph of $B(t) = (x(t), y(t))$ are shown in Fig. 39.

Practice 3: Take $P_0 = (0, 0)$ and $P_3 = (0, 1)$ to get the correct endpoints.

Take $P_1 = (-1, -1)$ and $P_2 = (-1, 2)$ to get a "C" shape. Then

$x(t) = (1-t)^3 \cdot 0 + 3(1-t)^2 \cdot t \cdot (-1) + 3(1-t) \cdot t^2 \cdot (-1) + t^3 \cdot 0$

$y(t) = (1-t)^3 \cdot 0 + 3(1-t)^2 \cdot t \cdot (-1) + 3(1-t) \cdot t^2 \cdot 2 + t^3 \cdot 1$.

The control points and the graph of $B(t) = (x(t), y(t))$ are shown in Fig. 40.

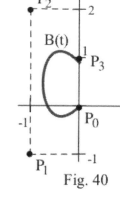

Fig. 40

Practice 4: Take $P_0 = (1, 3)$ and $P_3 = (5, 3)$ to get the correct endpoints.

Take $P_1 = (2, 1)$ and $P_2 = (4, 4)$ to get the correct slopes. Then

$x(t) = (1-t)^3 \cdot 1 + 3(1-t)^2 \cdot t \cdot 2 + 3(1-t) \cdot t^2 \cdot 4 + t^3 \cdot 5$

$y(t) = (1-t)^3 \cdot 3 + 3(1-t)^2 \cdot t \cdot 1 + 3(1-t) \cdot t^2 \cdot 4 + t^3 \cdot 3$.

The control points and the graph of

$B(t) = (x(t), y(t))$ are shown in Fig. 41.

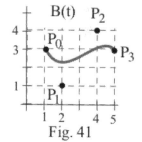

Fig. 41

Other choices for P_1 and P_2 can also yield correct slopes, and then we

have different formulas for $x(t)$ and $y(t)$

9.5 CONIC SECTIONS

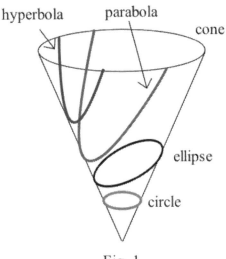

Fig. 1

The conic sections are the curves obtained when a cone is cut by a plane (Fig. 1). They have attracted the interest of mathematicians since the time of Plato, and they are still used by scientists and engineers. The early Greeks were interested in these shapes because of their beauty and their representations by sets of points that met certain distance definitions (e.g., the circle is the set of points at a fixed distance from a given point). Mathematicians and scientists since the 1600s have been interested in the conic sections because the planets, moons, and other celestial objects follow paths that are (approximately) conic sections, and the reflective properties of the conic sections are useful for designing telescopes and other instruments. Finally, the conic sections give the **complete** answer to the question, "what is the shape of the graph of the general quadratic equation $Ax^2 + Bxy + Cy^2 + Dx + Ey + F = 0$?"

This section discusses the "cut cone" and distance definitions of the conic sections and shows their standard equations in rectangular coordinate form. The section ends with a discussion of the discriminant, an easy way to determine the shape of the graph of any standard quadratic equation
$Ax^2 + Bxy + Cy^2 + Dx + Ey + F = 0$. Section 9.6 examines the polar coordinate definitions of the conic sections, some of the reflective properties of the conic sections, and some of their applications.

Cutting A Cone

When a (right circular double) cone is cut by a plane, only a few shapes are possible, and these are called the conic sections (Fig. 1). If the plane makes an angle of θ with the horizontal, and $\theta < \alpha$, then the set of points is an ellipse (Fig. 2). When $\theta = 0 < \alpha$, we have a circle, a special case of an ellipse (Fig. 3). If $\theta = \alpha$, a parabola is formed (Fig. 4), and if $\theta > \alpha$, a hyperbola is formed (Fig. 5). When the plane goes through the vertex of the cone, degenerate conics are formed: the degenerate ellipse ($\theta < \alpha$) is a point, the degenerate parabola ($\theta = \alpha$) is a line, and a degenerate hyperbola ($\theta > \alpha$) is a pair of intersecting lines.

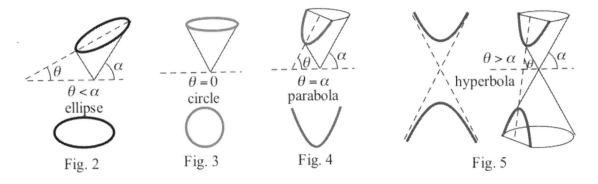

Fig. 2   Fig. 3   Fig. 4   Fig. 5

The conic sections are lovely to look at, but we will not use the conic sections as pieces of a cone because the "cut cone" definition of these shapes does not easily lead to formulas for them. To determine formulas for the conic sections it is easier to use alternate definitions of these shapes in terms of distances of points from fixed points and lines. Then we can use the formula for distance between two points and some algebra to derive formulas for the conic sections.

The Ellipse

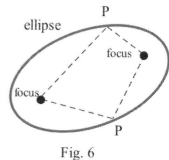
Fig. 6

Ellipse: An ellipse is the set of all points P for which the **sum** of the distances from P to two fixed points (called foci) **is a constant**: dist(P, one focus) + dist(P, other focus) = constant. (Fig. 6)

Example 1: Find the set of points whose distances from the foci $F_1 = (4,0)$ and $F_2 = (-4, 0)$ add up to 10.

Solution: If the point $P = (x, y)$ is on the ellipse, then the distances $PF_1 = \sqrt{(x-4)^2 + y^2}$ and $PF_2 = \sqrt{(x+4)^2 + y^2}$ must total 10 so we have the equation

$$PF_1 + PF_2 = \sqrt{(x-4)^2 + y^2} + \sqrt{(x+4)^2 + y^2} = 10 \text{ (Fig. 7)}$$

Moving the second radical to the right side of the equation, squaring both sides, and simplifying, we get

$$4x + 25 = 5\sqrt{(x+4)^2 + y^2} \ .$$

Squaring each side again and simplifying, we have $225 = 9x^2 + 25y^2$ so, after dividing each side by 225,

$$\frac{x^2}{25} + \frac{y^2}{9} = 1 \ .$$

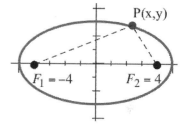
ellipse: $\overline{PF_1} + \overline{PF_2} = 10$
Fig. 7

Practice 1: Find the set of points whose distances from the foci $F_1 = (3,0)$ and $F_2 = (-3, 0)$ add up to 10.

Using the same algebraic steps as in Example 1, it can be shown (see the Appendix at the end of the problems) that the set of points $P = (x,y)$ whose distances from the foci $F_1 = (c,0)$ and $F_2 = (-c, 0)$ add up to 2a (a > c) is described by the formula

$$\frac{x^2}{a^2} + \frac{y^2}{b^2} = 1 \text{ where } b^2 = a^2 - c^2 \ .$$

Ellipse

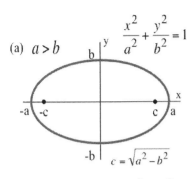

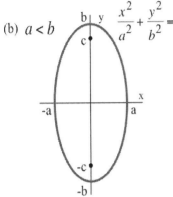

Fig. 8

The standard formula for an ellipse is $\dfrac{x^2}{a^2} + \dfrac{y^2}{b^2} = 1$.

a = b: The ellipse is a circle.

a > b: (Fig. 8a) The vertices are at $(\pm a, 0)$ on the x–axis,
the foci are at $(\pm c, 0)$ with $c = \sqrt{a^2 - b^2}$, and
for any point P on the ellipse,
dist(P, one focus) + dist(P, other focus) = 2a.
The length of the semimajor axis is a.

a < b: (Fig. 8b) The vertices are at $(0, \pm b)$ on the y–axis,
the foci are at $(0, \pm c)$ with $c = \sqrt{b^2 - a^2}$, and
for any point P on the ellipse,
dist(P, one focus) + dist(P, other focus) = 2b.
The length of the semimajor axis is b.

Practice 2: Use the information in the box to determine the vertices, foci, and length of the semimajor axis of the ellipse $\dfrac{x^2}{169} + \dfrac{y^2}{25} = 1$.

The Parabola

Parabola: A parabola is the set of all points P for which the distance from P to a fixed point (focus) **is equal to** the distance from P to a fixed line (directrix): dist(P, focus) = dist(P, directrix). (Fig. 9)

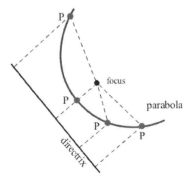

dist(P, focus) = dist(P, directix)
Fig. 9

Example 2: Find the set of points $P = (x,y)$ whose distance from the focus $F = (4,0)$ equals the distance from the directrix $x = -1$.

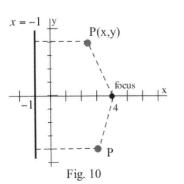

Fig. 10

Solution: The distance $PF = \sqrt{(x-4)^2 + y^2}$, and the distance from P to the directrix (Fig. 10) is $x+1$. If these two distances are equal then we have the equation $\sqrt{(x-4)^2 + y^2} = x + 1$.

Squaring each side,
$$(x-4)^2 + y^2 = (x+1)^2$$
so $x^2 - 8x + 16 + y^2 = x^2 + 2x + 1$.

This simplifies to $x = \frac{1}{10} y^2 + \frac{3}{2}$, the equation of a parabola opening to the right (Fig. 11).

Practice 3: Find the set of points $P = (x,y)$ whose distance from the focus $F = (0,2)$ equals the distance from the directrix $y = -2$.

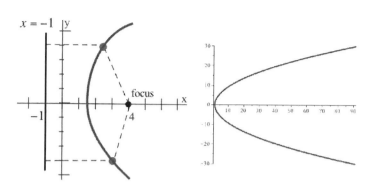

parabola: $x = \frac{3}{2} + \frac{1}{10} y^2$ (two scales)
Fig. 11

(a) $y = ax^2$ $(a > 0)$

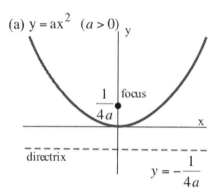

(b) $x = ay^2$ $(a > 0)$

Fig. 12

Parabola

The standard parabola $y = ax^2$ opens around the y–axis (Fig. 12a) with vertex = (0,0), focus = $(0, \frac{1}{4a})$, and directrix $y = -\frac{1}{4a}$.

The standard parabola $x = ay^2$ opens around the x–axis (Fig. 12b) with vertex = (0,0), focus = $(\frac{1}{4a}, 0)$, and directrix $x = -\frac{1}{4a}$.

Proof for the case $y = ax^2$:

The set of points $p = (x,y)$ that are equally distant from the focus $F = (0, \frac{1}{4a})$ and the directrix $y = -\frac{1}{4a}$ satisfy the distance equation

PF = PD so

$$\sqrt{x^2 + (y - \frac{1}{4a})^2} = (y + \frac{1}{4a}).$$ Squaring each side, we have

$x^2 + (y - \frac{1}{4a})^2 = (y + \frac{1}{4a})^2$ and $x^2 + y^2 - \frac{2}{4a} y + \frac{1}{16a^2} = y^2 + \frac{2}{4a} y + \frac{1}{16a^2}$.

Then $x^2 = \frac{2}{4a} y + \frac{2}{4a} y = \frac{1}{a} y$ and, finally, $y = ax^2$.

Practice 4: Prove that the set of points $P = (x,y)$ that are equally distant from the focus $F = (\frac{1}{4a}, 0)$, and directrix $x = -\frac{1}{4a}$ satisfy the equation $x = ay^2$.

Hyperbola

Hyperbola: A hyperbola is the set of all points P for which the **difference** of the distances from P to two fixed points (foci) **is a constant**:

dist(P, one focus) – dist(P, other focus) = constant. (Fig. 13)

Example 3: Find the set of points for which the **difference** of the distances from the points to the foci $F_1 = (5,0)$ and $F_2 = (-5, 0)$ is always 8.

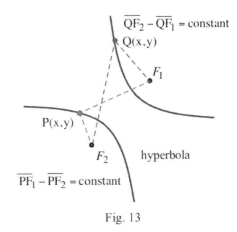

Fig. 13

Solution: If the point $P = (x, y)$ is on the hyperbola, then the difference of the distances $PF_1 = \sqrt{(x-5)^2 + y^2}$ and $PF_2 = \sqrt{(x+5)^2 + y^2}$ is 8 so we have the equation

$$PF_1 - PF_2 = \sqrt{(x-5)^2 + y^2} - \sqrt{(x+5)^2 + y^2} = 8 \quad \text{(Fig. 14)}.$$

Moving the second radical to the right side of the equation, squaring both sides, and simplifying, we get

$$5x + 16 = -4\sqrt{(x+5)^2 + y^2}.$$

Squaring each side again and simplifying, we have $9x^2 - 16y^2 = 144$.

After dividing each side by 144, $\frac{x^2}{16} - \frac{y^2}{9} = 1$.

If we start with the difference $PF_2 - PF_1 = 8$, we have the equation

$$\sqrt{(x+5)^2 + y^2} - \sqrt{(x-5)^2 + y^2} = 8.$$

Solving this equation, we again get $9x^2 - 16y^2 = 144$ and

$$\frac{x^2}{16} - \frac{y^2}{9} = 1.$$

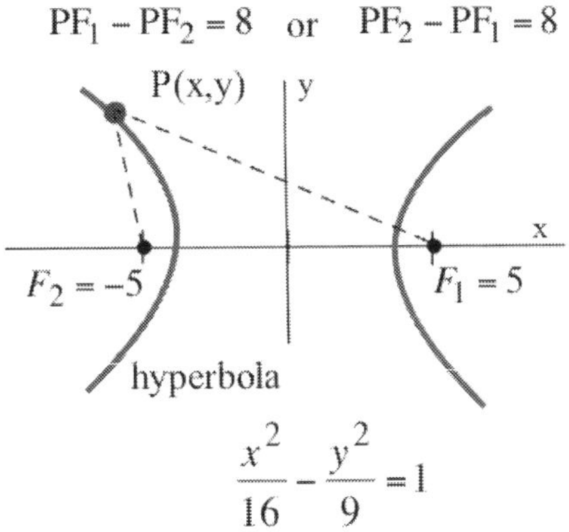

Fig. 14

Using the same algebraic steps as in the Example 3, it can be shown (see the Appendix at the end of the problems) that the set of points $P = (x,y)$ whose distances from the foci $F_1 = (c,0)$ and $F_2 = (-c, 0)$ differ by 2a ($a < c$) is described by the formula

$$\frac{x^2}{a^2} - \frac{y^2}{b^2} = 1 \quad \text{where } b^2 = c^2 - a^2.$$

(a) hyperbola $\frac{x^2}{a^2} - \frac{y^2}{b^2} = 1$

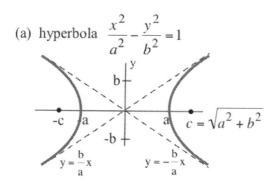

(b) hyperbola $\frac{y^2}{b^2} - \frac{x^2}{a^2} = 1$

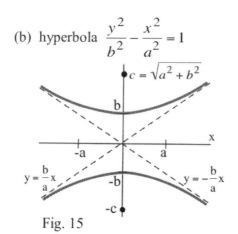

Fig. 15

Hyperbola

The standard hyperbola $\frac{x^2}{a^2} - \frac{y^2}{b^2} = 1$

opens around the x–axis (Fig. 15a)

with vertices at $(\pm a, 0)$, foci at $(\pm\sqrt{a^2 + b^2}, 0)$,

and linear asymptotes $y = \pm \frac{b}{a} x$.

The standard hyperbola $\frac{y^2}{b^2} - \frac{x^2}{a^2} = 1$

opens around the y–axis (Fig. 15b)

with vertices at $(0, \pm b)$, foci at $(0, \pm\sqrt{a^2 + b^2})$,

and linear asymptotes $y = \pm \frac{b}{a} x$.

Practice 5: Graph the hyperbolas $\frac{x^2}{25} - \frac{y^2}{16} = 1$ and $\frac{y^2}{25} - \frac{x^2}{16} = 1$ and find the linear asymptotes for each hyperbola.

Visually distinguishing the conic sections

If you only observe a small part of the graph of a conic section, it may be impossible to determine which conic section it is, and you may need to look at more of its graph. Near a vertex or in small pieces, all of the conic sections can be quite similar in appearance, but on a larger graph the ellipse is easy to distinguish from the other two. On a large graph, the hyperbola and parabola can be distinguished by noting that the hyperbola has two linear asymptotes and the parabola has no linear asymptotes.

The General Quadratic Equation and the Discriminant

Every equation that is quadratic in the variables x or y or both can be written in the form

$$Ax^2 + Bxy + Cy^2 + Dx + Ey + F = 0 \text{ where A through F are constants.}$$

The form $Ax^2 + Bxy + Cy^2 + Dx + Ey + F = 0$ is called the **general quadratic equation**.

In particular, each of the conic sections can be written in the form of a general quadratic equation by clearing all fractions and collecting all of the terms on one side of the equation. What is perhaps surprising is that the graph of a general quadratic equation is always a conic section or a degenerate form of a conic section. Usually the graph of a general quadratic equation is not centered at the origin and is not symmetric about either axis, but the shape is always an ellipse, parabola, hyperbola, or degenerate form of one of these.

Even more surprising, a quick and easy calculation using just the coefficients A, B, and C of the general quadratic equation tells us the shape of its graph: ellipse, parabola, or hyperbola. The value obtained by this simple calculation is called the discriminant of the general quadratic equation.

Discriminant

The **discriminant** of the the general quadratic form $Ax^2 + Bxy + Cy^2 + Dx + Ey + F = 0$

is the value $B^2 - 4AC$.

Example 4: Write each of the following in its general quadratic form and calculate its discriminant.

(a) $\dfrac{x^2}{25} + \dfrac{y^2}{9} = 1$ (b) $3y + 7 = 2x^2 + 5x + 1$ (c) $5x^2 + 3 = 7y^2 - 2xy + 4y + 8$

Solution: (a) $9x^2 + 25y^2 - 225 = 0$ so $A = 9, C = 25, F = -225,$ and $B = D = E = 0$. $B^2 - 4AC = -900$.

(b) $2x^2 + 5x - 3y - 6 = 0$ so $A = 2, D = 5, E = -3, F = -6,$ and $B = C = 0$. $B^2 - 4AC = 0$.

(c) $5x^2 + 2xy - 7y^2 - 4y - 5 = 0$ so $A = 5, B = 2, C = -7, D = 0, E = -4$ and $F = -5$.
$B^2 - 4AC = 4 - 4(5)(-7) = 144$.

Practice 6: Write each of the following in its general quadratic form and calculate its discriminant.

(a) $1 = \dfrac{x^2}{36} - \dfrac{y^2}{9}$ (b) $x = 3y^2 - 5$ (c) $\dfrac{x^2}{16} + \dfrac{(y-2)^2}{25} = 1$

One very important property of the discriminant is that it is invariant under translations and rotations, its value does not change even if the graph is rigidly translated around the plane and rotated. When a graph is shifted or rotated or both, its general quadratic equation changes, but the discriminant of the new quadratic equation is the same value as the discriminant of the original quadratic equation. And we can determine the shape of the graph simply from the sign of the discriminant.

Quadratic Shape Theorem

The graph of the general quadratic equation $Ax^2 + Bxy + Cy^2 + Dx + Ey + F = 0$ is

an ellipse if $\quad B^2 - 4AC < 0 \quad$ (degenerate forms: one point or no points)
a parabola if $\quad B^2 - 4AC = 0 \quad$ (degenerate forms: two lines, one line, or no points)
a hyperbola if $\quad B^2 - 4AC > 0 \quad$ (degenerate form: pair of intersecting lines).

The proofs of this result and of the invariance of the discriminant under translations and rotations are "elementary" and just require a knowledge of algebra and trigonometry, but they are rather long and are very computational. A proof of the invariance of the discriminant under translations and rotations and of the Quadratic Shape Theorem is given in the Appendix after the problem set.

Example 5: Use the discriminant to determine the shapes of the graphs of the following equations.
(a) $x^2 + 3xy + 3y^2 = -7y - 4$ (b) $4x^2 + 4xy + y^2 = 3x - 1$ (c) $y^2 - 4x^2 = 0$.

Solution: (a) $B^2 - 4AC = 3^2 - 4(1)(3) = -3 < 0$. The graph is an ellipse.
(b) $B^2 - 4AC = 4^2 - 4(4)(1) = 0$. The graph is a parabola.
(c) $B^2 - 4AC = 0^2 - 4(-4)(1) = 16 > 0$. The graph is a hyperbola — actually a degenerate hyperbola. The graph of $0 = y^2 - 4x^2 = (y + 2x)(y - 2x)$ consists of the two lines $y = -2x$ and $y = 2x$.

Practice 7: Use the discriminant to determine the shapes of the graphs of the following equations.
(a) $x^2 + 2xy = 2y^2 + 4x + 3$ (b) $y^2 + 2x^2 = xy - 3y + 7$ (c) $2x^2 - 4xy = 3 + 5y - 2y^2$.

Sketching Standard Ellipses and Hyperbolas

The graphs of general ellipses and hyperbolas require plotting lots of points (a computer or calculator can help), but it is easy to sketch good graphs of the standard ellipses and hyperbolas. The steps for doing so are given below.

Graphing the Standard Ellipse $\dfrac{x^2}{a^2} + \dfrac{y^2}{b^2} = 1$

1. Sketch short vertical line segments at the points $(\pm a, 0)$ on the x–axis and short horizontal line segments at the points $(0, \pm b)$ on the y–axis (Fig. 16a). Draw a rectangle whose sides are formed by extending the line segments.

2. Use the tangent line segments in step 1 as guide to sketching the ellipse (Fig. 16b). The graph of the ellipse is always inside the rectangle except at the 4 points that touch it.

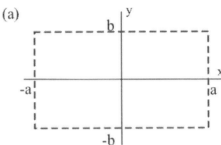

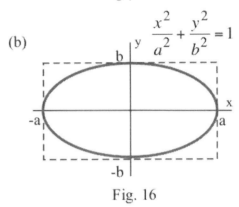

Fig. 16

Graphing the Standard Hyperbolas $\dfrac{x^2}{a^2} - \dfrac{y^2}{b^2} = 1$ and $\dfrac{y^2}{b^2} - \dfrac{x^2}{a^2} = 1$

1. Sketch the rectangle that intersects the x–axis at the points $(\pm a, 0)$ and the y–axis at the points $(0, \pm b)$. (Fig. 17a)

2. Draw the lines which go through the origin and the corners of the rectangle from step 1. (Fig. 17b) These lines are the asymptotes of the hyperbola.

3. For $\dfrac{x^2}{a^2} - \dfrac{y^2}{b^2} = 1$, plot the points $(\pm a, 0)$ on the hyperbola, and use the asymptotes from step 2 as a guide to sketching the rest of the hyperbola. (Fig. 17c)

3'. For $\dfrac{y^2}{b^2} - \dfrac{x^2}{a^2} = 1$, plot the points $(0, \pm b)$ on the hyperbola, and use the asymptotes from step 2 as a guide to sketching the rest of the hyperbola. (Fig. 17d)

The graph of the hyperbola is always outside the rectangle except at the 2 points which touch it.

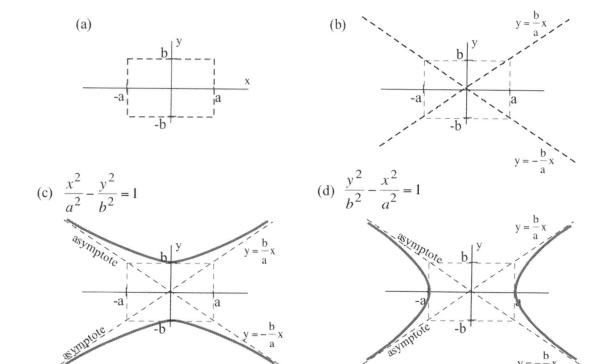

Fig. 17

Symmetry of the Conic Sections

Symmetry properties of the conic sections can simplify the task of graphing them. A parabola has one line of symmetry, so once we have graphed half of a parabola we can get the other half by folding along the line of symmetry. An ellipse and a hyperbola each have two lines of symmetry, so once we have graphed one fourth of an ellipse or hyperbola we can get the rest of the graph by folding along each line of symmetry.

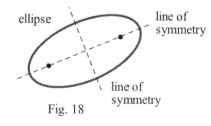

Fig. 18

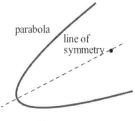

Fig. 19

- The parabola is symmetric about the line through the focus and the vertex (Fig. 18).
- The ellipse is symmetric about the line through the two foci. It is also symmetric about the perpendicular bisector of the line segment through the two foci (Fig. 19).
- The hyperbola is symmetric about the line through the two foci and about the perpendicular bisector of the line segment through the two foci (Fig. 20).

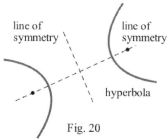

Fig. 20

The Conic Sections as "Shadows of Spheres"

There are a lot of different shapes at the beach on a sunny day, even conic sections. Suppose we have a sphere resting on a flat surface and a point radiating light.

- If the point of light is higher than the top of the sphere, then the shadow of the sphere is an ellipse (Fig. 21).
- If the point of light is exactly the same height as the top of the sphere, then the shadow of the sphere is a parabola (Fig. 22).
- If the point of light is lower than the top of the sphere, then the shadow of the sphere is one branch of a hyperbola (Fig. 23).

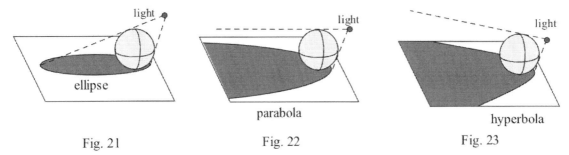

Fig. 21 Fig. 22 Fig. 23

PROBLEMS

1. What is the shape of the graph of the set of points whose distances from (6,0) and (–6,0) always add up to 20? Find an equation for the graph.

2. What is the shape of the graph of the set of points whose distances from (2,0) and (–2,0) always add up to 20? Find an equation for the graph.

3. What is the shape of the graph of the set of points whose distance from the point (0,5) is equal to the distance from the point to the line $y = -5$? Find an equation for the graph.

4. What is the shape of the graph of the set of points whose distance from the point (2,0) is equal to the distance from the point to the line $x = -4$? Find an equation for the graph.

5. Give the standard equation for the ellipse in Fig. 24.

6. Give the standard equation for the ellipse in Fig. 25.

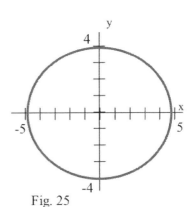

Fig. 24

Fig. 25

7. What lines are linear asymptotes for the hyperbola $4x^2 - 9y^2 = 36$, and where are the foci?

8. What lines are linear asymptotes for the hyperbola $25x^2 - 4y^2 = 100$, and where are the foci?

9. What lines are linear asymptotes for the hyperbola $5y^2 - 3x^2 = 15$, and where are the foci?

10. What lines are linear asymptotes for the hyperbola $5y^2 - 3x^2 = 120$, and where are the foci?

In problems 11–16, rewrite each equation in the form of the general quadratic equation $Ax^2 + Bxy + Cy^2 + Dx + Ey + F = 0$ and then calculate the value of the discriminant. What is the shape of each graph?

11. (a) $\dfrac{x^2}{4} + \dfrac{y^2}{25} = 1$ (b) $\dfrac{x^2}{a^2} + \dfrac{y^2}{b^2} = 1$ 12. (a) $\dfrac{x^2}{4} - \dfrac{y^2}{25} = 1$ (b) $\dfrac{x^2}{a^2} - \dfrac{y^2}{b^2} = 1$

13. $x + 2y = 1 + \dfrac{3}{x - y}$

14. $y = \dfrac{5 + 2y - x^2}{4x + 5y}$

15. $x = \dfrac{7x - 3 - 2y^2}{2x + 4y}$

16. $x = \dfrac{2y^2 + 7x - 3}{2x + 5y}$

Problems 17–20 illustrate that a small change in the value of just **one** coefficient in the quadratic equation $Ax^2 + Bxy + Cy^2 + Dx + Ey + F = 0$ can have a dramatic effect on the shape of the graph. Determine the shape of the graph for each formula.

17. (a) $2x^2 + \mathbf{3}xy + 2y^2 + $ (*terms for x, y, and a constant*) $= 0$.
    (b) $2x^2 + \mathbf{4}xy + 2y^2 + $ (*terms for x, y, and a constant*) $= 0$.
    (c) $2x^2 + \mathbf{5}xy + 2y^2 + $ (*terms for x, y, and a constant*) $= 0$.
    (d) What are the shapes if the coefficients of the xy term are $3.99, 4,$ and 4.01?

18. (a) $\mathbf{1}x^2 + 4xy + 2y^2 + $ (*terms for x, y, and a constant*) $= 0$.
    (b) $\mathbf{2}x^2 + 4xy + 2y^2 + $ (*terms for x, y, and a constant*) $= 0$.
    (c) $\mathbf{3}x^2 + 4xy + 2y^2 + $ (*terms for x, y, and a constant*) $= 0$.
    (d) What are the shapes if the coefficients of the x^2 term are $1.99, 2,$ and 2.01?

19. (a) $x^2 + 4xy + \mathbf{3}y^2 + $ (*terms for x, y, and a constant*) $= 0$.
    (b) $x^2 + 4xy + \mathbf{4}y^2 + $ (*terms for x, y, and a constant*) $= 0$.
    (c) $x^2 + 4xy + \mathbf{5}y^2 + $ (*terms for x, y, and a constant*) $= 0$.
    (d) What are the shapes if the coefficients of the y^2 term are $3.99, 4,$ and 4.01?

20. Just changing a single sign can also dramatically change the shape of the graph.

    (a) $x^2 + 2xy + y^2 + $ (*terms for x, y, and a constant*) $= 0$.

    (b) $x^2 + 2xy - y^2 + $ (*terms for x, y, and a constant*) $= 0$.

21. Find the volume obtained when the region enclosed by the ellipse $\dfrac{x^2}{2^2} + \dfrac{y^2}{5^2} = 1$ is rotated

    (a) about the x–axis, and (b) about the y–axis.

22. Find the volume obtained when the region enclosed by the ellipse $\dfrac{x^2}{a^2} + \dfrac{y^2}{b^2} = 1$ is rotated

    (a) about the x–axis, and (b) about the y–axis.

23. Find the volume obtained when the region enclosed by the hyperbola $\dfrac{x^2}{2^2} - \dfrac{y^2}{5^2} = 1$ and the vertical

    line $x = 10$ is rotated (a) about the x–axis, and (b) about the y–axis.

24. Find the volume obtained when the region enclosed by the hyperbola $\dfrac{x^2}{a^2} - \dfrac{y^2}{b^2} = 1$ and the vertical line $x = L$ (Fig. 26) is rotated (a) about the x–axis, and (b) about the y–axis. (Assume $a < L$.)

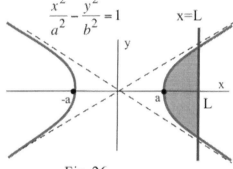

Fig. 26

25. Find the ratio of the area of the shaded parabolic region in Fig. 27 to the area of the rectangular region.

26. Find the ratio of the volumes obtained when the parabolic and rectangular regions in Fig. 27 are rotated about the y–axis.

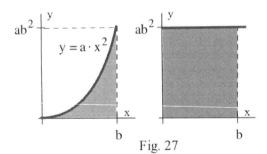

Fig. 27

String Constructions of Ellipses, Parabolas, and Hyperbolas   (Optional)

All of the conic sections can be drawn with the help of some pins and string, and the directions and figures show how it can be done. For each conic section, you are asked to determine and describe why each construction produces the desired shape.

Ellipse: Pin the two ends of the string to a board so the string is not taut. Put the point of a pencil in the bend in the string (Fig. 28), and, keeping the string taut, draw a curve.

27. How is the distance between the vertices of the ellipse related to the length of the string?

28. Explain why this method produces an ellipse, a set of points whose distances from the two fixed points (foci) always sum to a constant. What is the constant?

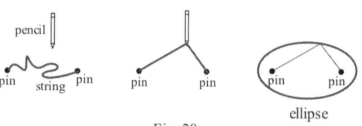

Fig. 28

29. What happens to the shape of the ellipse as the two foci are moved closer together (and the piece of string stays the same length)? Draw several ellipses using the same piece of string and different fixed points, and describe the results.

Parabola: Pin one end of the string to a board and the other end to the corner of a T–square bar that is the same length as the string. Put the point of a pencil in the bend in the string (Fig. 29) and keep the string taut. As the T–square is slid sideways, the pencil draws a curve.

30. Explain why this method produces a parabola, a set of points whose distance from a fixed point (one end of the string) is equal to the distance from a fixed line (the edge of the table).

31. What happens if the length of the string is slightly shorter than the length of the T–square bar? Draw several curves with several slightly shorter pieces of string and describe the results. What shapes are the curves?

32. Find a way to use pins, string and a pencil to sketch the graph of a hyperbola.

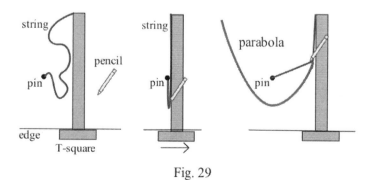

Fig. 29

Section 9.5 PRACTICE Answers

Practice 1: $F_1 = (3,0)$, $F_2 = (-3,0)$, and $P = (x,y)$. We want $\text{dist}(F_1, P) + \text{dist}(F_2, P) = 10$ so

$$\text{dist}((x,y),(3,0)) + \text{dist}((x,y),(-3,0)) = 10 \text{ and}$$
$$\sqrt{(x-3)^2 + y^2} + \sqrt{(x+3)^2 + y^2} = 10.$$

Moving the second radical to the right side and squaring, we get

$$(x-3)^2 + y^2 = 100 - 20\sqrt{(x+3)^2 + y^2} + (x+3)^2 + y^2 \text{ and}$$
$$x^2 - 6x + 9 + y^2 = 100 - 20\sqrt{(x+3)^2 + y^2} + x^2 + 6x + 9 + y^2 \text{ so}$$
$$-12x - 100 = -20\sqrt{(x+3)^2 + y^2}.$$

Dividing each side by -2 and then squaring, we have

$$36x^2 + 600x + 2500 = 100(x^2 + 6x + 9 + y^2) \text{ so}$$
$$1600 = 64x^2 + 100y^2 \text{ and}$$
$$1 = \frac{64x^2}{1600} + \frac{100y^2}{1600} = \frac{x^2}{25} + \frac{y^2}{16}.$$

Practice 2: $a = 13$ and $b = 5$ so the vertices of the ellipse are $(13, 0)$ and $(-13, 0)$. The value of c is $\sqrt{169 - 25} = 12$ so the foci are $(12, 0)$ and $(-12, 0)$. The length of the semimajor axis is 13.

Practice 3: $\text{dist}(P, \text{focus}) = \text{dist}(P, \text{directrix})$ so $\text{dist}((x,y),(0,2)) = \text{dist}((x,y), \text{line } y=-2)$:

$$\sqrt{(x-0)^2 + (y-2)^2} = y + 2.$$

Squaring, we get $x^2 + y^2 - 4y + 4 = y^2 + 4y + 4$ so $x^2 = 8y$ or $y = \frac{1}{8}x^2$.

Practice 4: This is similar to Practice 3: $\text{dist}(P, \text{focus}) = \text{dist}(P, \text{directrix})$ so $\text{dist}((x,y),(\frac{1}{4a}, 0)) = \text{dist}((x,y), \text{line } x = -\frac{1}{4a})$. Then

$$\sqrt{(x - \frac{1}{4a})^2 + (y - 0)^2} = x + \frac{1}{4a}. \text{ Squaring each side we get}$$

$$x^2 - 2x\frac{1}{4a} + \frac{1}{16a^2} + y^2 = x^2 + 2x\frac{1}{4a} + \frac{1}{16a^2} \text{ so } y^2 = \frac{1}{a}x \text{ and } x = ay^2.$$

Practice 5: The graphs are shown in Fig. 30. Both hyperbolas have the same linear asymptotes: $y = \frac{4}{5}x$ and $y = -\frac{4}{5}x$.

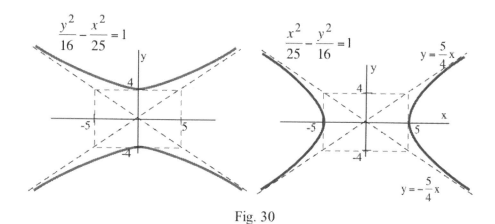

Fig. 30

Practice 6: (a) $324 = 9x^2 - 36y^2$ so $9x^2 - 36y^2 - 324 = 0$.
$A = 9$, $B = 0$, and $C = -36$ so $D = 0 - 4(9)(-36) = \mathbf{576}$.

(b) $0x^2 + 0xy + 3y^2 - x - 5 = 0$.
$A = 0$, $B = 0$, and $C = 3$ so $D = 0 - 4(0)(3) = \mathbf{0}$.

(c) $25x^2 + 16(y-2)^2 = 400$ so $25x^2 + 16y^2 - 64y + 48 - 400 = 0$.
$A = 25$, $B = 0$, and $C = 16$ so $D = 0 - 4(25)(16) = \mathbf{-1600}$.

Practice 7: (a) $x^2 + 2xy - 2y^2 - 4x - 3 = 0$.
$A = 1$, $B = 2$, $C = -2$ so $D = 4 - 4(1)(-2) = 12 > 0$: hyperbola.

(b) $2x^2 - 1xy + 1y^2 + 3y - 7 = 0$.
$A = 2$, $B = -1$, $C = 1$ so $D = 1 - 4(2)(1) = -7 < 0$: ellipse.

(c) $2x^2 - 4xy + 2y^2 - 5y - 3 = 0$.
$A = 2$, $B = -4$, and $C = 2$ so $D = 16 - 4(2)(2) = 0$: parabola.

Appendix for 9.5: Conic Sections

Deriving the Standard Forms from Distance Definitions of the Conic Sections

Ellipse

>   Ellipse   An ellipse is the set of all points P so the **sum** of the distances of P from two fixed points (called foci) **is a constant**.

If F_1 and F_2 are the foci (Fig. 40), then for every point P on the ellipse, the distance from P to F_1 PLUS the distance from P to F_2 is a constant: $PF_1 + PF_2 = $ constant. If the center of the ellipse is at the origin and the foci lie on the x–axis at $F_1 = (c, 0)$ and $F_2 = (-c, 0)$, we can translate the words into a formula:

$$PF_1 + PF_2 = \text{constant becomes} \quad \sqrt{(x-c)^2 + y^2} + \sqrt{(x+c)^2 + y^2} = 2a.$$

(Calling the constant 2a simply makes some of the later algebra easier.)

By moving the second radical to the right side of the equation, squaring each side, and simplifying, we get

$$\sqrt{(x-c)^2 + y^2} = 2a - \sqrt{(x+c)^2 + y^2}$$

$$(x-c)^2 + y^2 = 4a^2 - 4a\sqrt{(x+c)^2 + y^2} + (x+c)^2 + y^2$$

$$\text{so } x^2 - 2xc + c^2 + y^2 = 4a^2 - 4a\sqrt{(x+c)^2 + y^2} + x^2 + 2xc + c^2 + y^2$$

$$\text{and } xc + a^2 = a\sqrt{(x+c)^2 + y^2}.$$

Squaring each side again and simplifying, we get

$$(xc + a^2)^2 = a^2 \{ (x+c)^2 + y^2 \} \quad \text{so } x^2c^2 + 2xca^2 + a^4 = a^2x^2 + 2xca^2 + a^2c^2 + a^2y^2$$

$$\text{and } a^2(a^2 - c^2) = x^2(a^2 - c^2) + y^2 a^2.$$

Finally, dividing each side by $a^2(a^2 - c^2)$, we get $\dfrac{x^2}{a^2} + \dfrac{y^2}{a^2 - c^2} = 1$.

By setting $b^2 = a^2 - c^2$, we have $\dfrac{x^2}{a^2} + \dfrac{y^2}{b^2} = 1$, the standard form of the ellipse.

Hyperbola

> Hyperbola: A hyperbola is the set of all points P so the **difference** of the distances of P from the two fixed points (foci) **is a constant**.

If F_1 and F_2 are the foci (Fig. 9), then for every point P on the hyperbola, the distance from P to F_1 MINUS the distance from P to F_2 is a constant: $PF_1 - PF_2 =$ constant (Fig. 42). If the center of the hyperbola is at the origin and the foci lie on the x–axis at $F_1 = (c, 0)$ and $F_2 = (-c, 0)$, we can translate the words into a formula:

$$PF_1 - PF_2 = \text{constant} \quad \text{becomes} \quad \sqrt{(x-c)^2 + y^2} - \sqrt{(x+c)^2 + y^2} = 2a.$$

(Calling the constant 2a simply makes some of the later algebra easier.)

The algebra which follows is very similar to that used for the ellipse.

Moving the second radical to the right side of the equation, squaring each side, and simplifying, we get

$$\sqrt{(x-c)^2 + y^2} = 2a + \sqrt{(x+c)^2 + y^2}$$

$$(x-c)^2 + y^2 = 4a^2 + 4a\sqrt{(x+c)^2 + y^2} + (x+c)^2 + y^2$$

$$\text{so } x^2 - 2xc + c^2 + y^2 = 4a^2 + 4a\sqrt{(x+c)^2 + y^2} + x^2 + 2xc + c^2 + y^2$$

$$\text{and } xc + a^2 = -a\sqrt{(x+c)^2 + y^2}.$$

Squaring each side again and simplifying, we get

$$(xc + a^2)^2 = a^2 \{(x+c)^2 + y^2\} \quad \text{so} \quad x^2c^2 + 2xca^2 + a^2 = a^2x + 2xca^2 + a^2c^2 + a^2y^2$$

$$\text{and} \quad x^2(c^2 - a^2) - y^2a^2 = a^2(c^2 - a^2).$$

Finally, dividing each side by $a^2(c^2 - a^2)$, we get $\dfrac{x^2}{a^2} + \dfrac{y^2}{c^2 - a^2} = 1$.

By setting $b^2 = c^2 - a^2$, we have $\dfrac{x^2}{a^2} - \dfrac{y^2}{b^2} = 1$, the standard form of the hyperbola.

Invariance Properties of the Discriminant

The **discriminant** of $Ax^2 + Bxy + Cy^2 + Dx + Ey + F = 0$ is $d = \mathbf{B^2 - 4AC}$.

The Discriminant $B^2 - 4AC$ is invariant under translations (shifts):

If a point (x, y) is shifted h units up and k units to the right, then the coordinates of the new point are $(x', y') = (x+h, y+k)$. To show that the discriminant of $Ax^2 + Bxy + Cy^2 + Dx + Ey + F = 0$ is invariant under translations, we need to show that the discriminant of $Ax^2 + Bxy + Cy^2 + Dx + Ey + F = 0$ and the discriminant of $A(x')^2 + B(x')(y') + C(y')^2 + Dx' + Ey' + F = 0$ are equal for $x' = x + h$ and $y' = y + k$.

The discriminant of $Ax^2 + Bxy + Cy^2 + Dx + Ey + F = 0$ is equal to $B^2 - 4AC$.
Replacing x' with $x+h$ and y' with $y+k$,

$$A(x')^2 + B(x')(y') + C(y')^2 + Dx' + Ey' + F$$
$$= A(x+h)^2 + B(x+h)(y+k) + C(y+k)^2 + D(x+h) + E(y+k) + F$$
$$= A(x^2 + 2xh + h^2) + B(xy + xk + yh + hk) + C(y^2 + 2yk + k^2) + D(x+h) + E(y+k) + F$$
$$= Ax^2 + Bxy + Cy^2 + (2Ah + Bk + D)x + (Bh + 2Ck + E)y + (Ah^2 + Bhk + Ck^2 + Dh + Ek + F).$$

The discriminant of this final formula is $B^2 - 4AC$, the same as the discriminant of $Ax^2 + Bxy + Cy^2 + Dx + Ey + F$. In fact, a translation does not change the values of the coefficients of x^2, xy, and y^2 (the values of A, B, and C) so the discriminant is unchanged.

The Discriminant $B^2 - 4AC$ is invariant under rotation by an angle θ:

If a point (x, y) is rotated about the origin by an angle of θ, then the coordinates of the new point are $(x', y') = (x \cdot \cos(\theta) - y \cdot \sin(\theta), x \cdot \sin(\theta) + y \cdot \cos(\theta))$. To show that the discriminant of $Ax^2 + Bxy + Cy^2 + Dx + Ey + F = 0$ is invariant under rotations, we need to show that the discriminant of $Ax^2 + Bxy + Cy^2 + Dx + Ey + F = 0$ and the discriminant of $A(x')^2 + B(x')(y') + C(y')^2 + Dx' + Ey' + F = 0$ are equal when $x' = x \cdot \cos(\theta) - y \cdot \sin(\theta)$ and $y' = x \cdot \sin(\theta) + y \cdot \cos(\theta)$.

The discriminant of $Ax^2 + Bxy + Cy^2 + Dx + Ey + F = 0$ is equal to $B^2 - 4AC$.
Replacing x' with $x \cdot \cos(\theta) - y \cdot \sin(\theta) = x \cdot c - y \cdot s$ and y' with $x \cdot \sin(\theta) + y \cdot \cos(\theta) = x \cdot s + y \cdot c$

$$A(x')^2 + B(x')(y') + C(y')^2 + Dx' + Ey' + F = 0$$
$$= A(xc - ys)^2 + B(xc - ys)(xs + yc) + C(xs + yc)^2 + \ldots \text{(terms without } x^2, xy, \text{ and } y^2)$$
$$= A(x^2c^2 - 2xysc + y^2s^2) + B(x^2sc - xys^2 + xyc^2 - y^2sc) + C(x^2s^2 + 2xysc + y^2c^2) + \ldots$$
$$= (Ac^2 + Bsc + Cs^2)x^2 + (-2Asc - Bs^2 + Bc^2 + 2Csc)xy + (As^2 + Bsc + Cc^2)y^2 + \ldots$$

Then $A' = Ac^2 + Bsc + Cs^2$, $B' = -2Asc - Bs^2 + Bc^2 + 2Csc$, and $C' = As^2 + Bsc + Cc^2$, so the new discriminant is

$$(B')^2 - 4(A')(C')$$
$$= (-2Asc - Bs^2 + Bc^2 + 2Csc)^2 - 4(Ac^2 + Bsc + Cs^2)(As^2 + Bsc + Cc^2)$$
$$= \{ s^4(B^2) + s^3c(4AB - 4BC) + s^2c^2(4A^2 - 8AC - 2B^2 + 4C^2) + sc^3(-4AB + 4BC) + c^4(B^2) \}$$
$$\quad - 4\{ s^4(AC) + s^3c(AB - BC) + s^2c^2(A^2 - B^2 + C^2) + sc^3(-AB + BC) + c^4(AC) \}$$
$$= s^4(B^2 - 4AC) + s^2c^2(2B^2 - 8AC) + c^4(B^2 - 4AC)$$
$$= (B^2 - 4AC)(s^4 + 2s^2c^2 + c^4) = (B^2 - 4AC)(s^2 + c^2)(s^2 + c^2) = (B^2 - 4AC) \text{, the original discriminant.}$$

The invariance of the discriminant under translation and rotation shows that any conic section can be translated so its "center" is at the origin and rotated so its axis is the x–axis without changing the value of the discriminant: the value of the discriminant depends strictly on the shape of the curve, not on its location or orientation. When the axis of the conic section is the x–axis, the standard quadratic equation
$Ax^2 + Bxy + Cy^2 + Dx + Ey + F = 0$
does not have an xy term (B=0) so we only need to investigate the reduced form
$Ax^2 + Cy^2 + Dx + Ey + F = 0$.

1) $A = C = 0$ (discriminant d=0). A straight line. (special case: no graph)

2) $A = C \neq 0$ (d<0). A circle. (special cases: a point or no graph)

3) $A = 0, C \neq 0$ or $A \neq 0, C = 0$ (d=0): A parabola. (special cases: 2 lines, 1 line, or no graph)

4) A and C both positive or both negative (d<0): An Ellipse. (special cases: a point or no graph)

5) A and C have opposite signs (d>0): A Hyperbola. (special case: a pair of intersecting lines)

9.6 PROPERTIES OF THE CONIC SECTIONS

This section presents some of the interesting and important properties of the conic sections that can be proven using calculus. It begins with their reflection properties and considers a few ways these properties are used today. Then we derive the polar coordinate form of the conic sections and use that form to examine one of the reasons conic sections are still extensively used: the paths of planets, satellites, comets, baseballs, and even subatomic particles are often conic sections. The section ends with a specialized examination of elliptical orbits. To understand and describe the motions of the universe, at telescopic and microscopic levels, we need conic sections!

Reflections on the Conic Sections

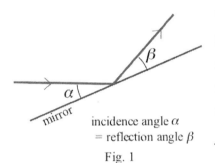

incidence angle α
= reflection angle β
Fig. 1

This discussion of reflection assumes that the angle of incidence of a light ray or billiard ball is equal to the angle of reflection of the ray or ball. The assumption is valid for light rays and mirrors (Fig. 1) but is not completely valid for balls: the spin of the ball before it hits the wall may make the reflection angle smaller than, greater than, or equal to the incidence angle.

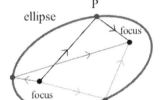

Fig. 2

Reflection Property of an Ellipse

An elliptical mirror reflects light from one focus to the other focus (Fig. 2) and all of the light rays take the same amount of time to be reflected to the other focus.

Outline of a proof: We can assume that the ellipse is oriented so its equation is

$$\frac{x^2}{a^2} + \frac{y^2}{b^2} = 1 \quad \text{(Fig. 3) and } a > b > 0.$$

Then the foci are at the points $F_1 = (-c, 0)$ and $F_2 = (c, 0)$ with $c = \sqrt{a^2 - b^2}$.

To show that the light rays from one focus are always reflected to the other focus, we need to show that angle α, the angle between the ray from F_1 and the tangent line to the ellipse, is equal to angle β, the angle between the tangent to the ellipse and the ray to F_2.

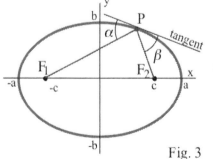

m_1 = slope of F_1P
m_2 = slope of F_2P
m_3 = slope of tangent line at P

Fig. 3

The most direct way to show that $\alpha = \beta$ is to start by calculating the slopes m_1 = slope of the line from F_1 to P, m_2 = the slope of the line from P to F_2, and m_3 = the slope of the tangent line:

$$m_1 = \frac{y}{x+c}, \quad m_2 = \frac{y}{x-c}, \text{ and, by implicit differentiation, } m_3 = \frac{-x}{y}\frac{b^2}{a^2}.$$

Then, from section 0.2, we know that $\tan(\alpha) = \dfrac{m_3 - m_1}{1 + m_1 m_3}$ and $\tan(\beta) = \dfrac{m_2 - m_3}{1 + m_2 m_3}$ so we just need to evaluate $\tan(\alpha)$ and $\tan(\beta)$ and show that they are equal. Since the process is algebraically tedious and not very enlightening, it has been relegated to an Appendix after the problem set.

Since straight paths from one focus to the ellipse and back to the other focus all have the same length (the definition of an ellipse), all of the light rays from the one focus take the same amount of time to reach the other focus. If a small stone is dropped into an elliptical pool at one focus (Fig. 4), then the waves radiate in all directions, reflect off the sides of the pool to the other focus and create a splash there because they all arrive at the same time. Similarly, if a room is in the shape of an (half) ellipsoid of revolution (Fig. 5), then the sound waves from a whisper at one focus will bounce off the walls and all arrive at the same time at the other focus where an eavesdropper can hear the conversation.

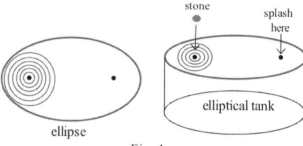
Fig. 4

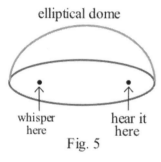
Fig. 5

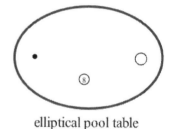
elliptical pool table
Fig. 6

Practice 1: What simple directions ensure that a ball shot from anywhere on an elliptical pool table (Fig. 6) will bounce off one wall and go into the single hole located at a focus of the ellipse?

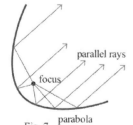
Fig. 7

Reflection Property of a Parabola

A parabolic mirror reflects light from the focus in a line parallel to the axis of the parabola (Fig. 7). In reverse, incoming light rays parallel to the axis are reflected to the focus.

Outline of a proof: If the parabolic mirror is given by $x = ay^2$ (Fig. 8), then its focus is at $F = (\frac{1}{4a}, 0)$ and the parabola is symmetric with respect to the x–axis. To prove the reflection property of the parabola, we need to show that $\alpha = \beta$ or, since α and β are both acute angles, that $\tan(\alpha) = \tan(\beta)$.

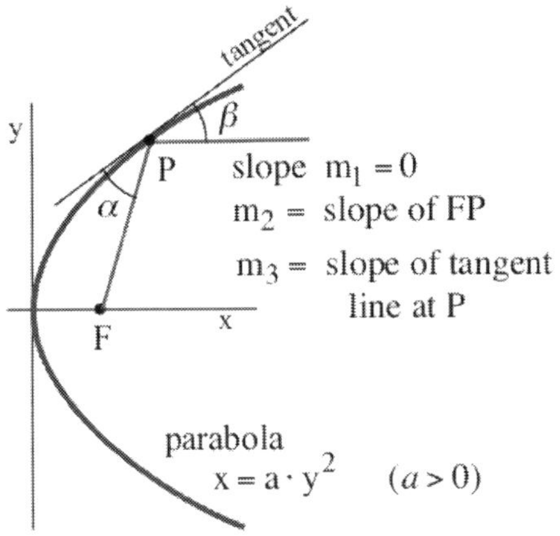

From Fig. 8 we calculate that

$$m_1 = 0, \quad m_2 = \frac{y}{x - \frac{1}{4a}} = \frac{4ay}{4ax - 1} = \frac{4ay}{4a^2y^2 - 1}, \text{ and, by implicit}$$

differentiation, $m_3 = \frac{1}{2ay}$.

Since $m_1 = 0$, we know that $\tan(\alpha) = \frac{m_3 - m_1}{1 + m_1 m_3} = m_3 = \frac{1}{2ay}$.

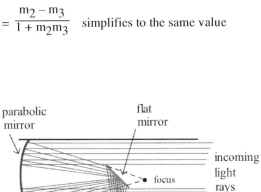

Fig. 8

An elementary but tedious algebraic argument shows that $\tan(\beta) = \frac{m_2 - m_3}{1 + m_2 m_3}$ simplifies to the same value as $\tan(\alpha)$ so $\tan(\alpha) = \tan(\beta)$ and $\alpha = \beta$.

Because of this reflection property, the parabola is used in a variety of instruments and devices. Mirrors in reflecting telescopes are parabolic (Fig. 9) so that the dim incoming (parallel) light rays from distant stars are all reflected to an eyepiece at the mirror's focus for viewing. Similarly, radio telescopes use a parabolic surface to collect weak signals. A well known scientific supply company sells an 18 inch diameter parabolic reflector "ideal for a broad range of applications including solar furnaces, solar energy collectors, and parabolic and directional microphones." For outgoing light, flashlights and automobile headlights use (almost) parabolic mirrors so a light source set at the focus of the mirror creates a tight beam of light (Fig. 10).

Fig. 9: Reflecting telescope

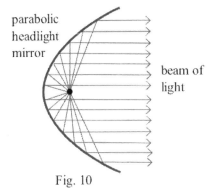

Fig. 10

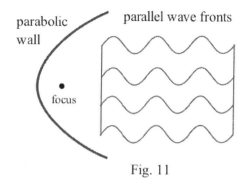

Fig. 11

Not only are incoming parallel rays reflected to the same point, but they reach that point at the same time. More precisely, if two objects start at the same distance from the y–axis and travel parallel to the x–axis, they both travel the same distance to reach the focus. If they are traveling at the same speed, they reach the focus together. An incoming linear wave front (Fig. 11) is reflected by a parabolic wall to create a splash at the focus. A small stone dropped into a wave tank at the focus of a parabola creates a linear outgoing wave. This "same distance" property of the reflection is something you can prove.

Practice 2: An object starts at the point (p,q), travels to the left until it encounters the parabola $x = ay^2$ (Fig. 12) and then goes straight to the focus at $(\frac{1}{4a}, 0)$. Show that the total distance traveled, L_1 plus L_2, equals $p + \frac{1}{4a}$ so the total distance is the same for all values of q. (Assume that $p > aq^2$ so the starting point is to the left of the parabola.)

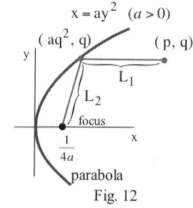

Fig. 12

The hyperbola also has a reflection property, but it is less useful than those for ellipses and parabolas.

Reflection Property of an Hyperbola

An hyperbolic mirror reflects light aimed at one focus to the other focus (Fig. 13).

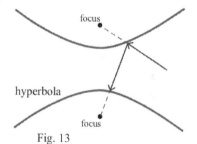

Fig. 13

Polar Coordinate Forms for the Conic Sections

In the rectangular coordinate system, the graph of the general quadratic equation $Ax^2 + Bxy + Cy^2 + Dx + Ey + F = 0$ is always a conic section, and the value of the discriminant $B^2 - 4AC$ tells us which type. In the polar coordinate system, an even simpler function describes all of the conic section shapes, and a single parameter in that function tells us the shape of the graph.

9.6 Properties of the Conic Sections

For $e \geq 0$, the polar coordinate graphs of $r = \dfrac{k}{1 \pm e \cdot \cos(\theta)}$ and $r = \dfrac{k}{1 \pm e \cdot \sin(\theta)}$

are conic sections with one focus at the origin.

If $e < 1$, the graph is an ellipse. (If $e = 0$, the graph is a circle.)

If $e = 1$, the graph is a parabola.

If $e > 1$, the graph is an hyperbola.

The number e is called the **eccentricity** of the conic section.

Fig. 14 shows graphs of $r = \dfrac{1}{1 + e \cdot \cos(\theta)}$ for several values of e.

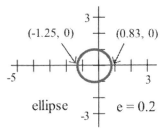
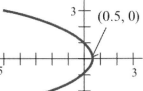
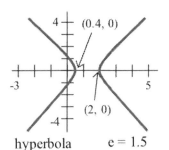
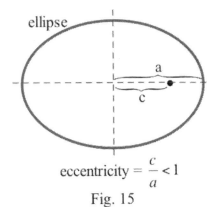
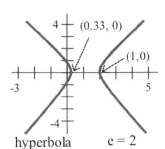

Fig. 14

For an ellipse, the eccentricity = $\dfrac{\text{dist(center, focus)}}{\text{dist(center, vertex)}}$ (Fig. 15).

If the eccentricity of an ellipse is close to zero, then the ellipse is "almost" a circle. If the eccentricity of an ellipse is close to 1, the ellipse is rather "narrow."

eccentricity = $\dfrac{c}{a} < 1$

Fig. 15

For a hyperbola, the eccentricity $= \frac{\text{dist(center, focus)}}{\text{dist(center, vertex)}}$ (Fig. 16). If the eccentricity of a hyperbola is close to 1, then the hyperbola is "narrow." If the eccentricity of a hyperbola is very large, the hyperbola "opens wide."

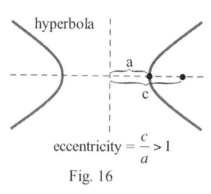

Fig. 16

Proof: The proof uses a strategy common in mathematics: move the problem into a system we know more about. In this case we move the problem from the polar coordinate system to the rectangular coordinate system, put the resulting equation into the form of a general quadratic equation, and then use the discriminant to determine the shape of the graph.

If $r = \frac{k}{1 + e \cdot \cos(\theta)}$, then $r + e \cdot r \cdot \cos(\theta) = k$. Replacing r with $\sqrt{x^2 + y^2}$ and $r \cdot \cos(\theta)$ with x, we get $\sqrt{x^2 + y^2} + e \cdot x = k$ and $\sqrt{x^2 + y^2} = k - e \cdot x$.

Squaring each side and collecting all of the terms on the right gives the equivalent general quadratic equation

$$(1 - e^2)x^2 + y^2 + 2kex - k = 0 \text{ so } A = 1 - e^2, B = 0, \text{ and } C = 1.$$

The discriminant of this general quadratic equation is $B^2 - 4AC = 0 - 4(1 - e^2)(1) = 4(e^2 - 1)$ so

if $e < 1$, then $B^2 - 4AC < 0$ and the graph is an ellipse,

if $e = 1$, then $B^2 - 4AC = 0$ and the graph is an parabola, and

if $e > 1$, then $B^2 - 4AC > 0$ and the graph is an hyperbola.

The graph of $r = \frac{k}{1 + e \cdot \cos(\theta)}$ is a conic section, and the value of the eccentricity tells which shape the graph has. We will not prove that one focus of the conic section is at the origin, but it's true.

Practice 3: Graph $r = \frac{k}{1 + (0.8) \cdot \cos(\theta)}$ for $k = 0.5, 1, 2,$ and 3. What effect does the value of k have on the graph?

Subtracting a constant α from θ rotates a polar coordinate graph counterclockwise about the origin by an angle of α, but does not change the shape of the graph, so the graphs of

$$r = \frac{k}{1 + e \cdot \cos(\theta - \alpha)}$$

are all conic sections whose shapes depend on the size of the parameter e.

In particular, the polar graphs of $r = \dfrac{1}{1 - e \cdot \cos(\theta)}$, $r = \dfrac{1}{1 + e \cdot \sin(\theta)}$, and $r = \dfrac{1}{1 - e \cdot \sin(\theta)}$

are conic sections, rotations about the origin of the graphs of $r = \dfrac{1}{1 + e \cdot \cos(\theta)}$, since

$-\cos(\theta) = \cos(\theta - \pi)$, $\sin(\theta) = \cos(\theta - \pi/2)$, and $-\sin(\theta) = \cos(\theta - -\pi/2)$. Fig. 17 shows several of these graphs for $e = 0.8$.

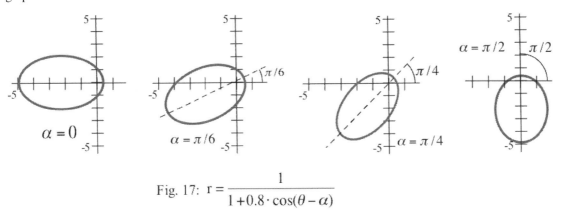

Fig. 17: $r = \dfrac{1}{1 + 0.8 \cdot \cos(\theta - \alpha)}$

The Path of Every Object in the Universe is a Conic Section (not really)

Rather than engage in endless philosophical discussion about how the planets ought to move, Tycho Brahe (1546–1601) had a better idea about how to find out how they actually do move — collect data! Even before the invention of the telescope, he built an observatory, and with the aid of devices like protractors he carefully cataloged the positions of the planets for 20 years. Just before his death, he passed this accumulated data to Johannes Kepler to edit and publish. From these remarkable data, Kepler deduced his three laws of planetary motion, the first of which says **each planet moves in an elliptical orbit with the sun at one focus** (Fig. 18). From Kepler's laws, Newton was able to deduce that a force, gravity, held the planets in orbits and that the force varied inversely as the square of the distance between the planet and the sun. From this "inverse square" fact it can be shown that the position of one object (e.g., a planet) with respect to another (e.g., the sun) is given by the polar coordinate formula

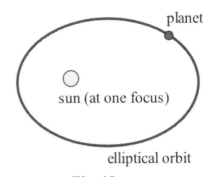

Fig. 18

$$r = \frac{h}{1 + (h-1)\cdot\cos(\theta)} \quad \text{where} \quad h = \frac{r_0 \cdot v_0^2}{GM}$$

r_0 = initial distance (m), v_0 = initial velocity (m/s)

G = universal gravitation constant ($6.7 \times 10^{-11} \frac{N\,m^2}{kg^2}$)

M = mass of the sun or "other object" (kg).

You should recognize the pattern of this formula as the pattern for the conic sections with eccentricity

$$e = |h - 1| = \left| \frac{r_0 \cdot v_0^2}{GM} - 1 \right|.$$

In a "two body" universe, all motion paths are conic sections, and the shape of the conic section is determined by the value of $r_0 \cdot v_0^2$. If the object is far away or moving very rapidly or both, then $r_0 \cdot v_0^2 > 2GM$, the eccentricity is greater than one, and the path is a hyperbola. If the objects are relatively close and/or moving slowly, then $r_0 \cdot v_0^2 < 2GM$ (the situation with each planet and the sun), the eccentricity is less than one and the path is an ellipse. If $r_0 \cdot v_0^2 = 2GM$, then the eccentricity is 1 and the path is a parabola. If $r_0 \cdot v_0^2 = GM$, then the eccentricity is 0 and the path is a circle. It is rare to encounter values of r_0 and v_0 so $r_0 \cdot v_0^2$ exactly equals GM or $2GM$.

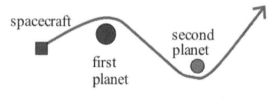

Fig. 19

The universe obviously contains more than two bodies and the paths of most objects are not conic sections, but there are still important situations in which the force on an object is due almost entirely to the gravitational attraction between it and one other body. For example, scientists and engineers use the position formula to determine the orbital position and velocity needed to put a satellite into an orbit with the desired eccentricity, and the position formula was used to help calculate how close Voyager 2 should come to Jupiter and Saturn so the gravity of those planets could be used to change the path of Voyager 2 to a hyperbola and send it on to other planets (Fig. 19). The conic sections even appear at less grand scales. In a vacuum, the path of a thrown baseball (or bat) is a parabola, unless it is thrown hard enough to achieve an elliptical orbit. And at the subatomic scale, Rutherford (1871 – 1937) discovered that alpha particles shot toward the nucleus of an atom are repelled away from the nucleus along hyperbolic paths (Fig. 20). Conic sections are everywhere.

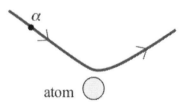

Fig. 20

Elliptical Orbits

When a planet orbits a sun, the orbit is an ellipse, and we can use information about ellipses to calculate information about these orbits.

The position of a planet in elliptical orbit around a sun is given by the polar equation

$$r = \frac{k}{1 + e \cdot \cos(\theta)}$$

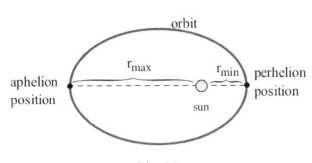

Fig. 21

for some value of the eccentricity $e < 1$ (Fig. 21). The planet is closest (at perhelion) to the sun when $\theta = 0$ and this minimum distance is $r_{min} = \frac{k}{1+e}$. The greatest distance (at aphelion) occurs when $\theta = \pi$ and that distance is $r_{max} = \frac{k}{1-e}$. If the width of the ellipse (technically, the length of the major axis) is 2a, then

$$2a = r_{min} + r_{max} = \frac{k}{1+e} + \frac{k}{1-e} = \frac{2k}{1-e^2}$$

so $k = a(1 - e^2)$ and the position of the planet is given by

$$r = \frac{a(1-e^2)}{1 + e \cdot \cos(\theta)} \quad \text{with} \quad r_{min} = \frac{a(1-e^2)}{1+e} = a(1-e) \quad \text{and} \quad r_{max} = \frac{a(1-e^2)}{1-e} = a(1+e).$$

Example 4: We want to put a satellite in an elliptical orbit around the earth (radius ≈ 6360 km) so the maximum height of the satellite is 20,000 km and the minimum height is 10,000 km (Fig. 22). Find the eccentricity of the orbit and give a polar formula for its position.

Solution: r_{max} = maximum height plus the radius of the earth = 26,360 km and r_{min} = minimum height plus the radius of the earth = 16,360 km so $a(1 + e) = 26,360$ and $a(1 - e) = 16,360$. Dividing these last two quantities, we have

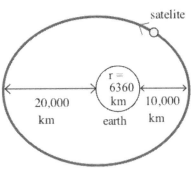

Fig. 22

$$\frac{r_{max}}{r_{min}} = \frac{26360}{16360} = \frac{a(1+e)}{a(1-e)} = \frac{1+e}{1-e} \quad \text{and} \quad e = \frac{10000}{42720} \approx \mathbf{0.234}$$

Using $r_{min} = a(1-e) = a(1 - 0.234) = 16360$ we have $a \approx 21358$.

Finally, $r = \dfrac{a(1-e^2)}{1 + e \cdot \cos(\theta)} = \dfrac{20189}{1 + (0.234) \cdot \cos(\theta)}$.

Practice 4: Pluto's orbit has an eccentricity of 0.2481 and its semimajor axis is 5,909 million kilometers. Find the minimum and maximum distance of Pluto from the sun during one orbit. The orbit of Neptune has an eccentricity of 0.0082 with a semimajor axis of 4,500 million kilometers. Is Neptune ever farther from the sun than Pluto?

PROBLEMS

Reflection properties

1. For the ellipse in Fig. 23, how far does a ball travel as it moves from one focus to any point on the ellipse and on to the other focus?

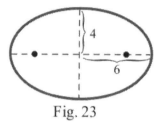

Fig. 23

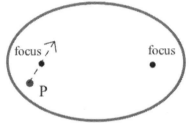

Fig. 24

2. In Fig. 24 a ball rolls from point P over the the focus at A and keeps rolling. Sketch the path of the ball for the first 5 bounces it makes off of the ellipse. What does the path of the ball look like after "a long time?"

3. In Fig. 25 a ball rolls from point P toward the the focus at A and bounces off of the hyperbola. Sketch the path of the ball for the first 5 bounces it makes off of the hyperbola. What does the path of the ball look like after "a long time?"

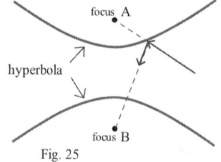

Fig. 25

4. An explosion is set off at one focus inside a very strong ellipsoidal shell. What might happen to a piece of graphite located at the other focus?

5. A straight wave front is approaching a parabolic jetty. Why wouldn't you want your boat to be at the focus of the parabola?

6. The members of a marching band are grouped near point A (Fig. 26), the focus of a parabola. At a signal from the director, the band members each march (at the same speed) in different directions toward the parabola, immediately turn and then march due west. What shape will the formation have after all of the marchers have made their turns?

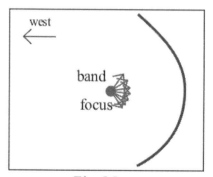

Fig. 26

7. A reflecting telescope is built with a parabolic mirror and a hyperbolic mirror (Fig. 27) so F_1 is the focus of the parabola and F_1 and F_2 are the foci of the hyperbola. Trace the paths of the incoming parallel light rays a, b, and c as they reflect off both mirrors. Why is the eyepiece located at F_2?

8. Make a slight design change in the telescope in Fig. 27 so the eyepiece can be located at the point off to the side of the parabolic mirror.

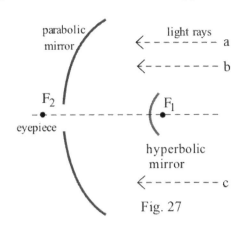

Fig. 27

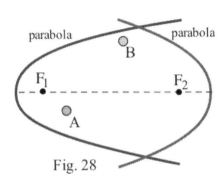

Fig. 28

9. The billiards table in Fig. 28 consists of parts of two parabolas: the right parabola has focus F_1 and the left parabola has focus F_2. (a) Determine a strategy for shooting the balls located at A and B to make a two–cushion (i.e., two bounce) shot into the hole located at F_2. (b) Are there any places on the table where your strategy in part (a) does not work? (Unfortunately for my ability to win at billiards, the angle of incidence does not necessarily equal the angle of reflection. But assume they are equal for these problems.)

10. The billiards table in Fig. 29 consists of parts of two ellipses: the short ellipse has foci F_1 and F_2 and the tall ellipse has foci F_2 and F_3.

    (a) Determine a strategy for shooting the balls located at A and B to make a two–cushion (i.e., two bounce) shot into the hole located at F_3.

    (b) Are there any places on the table where your strategy in part (a) does not work?

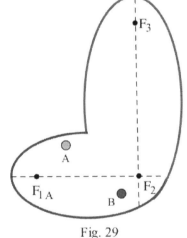

Fig. 29

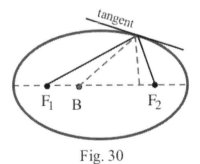

Fig. 30

11. Use Fig. 30 to help explain geometrically why a ball located on the major axis of an ellipse between the two foci is always reflected back to a point on the major axis between the two foci.

12. Is a rectangular reflection path (Fig. 31) possible for the ellipse

    (a) $\frac{x^2}{25} + \frac{y^2}{16} = 1$? (b) $\frac{x^2}{a^2} + \frac{y^2}{b^2} = 1$?

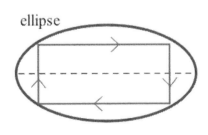

Fig. 31

Polar forms

In problems 13–18, determine the eccentricity of each conic section, identify the shape, and determine where it crosses the x and y axes.

13. $r = \dfrac{11}{3 + 5\cos(\theta)}$

14. $r = \dfrac{8}{7 + 3\sin(\theta + \pi/6)}$

15. $r = \dfrac{1}{2 + 2\sin(\theta - \pi/3)}$

16. $r = \dfrac{-4}{3 - 3\cos(\theta)}$

17. $r = \dfrac{17}{7 - 5\cos(\theta + 3\pi)}$

18. $r = \dfrac{3}{4 - 2\sin(\theta + \pi/11)}$

In problems 19–24, sketch each ellipse and determine the coordinates of the foci. (Reminder: In the standard polar coordinate form used here, one focus is always at the origin.)

19. $r = \dfrac{6}{2 + \cos(\theta)}$

20. $r = \dfrac{6}{2 + \sin(\theta)}$

21. $r = \dfrac{12}{3 - \sin(\theta)}$

22. $r = \dfrac{12}{3 - \cos(\theta)}$

23. $r = \dfrac{3}{2 + \sin(\theta - \pi/4)}$

24. $r = \dfrac{3}{2 + \cos(\theta + \pi/4)}$

In problems 25 and 26, represent the length and area of each ellipse as definite integrals and use Simpson's rule with n = 100 to approximate the values of the integrals.

25. $r = \dfrac{1}{1 + 0.5\cos(\theta)}$

26. $r = \dfrac{1}{1 + 0.9\cos(\theta)}$

Conic section paths:

Problems 27–30 refer to the two objects in Fig. 32. Determine the shape of the path of object B. (Object A has mass 10^{19} kg, r = 10^5 m)

27. v = 17.6 m/s

28. v = 115 m/s

29. v = 120 m/s

30. Determine a velocity for B so that the path is a parabola.

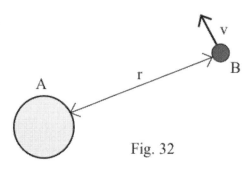

Fig. 32

31. An object at a distance r from the center of a planet of mass M (Fig. 33) has velocity v. Determine conditions on v as a function of r, M, and G so the resulting path is (a) circular,
(b) elliptical, (c) parabolic, and
(d) hyperbolic.

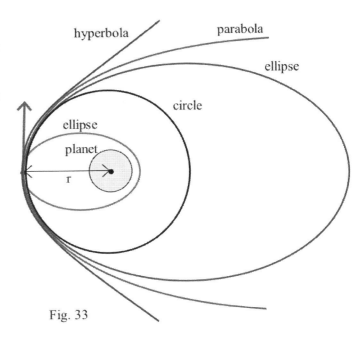

Fig. 33

Problems 32–34 refer to Earth and moon orbits: use
$G = 6.7 \times 10^{-11}$ N·m^2/kg^2,
$M_{Earth} = 5.98 \times 10^{24}$ kg,
$r_{Earth} = 6.360 \times 10^6$ m,
$M_{moon} = 7 \times 10^{22}$ kg, and
$r_{moon} = 1.738 \times 10^6$.

32. At the Earth's surface, determine a velocity so that the resulting path is circular. This is the minimum velocity needed to achieve "orbit," a very low orbit.

33. At the Earth's surface, determine a velocity so that the resulting path is parabolic. This is the minimum velocity needed to escape from orbit, and is called the "escape velocity".

34. At the moon's surface, determine the minimum orbital velocity and the (minimum) escape velocity.

Elliptical orbits

35. We want to put a satellite into orbit around Earth so the maximum altitude of the satellite is 1000 km and the minimum altitude is 800 km. Find the eccentricity of this orbit and give a polar coordinate formula for its position.

36. The Earth follows an elliptical orbit around the sun, and this ellipse has a semimajor axis of 149.6×10^6 km and an eccentricity of 0.017. (a) Determine the maximum and minimum distances of the Earth from the sun. (b) How far apart are the two foci of this ellipse?

37. Determine the altitude needed for an Earth satellite to make one orbit on a circular path every 24 hours ("geosyncronous"). (Since the orbit is circular and the satellite makes one orbit every 24 hours, you can determine the velocity v (in m/s) as a function of the distance r from the center of the Earth. Since the orbit is circular, $e = 0$ and $r \cdot v^2 = GM$.)

Section 9.6   **PRACTICE Answers**

Practice 1: "Shoot the ball toward the left focus." Since the table is an ellipse, the ball will roll over the focus, hit the wall, and be reflected into the hole at the other focus.

Practice 2: L_1 = distance from (p, q) to $(a \cdot q^2, q)$ = $p - a \cdot q^2$.

L_2 = distance from $(a \cdot q^2, q)$ to $(\frac{1}{4a}, 0)$

$= \sqrt{(a \cdot q^2 - \frac{1}{4a})^2 + (q-0)^2}$ $= \sqrt{a^2 q^4 - 2aq^2(\frac{1}{4a}) + \frac{1}{16a^2} + q^2}$

$= \sqrt{a^2 q^4 - \frac{1}{2} q^2 + \frac{1}{16a^2} + q^2}$

$= \sqrt{a^2 q^4 + \frac{1}{2} q^2 + \frac{1}{16a^2}} = \sqrt{(aq^2 + \frac{1}{4a})^2} = aq^2 + \frac{1}{4a}$.

Therefore, $L_1 + L_2 = (p - a \cdot q^2) + (aq^2 + \frac{1}{4a}) = p + \frac{1}{4a}$.

Practice 3: The graphs of $r = \dfrac{k}{1 + (0.8) \cdot \cos(\theta)}$

for $k = 0.5, 1,$ and 2 are shown in Fig. 34.
Each graph is an ellipse with eccentricity 0.8 and one focus at the origin. The value of k determines the size of the ellipse. The larger the magnitude of k, the larger the ellipse.

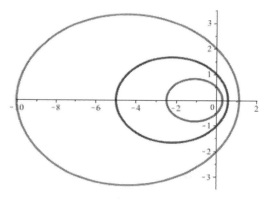

Fig. 34

Practice 4: For Pluto: $e = 0.2481$ and semimajor axis length = 5,909 km ($a = 5,909$ km).

$r_{min} = a(1 - e) = 5909(0.7519) \approx 4443$ km.

$r_{max} = a(1 + e) = 5909(1.2481) \approx 7375$ km.

For Neptune: $e = 0.0082$ and $a = 4,500$ km.

$r_{min} = a(1 - e) = 4500(0.9918) \approx 4463$ km, a distance closer than Pluto at its closest!

$r_{max} = a(1 + e) = 4500(1.0082) \approx 4537$ km.

In fact, Neptune is the farthest planet from the sun between January 1979 and March 1999. Now that Pluto has been reclassified as a "dwarf plane," Neptune is always the farthest planet from the sun. Poor Pluto.

Appendix: Reflection Property of the Ellipse

Let $P = (x,y)$ be on the ellipse $\dfrac{x^2}{a^2} + \dfrac{y^2}{b^2} = 1$ with foci at $(-c, 0)$ and $(c, 0)$ for $c = \sqrt{a^2 - b^2}$.

Then the slopes (Fig. 40) are m_1 = slope from P to $(-c,0)$ = $\dfrac{y}{x+c}$, m_2 = slope from P to $(c,0)$ = $\dfrac{y}{x-c}$,

and m_3 = slope of tangent line to ellipse at (x,y) = $\dfrac{-x}{y} \dfrac{b^2}{a^2}$ (by implicit differentiation).

We know that $\tan(\alpha) = \dfrac{m_3 - m_1}{1 + m_1 m_3}$ and $\tan(\beta) = \dfrac{m_2 - m_3}{1 + m_2 m_3}$, and we want to show that $\alpha = \beta$ or,

equivalently, that $\tan(\alpha) = \tan(\beta)$.

$$\tan(\alpha) = \frac{m_3 - m_1}{1 + m_1 m_3} = \frac{\dfrac{-x}{y}\dfrac{b^2}{a^2} - \dfrac{y}{x+c}}{1 + \dfrac{y}{x+c}\dfrac{-x}{y}\dfrac{b^2}{a^2}} \quad \text{multiply top \& bottom by } ya^2(x+c)$$

$$= \frac{-xb^2(x+c) - y^2 a^2}{ya^2(x+c) - xyb^2} = \frac{-x^2 b^2 - xb^2 c - y^2 a^2}{xya^2 + ya^2 c - xyb^2}$$

$$= \frac{-xb^2 c - (x^2 b^2 + y^2 a^2)}{xy(a^2 - b^2) + ya^2 c} \qquad x^2 b^2 + y^2 a^2 = a^2 b^2 \text{ and } a^2 - b^2 = c^2$$

$$= \frac{-xb^2 c - a^2 b^2}{xyc^2 + ya^2 c} = \frac{-b^2}{yc}\frac{xc + a^2}{xc + a^2} = \frac{-b^2}{yc}.$$

Similarly, $\tan(\beta) = \dfrac{m_2 - m_3}{1 + m_2 m_3} = \dfrac{\dfrac{y}{x-c} - \dfrac{-x}{y}\dfrac{b^2}{a^2}}{1 + \dfrac{y}{x-c}\dfrac{-x}{y}\dfrac{b^2}{a^2}}$ multiply top & bottom by $ya^2(x-c)$

$$= \frac{y^2 a^2 + xb^2(x-c)}{ya^2(x-c) - xyb^2} = \frac{x^2 b^2 - xb^2 c + y^2 a^2}{xya^2 - ya^2 c - xyb^2}$$

$$= \frac{-xb^2 c + (x^2 b^2 + y^2 a^2)}{xy(a^2 - b^2) - ya^2 c} \qquad (x^2 b^2 + y^2 a^2 = a^2 b^2 \text{ and } a^2 - b^2 = c^2)$$

$$= \frac{-xb^2 c + a^2 b^2}{xyc^2 - ya^2 c} = \frac{-b^2}{yc}\frac{xc - a^2}{xc - a^2} = \frac{-b^2}{yc} = \tan(\alpha). \text{ (Yes!)}$$

Reflection Property of the Parabola $x = ay^2$ with focus $(\frac{1}{4a}, 0)$

The slopes of the line segments in Fig. 41 are m_1 = slope of "incoming" ray = 0,

m_2 = slope from P to focus = $\dfrac{y}{x - \frac{1}{4a}}$ = $\dfrac{4ay}{4ax - 1}$ = $\dfrac{4ay}{4a^2y^2 - 1}$, and, by implicit differentiation,

m_3 = slope of the tangent line at (x,y) = $\dfrac{1}{2ay}$.

We know that $\tan(\alpha) = \dfrac{m_3 - m_1}{1 + m_1 m_3}$ and $\tan(\beta) = \dfrac{m_2 - m_3}{1 + m_2 m_3}$, and we want to show that $\alpha = \beta$ or, equivalently, that $\tan(\alpha) = \tan(\beta)$.

$\tan(\alpha) = \dfrac{m_3 - m_1}{1 + m_1 m_3} = m_3 = \dfrac{1}{2ay}$

$\tan(\beta) = \dfrac{m_2 - m_3}{1 + m_2 m_3} = \dfrac{\frac{4ay}{4a^2y^2 - 1} - \frac{1}{2ay}}{1 + \frac{4ay}{4a^2y^2 - 1} \cdot \frac{1}{2ay}}$ multiply top & bottom by $(4a^2y^2 - 1)(2ay)$

$= \dfrac{8a^2y^2 - 4a^2y^2 + 1}{(4a^2y^2 - 1)(2ay) + 4ay} = \dfrac{4a^2y^2 + 1}{8a^3y^3 - 2ay + 4ay}$

$= \dfrac{4a^2y^2 + 1}{8a^3y^3 + 2ay} = \dfrac{4a^2y^2 + 1}{2ay(4a^2y^2 + 1)} = \dfrac{1}{2ay} = \tan(\alpha)$. (Yes!)

There are other, more geometric ways to prove this result.

Chapter Nine

Section 9.1   Odd Answers

1. A: $(25, 0)$ B: $(25, 3\pi/2)$ C: $(15, \pi/6)$

3. A: $(30, \pi/3)$ B: $(20, 5\pi/6)$ C: $(25, 5\pi/4)$

5. Graph is given. Shape is almost rectangular.

7. Graph is given.

9. Graph is given.

11. Graph is given.

13. Graph is given.

15. Graphs are given.

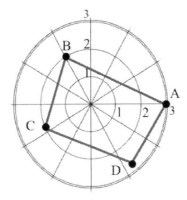
Prob. 5

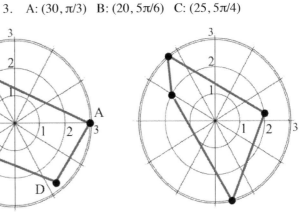
Prob. 7

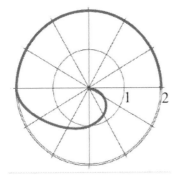
Prob. 9

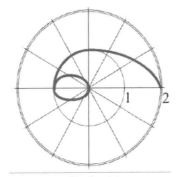
Prob. 11

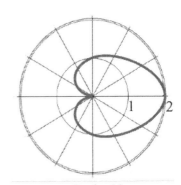
Prob. 13

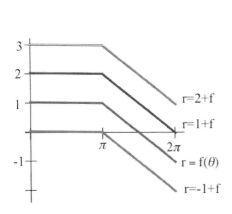
Prob. 15

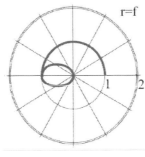

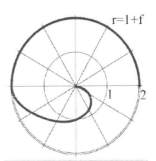

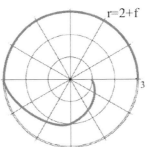

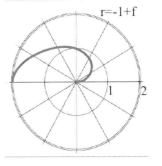

17. Graphs are given.

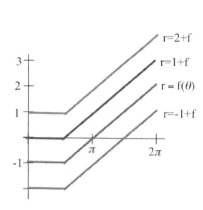

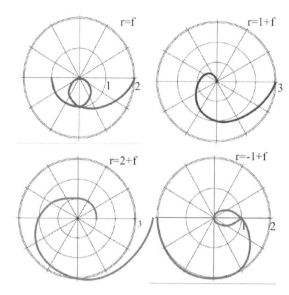

Prob. 17

19. The polar graph will approach (spiral in or out to) a circle of radius 3.

21. Circle centered at origin with radius 3.

23. Line through the origin making an angle of $\pi/6$ with the x–axis.

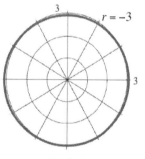

Prob. 21

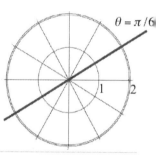

Prob. 23

25. A circle sitting atop the x–axis, touching the origin.

27. Graph is given.

29. Graph is given.

31. Graph is given.

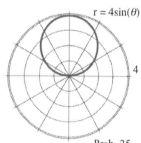

Prob. 25

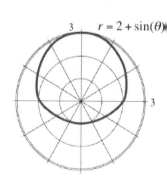

Prob. 27

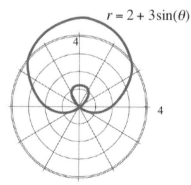

Prob. 29

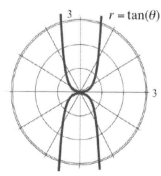

Prob. 31

33. Graph is given. A vertical line through the rectangular coordinate point (3,0).

35. Graph is given. A line: in rectangular coordinates y = 1 – x.

37. Graph is given: a fast growing spiral.

39. Graph is given.

41. {m = 1, n = 2}, {m = 2, n = 4}, {m = 3, n = 3}, {m = 4, n = 4} all have pleasing shapes. Find some others for yourself.

43. The graphs are given. The graph for a = 0 is rotated π/6, π/4, and π/2 counterclockwise about the origin as a = π/6, π/4, and π/2 respectively.

45.
n	1	2	3	4	5	6	7	8
# petals	1	4	3	8	5	12	7	16

47. Rectangular (–2, 3) is polar ($\sqrt{13}$, 2.159).
Rect (2, –3) is polar ($\sqrt{13}$, 5.300) or (–$\sqrt{13}$, 2.159).
Rect (0, –4) is polar (–4, π/2) or (4, 3π/2).

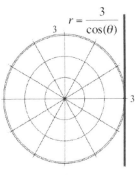
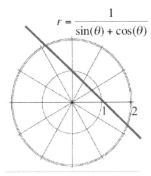
Prob. 33   Prob. 35

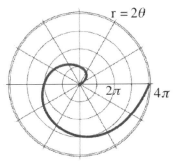
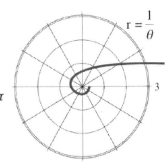
Prob. 37   Prob. 39

49. Rectangular (3, 4) is polar (5, 0.927).
Rect (–1, –3) is polar ($\sqrt{10}$, 4.391).
Rect (–7, 12) is polar ($\sqrt{193}$, –1.043) or ($\sqrt{193}$, –1.043 + π) ≈ ($\sqrt{193}$, 2.099).

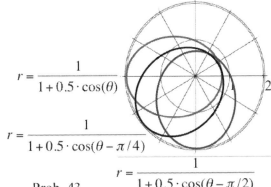
Prob. 43

51. Polar (–2, 3) is rect. (1.98, –0.282). Polar (2, –3) is rect. (–1.98, –0.282). Polar (0, –4) is rect. (0, 0).

53. Polar (2, 3) is rect. (–1.98, 0.282). Polar (–2, –3) is rect. (1.98, 0.282). Polar (0, 4) is rect. (0, 0).

55. x = 15·cos(–30°) + 10·cos(0°) ≈ 22.99 inches to the right of the shoulder.
y = 15·sin(–30°) + 10·sin(0°) ≈ –7.5 inches or 7.5 inches **below** the shoulder.
Polar location of hand: (24.18, –18.07°) or (24.18, –0.32).

57. x = 15·cos(–0.9) + 10·cos(–0.5) ≈ 18.10 inches to the right of the shoulder.
y = 15·sin(–0.9) + 10·sin(–0.5) ≈ 16.54 inches **below** the shoulder. Polar is (24.52, -0.74).

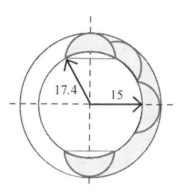

59. Figure is given: $-\pi/2 \leq \theta \leq \pi/2$ and $-\pi/2 \leq \phi \leq \pi/2$. The robot's hand can reach the shaded region.

60. Figure is given: $-\pi/2 \leq \theta \leq \pi/2$ and $-\pi \leq \phi \leq \pi$. The robot's hand can reach the shaded region.

61. On your own.

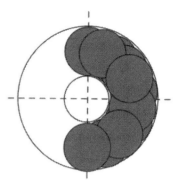

Prob. 59

Prob. 60

Section 9.2   Odd Answers

1. | Point | dr/dθ | dx/dθ | dy/dθ | dy/dx |
    |---|---|---|---|---|
    | A | − | − | + | − |
    | B | 0 | − | + | − |
    | C | + | − | + | − |
    | D | + | − | 0 | 0 |
    | E | − | + | − | − |

2. | Point | dr/dθ | dx/dθ | dy/dθ | dy/dx |
    |---|---|---|---|---|
    | A | + | + | + | + |
    | B | + | − | + | − |
    | C | − | − | − | + |
    | D | − | − | − | + |
    | E | + | − | − | + |

3. | Point | dr/dθ | dx/dθ | dy/dθ | dy/dx |
    |---|---|---|---|---|
    | A | 0 | 0 | + | Und. |
    | B | + | 0 | + | Und. |
    | C | U | U | U | U |
    | D | − | − | 0 | 0 |
    | E | + | − | 0 | 0 |

4. | Point | dr/dθ | dx/dθ | dy/dθ | dy/dx |
    |---|---|---|---|---|
    | A | − | + | − | − |
    | B | − | + | − | − |
    | C | + | + | − | − |
    | D | + | + | 0 | 0 |
    | E | + | + | + | + |

5. The polar graph is a circle centered at the origin with radius 5.

    At A: $dr/d\theta = 0$, $dy/dx = -1$. At B: $dr/d\theta = 0$, $dy/dx = 0$.

    At C: $dr/d\theta = 0$, $dy/dx =$ Und.

7. The graph is given.

    At A: $dr/d\theta = 0$, $dy/dx =$ Und. At B: $dr/d\theta = -1$, $dy/dx = -1/5$.

    At C: $dr/d\theta = 0$, $dy/dx = 0$.

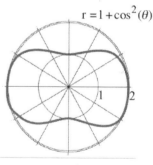

Prob. 7

9. When $\theta = 2\pi/3$, $dy/dx = -\sqrt{3}$. When $\theta = 4\pi/3$, $dy/dx = +\sqrt{3}$.

11. We have $\dfrac{dy}{dx} = \dfrac{r\cdot\cos(\theta) + \dfrac{dr}{d\theta}\cdot\sin(\theta)}{-r\cdot\sin(\theta) + \dfrac{dr}{d\theta}\cdot\cos(\theta)}$,

    so if $r(\theta) = 0$ (and $\dfrac{dr}{d\theta}$ exists and $\dfrac{dr}{d\theta} \neq 0$)

    then $\dfrac{dy}{dx} = \dfrac{\dfrac{dr}{d\theta}\cdot\sin(\theta)}{\dfrac{dr}{d\theta}\cdot\cos(\theta)} = \tan(\theta)$.

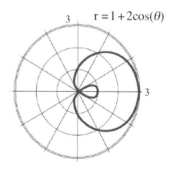

Prob. 9

13. $\frac{9}{16}\pi^2 \approx 5.552$

15. {area of cardioid in first quad.} – {area of circle in first quad.} = $\{1 + \frac{3\pi}{8}\} - \{\frac{\pi}{4}\} = 1 + \frac{\pi}{8} \approx 1.393$.
    This problem is worked out in Example 4.

17. 3–petal: $A = \frac{1}{2} \int_0^{\pi/3} \sin^2(3\theta)\, d\theta = \frac{\pi}{12} \approx 0.262$. 5–petal: $A = \frac{1}{2} \int_0^{\pi/5} \sin^2(5\theta)\, d\theta = \frac{\pi}{20} \approx 0.157$.

19. $A = \pi(2)^2 = 4\pi \approx 12.566$

21. (b) add several semicirular regions to get $\frac{1}{2}\pi(40)^2 + \frac{1}{2}\pi(30)^2 + \frac{1}{2}\pi(10)^2 = 1300\pi\ \text{ft}^2 \approx 4{,}084.1\ \text{ft}^2$

    (c) $\frac{3}{4}\pi(40)^2 + \frac{1}{2}\pi(20)^2 = 1400\pi\ \text{ft}^2 \approx 4{,}398.2\ \text{ft}^2 > 4{,}084.1\ \text{ft}^2$

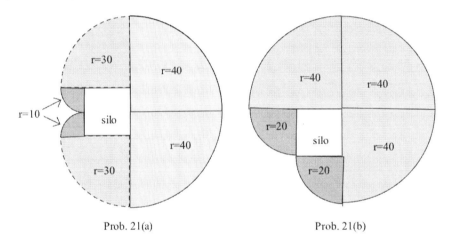

Prob. 21(a)   Prob. 21(b)

23. $r = \theta$, $dr/d\theta = 1$, $L = \int_0^{2\pi} \sqrt{\theta^2 + 1}\, d\theta \approx 21.256$ (using Simpson's rule with n = 20).

25. $r = 1 + \cos(\theta)$, $dr/d\theta = -\sin(\theta)$,

$$L = \int_0^{2\pi} \sqrt{\{1+\cos(\theta)\}^2 + \{-\sin(\theta)\}^2}\, d\theta = \int_0^{2\pi} \sqrt{2 + 2\cos(\theta)}\, d\theta = \int_0^{2\pi} \sqrt{4\cos^2(\theta/2)}\, d\theta$$

$$= 2\int_0^{2\pi} |\cos(\theta/2)|\, d\theta = 4\int_0^{\pi} \cos(\theta/2)\, d\theta = 8\sin(\theta/2)\Big|_0^{\pi} = 8.$$

(Simpson's rule with n= 20 gives the same result.)

27. 10π

29. $r = \sin(3\theta)$, $dr/d\theta = 3\cos(3\theta)$, $L = \int_0^{\pi/3} \sqrt{\sin^2(3\theta) + \{3\cos(3\theta)\}^2}\, d\theta \approx 2.227$ (Simpson, n = 20).

Section 9.3   Odd Answers

1. Graph is given.   3. Graph is given.

5. Graph is given.   7. Graph is given.

9. Graph is given: a straight line.

11. $(x(0), y(0)) = (b, d)$ and
    $(x(1), y(1)) = (a+b, c+d)$ so
    slope $= \dfrac{(c+d) - d}{(a+b) - b} = \dfrac{c}{a}$.

13. Graph is given.

15. Graphs are given.

    (a) is the entire line $y = x$.

    (b) is the "half–line" $y = x$ for $x \geq 0$.

    (c) is the line segment $y = x$ for
    $-1 \leq x \leq 1$: the location oscillates
    along the line between $(-1, -1)$ and $(1,1)$.

    All of these graphs satisfy the same realationship between x
    and y, $y = x$, but the graphs cover different parts of the
    graph of $y = x$ (different domains).

17. The graph is given. The graph begins
    at $\left(\dfrac{2\sqrt{2}}{\pi}, \dfrac{2\sqrt{2}}{\pi}\right)$ when $t = \pi/4$,
    and then it spirals counterclockwise
    around and in toward the origin.
    As t increases from $\pi/4$, the radial
    distance of a point on the graph
    decreases, approaching 0.

19. closer to

21. No. The new values give a point very close to the fixed point for these
    populations so the system will be in good balance.

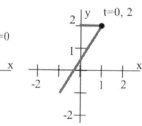
Prob. 1   Prob. 3

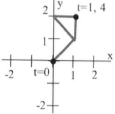
Prob. 5   Prob. 7

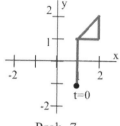
Prob. 9   Prob. 13

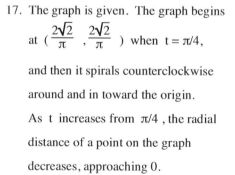
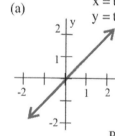
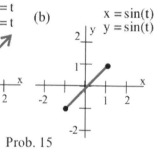
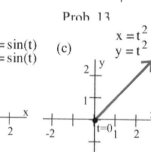
Prob. 15

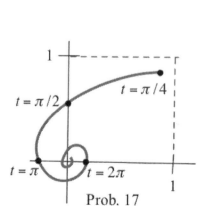
Prob. 17

23. Solve $x = at + b$ for t and substitute this value of t into the equation $y = cx + d$. Then $t = \dfrac{x-b}{a}$ and
$y = c(\dfrac{x-b}{t}) + d = \dfrac{c}{a} x + (d - \dfrac{b}{a})$, the equation of a straight line with slope $\dfrac{c}{a}$. The slope $\dfrac{c}{a}$ of y as a function of x is the slope of y as a function of t divided by the slope of x as a function of t.

25. Sketches of some possible paths are given.

   (a) walking slowly   (b) running (like a stretched spring)   (c) walking along a parabolic path

   parabola

   Prob. 25

27. (a) $x = R(t - \sin(t))$, $y = R(1 - \cos(t))$. If $t = 2\pi$, then $(x,y) = (2\pi R, 0) = (10\pi, 0)$ so $R = 5$.

   (b) Set $x = R(t - \sin(t)) = 5$ and $y = R(1 - \cos(t)) = 2$, divide y by x to eliminate R, solve (graphically or using Newton's method or some other way) to get $t \approx 3.820$. Substitute this value into the equation for x or y and solve for $R \approx 1.124$.

   (c) Set $x = R(t - \sin(t)) = 2$ and $y = R(1 - \cos(t)) = 3$, divide y by x to eliminate R, solve (graphically or using Newton's method or some other way) to get $t \approx 1.786$. Substitute this value into the equation for x or y and solve for $R \approx 2.472$.

   (d) Set $x = R(t - \sin(t)) = 4\pi$ and $y = R(1 - \cos(t)) = 8$ and solve to get $t = \pi$ and $R = 4$.

29. (a) $x = -50 + 30(\dfrac{1}{\sqrt{2}}) \cdot t = -50 + (15\sqrt{2}) \cdot t$ feet and
   $y = 6 + 30(\dfrac{1}{\sqrt{2}}) \cdot t - 16t^2 = 6 + (15\sqrt{2}) \cdot t - 16t^2$ feet.

   (b) $x = -50 + V(\dfrac{\sqrt{2}}{2}) \cdot t$ feet and $y = 6 + V(\dfrac{\sqrt{2}}{2}) \cdot t - 16t^2$ feet.

Section 9.4   Odd Answers

1. (a) The graph is given.

   (b) $dx/dt = 1 - 2t$, $dy/dt = 2$. When $t = 0$, $dy/dx = 2$. When $t = 1$, $dy/dx = -2$. When $t = 2$, $dy/dx = -2/3$.

   (c) dy/dx is never 0. dy/dx is undefined when $t = 1/2$: at $(x,y) = (1/4, 2)$.

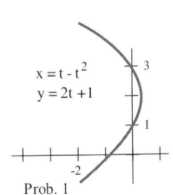

Prob. 1

3. (a) The graph is given.

   (b) $dx/dt = -\sin(t)$, $dy/dt = \cos(t)$. When $t = 0$, dy/dx is undefined. When $t = \pi/4$, $dy/dx = -1$. When $t = \pi/2$, $dy/dx = 0$.

   (c) $dy/dx = 0$ whenever $t = (k + \frac{1}{2})\pi$ for k an integer. dy/dx is undefined whenever $t = k\pi$ for k an integer.

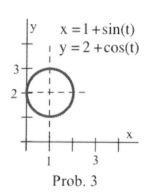
Prob. 3

5. (a) The graph is given.

   (b) $dx/dt = \cos(t)$, $dy/dt = -\sin(t)$. When $t = 0$, $dy/dx = 0$. When $t = \pi/4$, $dy/dx = -1$. When $t = \pi/2$, dy/dx is undefined. When $t = 17.3$, $dy/dx \approx 47.073$.

   (c) $dy/dx = 0$ whenever $t = k\pi$ for k an integer. dy/dx is undefined whenever $t = (k + \frac{1}{2})\pi$ for k an integer.

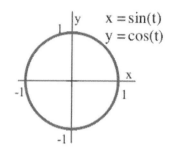
Prob. 5

7. (a) The graph is given.

   (b) $dx/dt = 1/t$, $dy/dt = -2t$. When $t = 1$, $dy/dx = -2$. When $t = 2$, $dy/dx = -8$. When $t = e$, $dy/dx = -2e^2$.

   (c) The function is only defined for $t > 0$, and for all $t > 0$ the slope of the tangent line dy/dx is defined and is not equal to 0.

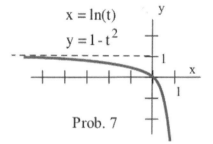
Prob. 7

9. (a) $m_0 \approx \frac{-1}{1} = -1$. $m_1 \approx \frac{0}{1} = 0$. $m_2 \approx \frac{0}{1} = 0$. m_3 is undefined.

   (b) $dy/dx = 0$ when $t = 1$ and $t = 2$.

11. (a) $m_0 \approx \frac{2}{-1} = -2$. $m_1 \approx \frac{1}{-1} = -1$. m_2 is undefined. $m_3 = \frac{-1}{1} = -1$.

    (b) dy/dx is undefined when $t = 2$.

13. $dx/dt = 1 - 2t$, $dy/dt = 2$ so $v = \sqrt{(1-2t)^2 + (2)^2} = \sqrt{4t^2 - 4t + 5}$. $v_0 = \sqrt{5} \approx 2.24$ ft/s, $v_1 = \sqrt{5} \approx 2.24$ ft/s, $v_2 = \sqrt{13} \approx 3.61$ ft/s.

15. $dx/dt = -\sin(t)$, $dy/dt = \cos(t)$ so
    $v = \sqrt{(-\sin(t))^2 + (\cos(t))^2} = \sqrt{\sin^2(t) + \cos^2(t)} = 1$ ft/s for all values of t.
    $v_0 = v_{\pi/4} = v_{\pi/2} = v_\pi = 1$ ft/s.

17. $v_0 \approx \sqrt{(1)^2 + (-1)^2} = \sqrt{2} \approx 1.41$ ft/s. $v_1 \approx \sqrt{(1)^2 + (0)^2} = 1$ ft/s.
    $v_2 \approx \sqrt{(1)^2 + (0)^2} = 1$ ft/s. $v_3 \approx \sqrt{(0)^2 + (-1)^2} = 1$ ft/s.
    $v_4 \approx \sqrt{(-1)^2 + (-1)^2} = \sqrt{2} \approx 1.41$ ft/s.

19. $v_0 \approx \sqrt{(-1)^2 + (2)^2} = \sqrt{5} \approx 2.24$ ft/s. $v_1 \approx \sqrt{(-1)^2 + (1)^2} = \sqrt{2} \approx 1.41$ ft/s.
    $v_2 \approx \sqrt{(0)^2 + (0)^2} = 0$ ft/s. $v_3 \approx \sqrt{(1)^2 + (-1)^2} = \sqrt{2} \approx 1.41$ ft/s.

21. (a) $dx/dt = R(1 - \cos(t))$ and $dy/dt = R\sin(t)$ so
    $v = \sqrt{R^2(1 - \cos(t))^2 + R^2(\sin(t))^2} = |R|\sqrt{2}\sqrt{1 - \cos(t)}$ ft/s.
    (b) v is maximum when $\cos(t) = -1$, when $t = (2k+1)\pi$ seconds for k an integer.
    (c) $v_{max} = 2R$ ft/s.

23. $L = \int_0^2 \sqrt{(1-2t)^2 + 2^2}\, dt \approx 4.939$ (using a calculator).

25. π (half the circumference of the circle).

27. $x = 3 - t, y = 1 + \frac{1}{2} t$. $L = \int_1^3 \sqrt{(-1)^2 + (1/2)^2}\, dt = 2 \cdot \frac{\sqrt{5}}{2} = \sqrt{5} \approx 2.24$.

    Alternately, the graph of (x, y) is a straight line, and we can calculate the distance from $(x(1), y(1)) = (2, 1.5)$ to the point $(x(3), y(3)) = (0, 2.5)$: distance $= \sqrt{2^2 + 1^2} = \sqrt{5}$.

29. $y = 4t^2 - t^4, dx/dt = 2t$. $A = \int_0^2 (4t^2 - t^4) \cdot 2t\, dt = 2t^4 - \frac{1}{3}t^6 \Big|_0^2 = 32 - \frac{64}{3} = \frac{32}{3}$.

31. $y = 1 + \cos(t), dx/dt = 2t$. $A = \int_0^2 (1 + \cos(t)) \cdot 2t\, dt \approx 4.805$ (using a calculator).

33. See Fig. 24. $A = \frac{\pi R^2}{4} + \frac{R^2}{2} + \frac{\pi R^2}{2} + \frac{R^2}{2} + \frac{\pi R^2}{4} = \pi R^2 + R^2 = R^2(\pi + 1)$.

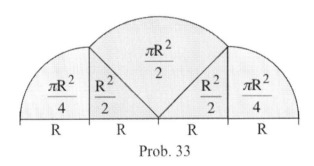

Prob. 33

Section 9.4.5 Odd Answers

1. $P_A = (0,4)$, $P_B = (5,2)$. $x(t) = 0 + 5t = (1-t) \cdot 0 + t \cdot 5$ and $y(t) = 4 + (-2)t = (1-t) \cdot 4 + t \cdot 2$.

3. $P_A = (4,3)$, $P_B = (1,2)$. $x(t) = 4 - 3t = (1-t) \cdot 4 + t \cdot 1$ and $y(t) = 3 - 1t = (1-t) \cdot 3 + t \cdot 2$.

5. $P_A = (1,4)$, $P_B = (5,1)$. $x(t) = 1 + 4t = (1-t) \cdot 1 + t \cdot 5$ and $y(t) = 4 + (-3)t = (1-t) \cdot 4 + t \cdot 1$

7. If we start with the equation $x(t) = x_0 + t \cdot \Delta x$ and replace Δx with $x_1 - x_0$ then
$x(t) = x_0 + t \cdot (x_1 - x_0) = x_0 + t \cdot x_1 - t \cdot x_0 = (1-t) \cdot x_0 + t \cdot x_1$ which is the pattern we wanted.
The algebra for y(t) is similar.

For problems 9-13 the Bezier pattern is
$$B(t) = (1-t)^3 \cdot P_0 + 3(1-t)^2 t \cdot P_1 + 3(1-t) t^2 \cdot P_2 + t^3 \cdot P_3$$

9. $x(t) = (1-t)^3 \cdot 0 + 3(1-t)^2 t \cdot 2 + 3(1-t)t^2 \cdot 1 + t^3 \cdot 4$
$y(t) = (1-t)^3 \cdot 5 + 3(1-t)^2 t \cdot 3 + 3(1-t)t^2 \cdot 4 + t^3 \cdot 2$

11. $x(t) = (1-t)^3 \cdot 6 + 3(1-t)^2 t \cdot 6 + 3(1-t)t^2 \cdot 2 + t^3 \cdot 2$
$y(t) = (1-t)^3 \cdot 5 + 3(1-t)^2 t \cdot 3 + 3(1-t)t^2 \cdot 5 + t^3 \cdot 0$

13. $P_0 = (5,1)$ and $B'(0) = 2$ tells us that P_1 could be (5+1, 1+2) or (5+2, 1+4) or (5+h, 1 + 2h)
We pick $P_1 = (6,3)$
$P_3 = (1,3)$ and $B'(1) = 3$ tells us P_2 could be (1+1,3+3) or (1+h, 3+3h).
We pick $P_2 = (2,6)$.
$x(t) = (1-t)^3 \cdot 5 + 3(1-t)^2 t \cdot 6 + 3(1-t)t^2 \cdot 2 + t^3 \cdot 1$
$y(t) = (1-t)^3 \cdot 1 + 3(1-t)^2 t \cdot 3 + 3(1-t)t^2 \cdot 6 + t^3 \cdot 3$

14. See Fig. 20.   15. See Fig. 21.

16. See Fig. 22.   17. See Fig. 23.

18. Violates Property (3): B'(1) does not equal the slope of the segment from P_2 to P_3

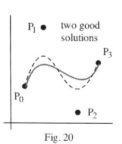

Fig. 20

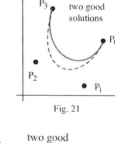

Fig. 21

19. Violates Property (3) at B(1). Also violates Property (4) since the B(t) graph goes outside the "rubber band" around the 4 control points.

20. Violates Property (2). Since the graph of B(t) has 3 "turns" then B(t) is not a cubic polynomial (which can only have 2 turns).

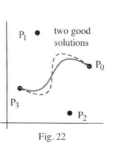

Fig. 22

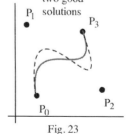

Fig. 23

21. Violates Property (1): B(1) = P_2 instead of P_3.

Section 9.5   Odd Answers

1. An ellipse. Figure is given Since $b^2 + 6^2 = 10^2$, $b = 8$. $c = \sqrt{a^2 - b^2}$ so $6 = \sqrt{a^2 - 64}$ and $a = 10$. The ellipse is given by $\dfrac{x^2}{10^2} + \dfrac{y^2}{8^2} = 1$.

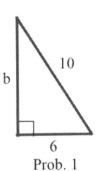
Prob. 1

3. A parabola with focus at (0,5) and vertex at (0,0). The parabola has an equation of the form $y = ax^2$ with $\dfrac{1}{4a} = 5$ so $a = \dfrac{1}{20}$. The parabola is given by $y = \dfrac{1}{20} x^2$.

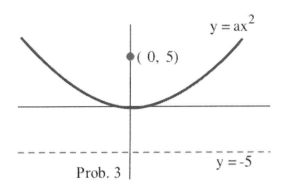
Prob. 3

5. $\dfrac{x^2}{4} + \dfrac{y^2}{9} = 1$.

7. Linear asymptotes: $y = \dfrac{2}{3} x$ and $y = -\dfrac{2}{3} x$. (Set $4x^2 - 9y^2 = 0$ and solve for y.) $c^2 = a^2 + b^2 = 4 + 9$ so $c = \sqrt{13}$: foci are at $(\sqrt{13}, 0)$ and $(-\sqrt{13}, 0)$.

9. Linear asymptotes: $y = \sqrt{\dfrac{3}{5}} x$ and $y = -\sqrt{\dfrac{3}{5}} x$. Foci: $(0, \sqrt{8})$ and $(0, -\sqrt{8})$.

11. (a) $25x^2 + 4y^2 + (-100) = 0$: discriminant = $(0)^2 - 4(25)(4) = -400 < 0$. The graph is an ellipse.
    (b) $b^2x^2 + a^2y^2 + (a^2b^2) = 0$: discriminant = $(0)^2 - 4(a^2)(b^2) = -4a^2b^2 < 0$. The graph is an ellipse.

13. $x^2 + xy - 2y^2 - x + y - 3 = 0$: discriminant = $(1)^2 - 4(1)(-2) = 9 > 0$. The graph is a hyperbola.

15. $2x^2 + 4xy + 2y^2 - 7x + 3 = 0$: discriminant = $(4)^2 - 4(2)(2) = 0$. The graph is a parabola.

17. (a) $B^2 - 4AC = 9 - 4(2)(2) < 0$: ellipse.
    (b) $B^2 - 4AC = 16 - 4(2)(2) = 0$: parabola.
    (c) $B^2 - 4AC = 25 - 4(2)(2) > 0$: hyperbola.
    (d) same answers as for parts (a), (b), and (c).

19. (a) $B^2 - 4AC = 16 - 4(1)(3) > 0$: hyperbola.
    (b) $B^2 - 4AC = 16 - 4(1)(4) = 0$: parabola.
    (c) $B^2 - 4AC = 16 - 4(1)(5) < 0$: ellipse.
    (d) same answers as for parts (a), (b), and (c).

21. Figure is shown.

(a) About the x-axis: $V = 2 \int_0^2 \pi y^2 \, dx = 2\pi \int_0^2 25(1 - \frac{x^2}{4}) \, dx$

$= 50\pi \{ x - \frac{x^3}{12} \} \Big|_0^2 = \frac{200\pi}{3}$.

(b) About the y-axis: $V = 2 \int_0^5 \pi x^2 \, dy$

$= 2\pi \int_0^5 4(1 - \frac{y^2}{25}) \, dx = 8\pi \{ y - \frac{y^3}{75} \} \Big|_0^5 = \frac{80\pi}{3}$.

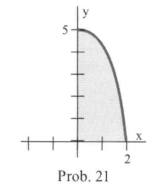

Prob. 21

23. Figure is shown.

(a) About the x-axis: $V = \int_2^{10} \pi y^2 \, dx$

$= \pi \int_2^{10} 25(\frac{x^2}{4} - 1) \, dx = 25\pi \{ \frac{x^3}{12} - x \} \Big|_2^{10}$

$= 25\pi \{ (\frac{1000}{12} - 10) - (\frac{8}{12} - 2) \} = \frac{5600\pi}{3} \approx 5864.3$.

(b) About the y-axis: $V = 2\pi \int_0^{10\sqrt{6}} \{ 100 - 4(\frac{y^2}{25} + 1) \} \, dy = 2\pi \{ 96y - \frac{4y^3}{75} \} \Big|_0^{10\sqrt{6}}$

$= 20\sqrt{6} \, \pi \{ 96 - \frac{2400}{75} \} = \frac{3840\sqrt{6}\,\pi}{3} \approx 9850$.

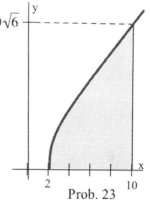

Prob. 23

25. $A_{parabolic} = \int_a^b ax^2 \, dx = \frac{1}{3} ax^3 \Big|_a^b = \frac{1}{3} ab^3$. $A_{rectangular} = ab^2 \cdot b = ab^3$. $\frac{A_{parabolic}}{A_{rectangular}} = \frac{1}{3}$.

27. The length of the string is the distance between the vertices.

29. When the pins (foci) are far apart, the ellipse tends to be long and narrow, cigar shaped. As the pins are moved closer together, the ellipse becomes more rounded and circular. In the limit, with the pins together, the ellipse is a cirle with diameter equal to the length of the string.

31. The curves are parabolas As the string is shortened, the vertex is moved up and the parabola narrows, approaching a vertical ray in the limit as the length of the string nears the vertical distance from the pin to the corner of the T-square where the string is attached.

Section 9.6   Odd Answers

1. 12 units, independent of where it bounces off of the ellipse.

3. After a "long time," the ball oscillates (almost) along a line between the vertices of the hyperbola.

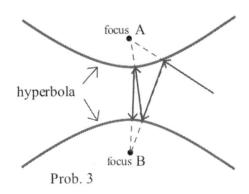
Prob. 3

5. All of the energy along the wave front is reflected to the focus of the parabolic jetty at the same time   The boat would be struck by a wave of considerable force.

7. The traced rays are shown. The eyepiece is located at F_2 because all of the incoming light is focused there.

9. (a) Roll the ball toward the focus F1 . The paths of A and B are shown.

   (b) The strategy in part (a) does not work for a ball in the shaded region. Why not?

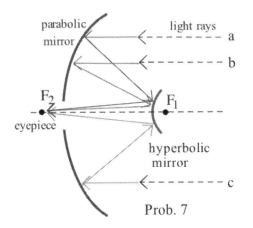
Prob. 7

11. At any point E on the ellipse, the angle of incidence equals the angle of reflection so the angles α and β are equal and the angle a and b are equal. Since {angle a} > {angle α} we have that {angle b} > {angle β} and the ball is reflected to a point C between the two foci.

13. $r = \dfrac{11/3}{1 + (5/3)\cos(\theta)}$ so $e = 5/3 > 1$ and the graph is a hyperbola. The hyperbola crosses the x–axis when $\theta = 0$ and π: at the points $(11/8, 0)$ and $(11/2, 0)$. It crosses the y–axis when $\theta = \pi/2$ and $3\pi/2$: at the points $(0, 11/3)$ and $(0, -11/3)$.

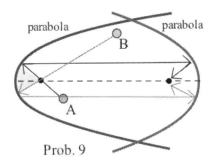
Prob. 9

15. $r = \dfrac{1/2}{1 + 1 \cdot \sin(\theta - \pi/3)}$ so $e = 1$ and the graph is a parabola.

The parabola crosses the x–axis when $\theta = 0$ and π: approximately at the points $(3.73, 0)$ and $(-0.27, 0)$. It crosses the y–axis when $\theta = \pi/2$ and $3\pi/2$: at the points $(0, 1/3)$ and $(0, -1)$.

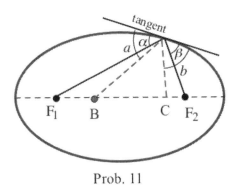
Prob. 11

17. $r = \dfrac{17/7}{1 - (5/7)\cdot\cos(\theta + 3\pi)}$ so $e = 5/7 \le 1$ and the graph is an ellipse.

The ellipse crosses the x–axis when $\theta = 0$ and π: approximately at the points $(1.42, 0)$ and $(-8.5, 0)$. It crosses the y–axis when $\theta = \pi/2$ and $3\pi/2$: at the points $(0, 2.43)$ and $(0, -2.43)$.

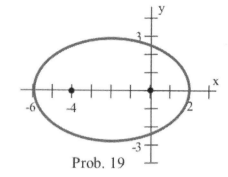
Prob. 19

19. One focus is at $(0,0)$ and, by symmetry, the other focus is at $(-4, 0)$.

21. One focus is at $(0,0)$ and, by symmetry, the other focus is at $(0, 3)$.

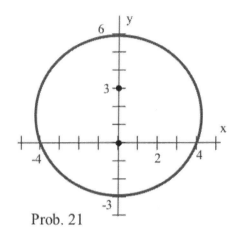
Prob. 21

23. See Fig. 41. This one is more difficult because the ellipse is tilted, but we can still use the symmetry of the ellipse and the fact that it is tilted at an angle of $\pi/4$ to the x–axis. One focus is at $(0,0)$ and the other focus is at $(1.41, -1.41)$.

25. $r = \dfrac{1}{1 + 0.5\cdot\cos(\theta)}$.

$\dfrac{dr}{d\theta} = -(1 + 0.5\cdot\cos(\theta))^{-2} \cdot \{-0.5\cdot\sin(\theta)\} = \dfrac{\sin(\theta)}{2 + \cos(\theta)}$

Length $= \displaystyle\int_0^{2\pi} \sqrt{(r)^2 + (dr/d\theta)^2}\, d\theta \approx 7.659$

Area $= \displaystyle\int_0^{2\pi} \tfrac{1}{2} r^2(\theta)\, d \approx 4.8368$

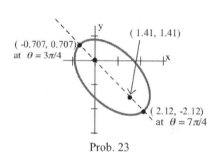
Prob. 23

Both integrals were approximated using Simpson's rule with $n = 20$)

27. $h = \dfrac{r_o v_o^2}{GM} = \dfrac{10^5 (17.6)^2}{(6.7)(10^{-11})(10^{19})} = \dfrac{3.097 \times 10^7}{6.7 \times 10^8} \approx 4.62 \times 10^{-2}$ so $e = |h - 1| \approx 0.95 < 1$

and the path is an ellipse (but a long narrow ellipse such as a comet might have).

29. $h = \dfrac{r_o v_o^2}{GM} = \dfrac{10^5 (120)^2}{(6.7)(10^{-11})(10^{19})} = \dfrac{1.44 \times 10^9}{6.7 \times 10^8} \approx 2.149$ so $e = |h - 1| \approx 1.149 > 1$ and the path is a hyperbola.

31. $e = \dfrac{r \cdot v^2}{GM} - 1$.

    (a) The path is circular if $e = 0$, so $\dfrac{r \cdot v^2}{GM} - 1 = 0$ and $v = \sqrt{\dfrac{GM}{r}}$.

    (b) The path is elliptical if $e < 1$, so $\dfrac{r \cdot v^2}{GM} - 1 < 1$ and

         $v < \sqrt{\dfrac{2GM}{r}} = \sqrt{2}\sqrt{\dfrac{GM}{r}} = \sqrt{2} \cdot \{\text{circular velocity}\}$.

    (c) The path is parabolic if $e = 1$, so $v = \sqrt{\dfrac{2GM}{r}} = \sqrt{2}\sqrt{\dfrac{GM}{r}} = \sqrt{2} \cdot \{\text{circular velocity}\}$.

    (d) The path is hyperbolic if $e > 1$, so $v > \sqrt{\dfrac{2GM}{r}} = \sqrt{2}\sqrt{\dfrac{GM}{r}} = \sqrt{2} \cdot \{\text{circular velocity}\}$.

33. We want $v = \sqrt{\dfrac{2GM}{r}} = \sqrt{\dfrac{2(6.7 \times 10^{-11})(5.98 \times 10^{24})}{6.36 \times 10^6}} \approx \sqrt{1.26 \times 10^8} \approx 11{,}225$ m/s or

    approximately 25,110 miles per hour.

35. $r_{max} = 1000 + r_{earth} \approx 7.36 \times 10^6$. $r_{min} = 800 + r_{earth} \approx 7.16 \times 10^6$.

    $2a = r_{min} + r_{max} \approx 14.520 \times 10^6$ so $a \approx 7.26 \times 10^6$.

    Also, $r_{max} = a(1 + e)$ so $7.36 \times 10^6 = 7.26 \times 10^6 (1 + e)$ and $e \approx 0.01377$.

    Finally, $k = a(1 - e^2) \approx 7.26 \times 10^6 (1 - (0.01377)^2) \approx 7.2586 \times 10^6$ so

    $r = \dfrac{k}{1 + e \cdot \cos(\theta)} \approx \dfrac{7.2586 \times 10^6}{1 + (0.01377) \cdot \cos(\theta)}$.

37. On your own.

10.0 INTRODUCTION TO SEQUENCES AND SERIES

Chapter 10 is an introduction to two special topics in calculus, sequences and series. The main idea underlying this chapter is that **polynomials are easy**, and that even the hard functions such as sin(x) and log(x) can be represented as "big polynomials."

Polynomials are easy. It is easy to do arithmetic (evaluate, add, subtract, multiply, and even divide) with polynomials. It is easy to do calculus (differentiate and integrate) with polynomials. And, strangely enough, every polynomial is <u>completely</u> determined by its value and the values of all of its derivatives at x = 0: if we know the values of P(0), P'(0), P''(0), etc., we can determine a formula for P(x) that is valid for all x.

Unfortunately, many of the important functions we need for applications (sin, cos, exp, log) are not polynomials: sin and cos have too many wiggles; exp grows too fast; and log has an asymptote. However, even these "hard" functions are "almost" polynomials and share many properties with polynomials:

For some values of x (to be specified in later sections), many important functions can be represented as "big polynomials" called power series:

$$\sin(x) = x - \frac{x^3}{2\cdot 3} + \frac{x^5}{2\cdot 3\cdot 4\cdot 5} - \frac{x^7}{2\cdot 3\cdot 4\cdot 5\cdot 6\cdot 7} + \ldots + (-1)^n \frac{x^{2n+1}}{(2n+1)!} + \ldots \quad \text{for } n = 0, 1, 2, \ldots$$

$$\cos(x) = 1 - \frac{x^2}{2} + \frac{x^4}{2\cdot 3\cdot 4} - \frac{x^6}{2\cdot 3\cdot 4\cdot 5\cdot 6} + \ldots + (-1)^n \frac{x^{2n}}{(2n)!} + \ldots$$

$$\exp(x) = 1 + x + \frac{x^2}{2} + \frac{x^3}{2\cdot 3} + \frac{x^4}{2\cdot 3\cdot 4} + \ldots + \frac{x^n}{n!} + \ldots$$

$$\frac{1}{1-x} = 1 + x + x^2 + x^3 + x^4 + \ldots + x^n + \ldots$$

the " ... " at the end means the pattern of the terms continues "forever"

In this chapter we examine
- what it means to sum an **infinite** number of terms,
- how to do algebra and calculus with series ("big polynomials"),
- how to represent functions as series ("big polynomials"), and
- how to use such series ("big polynomials") to calculate derivatives, integrals, and even solve differential equations.

First, however, we need to lay a foundation, and that foundation is the study of lists of numbers, their properties and behavior.

Section 10.1 focuses on this foundation material. It introduces lists of numbers, called sequences, and examines some specific sequences we will need later. It also introduces the idea of the convergence of a sequence and examines some ways we can determine whether or not a sequence converges.

Sections 10.2 to 10.7 focus on what it means to add up an infinite number of numbers, an infinite series, and on how we can determine whether the resulting sum is a finite number.

Sections 10.8 to 10.11 generalize the idea of an infinite series of numbers to infinite series that contain a variable. These series that contain powers of a variable are called power series. These sections discuss how we can represent and approximate functions such as $\sin(x)$ and e^x with power series, how accurate these approximations are for commonly needed functions, and how we can use them with derivatives and integrals.

PROBLEMS

These problems illustrate, at an elementary level, some of the problems and concepts we will examine more deeply in this chapter. They are intended to start you thinking in certain ways that are useful and necessary for Chapter 10.

Patterns in lists of numbers

For problems 1 – 6, the first four numbers $a_1, a_2, a_3,$ and a_4 in a list are given. (a) Write the next two numbers in the list, (b) write a formula for the 5th number a_5 in the list and (c) write a formula for the nth number a_n in the list.

1. 2, 4, 8, 16, ___ , ___
2. 3, 9, 27, 81, ___ , ___
3. –1, +1, –1, +1, ___ , ___
4. 1, 1/2, 1/3, 1/4, ___ , ___
5. 1, 2, 6, 24, ___ , ___
6. 1, 4, 9, 16, ___ , ___

For problems 7 – 11, evaluate each of the four given numbers and write the next two numbers in the list.

7. 1, 1 + 1/2, 1 + 1/2 + 1/3, 1 + 1/2 + 1/3 + 1/4, _____ , _____
8. 1, 1 + 1/2, 1 + 1/2 + 1/4, 1 + 1/2 + 1/4 + 1/6, _____ , _____
9. 1, 1 – 1/2, 1 – 1/2 + 1/4, 1 – 1/2 + 1/4 – 1/8, _____ , _____
10. 1, 1 + 2, 1 + 2 + 4, 1 + 2 + 4 + 8, ___ , ___
11. 1, 1 – 1, 1 – 1 + 1, 1 – 1 + 1 – 1, ___ , ___

Lists and graphs

For problems 12 – 15, (a) fill in the next two entries in the table and (b) graph the function for x = 1, 2, ... 6. These particular functions are defined only for integer values of x.

12.
x	f(x)
1	2
2	4
3	8
4	16
5	
6	

13.
x	g(x)
1	–1
2	+1
3	–1
4	+1
5	
6	

14.
x	s(x)
1	1 + 1/2
2	1 + 1/2 + 1/3
3	1 + 1/2 + 1/3 + 1/4
4	1 + 1/2 + 1/3 + 1/4 + 1/5
5	
6	

15.
x	t(x)
1	1 – 1/2
2	1 – 1/2 + 1/4
3	1 – 1/2 + 1/4 – 1/8
4	1 – 1/2 + 1/4 – 1/8 + 1/16
5	
6	

Polynomials and sine, cosine, and the exponential function

16. (a) Fill in the table for $P(x) = x - \dfrac{x^3}{2\cdot 3}$ and $\sin(x)$.

    (b) Graph $y = P(x)$ and $y = \sin(x)$ for $-2 \leq x \leq 2$.

    (c) Repeat parts (a) and (b) for $P(x) = x - \dfrac{x^3}{2\cdot 3} + \dfrac{x^5}{2\cdot 3\cdot 4\cdot 5}$.

x	P(x)	sin(x)	| P(x) – sin(x) |
0			
0.1			
0.2			
0.3			
1.0			
2.0			

17. (a) Fill in the table for $P(x) = 1 - \dfrac{x^2}{2}$ and $\cos(x)$.

    (b) Graph $y = P(x)$ and $y = \cos(x)$ for $-2 \leq x \leq 2$.

    (c) Repeat parts (a) and (b) for $P(x) = 1 - \dfrac{x^2}{2} + \dfrac{x^4}{2\cdot 3\cdot 4}$.

x	P(x)	cos(x)	| P(x) – cos(x) |
0			
0.1			
0.2			
0.3			
1.0			
2.0			

18. (a) Fill in the table for $P(x) = 1 + x + \dfrac{x^2}{2}$ and e^x.

    (b) Graph $y = P(x)$ and $y = e^x$ for $-2 \leq x \leq 2$.

    (c) Repeat parts (a) and (b) for $P(x) = 1 + x + \dfrac{x^2}{2} + \dfrac{x^3}{3}$.

x	P(x)	e^x	| P(x) – e^x |
0			
0.1			
0.2			
0.3			
1.0			
2.0			

Polynomials and their values at x = 0

These problems illustrate how we can determine a formula for a polynomial when we know the values of the polynomial and its derivatives at $x = 0$.

In problems 19 – 24, $P(x) = Ax + B$ is a linear polynomial, and the values of $P(0)$ and $P'(0)$ are given. Find the values of A and B and write a formula for $P(x)$.

19. $P(0) = 5$, $P'(0) = 3$
20. $P(0) = -2$, $P'(0) = 7$
21. $P(0) = 4$, $P'(0) = -1$

22. $P(0) = 8$, $P'(0) = 5$
23. $P(0) = 4$, $P'(0) = 0$
24. $P(0) = -3$, $P'(0) = -2$

25. How are the values of A and B related to the values of $P(0)$ and $P'(0)$?

In problems 26 – 31, $P(x) = Ax^2 + Bx + C$ is a quadratic polynomial, and the values of $P(0)$, $P'(0)$, and $P''(0)$ are given. Find the values of A, B, and C and write a formula for $P(x)$.

26. $P(0) = 5, P'(0) = 3, P''(0) = 4$
27. $P(0) = -2, P'(0) = 7, P''(0) = 6$

28. $P(0) = 4, P'(0) = -1, P''(0) = -2$
29. $P(0) = 8, P'(0) = 5, P''(0) = 10$

30. $P(0) = 4, P'(0) = 0, P''(0) = -4$
31. $P(0) = -3, P'(0) = -2, P''(0) = 4$

32. How are the values of A, B, and C related to the values of $P(0), P'(0)$, and $P''(0)$?

In problems 33 – 38, $P(x) = Ax^3 + Bx^2 + Cx + D$ is a cubic polynomial, and the values of $P(0)$, $P'(0)$, $P''(0)$, and $P'''(0)$ are given. Find the values of A, B, C, and D and write a formula for $P(x)$.

33. $P(0) = 5, P'(0) = 3, P''(0) = 4, P'''(0) = 6$
34. $P(0) = -2, P'(0) = 7, P''(0) = 6, P'''(0) = 18$

35. $P(0) = 4, P'(0) = -1, P''(0) = -2, P'''(0) = -12$
36. $P(0) = 8, P'(0) = 5, P''(0) = 10, P'''(0) = 12$

37. $P(0) = 4, P'(0) = 0, P''(0) = -4, P'''(0) = 36$
38. $P(0) = -3, P'(0) = -2, P''(0) = 4, P'''(0) = 36$

39. How are the values of A, B, C, and D related to the values of $P(0), P'(0), P''(0)$, and $P'''(0)$?

10.1 SEQUENCES

Sequences play important roles in several areas of theoretical and applied mathematics. As you study additional mathematics you will encounter them again. In this course, however, their role is primarily as a foundation for our study of series ("big polynomials"). In order to understand how and where it is valid to represent a function such as sine as a series, we need to examine what it means to add together an infinite number of values. And in order to understand this infinite addition we need to analyze lists of numbers (called sequences) and determine whether or not the numbers in the list are converging to a single value. This section examines sequences, how to represent sequences graphically, what it means for a sequence to converge, and several techniques to determine if a sequence converges.

Example 1: A person places $100 in an account that pays 8% interest at the end of each year. How much will be in the account at the end of 1 year, 2 years, 3 years, and n years?

Solution: After one year, the total is the principal plus the interest: $100 + (.08)100 = (\mathbf{1.08}) \cdot \mathbf{100} = \$108$.

At the end of the second year, the amount is 108% of the amount at the start of the second year:

$(1.08)\{(1.08)100\} = (\mathbf{1.08})^2 \cdot \mathbf{100} = \$116.64$.

At the end of the third year, the amount is 108% of the amount at the start of the third year:

$(1.08)\{(1.08)^2 \, 100\} = (\mathbf{1.08})^3 \cdot \mathbf{100} = \$125.97$.

These results are shown in Fig. 1. In general, at the end of the nth year, the amount in the account is $(\mathbf{1.08})^n \cdot \mathbf{100}$ dollars.

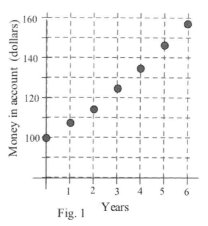
Fig. 1

Practice 1: A layer of protective film transmits two thirds of the light that reaches that layer. How much of the incoming light is transmitted through 1 layer, 2 layers, 3 layers, and n layers? (Fig. 2)

The Example and Practice each asked for a list of numbers in a definite order: a first number, then a second number, and so on. Such a list of numbers in a definite order is called a **sequence**. An infinite sequence is one that just keeps going and has no last number. Often the pattern of a sequence is clear from the first few numbers, but in order to precisely specify a sequence, a rule for finding the value of the nth term, a_n ("a sub n"), in the sequence is usually given.

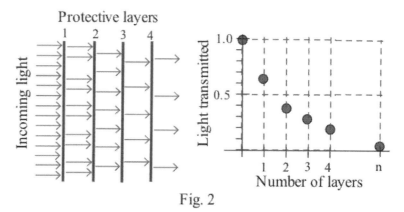
Fig. 2

Example 2: List the next two numbers in each sequence and give a rule for calculating the nth number, a_n:

(a) $1, 4, 9, 16, \ldots$ (b) $-1, 1, -1, 1, \ldots$ (c) $\frac{1}{2}, \frac{1}{4}, \frac{1}{8}, \frac{1}{16}, \ldots$

Solution: (a) $a_5 = 25$, $a_6 = 36$, and $a_n = n^2$. (b) $a_5 = -1$, $a_6 = 1$, and $a_n = (-1)^n$.

(c) $a_5 = \frac{1}{32}$, $a_6 = \frac{1}{64}$, and $a_n = (\frac{1}{2})^n = \frac{1}{2^n}$.

Practice 2: List the next two numbers in each sequence and give a rule for calculating the nth number, a_n:

(a) $1, \frac{1}{2}, \frac{1}{3}, \frac{1}{4}, \ldots$ (b) $\frac{-1}{2}, \frac{1}{4}, \frac{-1}{8}, \frac{1}{16}, \ldots$ (c) $2, 2, 2, 2, \ldots$

Definition and Notation

Since a sequence gives a single value for each integer n, a sequence is a function, but a function whose domain is restricted to the integers.

> **Definition**
>
> A **sequence** is a function whose domain is all integers greater than or equal to a starting integer.

Most of our sequences will have a starting integer of 1, but sometimes it is convenient to start with 0 or another integer value.

Notation: The symbol a_n represents a single number called the nth term.

The symbol $\{ a_n \}$ represents the entire sequence of numbers, the set of all terms.

The symbol $\{ rule \}$ represents the sequence generated by the rule.

The symbol $\{ a_n \}_{n=3}$ represents the sequence that starts with $n = 3$.

Because sequences are functions, we can add, subtract, multiply, and divide them, and we can combine them with other functions to form new sequences. We can also graph sequences, and their graphs can sometimes help us describe and understand their behavior.

Example 3: For the sequences given by $a_n = 3 - \frac{1}{n}$ and $b_n = \frac{1}{2^n}$, graph the points (n, a_n) and (n, b_n) for $n = 1$ to 5. Calculate the first 5 terms of $c_n = a_n + b_n$ and graph the points (n, c_n).

Solution: $c_1 = (3 - \frac{1}{1}) + (\frac{1}{2^1}) = 2.5$, $c_2 = 2.75$, $c_3 \approx 2.792$,

$c_4 = 2.8125$, $c_5 = 2.83125$. The graphs of (n, a_n), (n, b_n), and (n, c_n) are shown in Fig. 3.

Practice 3: For a_n and b_n in the previous example, calculate the first 5 terms of $c_n = a_n - b_n$ and $d_n = (-1)^n b_n$ and graph the points (n, c_n) and (n, d_n).

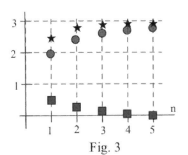

Fig. 3

Recursive Sequences

A recursive sequence is a sequence defined by a rule that gives each new term in the sequence as a combination of some of the previous terms. We already encountered a recursive sequence when we studied Newton's Method for approximating roots of a function (Section 2.7). Newton's method for finding the roots of a function generates a recursive sequence $\{ x_1, x_2, x_3, x_4, ...\}$, as do successive iterations of a function and other operations.

Example 4: Let $f(x) = x^2 - 4$. Take $x_1 = 3$ and apply Newton's method (Section 2.7) to calculate x_2 and x_3. Give a rule for x_n.

Solution: $f(x) = x^2 - 4$ so $f'(x) = 2x$, and, by Newton's method,

$$x_2 = x_1 - \frac{f(x_1)}{f'(x_1)} = 3 - \frac{f(3)}{f'(3)} = 3 - \frac{5}{6} = \frac{13}{6} \approx 2.1667.$$

$$x_3 = x_2 - \frac{f(x_2)}{f'(x_2)} = \frac{13}{6} - \frac{f(13/6)}{f'(13/6)} = \frac{13}{6} - \frac{25}{156} = \frac{313}{156} \approx 2.0064.$$

In general, $x_n = x_{n-1} - \frac{f(x_{n-1})}{f'(x_{n-1})} = x_{n-1} - \frac{(x_{n-1})^2 - 4}{2x_{n-1}}$.

The terms $x_1, x_2, ...$ approach the value 2, one solution of $x^2 - 4 = 0$. The sequence $\{ x_n \}$ is a recursive sequence since each term x_n is defined as a function of the previous term x_{n-1}.

Practice 4: Let $f(x) = 2x - 1$, and define $a_n = f(f(f(...f(a_0)...)))$ where the function is applied n times. Put $a_0 = 3$ and calculate $a_1, a_2,$ and a_3. Note that a_n can be defined recursively as $a_n = f(a_{n-1})$.

Example 5: Let $a_n = 1/2^n$, and define a second sequence $\{ s_n \}$ by the rule that s_n is the **sum** of the first n terms of a_n. Calculate the values of s_n for n = 1 to 5.

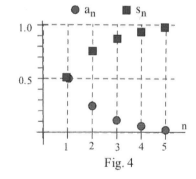

Solution: $s_1 = a_1 = 1/2,$ $s_2 = a_1 + a_2 = 1/2 + 1/4 = 3/4$,
$s_3 = a_1 + a_2 + a_3 = 1/2 + 1/4 + 1/8 = 7/8$, $s_4 = 15/16$, and $s_5 = 31/32$. (Fig. 4)

Fig. 4

You should notice two patterns in these sums.

First, it appears that $s_n = (2^n - 1)/2^n$.

Second, you can simplify the addition process: each term s_n is the sum of the previous term s_{n-1} and the a_n term: $s_n = s_{n-1} + a_n$. We will meet this second pattern again in the next section.

Practice 5: Let $b_0 = 0$ and, for n > 0, define $b_n = b_{n-1} + 1/3^n$. Calculate b_n for n = 1 to 4.

Limits of Sequences: Convergence

Since sequences are discrete functions defined only on integers, some calculus ideas for continuous functions are not applicable to sequences. One type of limit, however, is used: the limit as n approaches infinity. Do the values a_n eventually approach (or equal) some number?

We say that the limit of a sequence $\{a_n\}$ is L if the terms a_n are arbitrarily close to L for sufficiently large values of n — the terms at the beginning of the sequence can be any values, but for large values of n, the a_n terms are all close to L. The following definition puts this idea more precisely.

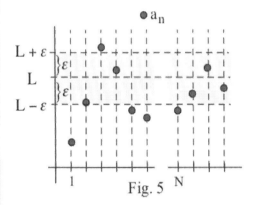

Definition

$\lim\limits_{n \to \infty} a_n = L$ if for any $\varepsilon > 0$ ("epsilon > 0")

there is an index N (typically depending on ε)

so that a_n is within ε of L whenever

n is larger than N :

$n > N$ implies $|a_n - L| < \varepsilon$ (Fig. 5)

If a sequence has a finite limit L, we say that the sequence "converges to L." If a sequence does not have a finite limit, we say the sequence "diverges." Typically a sequence diverges because its terms grow infinitely large (positively or negatively) or because the terms oscillate and do not approach a single number.

Example 7: For $a_n = 3 + 1/n^2$ show that $\lim\limits_{n \to \infty} a_n = 3$.

Solution: We need to show that for any positive ε, there is a number N so that the distance from a_n to L, $|a_n - L|$, is less than ε whenever n is larger than N. For this particular sequence and limit we need to show that for any positive ε, there is a number N so that (Fig. 6)

$|(3 + 1/n^2) - 3| < \varepsilon$ whenever $n > N$.

To determine what N might be, we solve the inequality

$|(3 + 1/n^2) - 3| < \varepsilon$ for n in terms of ε:

$|1/n^2| < \varepsilon$ so $1/\varepsilon < n^2$ and $n > 1/\sqrt{\varepsilon}$.

So for **any** positive ε, we can take $N = 1/\sqrt{\varepsilon}$ (or the next larger integer). Then for $n > N$ we know that

$n > 1/\sqrt{\varepsilon}$ so $1/n^2 < \varepsilon$ and $|(3 + 1/n^2) - 3| < \varepsilon$.

Practice 6: For $a_n = (n+1)/n$ show that $\lim_{n \to \infty} a_n = 1$. (Fig. 7)

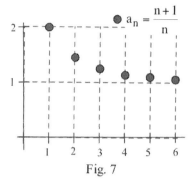

Fig. 7

The limit of a sequence, as n approaches infinity, depends only on the behavior of the terms of the sequence for large values of n (the "tail end") and not on the values of the first few (or few thousand) terms. As a consequence, we can insert or delete any **finite** number of terms without changing the convergence behavior of the sequence.

Sequences are functions, so limits of sequences share many properties with limits of other functions, and we state only a few of them.

Uniqueness Theorem

If a sequence converges to a limit, then the limit is unique.

A sequence can not converge to two different values.

A proof of the Uniqueness Theorem is given after the problem set.

Sometimes it is useful to replace a sequence $\{a_n\}$, a function whose domain is integers, with a function f whose domain is the real numbers so $a_n = f(n)$. If f(x) has a limit as "$x \to \infty$," as x gets arbitrarily large, then f(n) has the same limit as "$n \to \infty$" (Fig. 8). This replacement of "x" with "n" allows us to use earlier results about functions, particularly L'Hopital's Rule, to calculate limits of sequences.

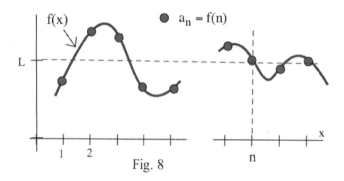

Fig. 8

Theorem: If $a_n = f(n)$ and $\lim_{x \to \infty} f(x) = L$

then $\{a_n\}$ converges to L: $\lim_{n \to \infty} a_n = L$

Example 8: Calculate $\lim_{n \to \infty} \left(1 + \dfrac{2}{n}\right)^n$.

Solution: The terms of the sequence are $a_n = \left(1 + \dfrac{2}{n}\right)^n$, so we can define $f(x) = \left(1 + \dfrac{2}{x}\right)^x$ by replacing the integer values n with real number values x. Then $a_n = f(n)$, and we can use

L'Hopital's rule to get $\lim_{x \to \infty} \left(1 + \frac{2}{x}\right)^x = e^2$ (Section 3.7, Example 6). Finally, we can conclude that

$$\lim_{n \to \infty} \left(1 + \frac{2}{n}\right)^n = e^2 \approx 7.389 \quad \text{(Fig. 9)}.$$

Practice 7: Calculate $\lim_{n \to \infty} \frac{\ln(n)}{n}$.

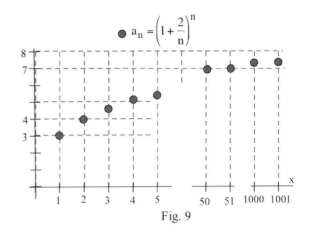

Fig. 9

A **subsequence** is an infinite set of terms from a sequence that occur in the same order as they appear in the original sequence. The sequence of even integers $\{2, 4, 6, \ldots\}$ is a subsequence of the sequence of all positive integers $\{1, 2, 3, 4, \ldots\}$. The sequence of reciprocals of primes $\{1/2, 1/3, 1/5, 1/7, \ldots\}$ is a subsequence of the sequence of the reciprocals of all positive integers $\{1, 1/2, 1/3, 1/4, 1/5, \ldots\}$. Subsequences inherit some properties from their original sequences.

Subsequence Theorem

Every subsequence of a convergent sequence converges to the same limit as the original sequence:

if $\quad \lim_{n \to \infty} a_n = L$ and $\{b_n\}$ is a subsequence of $\{a_n\}$,

then $\quad \lim_{n \to \infty} b_n = L$. (Fig. 10)

If the sequence $\{a_n\}$ does not converge, then the subsequence $\{b_n\}$ may or may not converge.

Corollary: If two subsequences of the same sequence converge to two different limits, then the original sequence diverges.

Example 9: Show that the sequence $\left\{ \frac{(-1)^n n}{n+1} \right\}$ diverges.

Solution: If n is even, then the even terms $a_n = \frac{(-1)^{even} n}{n+1} = \frac{n}{n+1}$ converge to $+1$ so the subsequence of even terms converges to 1.

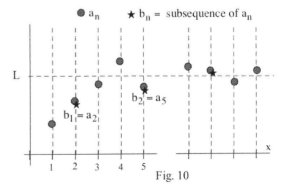

Fig. 10

If n is odd, then the odd terms $a_n = \frac{(-1)^{odd} n}{n+1} = \frac{-n}{n+1}$ converge to -1 so the subsequence of odd terms converges to -1. Finally, since the two subsequences converge to different values, we can conclude that the original sequence $\left\{ \frac{(-1)^n n}{n+1} \right\}$ diverges.

Practice 8: Show that the sequence $\{\sin(n\pi/2)\}$ diverges.

Bounded and Monotonic Sequences

A sequence $\{a_n\}$ is **bounded above** if there is a value A so that $a_n \leq A$ for all values of n: A is called an **upper bound** of the sequence (Fig. 11). Similarly, $\{a_n\}$ is **bounded below** if there is a value B so that $B \leq a_n$ for all n: B is called a **lower bound** of the sequence. A sequence is **bounded** if it has an upper bound and a lower bound. All of the terms of a bounded sequence are between (or equal to) the upper and lower bounds (Fig. 12)

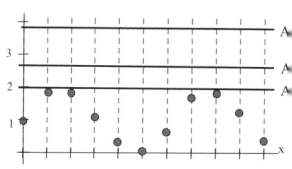

Fig. 11: Several upper bounds of the sequence $a_n = 1 + \sin(n)$

A **monotonically increasing** sequence is a sequence in which each term is greater than or equal to the previous term, $a_1 \leq a_2 \leq a_3 \leq \ldots$ (Fig. 13); a **monotonically decreasing** sequence is one in which each term is less than or equal to the previous term, $a_1 \geq a_2 \geq a_3 \geq \ldots$ (Fig. 14). A monotonic sequence does not oscillate: if one term is larger than a previous term and another term is smaller than a previous term, then the sequence is not monotonic increasing or decreasing.

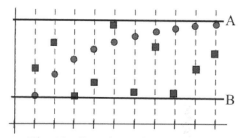

Fig. 12: Two bounded sequences

There are three basic ways to show that a sequence is monotonically increasing:

(i) by showing that $a_{n+1} \geq a_n$ for all n,

(ii) by showing that all the a_n are positive and
$$\frac{a_{n+1}}{a_n} \geq 1 \text{ for all } n, \text{ or}$$

(iii) by showing that $a_n = f(n)$ for integer values n and $f'(x) \geq 0$ for all $x > 0$.

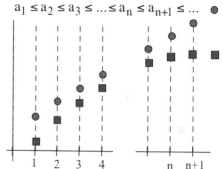

$a_1 \leq a_2 \leq a_3 \leq \ldots \leq a_n \leq a_{n+1} \leq \ldots$

Two monotonic increasing sequences

Fig. 13

Practice 9: List three ways you can show that a sequence is monotonically decreasing.

Example 10: Show that the sequence $a_n = \dfrac{2^n}{n!}$ is monotonically decreasing by showing that $\dfrac{a_{n+1}}{a_n} \leq 1$ for all n.

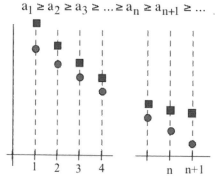

$a_1 \geq a_2 \geq a_3 \geq \ldots \geq a_n \geq a_{n+1} \geq \ldots$

Two monotonic decreasing sequences

Fig. 14

Solution: $a_n = \dfrac{2^n}{n!} = \dfrac{2^n}{1 \cdot 2 \cdot 3 \cdot \ldots \cdot n}$ and $a_{n+1} = \dfrac{2^{n+1}}{(n+1)!} = \dfrac{2^n \cdot 2}{1 \cdot 2 \cdot 3 \cdot \ldots \cdot n \cdot (n+1)}$.

Then $\dfrac{a_{n+1}}{a_n} = \dfrac{2^n \cdot 2}{1 \cdot 2 \cdot 3 \cdot \ldots \cdot n \cdot (n+1)} \cdot \dfrac{1 \cdot 2 \cdot 3 \cdot \ldots \cdot n}{2^n}$ by inverting and multiplying

$= \dfrac{2^n \cdot 2}{2^n} \cdot \dfrac{1 \cdot 2 \cdot 3 \cdot \ldots \cdot n}{1 \cdot 2 \cdot 3 \cdot \ldots \cdot n \cdot (n+1)}$ by reorganizing the top and bottom

$= \dfrac{2}{n+1} \leq 1$ for all positive integers n.

Practice 10: Show that $\left\{ \left(\dfrac{2}{3} \right)^n \right\}$ is monotonically decreasing.

Since the behavior of a monotonic sequence is so regular, it is usually easy to determine if a monotonic sequence has a finite limit: all we need to do is show that it is bounded.

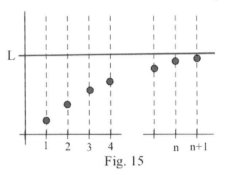
Fig. 15

Monotone Convergence Theorem

If   a monotonic sequence is bounded,

then   the sequence converges. (Fig. 15)

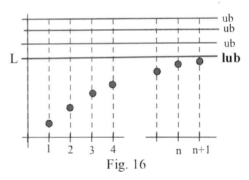
Fig. 16

Idea for a proof for a monotonically increasing sequence that is bounded above:

If the sequence $\{ a_n \}$ is bounded above, then $\{ a_n \}$ has an infinite number of upper bounds (Fig. 16), and each of these upper bounds is larger than every a_n. If L is the smallest of the upper bounds (the least upper bound) of $\{ a_n \}$, then there is a value a_N as close as we want to L (otherwise there would be an upper bound smaller than L). Finally, if a_N is close to L, then the later values, a_n with $n \geq N$, are even closer to L because $\{ a_n \}$ is monotonically increasing, so L is the limit of $\{ a_n \}$.

Cauchy and other mathematicians accepted this theorem on intuitive and geometric grounds similar to the "idea for a proof" given above, but later mathematicians felt more rigor was needed. However, even the mathematician Dedekind who supplied much of that rigor recognized the usefulness of geometric intuition.

"Even now such resort to geometric intuition in a first presentation of differential calculus, I regard as exceedingly useful, from a didactic standpoint, and indeed indispensable if one does not wish to lose too much time." (Dedekind, Essays on the Theory of Numbers, 1901 (Dover, 1963), pp. 1–2)

PROBLEMS In problems 1 – 6, find a rule which describes the given numbers in the sequence.

1. $1, 1/4, 1/9, 1/16, 1/25, \ldots$
2. $1, 1/8, 1/27, 1/64, 1/125, \ldots$
3. $0, 1/2, 2/3, 3/4, 4/5, \ldots$
4. $-1, 1/3, -1/9, 1/27, -1/81, \ldots$
5. $1/2, 2/4, 3/8, 4/16, 5/32, \ldots$
6. $7, 7, 7, 7, 7, \ldots$

(Bonus: O, T, T, F, F, S, S, E, ?, ?)

In problems 7 – 18, calculate the first 6 terms (starting with n = 1) of each sequence and graph these terms.

7. $\left\{ 1 - \dfrac{2}{n} \right\}$
8. $\left\{ 3 + \dfrac{1}{n^2} \right\}$
9. $\left\{ \dfrac{n}{2n-1} \right\}$
10. $\left\{ \dfrac{\ln(n)}{n} \right\}$

11. $\left\{ 3 + \dfrac{(-1)^n}{n} \right\}$
12. $\{ 4 + (-1)^n \}$
13. $\left\{ (-1)^n \dfrac{n-1}{n} \right\}$
14. $\{ \cos(n\pi/2) \}$

15. $\left\{ \dfrac{1}{n!} \right\}$
16. $\left\{ \dfrac{n+1}{n!} \right\}$
17. $\left\{ \dfrac{2^n}{n!} \right\}$
18. $\left\{ \left(1 + \dfrac{1}{n}\right)^n \right\}$

In problems 19 – 24, calculate the first 10 terms (starting with n = 1) of each sequence.

19. $a_1 = 2$ and $a_{n+1} = -a_n$
20. $b_1 = 3$ and $b_{n+1} = 1/b_n$

21. $\left\{ \sin\left(\dfrac{2\pi n}{3}\right) \right\}$
22. $a_1 = 2, a_2 = 3$, and, for $n \geq 3$, $a_n = a_{n-1} - a_{n-2}$

23. c_n = the sum of the first n positive integers

24. d_n = the sum of the first n prime numbers (2 is the first prime)

In problems 25 – 28, state whether each sequence appears to be converging or diverging. If you think the sequence is converging, mark its limit as a value on the vertical axis. (***Important Note:*** *The behavior of a sequence can change drastically after awhile, and the first terms have **no** influence on whether or not the sequence converges. However, sometimes the first few terms are the only values we have, and we need to reach a **tentative** conclusion based on those values*.)

25. Sequences A and B in Fig. 17.
26. Sequences C and D in Fig. 18.

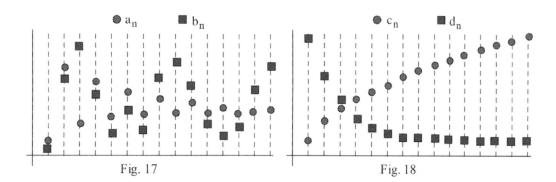

Fig. 17   Fig. 18

27. Sequences E and F in Fig. 19.   28. Sequences G and H in Fig. 20.

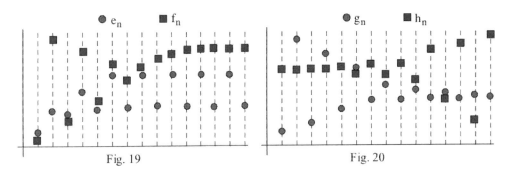

Fig. 19   Fig. 20

In problems 29 – 43, state whether each sequence converges or diverges. If the sequence converges, find its limit.

29. $\{ 1 - \frac{2}{n} \}$ 30. $\{ n^2 \}$ 31. $\{ \frac{n^2}{n+1} \}$ 32. $\{ 3 + \frac{1}{n^2} \}$

33. $\{ \frac{n}{2n-1} \}$ 34. $\{ \frac{\ln(n)}{n} \}$ 35. $\{ \ln(3 + \frac{7}{n}) \}$ 36. $\{ 3 + \frac{(-1)^{n+1}}{n} \}$

37. $\{ 4 + (-1)^n \}$ 38. $\{ (-1)^n \frac{n-1}{n} \}$ 39. $\{ \frac{1}{n!} \}$ 40. $\{ (1 + \frac{3}{n})^n \}$

41. $\{ (1 - \frac{1}{n})^n \}$ 42. $\{ \frac{\sqrt{n} - 3}{\sqrt{n} + 3} \}$ 43. $\{ \frac{(n+2)(n-5)}{n^2} \}$

In problems 44 – 47, prove that the sequence converges to the given limit by showing that for any $\varepsilon > 0$, you can find an N which satisfies the conditions of the definition of convergence.

44. $\lim_{n \to \infty} 2 - \frac{3}{n} = 2$ 45. $\lim_{n \to \infty} \frac{3}{n^2} = 0$

46. $\lim_{n \to \infty} \frac{7}{n+1} = 0$ 47. $\lim_{n \to \infty} \frac{3n-1}{n} = 3$

In problems 48 – 53, use subsequences to help determine whether the sequence converges or diverges. If the sequence converges, find its limit.

48. $\{ (-1)^n 3 \}$ 49. $\{ \frac{1}{n^{\text{th}} \text{ prime}} \}$ 50. $\{ (-1)^n \frac{n+1}{n} \}$

51. $\{ (-2)^n (\frac{1}{3})^n \}$ 52. $\{ (1 + \frac{1}{3n})^{3n} \}$ 53. $\{ (1 + \frac{5}{n^2})^{(n^2)} \}$

In problems 54 – 58, calculate $a_{n+1} - a_n$ and use that value to determine whether $\{ a_n \}$ is monotonic increasing, monotonic decreasing, or neither.

54. $\{ \frac{3}{n} \}$ 55. $\{ 7 - \frac{2}{n} \}$ 56. $\{ \frac{n-1}{2n} \}$ 57. $\{ 2^n \}$ 58. $\{ 1 - \frac{1}{2^n} \}$

In problems 59 – 63, calculate a_{n+1}/a_n and use that value to determine whether each sequence is monotonic increasing, monotonic decreasing, or neither.

59. $\left\{ \dfrac{n+1}{n!} \right\}$ 60. $\left\{ \dfrac{n}{n+1} \right\}$ 61. $\left\{ \left(\dfrac{5}{4}\right)^n \right\}$ 62. $\left\{ \dfrac{n^2}{n!} \right\}$ 63. $\left\{ \dfrac{n}{e^n} \right\}$

In problems 64 – 68, use derivatives to determine whether each sequence is monotonic increasing, monotonic decreasing, or neither.

64. $\left\{ \dfrac{n+1}{n} \right\}$ 65. $\left\{ 5 - \dfrac{3}{n} \right\}$ 66. $\left\{ n \cdot e^{-n} \right\}$ 67. $\left\{ \cos(1/n) \right\}$ 68. $\left\{ \left(1 + \dfrac{1}{n}\right)^3 \right\}$

In problems 69 – 73, show that each sequence is monotonic.

69. $\left\{ \dfrac{n+3}{n!} \right\}$ 70. $\left\{ \dfrac{n}{n+1} \right\}$ 71. $\left\{ 1 - \dfrac{1}{2^n} \right\}$ 72. $\left\{ \sin(1/n) \right\}$ 73. $\left\{ \dfrac{n+1}{e^n} \right\}$

74. The Fibonacci sequence (after Leonardo Fibonacci (1170–1250) who used it to model a population of rabbits) is obtained by setting the first two terms equal to 1 and then defining each new term as the sum of the two previous terms: $a_n = a_{n-1} + a_{n-2}$ for $n \geq 3$. (a) Write the first 7 terms of this sequence. (b) Calculate the successive ratios of the terms, a_n/a_{n-1}. (These ratios approach the "golden mean," approximately 1.618)

75. Heron's method for approximating roots: To approximate the square root of a positive number N, put $a_1 = N$ and let $a_{n+1} = \dfrac{1}{2}\left(a_n + \dfrac{N}{a_n} \right)$. Then $\{ a_n \}$ converges to $\sqrt{N}$. Calculate a_1 through a_4 for $N = 4, 9,$ and 5. (Heron's method is equivalent to Newton's method applied to the function $f(x) = x^2 - N$.)

76. **Hailstone Sequence** For the initial or "seed" value h_0, define the hailstone sequence by the rule

$$h_n = \begin{cases} 3 \cdot h_{n-1} + 1 & \text{if } h_{n-1} \text{ is odd} \\ \dfrac{1}{2} \cdot h_{n-1} & \text{if } h_{n-1} \text{ is even} \end{cases}$$

Define the **length** of the sequence to be the first value of n so that $h_n = 1$. If the seed value is $h_0 = 3$, then $h_1 = 3(3) + 1 = 10, h_2 = (10)/2 = 5, h_3 = 3(5) + 1 = 16, h_4 = 16/2 = 8, h_5 = 4, h_6 = 2,$ and $h_7 = 1$ so the **length** of the hailstone sequence is 7 for the seed value $h_0 = 3$.

(a) Find the length of the hailstone sequence for each seed value from 2 to 10.
(b) Find the length of the hailstone sequence for a seed value $h_0 = 2^n$.

** (c) Open question (no one has been able to answer the this question): Is the length of the hailstone sequence finite for every seed value?

(This is called the hailstone sequence because for some seed values, the terms of the sequence rise and drop just like the path of a hailstone as it forms. This sequence is attributed to Lothar Collatz and (c) is also called Ulam's conjecture, Syracuse's problem, Kakutani's problem and Hasse's algorithm. "The 3n+1 sequence has probably consumed more CPU time than any other number theoretic conjecture," says Gaston Gonnett of Zurich.)

77. **Negative Eugenics:** Suppose that individuals with the gene combination "aa" do not reproduce and those with the combinations "aA" and "AA" do reproduce. When the initial proportion of individuals with "aa" is $a_0 = p$ (typically a small number), then the proportion of individuals with "aa" in the kth generation is $a_k = \dfrac{p}{kp + 1}$. Use this formula for a_k to answer the following questions.

 (a) If 2% of a population initially have the combination "aa" and these individuals do not reproduce, then how many generations will it take for the proportion of individuals with "aa" to drop to 1%?

 (b) In general, find the number of generations until the proportion of individuals with "aa" is half of the initial proportion.

 ("Negative eugenics" is a strategy in which individuals with an undesirable trait are prevented from reproducing. It is not an effective strategy for traits carried by recessive genes (the above example) which are uncommon (p small) in a species which reproduces slowly (people). Mathematics shows that the social strategy of sterilizing people with some undesirable trait, as proposed in the early 20th century, won't effectively reduce the trait in the population.)

78. The fractional part of a number is the number minus its integer part: $x - \text{INT}(x)$. The sequence of fractional parts of multiples of a number x is the sequence with terms $a_n = n \cdot x - \text{INT}(n \cdot x)$. The behavior of the sequence of fractional parts of the multiples of a number is one way in which rational numbers differ from irrational numbers.

 (a) Let $a_n = nx - \text{INT}(nx)$ be the fractional part of the nth multiple of x. Calculate a_1 through a_6 for $x = 1/3$. These are the fractional parts of the first 6 multiples of $1/3$.

 (b) Calculate the fractional parts of the first 9 multiples of 3/4, and 2/5.

 (c) Calculate the fractional parts of the first 5 multiples of π.

 * (d) Let $a_n = n \cdot \pi - \text{INT}(n \cdot \pi)$ be the fractional part of the nth multiple of π. Is it possible for two different multiples of π to have the same fractional part? (Suggestion: Assume the answer is yes and derive a contradiction. Assume that $a_n = a_m$ for some $m \neq n$, and derive the contradiction that $\pi = \dfrac{\text{INT}(n\pi) - \text{INT}(m\pi)}{n-m}$. Why is this a contradiction?)

An Alternate Way to Visualize Sequences and Convergence

A sequence is a function, and we have graphed sequences in the xy plane in the same way we graphed other functions: since $a_n = f(n)$, we plotted the point (n, a_n). If the sequence $\{a_n\}$ converges to L, then the points (n, a_n) are eventually (for big values of n) close to or on the horizontal line $y = L$.

We can also graph a sequence $\{a_n\}$ in one dimension, on the x–axis. For each value of n, we plot the point $x = a_n$. Then the graph of $\{a_n\}$ consists of a collection of points on the x–axis. Fig. 21 shows the one dimensional graphs of $a_n = \frac{1}{n}$, $b_n = 2 + \frac{(-1)^n}{n}$, and $c_n = (-1)^n$.

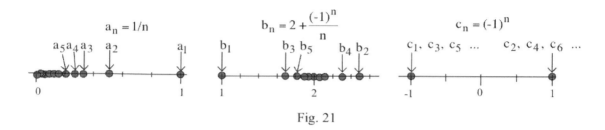

Fig. 21

If $\{a_n\}$ converges to L, then the points $x = a_n$ are eventually (for big values of n) close to or on the point $x = L$. If we build a narrow box, with width $2\varepsilon > 0$, and center the box at the point $x = L$, then all of the points a_n will fall into the box once n is larger than some value N.

79. Suppose that the sequence $\{a_n\}$ converges to 3 and that you place a single grain of sand at each point $x = a_n$ on the x–axis. Describe the likely result (a) after a few grains have been placed and (b) after a lot (thousands or millions) of grains have been placed.

80. Suppose the sequence $\{a_n\}$ converges to 3, $\{b_n\}$ converges to 1, and that you place a single grain of sand at each point $(x, y) = (a_n, b_n)$ on the xy–plane. Describe the likely result (a) after a few grains have been placed and (b) after a lot (thousands or millions) of grains have been placed.

81. Suppose that $a_n = \sin(n)$ for positive integers n. If you place a single grain of sand at each point $x = a_n$ on the x–axis. Describe the likely result (a) after a few grains have been placed and (b) after a lot (thousands or millions) of grains have been placed. (c) Do two grains ever end up on the same point?

82. Suppose that $a_n = \cos(n)$, and $b_n = \sin(n)$ for positive integers n. If you place a single grain of sand at each point $(x, y) = (a_n, b_n)$ on the xy–plane. Describe the likely result (a) after a few grains have been placed and (b) after a lot (thousands or millions) of grains have been placed.

Practice Answers

Practice 1: One layer transmits 2/3 of the original light. Two layers transmit $(\frac{2}{3})(\frac{2}{3}) = (\frac{2}{3})^2$ of the original light. Three layers transmit $(\frac{2}{3})^3$, and, in general, n layers transmit $(\frac{2}{3})^n$ of the original light.

Practice 2: (a) $1, 1/2, 1/3, 1/4, 1/5, 1/6, \ldots, 1/n, \ldots$

(b) $-1/2, 1/4, -1/8, 1/16, -1/32, 1/64, \ldots, (-1/2)^n$ or $(-1)^n (1/2)^n, \ldots$

(c) $2, 2, 2, 2, 2, \ldots, 2, \ldots$

Practice 3: $c_n = a_n - b_n$: $c_1 = (3 - \frac{1}{1}) - (\frac{1}{2}) = \frac{3}{2} = 1.5$, $c_2 = (3 - \frac{1}{2}) - (\frac{1}{2^2}) = \frac{9}{4} = 2.25$,

$c_3 = (3 - \frac{1}{3}) - (\frac{1}{2^3}) = \frac{61}{24} \approx 2.542$, $c_4 = (3 - \frac{1}{4}) - (\frac{1}{2^4}) = \frac{43}{16} \approx 2.687$,

$c_5 = (3 - \frac{1}{5}) - (\frac{1}{2^5}) = \frac{443}{160} \approx 2.769$

$d_n = (-1)^n b_n$: $d_1 = (-1)^1 (\frac{1}{2}) = -\frac{1}{2}$, $d_2 = (-1)^2 \frac{1}{2^2} = \frac{1}{4}$, $d_3 = (-1)^3 (\frac{1}{2^3}) = -\frac{1}{8}$,

$d_4 = \frac{1}{16}$, $d_2 - \frac{1}{32}$

Practice 4: $a_0 = 3$: $a_1 = f(a_0) = 2(3) - 1 = 5$, $a_2 = f(a_1) = 2(5) - 1 = 9$, $a_3 = f(a_2) = 2(9) - 1 = 17$

Practice 5: $b_1 = b_0 + \frac{1}{3} = 0 + \frac{1}{3} = \frac{1}{3}$, $b_2 = b_1 + \frac{1}{9} = \frac{4}{9}$, $b_3 = b_2 + \frac{1}{27} = \frac{13}{27}$, $b_4 = b_3 + \frac{1}{81} = \frac{40}{81}$

Practice 6: For $a_n = (n+1)/n$ show that $\lim_{n \to \infty} a_n = 1$:

We need to show that for any positive ε, there is a number N so that the distance from $a_n = \frac{n+1}{n}$ to L, $|a_n - L|$, is less than ε whenever n is larger than N.

For this particular $|a_n - L| = |\frac{n+1}{n} - 1| = |\frac{n}{n} + \frac{1}{n} - 1| = |\frac{1}{n}|$.

To determine what N might be, we solve the inequality

$|\frac{1}{n}| < \varepsilon$ for n in terms of ε: $|1/n| < \varepsilon$ so $1/\varepsilon < n$ and $n > 1/\varepsilon$.

So for **any** positive ε, we can take $N = 1/\varepsilon$ (or any larger number).

Then $n > N = 1/\varepsilon$ implies that $\varepsilon > \frac{1}{n} = |\frac{1}{n}| = |\frac{n+1}{n} - 1| = |a_n - L|$.

Practice 7: $\lim\limits_{n\to\infty} \dfrac{\ln(n)}{n}$: Since $\lim\limits_{n\to\infty} \ln(x) = \infty$ and $\lim\limits_{n\to\infty} x = \infty$, we can use L'Hopital's rule

(Section 3.7). $D(\ln(x)) = \dfrac{1}{x}$ and $D(x) = 1$, so $\lim\limits_{n\to\infty} \dfrac{\ln(n)}{n} = \lim\limits_{x\to\infty} \dfrac{1/x}{1} = 0$.

Then $\lim\limits_{x\to\infty} \dfrac{\ln(x)}{x} = 0$ and $\lim\limits_{n\to\infty} \dfrac{\ln(n)}{n} = 0$.

Practice 8: We can show that $a_n = \{\sin(\dfrac{\pi n}{2})\}$ diverges by finding two subsequences b_n and c_n of a_n so that the subsequences b_n and c_n converge to different limiting values.

Let $\{b_n\}$ consist of the terms $\{a_1, a_5, a_9, \ldots, a_{4n-3}, \ldots\} = \{\sin(\pi/2), \sin(5\pi/2),$ $\sin(9\pi/2), \ldots, \sin(\dfrac{(4n-3)\pi}{2}), \ldots\} = \{1, 1, 1, \ldots, 1, \ldots\}$. Then $\lim\limits_{n\to\infty} b_n = 1$.

Let $\{c_n\}$ consist of the terms $\{a_2, a_4, a_6, \ldots, a_{2n}, \ldots\} = \{\sin(2\pi/2), \sin(4\pi/2),$ $\sin(6\pi/2), \ldots, \sin(\dfrac{2n\pi}{2}), \ldots\} = \{0, 0, 0, \ldots, 0, \ldots\}$. Then $\lim\limits_{n\to\infty} c_n = 0$.

Since the subsequences $\{b_n\}$ and $\{c_n\}$ have different limits, we can conclude that the original sequence $\{a_n\}$ diverges. (Note: Many other pairs of subsequences also work in place of the $\{b_n\}$ and $\{c_n\}$ that we used.)

Practice 9: (i) by showing that $a_{n+1} \leq a_n$ for all n,

(ii) by showing that all the a_n are positive and $\dfrac{a_{n+1}}{a_n} \leq 1$ for all n, or

(iii) by showing that $a_n = f(n)$ for integer values n and $f'(x) \leq 0$ for all $x \geq 1$.

Practice 10: Show that $\{(\dfrac{2}{3})^n\}$ is monotonically decreasing.

Using method (i) of Practice 9: $(\dfrac{2}{3}) < 1$ so multiplying each side by $(\dfrac{2}{3})^n > 0$ we have $(\dfrac{2}{3})(\dfrac{2}{3})^n < 1(\dfrac{2}{3})^n$ and $(\dfrac{2}{3})^{n+1} < (\dfrac{2}{3})^n$ so $a_{n+1} < a_n$.

We could have used method (ii) of Practice 9 instead: $a_{n+1} = (\dfrac{2}{3})^{n+1}$ and $a_n = (\dfrac{2}{3})^n$ so $\dfrac{a_{n+1}}{a_n} = \dfrac{(2/3)^{n+1}}{(2/3)^n} = \dfrac{2}{3} < 1$ so $\{(\dfrac{2}{3})^n\}$ is monotonically decreasing.

Appendix: Proof of the Uniqueness Theorem

The proof starts by assuming that the limit is not unique. Then we show that this assumption of nonuniqueness leads to a contradiction so the assumption is false and the limit is unique.

Suppose that a sequence $\{a_n\}$ converges to two different limits L_1 and L_2. Then the distance between L_1 and L_2 is $d = |L_1 - L_2| > 0$. Since $\{a_n\}$ converges to L_1, then for any $\varepsilon > 0$ there is an N_1 so that $n > N_1$ implies $|a_n - L_1| < \varepsilon$. Take $\varepsilon = d/3$. Then there is an N_1 so $n > N_1$ implies $|a_n - L_1| < \varepsilon = d/3$. Similarly, there is an N_2 so that $n > N_2$ implies $|a_n - L_2| < \varepsilon = d/3$. Finally, if n is larger than both N_1 and N_2, then both conditions are satisfied and

$$d = |L_1 - L_2| = |(a_n - L_2) - (a_n - L_1)| \quad \text{by adding } 0 = a_n - a_n \text{ inside the absolute value}$$
$$\leq |a_n - L_2| + |a_n - L_1| \quad \text{by the Triangle Inequality}$$
$$< d/3 + d/3 \quad \text{since } |a_n - L_1| < \varepsilon = d/3 \text{ and } |a_n - L_2| < \varepsilon = d/3$$
$$= 2d/3.$$

We have found that $0 < d < \frac{2}{3}d$, a contradiction, and we can conclude the assumption, $L_1 \neq L_2$, is false. A sequence **can not** converge to two different values.

(Note: *We chose $\varepsilon = d/3$ because it "works" for our purpose by leading to a contradiction. Several other choices, any ε less than $d/2$, also lead to the contradiction. Since the definition says "for **any** $\varepsilon > 0$," we picked one we wanted.*)

10.2 INFINITE SERIES

Our goal in this section is to add together the numbers in a sequence. Since it would take a "very long time" to add together the infinite number of numbers, we first consider finite sums, look for patterns in these finite sums, and take limits as more and more numbers are included in the finite sums.

What does it mean to add together an infinite number of terms? We will define that concept carefully in this section. Secondly, is the sum of all the terms a finite number? In the next few sections we will examine a variety of techniques for determining whether an infinite sum is finite. Finally, if we know the sum is finite, can we determine the value of the sum? The difficulty of finding the exact value of the sum varies from very easy to very, very difficult.

Example 1: A golf ball is thrown 9 feet straight up into the air, and on each bounce it rebounds to two thirds of its previous height (Fig. 1). Find a sequence whose terms give the distances the ball travels during each successive bounce. Represent the **total** distance traveled by the ball as a sum.

(This is **not** the ball's path. The actual motion of the ball is straight up and down.)

Solution: The heights of the successive bounces are 9 feet, $(\frac{2}{3}) \cdot 9$ feet, $(\frac{2}{3}) \cdot [(\frac{2}{3}) \cdot 9]$ feet, $(\frac{2}{3})^3 \cdot 9$ feet, and so forth. On each bounce, the ball rises and falls so the distance traveled is twice the height of that bounce:

$$18 \text{ feet}, (\frac{2}{3}) \cdot 18 \text{ feet}, (\frac{2}{3}) \cdot (\frac{2}{3}) \cdot 18 \text{ feet}, (\frac{2}{3})^3 \cdot 18 \text{ feet}, (\frac{2}{3})^4 \cdot 18 \text{ feet}, \ldots .$$

The total distance traveled is the sum of the bounce–distances:

$$\text{total distance} = 18 + (\frac{2}{3}) \cdot 18 + (\frac{2}{3}) \cdot (\frac{2}{3}) \cdot 18 + (\frac{2}{3})^3 \cdot 18 + (\frac{2}{3})^4 \cdot 18 + \ldots$$

$$= 18 \{ 1 + \frac{2}{3} + (\frac{2}{3})^2 + (\frac{2}{3})^3 + (\frac{2}{3})^4 + \ldots \}$$

At the completion of the first bounce the ball has traveled 18 feet. After the second bounce, it has traveled 30 feet, a total of 38 feet after the third bounce, $43\frac{1}{3}$ feet after the fourth, and so on. With a calculator and some patience, we see that after the 20th bounce the ball has traveled a total of approximately 53.996 feet, after the 30th bounce approximately 53.99994 feet, and after the 40th bounce approximately 53.9999989 feet.

Practice 1: A tennis ball is thrown 10 feet straight up into the air, and on each bounce it rebounds to 40% of its previous height. Represent the total distance traveled by the ball as a sum, and find the total distance traveled by the ball after the completion of its third bounce. (Fig. 2)

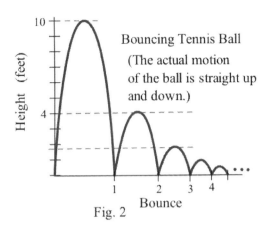

Fig. 2

Infinite Series

The infinite sums in the Example and Practice are called infinite series, and they are the objects we will start to examine in this section.

Definitions

An **infinite series** is an expression of the form

$$a_1 + a_2 + a_3 + a_4 + \ldots \quad \text{or} \quad \sum_{k=1}^{\infty} a_k .$$

The numbers $a_1, a_2, a_3, a_4, \ldots$ are called the **terms** of the series. (Fig. 3)

Example 2: Represent the following series using the sigma notation. (a) $1 + 1/3 + 1/9 + 1/27 + \ldots$,
(b) $-1 + 1/2 - 1/3 + 1/4 - 1/5 + \ldots$, (c) $18(2/3 + 4/9 + 8/27 + 16/81 + \ldots)$
(d) $0.777\ldots = 7/10 + 7/100 + 7/1000 + \ldots$, and (e) $0.222\ldots$

Solution: (a) $1 + 1/3 + 1/9 + 1/27 + \ldots = \sum_{k=0}^{\infty} (\tfrac{1}{3})^k \quad \text{or} \quad \sum_{k=1}^{\infty} (\tfrac{1}{3})^{k-1}$

(b) $-1 + 1/2 - 1/3 + 1/4 - 1/5 + \ldots = \sum_{k=1}^{\infty} (-1)^k \tfrac{1}{k}$ \qquad (c) $18 \sum_{k=1}^{\infty} (\tfrac{2}{3})^k$

(d) $0.777\ldots = 7/10 + 7/100 + 7/1000 + \ldots = \sum_{k=1}^{\infty} \tfrac{7}{10^k}$ \qquad (e) $\sum_{k=1}^{\infty} \tfrac{2}{10^k}$

Practice 2: Represent the following series using the sigma notation. (a) $1 + 2 + 3 + 4 + \ldots$,
(b) $-1 + 1 - 1 + 1 - \ldots$ \qquad (c) $2 + 1 + 1/2 + 1/4 + \ldots$
(d) $1/2 + 1/4 + 1/6 + 1/8 + 1/10 + \ldots$ \qquad (e) $0.111\ldots$

In order to determine if the infinite series adds up to a finite value, we examine the sums as more and more terms are added.

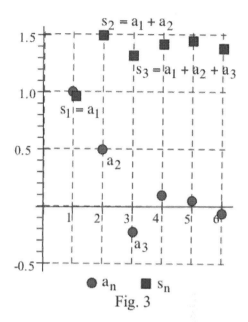

Definition

The **partial sums** s_n of the infinite series $\sum_{k=1}^{\infty} a_k$ are the numbers

$s_1 = a_1$,
$s_2 = a_1 + a_2$,
$s_3 = a_1 + a_2 + a_3$,
...

In general, $s_n = a_1 + a_2 + a_3 + \ldots + a_n = \sum_{k=1}^{n} a_k$

or, recursively, as $s_n = s_{n-1} + a_n$.

The partial sums form the **sequence of partial sums** $\{s_n\}$.

Fig. 3

Example 3: Calculate the first 4 partial sums for the following series.

(a) $1 + 1/2 + 1/4 + 1/8 + 1/16 + \ldots$, (b) $\sum_{k=1}^{\infty} (-1)^k$, and (c) $\sum_{n=1}^{\infty} \frac{1}{n}$.

Solution: (a) $s_1 = 1, s_2 = 1 + 1/2 = 3/2, s_3 = 1 + 1/2 + 1/4 = 7/4, s_4 = 1 + 1/2 + 1/4 + 1/8 = 15/8$

It is usually easier to use the recursive version of s_n:

$s_3 = s_2 + a_3 = 3/2 + 1/4 = 7/4$ and $s_4 = s_3 + a_4 = 7/4 + 1/8 = 15/8$.

(b) $s_1 = (-1)^1 = -1$, $s_2 = s_1 + a_2 = -1 + (-1)^2 = 0$, $s_3 = s_2 + a_3 = 0 + (-1)^3 = -1$, $s_4 = 0$.

(c) $s_1 = 1, s_2 = 3/2, s_3 = 11/6, s_4 = 25/12$.

Practice 3: Calculate the first 4 partial sums for the following series.

(a) $1 - 1/2 + 1/4 - 1/8 + 1/16 - \ldots$, (b) $\sum_{k=1}^{\infty} (\frac{1}{3})^k$, and (c) $\sum_{n=2}^{\infty} \frac{(-1)^n}{n}$.

If we know the values of the partial sums s_n, we can recover the values of the terms a_n used to build the s_n.

Example 4: Suppose $s_1 = 2.1$, $s_2 = 2.6$, $s_3 = 2.84$, and $s_4 = 2.87$ are the first partial sums of $\sum_{k=1}^{\infty} a_k$. Find the values of the first four terms of $\{a_n\}$.

Solution: $s_1 = a_1$ so $a_1 = 2.1$. $s_2 = a_1 + a_2$ so $2.6 = 2.1 + a_2$ and $a_2 = 0.5$.

Similarly, $s_3 = a_1 + a_2 + a_3$ so $2.84 = 2.1 + 0.5 + a_3$ and $a_3 = 0.24$. Finally, $a_4 = 0.03$.

An alternate solution method starts with $a_1 = s_1$ and then uses the fact that $s_n = s_{n-1} + a_n$ so $a_n = s_n - s_{n-1}$. Then

$$a_2 = s_2 - s_1 = 2.6 - 2.1 = 0.5 .$$
$$a_3 = s_3 - s_2 = 2.84 - 2.6 = 0.24, \text{ and}$$
$$a_4 = s_4 - s_3 = 2.87 - 2.84 = 0.03 .$$

Practice 4: Suppose $s_1 = 3.2$, $s_2 = 3.6$, $s_3 = 3.5$, $s_4 = 4$, $s_{99} = 7.3$, $s_{100} = 7.6$, and $s_{101} = 7.8$ are partial sums of $\sum_{k=1}^{\infty} a_k$. Find the values of a_1, a_2, a_3, a_4, and a_{100}.

Example 5: Graph the first five **terms** of the series $\sum_{k=1}^{\infty} (\frac{1}{2})^k$. Then graph the first five **partial sums**.

Solution:
$$a_1 = \left(\frac{1}{2}\right)^1 = \frac{1}{2}, \; a_2 = \left(\frac{1}{2}\right)^2 = \frac{1}{4}, \; a_3 = \left(\frac{1}{2}\right)^3 = \frac{1}{8}, \; a_4 = \frac{1}{16}, \; a_4 = \frac{1}{32}$$

$$s_1 = \frac{1}{2}, \; s_2 = \frac{1}{2} + \frac{1}{4} = \frac{3}{4}, \; s_3 = \frac{1}{2} + \frac{1}{4} + \frac{1}{8} = \frac{7}{8}, \; s_4 = \frac{15}{16}, \text{ and } s_5 = \frac{31}{32}.$$

These values are graphed in Fig. 4.

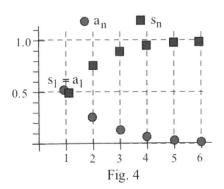

Fig. 4

Practice 5: Graph the first five terms of the series

$$\sum_{k=1}^{\infty} \left(-\frac{1}{2}\right)^k .$$ Then graph the first five partial sums.

Convergence of a Series

The convergence of a series is defined in terms of the behavior of the sequence of partial sums. If the partial sums, the sequence obtained by adding more and more of the terms of the series, approaches a finite number, we say the **series** converges to that finite number. If the sequence of partial sums diverges (does not approach a single finite number), we say that the **series** diverges.

Definitions

Let $\{s_n\}$ be the sequence of partial sums

of the series $\sum_{k=1}^{\infty} a_k$: $s_n = \sum_{k=1}^{n} a_k$

If $\{s_n\}$ is a convergent sequence, (Fig. 5)

we say the series $\sum_{k=1}^{\infty} a_k$ **converges**.

If the sequence of partial sums $\{s_n\}$ converges to A,

we say the series $\sum_{k=1}^{\infty} a_k$ **converges to A**

or **the sum of the series is A**,

and we write $\sum_{k=1}^{\infty} a_k = A$.

If the sequence of partial sums $\{s_n\}$ diverges,

we say the series $\sum_{k=1}^{\infty} a_k$ **diverges**.

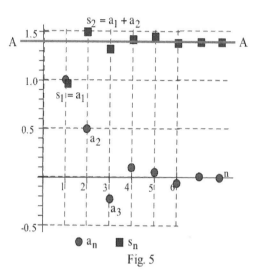

Fig. 5

Example 6: In the next section we present a method for determining that the nth partial sum

of $\sum_{k=1}^{\infty} (\frac{1}{2})^k$ is $s_n = \frac{2^n - 1}{2^n} = 1 - \frac{1}{2^n}$. Use this result to evaluate the limit of $\{s_n\}$.

Does the series $\sum_{k=1}^{\infty} (\frac{1}{2})^k$ converge? If so, to what value?

Solution: $\lim_{n \to \infty} s_n = \lim_{n \to \infty} 1 - \frac{1}{2^n} = 1$ (Fig. 6), so $\sum_{k=1}^{\infty} (\frac{1}{2})^k$ converges to 1: $\sum_{k=1}^{\infty} (\frac{1}{2})^k = 1$.

Practice 6: The nth partial sum of $\sum_{k=1}^{\infty} (-\frac{1}{2})^k$ is $s_n = -\frac{1}{3} + \frac{1}{3} \cdot (-\frac{1}{2})^n$. Use this result to

evaluate the limit of $\{s_n\}$. Does the series $\sum_{k=1}^{\infty} (-\frac{1}{2})^k$ converge? If so, to what value?

The next theorem says that if a series converges, then the terms of the series must approach 0. When a series is convergent then the partial sums s_n approach a finite limit (Fig. 3) so all of the s_n must be close to that limit when n is large. Then s_n and s_{n-1} must be close to each other (why?) and $a_n = s_n - s_{n-1}$ must be close to 0.

Theorem: If the series $\sum_{k=1}^{\infty} a_k$ converges, then $\lim_{k \to \infty} a_k = 0$.

We can NOT use this theorem to conclude that a series converges. If the terms of the series do approach 0, then the series may or may not converge — more information is needed to draw a conclusion. However, an alternate form of the theorem, called the nth Term Test for Divergence, is **very useful for quickly concluding that some series diverge.**

---

n^{th} Term Test for Divergence of a Series

If the terms a_n of a series do not approach 0 (as "$n \to \infty$")

then the series diverges:

if $\lim_{n \to \infty} a_n \neq 0$, then $\sum_{n=1}^{\infty} a_n$ diverges.

Example 7: Which of these series diverge **by the nth Term Test**?

(a) $\sum_{n=1}^{\infty} (-1)^n$ (b) $\sum_{n=1}^{\infty} (\frac{3}{4})^n$ (c) $\sum_{n=1}^{\infty} (1 + \frac{1}{n})^n$ (d) $\sum_{n=1}^{\infty} \frac{1}{n}$

Solution: (a) $a_n = (-1)^n$ oscillates between -1 and $+1$ and does not approach 0. $\sum_{n=1}^{\infty} (-1)^n$ **diverges**.

(b) $a_n = (\frac{3}{4})^n$ approaches 0 so $\sum_{n=1}^{\infty} (\frac{3}{4})^n$ **may or may not** converge.

(c) $a_n = (1 + \frac{1}{n})^n$ approaches $e \neq 0$, so $\sum_{n=1}^{\infty} (1 + \frac{1}{n})^n$ **diverges**.

(d) $a_n = \frac{1}{n}$ approaches 0 so $\sum_{n=1}^{\infty} \frac{1}{n}$ **may or may not** converge.

We can be certain that (a) and (c) diverge. We don't have enough information yet to decide about (b) and (d). (In the next section we show that (b) converges and (d) diverges.)

Practice 7: Which of these series diverge by the nth Term Test?

(a) $\sum_{n=1}^{\infty} (-0.9)^n$ (b) $\sum_{n=1}^{\infty} (1.1)^n$ (c) $\sum_{n=1}^{\infty} \sin(n\pi)$ (d) $\sum_{n=1}^{\infty} \frac{1}{\sqrt{n}}$

New Series From Old

If we know about the convergence of a series, then we also know about the convergence of several related series.

- Inserting or deleting a "few" terms, any **finite** number of terms, does not change the convergence or divergence of a series. The insertions or deletions typically change the sum (the limit of the partial sums), but they do not change whether or not the series converges. (Inserting or deleting an infinite number of terms can change the convergence or divergence.)

- Multiplying each term in a series by a nonzero constant does not change the convergence or divergence of a series:

  Suppose $c \neq 0$. $\sum_{n=1}^{\infty} a_n$ converges if and only if $\sum_{n=1}^{\infty} c \cdot a_n$ converges.

- Term–by–term addition and subtraction of the terms of two convergent series result in convergent series. (Term by term multiplication and division of series do not have such nice results.)

Theorem:

If $\sum_{n=1}^{\infty} a_n$ and $\sum_{n=1}^{\infty} b_n$ converge with $\sum_{n=1}^{\infty} a_n = A$ and $\sum_{n=1}^{\infty} b_n = B$,

then $\sum_{n=1}^{\infty} C \cdot a_n = C \cdot A$,

$\sum_{n=1}^{\infty} (a_n + b_n) = A + B$, and

$\sum_{n=1}^{\infty} (a_n - b_n) = A - B$.

The proofs of these statements follow directly from the definition of convergence of a series and from results about convergence of sequences (of partial sums).

PROBLEMS

In problems 1 – 6, rewrite each sum using sigma notation starting with $k = 1$.

1. $1 + \frac{1}{2} + \frac{1}{3} + \frac{1}{4} + \frac{1}{5} + ...$

2. $1 + \frac{1}{4} + \frac{1}{9} + \frac{1}{16} + \frac{1}{25} + ...$

3. $\frac{2}{3} + \frac{2}{6} + \frac{2}{9} + \frac{2}{12} + \frac{2}{15} + \frac{2}{18} + ...$

4. $\sin(1) + \sin(8) + \sin(27) + \sin(64) + \sin(125) + ...$

5. $(-\frac{1}{2}) + (\frac{1}{4}) + (-\frac{1}{8}) + (\frac{1}{16}) + (-\frac{1}{32}) + ...$

6. $(-\frac{1}{3}) + (\frac{1}{9}) + (-\frac{1}{27}) + (\frac{1}{81}) + (-\frac{1}{243}) + ...$

In problems 7 – 14, calculate and graph the first four partial sums s_1 to s_4 of the given series $\sum_{n=1}^{\infty} a_n$.

7. $\sum_{n=1}^{\infty} n^2$

8. $\sum_{n=1}^{\infty} (-1)^n$

9. $\sum_{n=1}^{\infty} \frac{1}{n+2}$

10. $\sum_{n=1}^{\infty} \left\{ \frac{1}{n} - \frac{1}{n+1} \right\}$

11. $\sum_{n=1}^{\infty} \frac{1}{2^n}$

12. $\sum_{n=1}^{\infty} (-\frac{1}{2})^n$

In problems 13 – 18, the first five partial sums s_1 to s_5 are given. Find the first four terms a_1 to a_4 of the series.

13. $s_1 = 3$, $s_2 = 2$, $s_3 = 4$, $s_4 = 5$, $s_5 = 3$

14. $s_1 = 3$, $s_2 = 5$, $s_3 = 4$, $s_4 = 6$, $s_5 = 5$

15. $s_1 = 4$, $s_2 = 4.5$, $s_3 = 4.3$, $s_4 = 4.8$, $s_5 = 5$

16. $s_1 = 4$, $s_2 = 3.7$, $s_3 = 3.9$, $s_4 = 4.1$, $s_5 = 4$

17. $s_1 = 1$, $s_2 = 1.1$, $s_3 = 1.11$, $s_4 = 1.111$, $s_5 = 1.1111$

18. $s_1 = 1$, $s_2 = 0.9$, $s_3 = 0.93$, $s_4 = 0.91$, $s_5 = 0.92$

In problems 19 – 28, represent each repeating decimal as a series using the sigma notation.

19. 0.888 ...
20. 0.333 ...
21. 0.555 ...
22. 0.111 ...
23. 0.aaa ...

24. 0.232323 ...
25. 0.171717 ...
26. 0.838383 ...
27. 0.070707 ...
28. 0.ababab ...

29. Find a pattern for a fraction representation of the repeating decimal 0.abcabcabc ... .

30. A golf ball is thrown 20 feet straight up into the air, and on each bounce it rebounds to 60% of its previous height. Represent the total distance traveled by the ball as a sum.

31. A "super ball" is thrown 15 feet straight up into the air, and on each bounce it rebounds to 80% of its previous height. Represent the total distance traveled by the ball as a sum.

32. Each special washing of a pair of overalls removes 80% of the radioactive particles attached to the overalls. Represent, as a sequence of numbers, the percent of the original radioactive particles that remain after each washing.

33. Each week, 20% of the argon gas in a container leaks out of the container. Represent, as a sequence of numbers, the percent of the original argon gas that remains in the container at the end of the 1st, 2nd, 3rd, and n^{th} weeks.

34. Eight people are going on an expedition by horseback through desolate country. The people and scientific equipment (fishing gear) require 12 horses, and additional horses are needed to carry food for the horses. Each horse can carry enough food to feed 2 horses for the trip. Represent the number of horses needed to carry food as a sum. (Start of a solution: The original 12 horses will require 6 new horses to carry their food. The 6 new horses require 3 additional horses to carry their food. The 3 additional horses require another 1.5 horses to carry food for them, etc.)

Which of the series in problems 35 – 43, definitely diverge by the nth Term Test? What can we conclude about the other series in these problems?

35. $\sum_{n=1}^{\infty} \left(\frac{1}{4}\right)^n$

36. $\sum_{n=1}^{\infty} \frac{7}{n}$

37. $\sum_{n=1}^{\infty} \left(\frac{4}{3}\right)^n$

38. $\sum_{n=1}^{\infty} \left(-\frac{7}{4}\right)^2$

39. $\sum_{n=1}^{\infty} \frac{\sin(n)}{n}$

40. $\sum_{n=1}^{\infty} \frac{\ln(n)}{n}$

41. $\sum_{n=1}^{\infty} \cos\left(\frac{1}{n}\right)$

42. $\sum_{n=1}^{\infty} \frac{n^2-20}{n^2+4}$

43. $\sum_{n=1}^{\infty} \frac{n^2-20}{n^5+4}$

Practice Answers

Practice 1: The heights of the bounces are $10, (0.4) \cdot 10, (0.4) \cdot (0.4) \cdot 10, (0.4)^3 \cdot 10, \ldots$ so the distances traveled (up and down) by the ball are $20, (0.4) \cdot 20, (0.4) \cdot (0.4) \cdot 20, (0.4)^3 \cdot 20, \ldots$
The total distance traveled is

$$20 + (0.4) \cdot 20 + (0.4)^2 \cdot 20 + (0.4)^3 \cdot 20 + \ldots = 20\{1 + 0.4 + (0.4)^2 + (0.4)^3 + \ldots\} = 20 \sum_{k=0}^{\infty} (0.4)^k$$

After 3 bounces the ball has traveled $20 + (0.4)(20) + (0.4)^2(20) = 20 + 8 + 3.2 = 31.2$ feet.

Practice 2: (a) $\sum_{k=1}^{\infty} k$ (b) $\sum_{k=1}^{\infty} (-1)^k$

(c) $2(1 + \frac{1}{2} + \frac{1}{4} + ...) = 2\sum_{k=0}^{\infty}(\frac{1}{2})^k$ or $\sum_{k=0}^{\infty}(\frac{1}{2})^{k-1}$ or $\sum_{k=1}^{\infty}(\frac{1}{2})^{k-2}$

(d) $\sum_{k=1}^{\infty} \frac{1}{2k}$ (e) $\frac{1}{10} + \frac{1}{100} + \frac{1}{1000} + ... = \frac{1}{10} + \frac{1}{10^2} + \frac{1}{10^3} + ... = \sum_{k=1}^{\infty}(\frac{1}{10})^k$ or $\sum_{k=1}^{\infty} \frac{1}{10^k}$

Practice 3: (a) Partial sums: 1, 1/2, 3/4, 5/8 (b) $\frac{1}{3} + \frac{1}{9} + \frac{1}{27} + ...$; partial sums: $\frac{1}{3}, \frac{4}{9}, \frac{13}{27}, \frac{40}{81}$

(c) $\frac{1}{2} - \frac{1}{3} + \frac{1}{4} - ...$; partial sums: $\frac{1}{2}, \frac{1}{6}, \frac{10}{24} = \frac{5}{12}, \frac{13}{60}$

Practice 4: $a_1 = s_1 = 3.2$, $a_2 = s_2 - s_1 = (3.6) - (3.2) = 0.4$, $a_3 = s_3 - s_2 = (3.5) - (3.6) = -0.1$,

$a_4 = s_4 - s_3 = (4) - (3.5) = 0.5$, $a_{100} = s_{100} - s_{99} = (7.6) - (7.3) = 0.3$

Practice 5: $a_1 = -1/2, a_2 = 1/4, a_3 = -1/8, a_4 = 1/16, a_5 = -1/32$

$s_1 = a_1 = -\frac{1}{2}$, $s_2 = a_1 + a_2 = -\frac{1}{2} + \frac{1}{4} = -\frac{1}{4}$,

$s_3 = s_2 + a_3 = -\frac{1}{4} - \frac{1}{8} = -\frac{3}{8} \approx -0.375$,

$s_4 = s_3 + a_4 = -\frac{3}{8} + \frac{1}{16} = -\frac{5}{16} \approx -0.3125$,

$s_5 = s_4 + a_5 = -\frac{5}{16} - \frac{1}{32} = -\frac{11}{32} \approx -0.34375$,

The graphs of a_n and s_n are shown in Fig. 6.

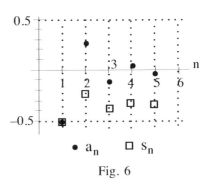

Fig. 6

Practice 6: $\lim_{n \to \infty} s_n = \lim_{n \to \infty} -\frac{1}{3} + \frac{1}{3} \cdot (-\frac{1}{2})^n = -\frac{1}{3} + \frac{1}{3} \cdot 0$

$= -\frac{1}{3}$.

The limit, as "$n \to \infty$", of s_n is a finite number, so $\sum_{k=1}^{\infty}(-\frac{1}{2})^k$ **converges to** $-\frac{1}{3}$.

Practice 7: (a) $\sum_{n=1}^{\infty}(-0.9)^n$: $a_n = (-0.9)^n$ approaches 0 so $\sum_{n=1}^{\infty}(-0.9)^n$ **may converge**.

(b) $\sum_{n=1}^{\infty}(1.1)^n$: $a_n = (1.1)^n$ "approaches infinity" so $\sum_{n=1}^{\infty}(1.1)^n$ **diverges.**

(c) $\sum_{n=1}^{\infty} \sin(n\pi) = 0 + 0 + 0 + ...$: $\sin(n\pi) = 0$ "approaches 0" so $\sum_{n=1}^{\infty} \sin(n\pi)$ **may converge**.

(d) $\sum_{n=1}^{\infty} \frac{1}{\sqrt{n}}$: $a_n = \frac{1}{\sqrt{n}}$ approaches 0 so $\sum_{n=1}^{\infty} \frac{1}{\sqrt{n}}$ **may converge**.

(Later in this chapter we will show that series (a) and (c) converge and series (d) diverges.)

Appendix: Programming Partial Sums of Numerical Series

MAPLE commands for $\sum_{n=1}^{100}\frac{1}{n}$, $\sum_{n=1}^{100}\frac{1}{n^2}$, $\sum_{n=1}^{100}\frac{1}{n!}$:

> sum(1/n , n=1..100) ;   (then press ENTER key)

> sum(1/n^2 , n=1..100) ;

> sum(1/(n!) , n=1..100) ;

TI–85 program for $\sum_{n=1}^{M}\frac{1}{n}$

```
Prgm1:NUMSUM
Disp "NUM TERMS ="
Input M
1 → A
0 → N
Lbl ONE
N+1 → N
1/N → A          (value of the new term)            1/N → A
S + A → S        (add new term to partial sum)      S + A → S
If N<M           (test if need next term)           If N<M
Goto ONE
Disp S
Stop                                                 Stop
```

For $\sum_{n=1}^{M}\frac{1}{n^2}$ change the bold line to **1/(N*N) → A** . For $\sum_{n=1}^{M}\frac{1}{n!}$ change the bold line to **A/N → A** .

10.3 GEOMETRIC AND HARMONIC SERIES

This section uses ideas from Section 10.2 about series and their convergence to investigate some special types of series. Geometric series are very important and appear in a variety of applications. Much of the early work in the 17th century with series focused on geometric series and generalized them. Many of the ideas used later in this chapter originated with geometric series. It is easy to determine whether a geometric series converges or diverges, and when one does converge, we can easily find its sum. The harmonic series is important as an example of a divergent series whose terms approach zero. A final type of series, called "telescoping," is discussed briefly. Telescoping series are relatively uncommon, but their partial sums exhibit a particularly nice pattern.

Geometric Series: $\sum_{k=0}^{\infty} C \cdot r^k = C + C \cdot r + C \cdot r^2 + C \cdot r^3 + \ldots$

Example 1: Bouncing Ball: A "super ball" is thrown 10 feet straight up into the air. On each bounce, it rebounds to four fifths of its previous height (Fig. 1) so the sequence of heights is 10 feet, 8 feet, 32/5 feet, 128/25 feet, etc.
(a) How far does the ball travel (up and down) during its nth bounce? (b) Use a sum to represent the total distance traveled by the ball.

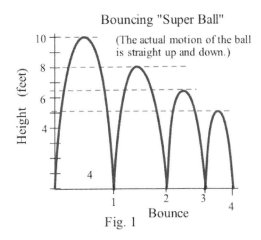

Fig. 1

Solution: Since the ball travels up and down on each bounce, the distance traveled during each bounce is twice the height of the ball on that bounce so $d_1 = 2(10 \text{ feet}) = 20$ feet, $d_2 = 16$ feet, $d_3 = 64/5$ feet, and, in general, $d_n = \frac{4}{5} \cdot d_{n-1}$. Looking at these values in another way,

$d_1 = 20$, $d_2 = \frac{4}{5} \cdot (20)$, $d_3 = \frac{4}{5} d_2 = \frac{4}{5} \cdot \frac{4}{5} \cdot 20 = (\frac{4}{5})^2 (20)$, $d_4 = \frac{4}{5} \cdot d_3 = \frac{4}{5} \cdot ((\frac{4}{5})^2 \cdot (20)) = (\frac{4}{5})^3 \cdot (20)$,

and, in general, $d_n = (\frac{4}{5})^{n-1} \cdot (20)$.

In theory, the ball bounces up and down forever, and the total distance traveled by the ball is the sum of the distances traveled during each bounce (an up and down flight):

(first bounce) + (second bounce) + (third bounce) + (forth bounce) + . . .

$$= 20 + \frac{4}{5}(20) + (\frac{4}{5})^2 (20) + (\frac{4}{5})^3 (20) + \ldots$$

$$= 20 \cdot (1 + \frac{4}{5} + (\frac{4}{5})^2 + (\frac{4}{5})^3 + \ldots) = 20 \cdot \sum_{k=0}^{\infty} (\frac{4}{5})^k .$$

Practice 1: **Cake:** Three calculus students want to share a small square cake equally, but they go about it in a rather strange way. First they cut the cake into 4 equal square pieces, each person takes one square, and one square is left (Fig. 2). Then they cut the leftover piece into 4 equal square pieces, each person takes one square and one square is left. And they keep repeating this process. (a) What fraction of the total cake does each person "eventually" get? (b) Represent the amount of cake each person gets as a geometric series: (amount of first piece) + (amount of second piece) + . . .

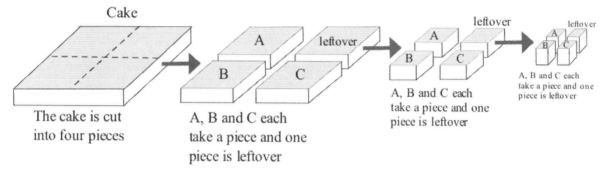

Fig. 2: Sharing a cake evenly among three students

Each series in the previous Example and Practice problems is a Geometric series, a series in which each term is a fixed multiple of the previous term. Geometric series have the form

$$\sum_{k=0}^{\infty} C \cdot r^k = C + C \cdot r + C \cdot r^2 + C \cdot r^3 + \ldots = C \cdot \sum_{k=0}^{\infty} r^k$$

with $C \neq 0$ and $r \neq 0$ representing fixed numbers. Each term in the series is r times the previous term. Geometric series are among the most common and easiest series we will encounter. A simple test determines whether a geometric series converges, and we can even determine the "sum" of the geometric series.

Geometric Series Theorem

The geometric series $\sum_{k=0}^{\infty} r^k = 1 + r + r^2 + r^3 + \ldots$ $\begin{cases} \text{converges to } \frac{1}{1-r} & \text{if } |r| < 1 \\ \text{diverges} & \text{if } |r| \geq 1 \end{cases}$

Proof: **If $|r| \geq 1$**, then $|r^k|$ approaches 1 or $+\infty$ as k becomes arbitrarily large, so the terms $a_k = r^k$ of the geometric series do not approach 0. Therefore, by the nth term test for divergence, the series diverges.

If $|r| < 1$, then the terms $a_k = r^k$ of the geometric series approach 0 so the series may or may not converge, and we need to examine the limit of the partial sums $s_n = 1 + r + r^2 + r^3 + \ldots + r^n$ of the series. For a geometric series, a clever insight allows us to calculate those partial sums:

$$(1-r) \cdot s_n = (1-r) \cdot (1 + r + r^2 + r^3 + \ldots + r^n)$$

$$= 1 \cdot (1 + r + r^2 + r^3 + \ldots + r^n) - r \cdot (1 + r + r^2 + r^3 + \ldots + r^n)$$

$$= (1 + r + r^2 + r^3 + \ldots + r^n) - (r + r^2 + r^3 + r^4 + \ldots + r^n + r^{n+1})$$

$$= 1 - r^{n+1}.$$

Since $|r| < 1$ we know $r \neq 1$ so we can divide the previous result by $1 - r$ to get

$$s_n = 1 + r + r^2 + r^3 + \ldots + r^n = \frac{1 - r^{n+1}}{1 - r} = \frac{1}{1-r} - \frac{r^{n+1}}{1-r}.$$

This formula for the nth partial sum of a geometric series is sometimes useful, but now we are interested in the limit of s_n as n approaches infinity. Since $|r| < 1$, r^{n+1} approaches 0 as n approaches infinity, so we can conclude that the partial sums $s_n = \frac{1}{1-r} - \frac{r^{n+1}}{1-r}$ approach $\frac{1}{1-r}$ (as "$n \to \infty$").

The geometric series $\sum_{k=0}^{\infty} r^k$ converges to the value $\frac{1}{1-r}$ when $-1 < r < 1$.

Finally, $\sum_{k=0}^{\infty} C \cdot r^k = C \cdot \sum_{k=0}^{\infty} r^k$ so we can easily determine whether or not $\sum_{k=0}^{\infty} C \cdot r^k$ converges and to what number.

Example 2: How far did the ball in Example 1 travel?

Solution: The distance traveled, $20(1 + \frac{4}{5} + (\frac{4}{5})^2 + (\frac{4}{5})^3 + \ldots)$, is a geometric series with $C = 20$ and $r = 4/5$. Since $|r| < 1$, the series $1 + \frac{4}{5} + (\frac{4}{5})^2 + (\frac{4}{5})^3 + \ldots$ converges to $\frac{1}{1-r} = \frac{1}{1 - 4/5} = 5$, so the total distance traveled is

$$20(1 + \frac{4}{5} + (\frac{4}{5})^2 + (\frac{4}{5})^3 + \ldots) = 20(5) = \textbf{100 feet}.$$

Repeating decimal numbers are really geometric series in disguise, and we can use the Geometric Series Theorem to represent the exact value of the sum as a fraction.

Example 3: Represent the repeating decimals $0.\overline{4}$ and $0.\overline{13}$ as geometric series and find their sums.

Solution: $0.\overline{4} = 0.444\ldots = \frac{4}{10} + \frac{4}{100} + \frac{4}{1000} + \ldots = \frac{4}{10} \cdot (1 + \frac{1}{10} + (\frac{1}{10})^2 + (\frac{1}{10})^3 + \ldots)$

which is a geometric series with $a = 4/10$ and $r = 1/10$. Since $|r| < 1$, the geometric series converges to $\frac{1}{1-r} = \frac{1}{1-1/10} = \frac{10}{9}$, and $0.\overline{4} = \frac{4}{10}(\frac{10}{9}) = \frac{4}{9}$.

Similarly, $0.\overline{13} = 0.131313... = \frac{13}{100} + \frac{13}{10000} + \frac{13}{1000000} + ...$

$= \frac{13}{100} \cdot (1 + \frac{1}{100} + (\frac{1}{100})^2 + (\frac{1}{100})^3 + ...)$

$= \frac{13}{100} \cdot (\frac{1}{1 - 1/100}) = \frac{13}{100}(\frac{100}{99}) = \frac{13}{99}$.

Practice 2: Represent the repeating decimals $0.\overline{3}$ and $0.\overline{432}$ as geometric series and find their sums.

One reason geometric series are important for us is that some series involving powers of x are geometric series.

Example 4: $\sum_{k=0}^{\infty} 3x^k = 3 + 3x + 3x^2 + ...$ and $\sum_{k=0}^{\infty} (2x-5)^k = 1 + (2x-5) + (2x-5)^2 + ...$

are geometric series with $r = x$ and $r = 2x - 5$, respectively. Find the values of x for each series so that the series converges.

Solution: A geometric series converges if and only if $|r| < 1$, so the first series converges if and only if $|x| < 1$, or, equivalently, $-1 < x < 1$. The sum of the first series is $\frac{3}{1-x}$.

In the second series $r = 2x - 5$ so the series converges if and only if $|2x - 5| < 1$. Removing the absolute value and solving for x, we get $-1 < 2x - 5 < 1$, and (adding 5 to each side and then dividing by 2) $2 < x < 3$. The second series converges if and only if $2 < x < 3$. The sum of the second series is $\frac{1}{1-(2x-5)}$ or $\frac{1}{6-2x}$.

Practice 3: The series $\sum_{k=0}^{\infty} (2x)^k$ and $\sum_{k=0}^{\infty} (3x-4)^k$ are geometric series. Find the ratio r for each series, and find all values of x for each series so that the series converges.

The series in the previous Example and Practice are called "power series" because they involve powers of the variable x. Later in this chapter we will investigate other power series which are not geometric series (e.g., $1 + x + x^2/2 + x^3/3 + ...$), and we will try to find values of x which guarantee that the series converge.

Harmonic Series: $\sum_{k=1}^{\infty} \frac{1}{k}$

The series $\sum_{k=1}^{\infty} \frac{1}{k}$ is one of the best known and most important **divergent** series. It is called the **harmonic series** because of its ties to music (Fig. 3).

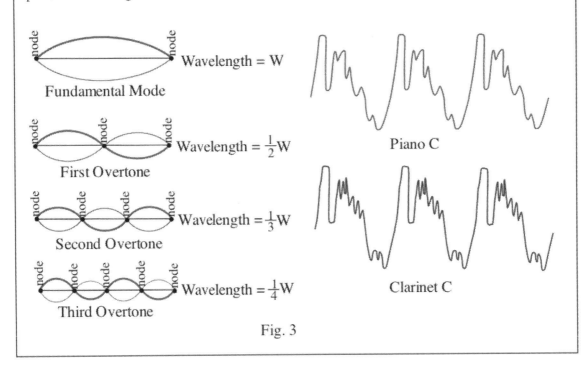

A taut piece of string such as a guitar string or a piano wire can only vibrate so that an integer number of waves are formed. The fundamental mode determines the note being played, and the number and intensity of the harmonics (overtones) present determine the characteristic quality of the sound. Because of these different characteristic qualities, a middle C (264 vibrations per second) played on a piano can be distinguished from a C on a clarinet.

Fig. 3

If we simply calculate partial sums of the harmonic series, it is not clear that the series diverges — the partial sums s_n grow, but as n becomes large, the values of s_n grow very, very slowly. Fig. 4 shows the values of n needed for the partial sums s_n to finally exceed the integer values 4, 5, 6, 8, 10, and 15. To examine the divergence of the harmonic series, brain power is much more effective than a lot of computing power.

n	s_n
31	4.0224519544
83	5.00206827268
227	6.00436670835
1,674	8.00048557200
12,367	10.00004300827
1,835,421	15.00000378267

Fig. 4

We can show that the harmonic series is divergent by showing that the terms of the harmonic series can be grouped into an infinite number of disjoint "chunks" each of which has a sum larger than 1/2. The series $\sum_{k=1}^{\infty} \frac{1}{2}$ is clearly divergent because the partial sums grow arbitrarily large: by adding enough of the terms together we can make the partial sums, $s_n > n/2$, larger than any predetermined number. Then we can conclude that the partial sums of the harmonic series also approach infinity so the harmonic series diverges.

> Theorem: The harmonic series $\sum_{k=1}^{\infty} \frac{1}{k} = 1 + \frac{1}{2} + \frac{1}{3} + \frac{1}{4} + \ldots$ **diverges**.

Proof: (This proof is essentially due to Oresme in 1630, twelve years before Newton was born. In 1821 Cauchy included Oresme's proof in a "Course in Analysis" and it became known as Cauchy's argument.)

Let S represent the sum of the harmonic series, $S = 1 + \frac{1}{2} + \frac{1}{3} + \frac{1}{4} + \ldots$, and group the terms of the series as indicated by the parentheses:

$$S = 1 + \frac{1}{2} + \left(\frac{1}{3} + \frac{1}{4}\right) + \left(\frac{1}{5} + \frac{1}{6} + \frac{1}{7} + \frac{1}{8}\right) + \left(\frac{1}{9} + \frac{1}{10} + \ldots + \frac{1}{16}\right) + \left(\frac{1}{17} + \ldots + \frac{1}{32}\right) + \ldots$$

    ↑   ↑   ↑   ↑

2 terms, each greater   4 terms, each greater   8 terms, each greater   16 terms, each greater
than or equal to 1/4   than or equal to 1/8   than or equal to 1/16   than or equal to 1/32

Each group in parentheses has a sum greater than 1/2, so

$$S > 1 + \frac{1}{2} + \left(\frac{1}{2}\right) + \left(\frac{1}{2}\right) + \left(\frac{1}{2}\right) + \ldots \text{ and}$$

the sequence of partial sums $\{s_n\}$ does not converge to a finite number. Therefore, the harmonic series diverges.

The harmonic series is an example of a **divergent** series whose terms, $a_k = 1/k$, approach 0. If the terms of a series approach 0, the series may or may not converge — we need to investigate further.

Telescoping Series

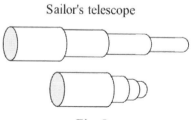

Sailor's telescope

Fig. 5

Sailors in the seventeenth and eighteenth centuries used telescopes (Fig. 5) which could be extended for viewing and collapsed for storing. Telescoping series get their name because they exhibit a similar "collapsing" property. Telescoping series are rather uncommon. But they are easy to analyze, and it can be useful to recognize them.

Example 5: Determine a formula for the partial sum s_n of the series $\sum_{k=1}^{\infty}\left[\dfrac{1}{k}-\dfrac{1}{k+1}\right]$.

Then find $\lim_{n\to\infty} s_n$. (Suggestion: It is tempting to algebraically consolidate terms, but the pattern is clearer in this case if you first write out all of the terms.)

Solution: $s_1 = a_1 = 1 - \dfrac{1}{2}$. In later values of s_n, part of each term cancels part of the next term:

$$s_2 = a_1 + a_2 = \left(1 - \dfrac{1}{2}\right) + \left(\dfrac{1}{2} - \dfrac{1}{3}\right) = 1 - \dfrac{1}{3}$$

$$s_3 = a_1 + a_2 + a_3 = \left(1 - \dfrac{1}{2}\right) + \left(\dfrac{1}{2} - \dfrac{1}{3}\right) + \left(\dfrac{1}{3} - \dfrac{1}{4}\right) = 1 - \dfrac{1}{4}$$

In general, many of the pieces in each partial sum "collapse" and we are left with a simple form of s_n:

$$s_n = a_1 + a_2 + \ldots + a_{n-1} + a_n = \left(1 - \dfrac{1}{2}\right) + \left(\dfrac{1}{2} - \dfrac{1}{3}\right) + \ldots + \left(\dfrac{1}{n-1} - \dfrac{1}{n}\right) + \left(\dfrac{1}{n} - \dfrac{1}{n+1}\right) = 1 - \dfrac{1}{n+1}$$

Finally, $\lim_{n\to\infty} s_n = \lim_{n\to\infty} 1 - \dfrac{1}{n+1} = 1$ so the series converges to 1: $\sum_{k=1}^{\infty}\left[\dfrac{1}{k} - \dfrac{1}{k+1}\right] = 1$.

Practice 4: Find the sum of the series $\sum_{k=3}^{\infty}\left[\sin\left(\dfrac{1}{k}\right) - \sin\left(\dfrac{1}{k+1}\right)\right]$.

PROBLEMS

In problems 1 – 6, rewrite each geometric series using the sigma notation and calculate the value of the sum.

1. $1 + \dfrac{1}{3} + \dfrac{1}{9} + \dfrac{1}{27} + \ldots$

2. $1 + \dfrac{2}{3} + \dfrac{4}{9} + \dfrac{8}{27} + \ldots$

3. $\dfrac{1}{8} + \dfrac{1}{16} + \dfrac{1}{32} + \dfrac{1}{64} + \ldots$

4. $1 - \dfrac{1}{2} + \dfrac{1}{4} - \dfrac{1}{8} + \dfrac{1}{16} - \ldots$

5. $-\dfrac{2}{3} + \dfrac{4}{9} - \dfrac{8}{27} + \ldots$

6. $1 + \dfrac{1}{e} + \dfrac{1}{e^2} + \dfrac{1}{e^3} + \ldots$

7. Rewrite each series in the form of a sum of r^k, and then show that

   (a) $\dfrac{1}{2} + \dfrac{1}{4} + \dfrac{1}{8} + \ldots = 1$, $\dfrac{1}{3} + \dfrac{1}{9} + \dfrac{1}{27} + \ldots = \dfrac{1}{2}$, and (b) for $a > 1$, $\dfrac{1}{a} + \dfrac{1}{a^2} + \dfrac{1}{a^3} + \ldots = \dfrac{1}{a-1}$.

8. A ball is thrown 10 feet straight up into the air, and on each bounce, it rebounds to 60% of its previous height. (a) How far does the ball travel (up and down) during its nth bounce. (b) Use a sum to represent the total distance traveled by the ball. (c) Find the total distance traveled by the ball.

9. An old tennis ball is thrown 20 feet straight up into the air, and on each bounce, it rebounds to 40% of its previous height. (a) How far does the ball travel (up and down) during its nth bounce? (b) Use a sum to represent the total distance traveled by the ball. (c) Find the total distance traveled by the ball.

10. Eighty people are going on an expedition by horseback through desolate country. The people and gear require 90 horses, and additional horses are needed to carry food for the original 90 horses. Each additional horse can carry enough food to feed 3 horses for the trip. How many additional horses are needed? (The original 90 horses will require 30 extra horses to carry their food. The 30 extra horses require 10 more horses to carry their food. etc.)

11. The mathematical diet you are following says you can eat "half of whatever is on the plate," so first you bite off one half of the cake and put the other half back on the plate. Then you pick up the remaining half from the plate (it's "on the plate"), bite off half of that, and return the rest to the plate. And you continue this silly process of picking up the piece from the plate, biting off half, and returning the rest to the plate. (a) Represent the total amount you eat as a series. (b) How much of the cake is **left** after 1 bite, 2 bites, n bites? (c) "Eventually," how much of the cake do you eat?

12. Suppose in Fig. 6 we begin with a square with sides of length 1 (area = 1) and construct another square inside by connecting the midpoints of the sides. Then the new square has area 1/2. If we continue the process of constructing each new square by connecting the midpoints of the sides of the previous square, we get a sequence of squares each of which has 1/2 half the area of the previous square. Find the total area of all of the squares.

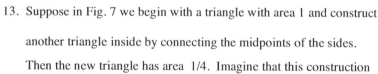

Fig. 6

13. Suppose in Fig. 7 we begin with a triangle with area 1 and construct another triangle inside by connecting the midpoints of the sides. Then the new triangle has area 1/4. Imagine that this construction process is continued and find the total area of all of the triangles.

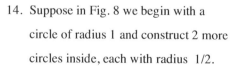

Fig. 7

14. Suppose in Fig. 8 we begin with a circle of radius 1 and construct 2 more circles inside, each with radius 1/2. Continue the process of constructing two new circles inside each circle from the previous step and find the total area of the circles.

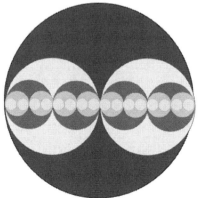

Fig. 8

15. The construction of the Helga von Koch snowflake begins with an equilateral triangle of area 1 (Fig. 9). Then each edge is subdivided into three equal lengths, and three equilateral triangles, each with area 1/9, are built on these "middle thirds" adding a total of 3(1/9) to the original area. The process is repeated: at the next stage, 3·4 equilateral triangles, each with area 1/81, are built on the new "middle thirds" adding 3·4·(1/81) more area. (a) Find the total area that results when this process is repeated forever.

number of edges	# triangles added	area of each triangle	total area added
3	0	0	0
3·4	3	1/9	3/9
$3 \cdot 4^2$	3·4	$1/9^2$	$3 \cdot 4/9^2 = (3/9)(4/9)$
$3 \cdot 4^3$	$3 \cdot 4^2$	$1/9^3$	$3 \cdot 4^2/9^3 = (3/9)(4^2/9^2)$
$3 \cdot 4^4$	$3 \cdot 4^3$	$1/9^4$	$3 \cdot 4^3/9^4 = (3/9)(4^3/9^3)$

so the total added area of the snowflake is

$$1 + \frac{3}{9} + (\frac{3}{9})(\frac{4}{9}) + (\frac{3}{9})(\frac{4^2}{9^2}) + (\frac{3}{9})(\frac{4^3}{9^3}) + \ldots$$

(b) Express the perimeter of the Koch Snowflake as a geometric series and find its sum.

(The area is finite, but the perimeter is infinite.)

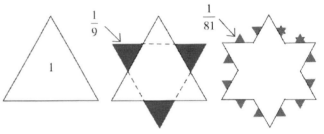

Fig. 9: Helga von Koch Snowflake

16. **Harmonic Tower:** The base of a tower is a cube whose edges are each one foot long. On top of it are cubes with edges of length 1/2, 1/3, 1/4, ... (Fig. 10).

    (a) Represent the total height of the tower as a series. Is the height finite?

    (b) Represent the total surface area of the cubes as a series.

    (c) Represent the total volume of the cubes as a series.

    (In the next section we will be able to determine if this surface area and volume are finite or infinite.)

17. The base of a tower is a sphere whose radius is each one foot long. On top of each sphere is another sphere with radius one half the radius of the sphere immediately beneath it (Fig 11). (a) Represent the total height of the tower as a series and find its sum. (b) Represent the total surface area of the spheres as a series and find its sum. (c) Represent the total volume of the spheres as a series and find its sum.

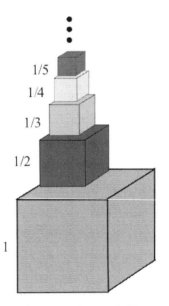

Fig. 10: Harmonic Tower

18. Represent the repeating decimals $0.\overline{6}$ and $0.\overline{63}$ as geometric series and find the value of each series as a simple fraction.

19. Represent the repeating decimals $0.\overline{8}$, $0.\overline{9}$, and $0.\overline{285714}$ as geometric series and find the value of each series as a simple fraction.

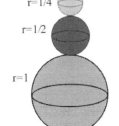

Fig. 11

20. Represent the repeating decimals $0.\overline{a}$, $0.\overline{ab}$, and $0.\overline{abc}$ as geometric series and find the value of each series as a simple fraction. What do you think the simple fraction representation is for $0.\overline{abcd}$?

In problems 21 – 32, find all values of x for which each geometric series converges.

21. $\sum_{k=1}^{\infty} (2x+1)^k$ 22. $\sum_{k=1}^{\infty} (3-x)^k$ 23. $\sum_{k=1}^{\infty} (1-2x)^k$

24. $\sum_{k=1}^{\infty} 5x^k$ 25. $\sum_{k=1}^{\infty} (7x)^k$ 26. $\sum_{k=1}^{\infty} (x/3)^k$

27. $1 + \frac{x}{2} + \frac{x^2}{4} + \frac{x^3}{8} + \ldots$ 28. $1 + \frac{2}{x} + \frac{4}{x^2} + \frac{8}{x^3} + \ldots$ 29. $1 + 2x + 4x^2 + 8x^3 + \ldots$

30. $\sum_{k=1}^{\infty} (2x/3)^k$ 31. $\sum_{k=1}^{\infty} \sin^k(x)$ 32. $\sum_{k=1}^{\infty} e^{kx} = \sum_{k=1}^{\infty} (e^x)^k$

33. One student thought the formula was $1 + x + x^2 + x^3 + \ldots = \frac{1}{1-x}$. The second student said "That can't be right. If we replace x with 2, then the formula says the sum of the positive numbers $1 + 2 + 4 + 8 + \ldots$ is a negative number $\frac{1}{1-2} = -1$." Who is right? Why?

34. The Classic Board Problem: If you have identical 1 foot long boards, they can be arranged to hang over the edge of a table. One board can extend 1/2 foot beyond the edge (Fig. 12), two boards can extend 1/2 + 1/4 feet, and, in general, n boards can extend 1/2 + 1/4 + 1/6 + ... + 1/(2n) feet beyond the edge.

    (a) How many boards are needed for an arrangement in which the entire top board is beyond the edge of the table?

    (b) How many boards are needed for an arrangement in which the entire top two boards are beyond the edge of the table?

    (c) How far can an arrangement extend beyond the edge of the table?

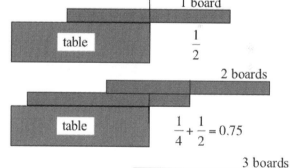
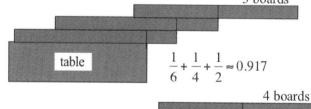
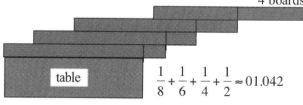

Fig. 12

In problems 35 – 40, calculate the value of the partial sum for $n = 4$ and $n = 5$ and find a formula for s_n. (The patterns may be more obvious if you do not simplify each term.)

35. $\sum_{k=3}^{\infty} \left[\frac{1}{k} - \frac{1}{k+1} \right]$

36. $\sum_{k=1}^{\infty} \left[\frac{1}{k} - \frac{1}{k+2} \right]$

37. $\sum_{k=1}^{\infty} [k^3 - (k+1)^3]$

38. $\sum_{k=1}^{\infty} \ln\left(\frac{k}{k+1}\right)$

39. $\sum_{k=3}^{\infty} [f(k) - f(k+1)]$

40. $\sum_{k=1}^{\infty} [g(k) - g(k+2)]$

In problems 41 – 44, calculate s_4 and s_5 for each series and find the limit of s_n as n approaches infinity. If the limit is a finite value, it represents the value of the infinite series.

41. $\sum_{k=1}^{\infty} \sin(\frac{1}{k}) - \sin(\frac{1}{k+1})$

42. $\sum_{k=2}^{\infty} \cos(\frac{1}{k}) - \cos(\frac{1}{k+1})$

43. $\sum_{k=2}^{\infty} \frac{1}{k^2} - \frac{1}{(k+1)^2}$

44. $\sum_{k=3}^{\infty} \ln(1 - \frac{1}{k^2})$ (Suggestion: Rewrite $1 - \frac{1}{k^2}$ as $\frac{1-\frac{1}{k}}{1-\frac{1}{k+1}}$)

Problems 45 and 46 are outlines of two "proofs by contradiction" that the harmonic series is divergent. Each proof starts with the assumption that the "sum" of the harmonic series is a finite number, and then an obviously false conclusion is derived from the assumption. Verify that each step follows from the assumption and previous steps, and explain why the conclusion is false.

45. Assume that $H = 1 + \frac{1}{2} + \frac{1}{3} + \frac{1}{4} + \frac{1}{5} + \frac{1}{6} + \frac{1}{7} + \ldots$ is a finite number, and let

$O = 1 + \frac{1}{3} + \frac{1}{5} + \frac{1}{7} + \ldots$ be the sum of the "odd reciprocals," and $E = \frac{1}{2} + \frac{1}{4} + \frac{1}{6} + \frac{1}{8} + \ldots$

be the sum of the "even reciprocals." Then

(i) $H = O + E$, (ii) each term of O is larger than the corresponding term of E so $O > E$,

and (iii) $E = \frac{1}{2} + \frac{1}{4} + \frac{1}{6} + \frac{1}{8} + \ldots = \frac{1}{2} \{ 1 + \frac{1}{2} + \frac{1}{3} + \frac{1}{4} + \ldots \} = \frac{1}{2} H$.

Therefore $H = O + E > \frac{1}{2} H + \frac{1}{2} H = H$ (so "H is strictly bigger than H," a contradiction).

II. Assume that $H = 1 + \frac{1}{2} + \frac{1}{3} + \frac{1}{4} + \frac{1}{5} + \frac{1}{6} + \frac{1}{7} + \ldots$ is a finite number, and, starting with the second term, group the terms into groups of three. Then, using $\frac{1}{n-1} + \frac{1}{n} + \frac{1}{n+1} > 3 \cdot \frac{1}{n}$, we have

$H = 1 + (\frac{1}{2} + \frac{1}{3} + \frac{1}{4}) + (\frac{1}{5} + \frac{1}{6} + \frac{1}{7}) + (\frac{1}{8} + \frac{1}{9} + \frac{1}{10}) + \ldots$

$> 1 + (\quad 1 \quad) + (\quad \frac{1}{2} \quad) + (\quad \frac{1}{3} \quad) + \ldots$

$= 1 + (1 + \frac{1}{2} + \frac{1}{3} + \frac{1}{4} + \ldots) = 1 + H$. Therefore, $H > 1 + H$ (so "H is bigger than 1 + H").

46. Jacob Bernoulli (1654–1705) was a master of understanding and manipulating series by breaking a difficult series into a sum of easier series. He used that technique to find the sum of the non–geometric series

$$\sum_{k=1}^{\infty} \frac{k}{2^k} = 1/2 + 2/4 + 3/8 + 4/16 + 5/32 + \ldots + k/2^k + \ldots \text{ in his book } \underline{\text{Ars Conjectandi}}, 1713.$$

Show that $1/2 + 2/4 + 3/8 + 4/16 + 5/32 + \ldots + n/2^n + \ldots$ can be written as

$\phantom{\text{plus}}$ $1/2 + 1/4 + 1/8 + 1/16 + 1/32 + \ldots + 1/2^n + \ldots$ (a)

plus $1/4 + 1/8 + 1/16 + 1/32 + \ldots + 1/2^n + \ldots$ (b)

plus $1/8 + 1/16 + 1/32 + \ldots + 1/2^n + \ldots$ (c)

plus . . . etc.

Find the values of the geometric series (a), (b), (c), etc. and then find the sum of these values (another geometric series).

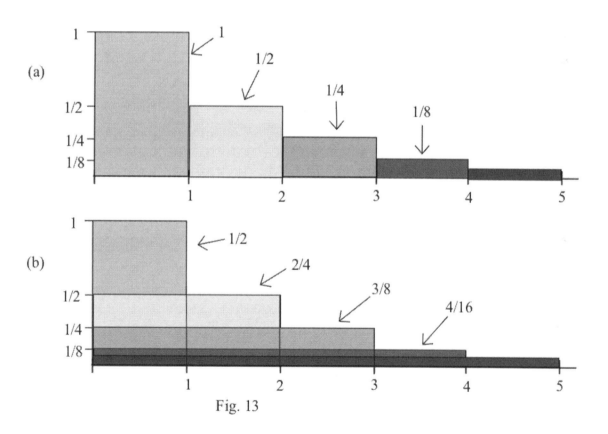

Fig. 13

47. Bernoulli's approach in problem 46 can also be interpreted as a geometric argument for representing the area in Fig. 13 in two different ways. (a) Represent the total area in Fig. 13a as a (geometric) sum of the areas of the side–by–side rectangles, and find the sum of the series. (b) Represent the total area of the stacked rectangles in Fig. 13b as a sum of the areas of the horizontal slices.

Since both series represent the same total area, the values of the series are equal.

48. Use the approach of problem 46 to find a formula for the sum of

(a) the value of $\displaystyle\sum_{k=1}^{\infty} \frac{k}{3^k} = 1/3 + 2/9 + 3/27 + 4/81 + \ldots + k/3^k + \ldots$ and

(b) a formula for the value of $\displaystyle\sum_{k=1}^{\infty} \frac{k}{c^k} = 1/c + 2/c^2 + 3/c^3 + 4/c^4 + \ldots + k/c^k + \ldots$ for $c > 1$.

(answers: (a) 3/4, (b) $c/(c-1)^2$)

Practice Answers

Practice 1: (a) Since they each get equal shares, and the whole cake is distributed, they each get 1/3 of the cake. More precisely, after step 1, 1/4 of the cake remains and 3/4 was shared. After step 2, $(1/4)^2$ of the cake remains and $1 - (1/4)^2$ was shared. After step n, $(1/4)^n$ of the cake remains and $1 - (1/4)^n$ was shared. So after step n, each student has $(\frac{1}{3})(1 - (1/4)^n)$ of the cake. "Eventually," each student gets (almost) $\frac{1}{3}$ of the cake.

(b) $(\frac{1}{4}) + (\frac{1}{4})^2 + (\frac{1}{4})^3 + \ldots = (\frac{1}{4})\{1 + (\frac{1}{4}) + (\frac{1}{4})^2 + \ldots\}$

Practice 2: $0.\overline{3} = 0.333\ldots = \frac{3}{10} + \frac{3}{100} + \frac{3}{1000} + \ldots = \frac{3}{10} \cdot (1 + \frac{1}{10} + (\frac{1}{10})^2 + (\frac{1}{10})^3 + \ldots)$

which is a geometric series with $a = 3/10$ and $r = 1/10$. Since $|r| < 1$, the geometric series

converges to $\frac{1}{1-r} = \frac{1}{1 - 1/10} = \frac{10}{9}$, and $0.\overline{3} = \frac{3}{10}(\frac{10}{9}) = \frac{3}{9} = \frac{1}{3}$.

Similarly, $0.\overline{432} = 0.432432432\ldots = \frac{432}{1000} + \frac{432}{1000000} + \frac{432}{1000000000} + \ldots$

$= \frac{432}{1000} \cdot (1 + \frac{1}{1000} + (\frac{1}{1000})^2 + (\frac{1}{100})^3 + \ldots)$

$= \frac{432}{1000} \cdot (\frac{1}{1 - 1/1000}) = \frac{432}{1000}(\frac{1000}{999}) = \frac{432}{999} = \frac{16}{37}$.

Practice 3: $r = 2x$: If $|2x| < 1$, then $-1 < 2x < 1$ so $-1/2 < x < 1/2$.

$\displaystyle\sum_{k=0}^{\infty} (2x)^k$ converges (to $\frac{1}{1 - 2x}$) when $-1/2 < x < 1/2$.

$r = 3x - 4$: If $|3x - 4| < 1$, then $-1 < 3x - 4 < 1$ so $3 < 3x < 5$ and $1 < x < 5/3$.

$\displaystyle\sum_{k=0}^{\infty} (3x - 4)^k$ converges (to $\frac{1}{1 - (3x - 4)} = \frac{1}{5 - 3x}$) when $1 < x < 5/3$.

Practice 4: Let $s_n = \sum_{k=3}^{n} \sin(\frac{1}{k}) - \sin(\frac{1}{k+1})$

$= \{\sin(\frac{1}{3}) - \sin(\frac{1}{4})\} + \{\sin(\frac{1}{4}) - \sin(\frac{1}{5})\} + \{\sin(\frac{1}{5}) - \sin(\frac{1}{6})\} + \ldots + \{\sin(\frac{1}{n}) - \sin(\frac{1}{n+1})\}$

$= \sin(\frac{1}{3}) - \sin(\frac{1}{n+1})$.

Then $\lim_{n \to \infty} s_n = \lim_{n \to \infty} \{\sin(\frac{1}{3}) - \sin(\frac{1}{n+1})\} = \sin(\frac{1}{3})$ so the series converges to $\sin(\frac{1}{3})$:

$\sum_{k=1}^{\infty} \sin(\frac{1}{k}) - \sin(\frac{1}{k+1}) = \sin(\frac{1}{3}) \approx 0.327$.

Appendix: MAPLE and WolframAlpha for Partial Sums of Geometric Series

MAPLE command for

$\sum_{n=0}^{100} \frac{3}{2^n}$: sum(3*(1/2)^n , n=0..100) ; (then press ENTER key)

$\sum_{n=0}^{\infty} \frac{3}{2^n}$: sum(3*(1/2)^n , n=0..infinity) ; (then press ENTER key)

WolframAlpha (free at http://www.wolframalpha.com)

Asking   sum 3/2^k , k = 1 to infinity

gives the sum of the infinite series and a graph of the partial sums.

Asking   sum 3/2^k , k = 1 to 100

gives the sum as an exact fraction (strange) and as a decimal.

10.3.5   AN INTERLUDE

The previous three sections introduced the topics of sequences and series, discussed the meaning of convergence of series, and examined geometric series in some detail. The ideas, definitions, and results in those sections are fundamental for understanding and working with the material in the rest of this chapter and for later work in theoretical and applied mathematics.

The material in the next several sections is of a different sort — it is more technical and specialized. In order to work effectively with power series we need to know where (for which values of x) the power series converge. And to determine that convergence we need additional methods. In the next several sections we examine several methods for determining where particular series converge or diverge. These methods are called "convergence tests."

- The Integral Test in Section 10.4 says that a series converges if and only if a certain related improper integral is finite. This result lets us change a question about convergence of a series into a question about the convergence of an integral. Sometimes the related integral is easy to evaluate so it is easy to determine the convergence of the series. (Sometimes the related integral is very difficult to evaluate.) The integral test is then used to determine the convergence of P–series, the whole family of series of the form $\sum_{k=1}^{\infty} \frac{1}{k^p}$.

- Section 10.5 introduces some methods for determining the convergence of a new series by comparing the new series with some series which we already know converge or diverge. These comparison tests can be very powerful and useful, but their power and usefulness depends on already knowing about the convergence of some particular series to compare against the new series. Typically we will compare new series against two types of series, geometric series and P–series.

- In Section 10.6 we derive a result about the convergence of a series $\sum_{k=1}^{\infty} a_k$ by examining the ratios of successive terms of the series, a_{k+1}/a_k . If this ratio is small enough, then we will be able to conclude that the series converges. If the ratio is large enough, then we will be able to conclude that the series diverges. Unfortunately, sometimes the value of the ratio will not allow us to conclude anything.

- Section 10.7 examines series whose terms alternate in sign, such as the "alternating harmonic series," $1 - 1/2 + 1/3 - 1/4 + \ldots$, and discusses methods to determine whether these alternating series converge.

Each of these sections is rather short and focuses on one or two tests of convergence. As you study the material in each section by itself, you need to be able to use the method discussed in that section. When you finish all of the sections, you also need to be able to decide which convergence test to use.

PROBLEMS

These problems illustrate some of the reasoning that is used in sections 10.4 – 10.7, but they do not assume any information from those sections.

Integrals and sums

1. Which shaded region in Fig. 1 has the larger area, the sum or the integral?

2. Which shaded region in Fig. 2 has the larger area, the sum or the integral?

3. Represent the area of the shaded region in Fig. 3 as an infinite series.

4. Represent the area of the shaded region in Fig. 4 as an infinite series.

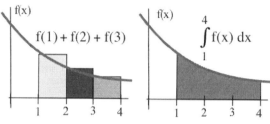

Fig. 1

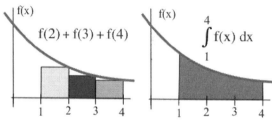

Fig. 2

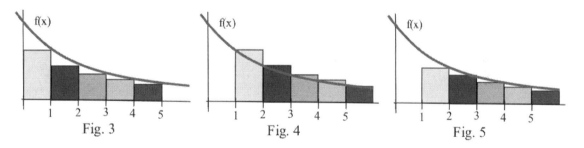

Fig. 3   Fig. 4   Fig. 5

5. Represent the area of the shaded region in Fig. 5 as an infinite series.

6. Which of the following represents the shaded area in Fig. 6?

   (a) $f(0) + f(1)$ (b) $f(1) + f(2)$

   (c) $f(2) + f(3)$ (d) $f(3) + f(4)$

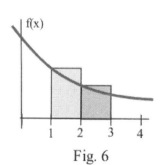

Fig. 6

7. Which of the following represents the shaded area in Fig. 7?

   (a) f(0) + f(1)   (b) f(1) + f(2)   (c) f(2) + f(3)   (d) f(3) + f(4)

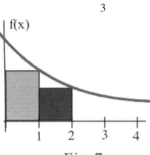
Fig. 7

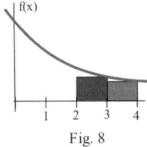
Fig. 8

8. Which of the following represents the shaded area in Fig. 8?

   (a) f(0) + f(1)   (b) f(1) + f(2)   (c) f(2) + f(3)   (d) f(3) + f(4)

9. Which of the following represents the shaded area in Fig. 9?

   (a) f(0) + f(1)   (b) f(1) + f(2)   (c) f(2) + f(3)   (d) f(3) + f(4)

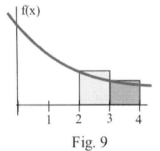
Fig. 9

Fig. 10

10. Arrange the following four values in increasing order (Fig. 10):

    (a) $\int_1^3 f(x)\,dx$ (b) $\int_2^4 f(x)\,dx$ (c) f(1) + f(2)   (d) f(2) + f(3)

11. Arrange the following four values in increasing order (Fig. 11):

    (a) $\int_1^4 f(x)\,dx$ (b) $\int_2^5 f(x)\,dx$

    (c) f(1) + f(2) + f(3)   (d) f(2) + f(3) + f(4)

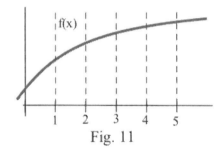
Fig. 11

Comparisons

12. You want to get a summer job operating a type of heavy equipment, and you know there are certain height requirements in order for the operator to fit safely in the cab of the machine. You don't remember what the requirements are, but three of your friends applied. Tom was rejected as too tall. Sam was rejected as too short. Justin got a job. Should you apply for the job if

    (a) you are taller than Tom? Why?   (b) you are taller than Sam? Why?

    (c) you are shorter than Sam? Why?   (d) you are shorter than Justin? Why?

    (e) List the comparisons which indicate that you are the wrong height for the job.

    (f) List the comparisons which do not give you enough information about whether you are an acceptable height for the job.

13. You know Wendy did well on the Calculus test and Paula did poorly, but you haven't received your test back yet. If the instructor tells you the following, what can you conclude?
    (a) "You did better than Wendy." (b) "You did better than Paula."
    (c) "You did worse than Wendy."   (d) "You did worse than Paula."

14. You have recently taken up mountain climbing and are considering a climb of Mt. Baker. You know that Mt. Index is too easy to be challenging, but that Mt. Liberty Bell is too difficult for you. Should you plan a climb of Mt. Baker if an experienced climber friend tells you that
    (a) "Baker is easier than Index."   (b) "Baker is more difficult than Index."
    (c) "Baker is easier than Liberty Bell." (d) "Baker is more difficult than Liberty Bell."
    (e) Which comparisons indicate that Baker is appropriate: challenging but not too difficult?
    (f) Which comparisons indicate that Baker is not appropriate?

15. As a student you have had Professors Good and Bad for classes, and they each lived up to their names. Now you are considering taking a class from Prof. Unknown whom you don't know. What can you expect if
    (a) a classmate who had Good and Unknown says "Unknown was better than Good" ?
    (b) a classmate who had Good and Unknown says "Good was better than Unknown" ?
    (c) a classmate who had Bad and Unknown says "Unknown was better than Bad" ?
    (d) a classmate who had Bad and Unknown says "Bad was better than Unknown" ?

In problems 16 – 19, all of the series converge. In each pair, which series has the larger sum?

16. $\sum_{k=1}^{\infty} \frac{1}{k^2+1}$, $\sum_{k=1}^{\infty} \frac{1}{k^2}$

17. $\sum_{k=2}^{\infty} \frac{1}{k^3-5}$, $\sum_{k=2}^{\infty} \frac{1}{k^3}$

18. $\sum_{k=1}^{\infty} \frac{1}{k^2+3k-1}$, $\sum_{k=1}^{\infty} \frac{1}{k^2}$

19. $\sum_{k=3}^{\infty} \frac{1}{k^2+5k}$, $\sum_{k=3}^{\infty} \frac{1}{k^3+k-1}$

Ratios of successive terms

In problems 20 – 28, a formula is given for each term a_k of a sequence. For each sequence (a) write a formula for a_{k+1}, (b) write the ratio a_{k+1}/a_k, and (c) simplify the ratio a_{k+1}/a_k.

20. $a_k = 3k$ 21. $a_k = k+3$ 22. $a_k = 2k+5$

23. $a_k = 3/k$ 24. $a_k = k^2$ 25. $a_k = 2^k$

26. $a_k = (1/2)^k$ 27. $a_k = x^k$ 28. $a_k = (x-1)^k$

In problems 29–36, state whether the series converges or diverges and calculate the ratio a_{k+1}/a_k.

29. $\sum_{k=1}^{\infty} (\frac{1}{2})^k$
30. $\sum_{k=1}^{\infty} (\frac{1}{5})^k$
31. $\sum_{k=1}^{\infty} 2^k$
32. $\sum_{k=1}^{\infty} (-3)^k$

33. $\sum_{k=1}^{\infty} 4$
34. $\sum_{k=1}^{\infty} (-1)^k$
35. $\sum_{k=1}^{\infty} \frac{1}{k}$
36. $\sum_{k=1}^{\infty} \frac{7}{k}$

Alternating terms

For problems 37 – 40, s_n represents the nth partial sum of the series with terms a_k (e.g., $s_3 = a_1 + a_2 + a_3$).

In each problem, circle the appropriate symbol: "<" or "=" or ">."

37. If $a_5 > 0$, then s_4 < = > s_5.

38. If $a_5 = 0$, then s_4 < = > s_5.

39. If $a_5 < 0$, then s_4 < = > s_5.

40. If $a_{n+1} > 0$ for all n, then s_n < = > s_{n+1} for all n.

41. If $a_{n+1} < 0$ for all n, then s_n < = > s_{n+1} for all n.

42. If $a_4 > 0$ and $a_5 < 0$, then how do s_3, s_4, and s_5 compare?

43. If $a_4 = 0.2$ and $a_5 = -0.1$ and $a_6 = 0.2$, then how do s_3, s_4, and s_5 compare?

44. If $a_4 = -0.3$ and $a_5 = 0.2$ and $a_6 = -0.1$, then how do s_3, s_4, and s_5 compare?

45. If $a_4 = -0.3$ and $a_5 = -0.2$ and $a_6 = 0.1$, then how do s_3, s_4, and s_5 compare?

In problems 46 – 50, the first 8 terms $a_1, a_2, \ldots, a_8$ of a series are given. Calculate and graph the first 8 partial sums $s_1, s_2, \ldots, s_8$ of the series and describe the pattern of the graph of the partial sums.

46. $a_1 = 2, a_2 = -1, a_3 = 2, a_4 = -1, a_5 = 2, a_6 = -1, a_7 = 2, a_8 = -1$.

47. $a_1 = 2, a_2 = -1, a_3 = 0.9, a_4 = -0.8, a_5 = 0.7, a_6 = -0.6, a_7 = 0.5, a_8 = -0.4$.

48. $a_1 = 2, a_2 = -1, a_3 = 1, a_4 = -1, a_5 = 1, a_6 = -1, a_7 = 1, a_8 = -1$.

49. $a_1 = -2, a_2 = 1.5, a_3 = -0.8, a_4 = 0.6, a_5 = -0.4, a_6 = 0.2, a_7 = 2, a_8 = -0.1$.

50. $a_1 = 5, a_2 = 1, a_3 = -0.6, a_4 = -0.4, a_5 = 0.2, a_6 = 0.1, a_7 = 0.1, a_8 = -0.2$.

51. What condition on the terms a_k guarantees that the graph of the partial sums s_n follows an "up–down–up–down" pattern?

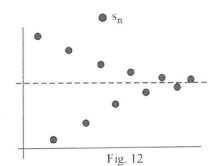

Fig. 12

52. What condition on the terms a_k guarantees that the graph of the partial sums s_n forms a "narrowing funnel" pattern in Fig. 12 ?

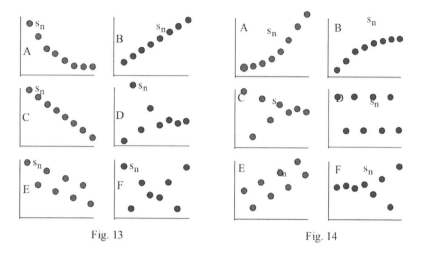

Fig. 13   Fig. 14

53. Fig. 13 shows the graphs of several partial sums s_n .
    (a) For which graphs do the terms a_k alternate in sign?
    (b) For which graphs do the $|a_k|$ decrease?
    (c) For which graphs do the terms a_k alternate in sign and decrease in absolute value?

54. Fig. 14 shows the graphs of several partial sums s_n .
    (a) For which graphs do the terms a_k alternate in sign?
    (b) For which graphs do the $|a_k|$ decrease?
    (c) For which graphs do the terms a_k alternate in sign and decrease in absolute value?

55. The geometric series $\sum_{k=0}^{\infty} (-\frac{1}{2})^k = 1 - \frac{1}{2} + \frac{1}{4} - \ldots$ converges to $\frac{1}{1-(-1/2)} = \frac{1}{3/2} = \frac{2}{3}$.

    Graph the horizontal line $y = \frac{2}{3}$ and then graph the partial sums $s_0, \ldots, s_8$ of $\sum_{k=0}^{\infty} (-\frac{1}{2})^k$.

56. The geometric series $\sum_{k=0}^{\infty} (-0.6)^k = 1 - 0.6 + 0.36 - \ldots$ converges to $\frac{1}{1-(-0.6)} = \frac{1}{1.6} = 0.625$.

    Graph the horizontal line $y = 0.625$ and then graph the partial sums $s_0, \ldots, s_8$ of $\sum_{k=0}^{\infty} (-0.6)^k$.

57. The geometric series $\sum_{k=0}^{\infty} (-2)^k = 1 - 2 + 4 - \ldots$ diverges ($|r| = |-2| = 2 > 1$). Graph the partial sums $s_0, \ldots, s_8$ of $\sum_{k=0}^{\infty} (-2)^k$.

10.4 POSITIVE TERM SERIES: INTEGRAL TEST & P–TEST

This section discusses two methods for determining whether some series are convergent. The first, the integral test, says that a given series converges if and only if a related improper integral converges. This lets us trade a question about the convergence of a series for a question about the convergence of an improper integral. The second convergence test, the P–test, says that the convergence of one particular type of series, the sum $\sum_{k=1}^{\infty} \frac{1}{k^p}$, depends only on the value of p. These tests only apply to series whose terms are positive. And, unfortunately, the tests only tell us if the series converge or diverge, but they do not tell us the actual sum of the series.

The Integral Test is the more fundamental and general of the two tests examined in this section, and it is used to prove the P–Test. The P–Test, however, is easier to apply and is likely to be the test you use more often.

Integral Test

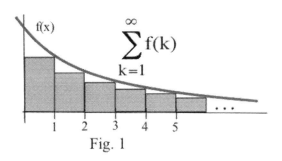
Fig. 1

A series can be thought of as a sum of areas of rectangles each having a base of one unit (Fig. 1). With this area interpretation of series there is a natural connection between series and integrals and between the convergence of a series and the convergence of an appropriate improper integral.

Example 1: Suppose the shaded region in Fig. 2a can be painted using 3 gallons of paint. How much paint is needed for the shaded region in Fig. 2b?

Solution: We don't have enough information to determine the exact amount of paint needed for the region in Fig. 2b, but the total of the rectangular areas is smaller than the area in Fig. 2a so less than 3 gallons of paint are needed for the region in Fig. 2b.

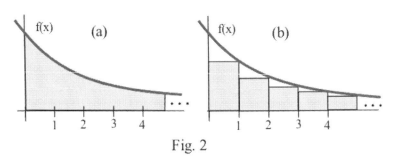
Fig. 2

Practice 1: Suppose the area of the shaded region in Fig. 3a is infinite. What can you say about the total area of the rectangular regions in Fig. 3b?

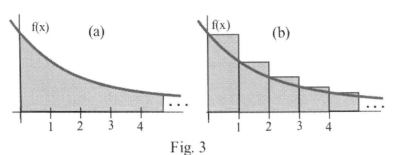
Fig. 3

The geometric reasoning used in Example 1 and Practice 1 can also be used to determine the convergence and divergence of some series.

Example 2: (a) Which is larger: $\sum_{k=2}^{\infty} \frac{1}{k^2}$ or $\int_{1}^{\infty} \frac{1}{x^2} dx$?

(b) Use the result of (a) to show that $\sum_{k=2}^{\infty} \frac{1}{k^2}$ is convergent.

Solution: Fig. 4 illustrates that the area of the rectangles, $\frac{1}{2^2} + \frac{1}{3^2} + \frac{1}{4^2} + \ldots + \frac{1}{n^2}$, is less than the area under the graph of the function $f(x) = \frac{1}{x^2}$ for $1 \leq x \leq n$:

$$\sum_{k=2}^{n} \frac{1}{k^2} < \int_{1}^{n} \frac{1}{x^2} dx \quad \text{so} \quad \sum_{k=2}^{n} \frac{1}{k^2} < \int_{1}^{\infty} \frac{1}{x^2} dx = 1 \text{ for every } n \geq 2.$$

Therefore, the partial sums of $\sum_{k=2}^{\infty} \frac{1}{k^2}$ are bounded. Also, each term $a_k = \frac{1}{k^2}$ is positive, so the partial sums of $\sum_{k=2}^{\infty} \frac{1}{k^2}$ are monotonically increasing. So, by the Monotonic Theorem of Section 10.1, the sequence of partial sums converges, so the series $\sum_{k=2}^{\infty} \frac{1}{k^2}$ is convergent.

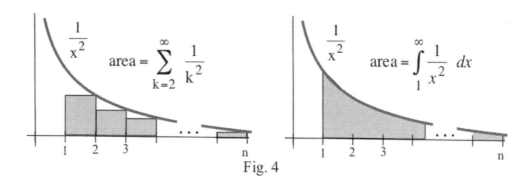
Fig. 4

The reasoning of Example 2 can be extended to the comparison of other series and the appropriate integrals.

Integral Test

Suppose f is a **continuous, positive, decreasing** function on $[1, \infty)$ and $a_k = f(k)$.

The series $\sum_{k=1}^{\infty} a_k$ converges **if and only if** the integral $\int_{1}^{\infty} f(x)\, dx$ converges.

Equivalently, (a) if $\int_{1}^{\infty} f(x)\, dx$ converges, then $\sum_{k=1}^{\infty} a_k$ converges,

and (b) if $\int_{1}^{\infty} f(x)\, dx$ diverges, then $\sum_{k=1}^{\infty} a_k$ diverges.

The proof is simply a careful use of the reasoning in the previous Examples.

Proof: Assume that f is a continuous, positive, decreasing function on $[1, \infty)$ and that $a_k = f(k)$.

Part (a): Assume that $\int_{1}^{\infty} f(x)\, dx$ converges: $\lim_{n \to \infty} \int_{1}^{n} f(x)\, dx$ is a finite number.

Since each $a_k > 0$, the sequence of partial sums s_n is increasing. If we arrange the rectangles under the graph of f as in Fig. 5, it is clear that

$$s_n = \sum_{k=1}^{n} a_k = a_1 + \sum_{k=2}^{n} a_k \leq a_1 + \int_{1}^{n} f(x)\, dx \leq a_1 + \int_{1}^{\infty} f(x)\, dx .$$

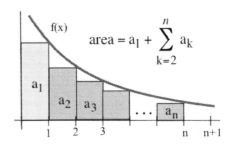

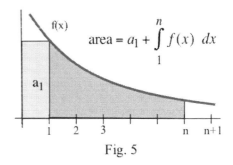

$\{ s_n \}$ is a bounded, increasing sequence so, by the Monotone Convergence Theorem of Section 10.1, $\{ s_n \}$ converges and $\sum_{k=1}^{\infty} a_k$ is convergent.

Fig. 5

Part (b): Assume that $\int_1^\infty f(x)\,dx$ diverges: $\lim_{n\to\infty} \int_1^n f(x)\,dx = \infty$.

If we arrange the rectangles under the graph of f as in Fig. 6, it is clear that

$$s_n = \sum_{k=1}^n a_k \geq \int_1^{n+1} f(x)\,dx \quad \text{for all } n,$$

so $\lim_{n\to\infty} s_n \geq \lim_{n\to\infty} \int_1^n f(x)\,dx = \infty$.

In other words, $\lim_{n\to\infty} s_n = \infty$ and $\sum_{k=1}^\infty a_k$ diverges.

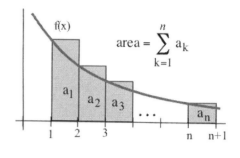

The inequalities in the proof relating the partial sums of the series to the values of integrals are sometimes used to approximate the values of the partial sums of a series:

$$\int_1^{n+1} f(x)\,dx \leq \sum_{k=1}^n a_k \leq a_1 + \int_1^n f(x)\,dx.$$

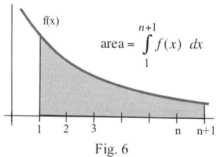

Fig. 6

We can use this last set of inequalities with $a_k = \frac{1}{k}$ and $n = 1{,}000$ to conclude that

$$\int_1^{10^3+1} \frac{1}{x}\,dx \leq \sum_{k=1}^{10^3} \frac{1}{k} \leq 1 + \int_1^{10^3} \frac{1}{x}\,dx \quad \text{so} \quad 7.48646986155 \leq \sum_{k=1}^{10^3} \frac{1}{k} \leq 8.48547086055.$$

(If $n = 1{,}000{,}000$, then the same type of reasoning shows that the partial sum of $1/k$ from $k = 1$ to $k = 10^6$ is between 13.815511 and 14.815510 .)

If the series does not start with $k = 1$, a Corollary of the Integral Test can be used.

Corollary: If f satisfies the hypotheses of the Integral Test on $[N, \infty)$ and $a_k = f(k)$,

then $\sum_{k=N}^\infty a_k$ and $\int_N^\infty f(x)\,dx$ both converge or both diverge.

Example 3: Use the Integral Test to determine whether (a) $\sum_{k=1}^\infty \frac{1}{k^3}$ and (b) $\sum_{k=2}^\infty \frac{1}{k \cdot \ln(k)}$ converge.

Solution: (a) If $f(x) = \frac{1}{x^3}$, then $a_k = f(k)$ and f is continuous, positive and decreasing on $[1, \infty)$.

Then $\int_1^\infty \frac{1}{x^3} dx = \lim_{n \to \infty} \int_1^n \frac{1}{x^3} dx = \lim_{n \to \infty} \left(-\frac{1}{2x^2} \right) \Big|_1^n$

$= \lim_{n \to \infty} \left(-\frac{1}{2n^2} \right) - \left(-\frac{1}{2 \cdot 1^2} \right) = \frac{1}{2}$.

The integral $\int_1^\infty \frac{1}{x^3} dx$ converges so the series $\sum_{k=1}^\infty \frac{1}{k^3}$ converges.

(b) If $f(x) = \frac{1}{x \cdot \ln(x)}$, then $a_k = f(k)$ and f is continuous, positive and decreasing on $[2, \infty)$.

Then $\int_2^\infty \frac{1}{x \cdot \ln(x)} dx = \lim_{n \to \infty} \int_2^n \frac{1}{x \cdot \ln(x)} dx = \lim_{n \to \infty} \ln(\ln(x)) \Big|_2^n$

$= \lim_{n \to \infty} \ln(\ln(n)) - \ln(\ln(2)) = \infty$.

The integral $\int_2^\infty \frac{1}{x \cdot \ln(x)} dx$ diverges so the series $\sum_{k=2}^\infty \frac{1}{k \cdot \ln(k)}$ diverges.

Practice 2: Use the Integral Test to determine whether (a) $\sum_{k=4}^\infty \frac{1}{\sqrt{k}}$ and (b) $\sum_{k=1}^\infty e^{-k}$ converge.

Note: The Integral Test does not give the value of the sum, it only answers the question of whether the series converges or diverges. Typically the value of the improper integral is not equal to the sum of the series.

P–Test for Convergence of $\sum_{k=1}^\infty \frac{1}{k^p}$

The P–Test is very easy to use. And it answers the convergence question for a whole family of series.

P–Test

The series $\sum_{k=1}^\infty \frac{1}{k^p}$ $\begin{cases} \text{converges} & \text{if } p > 1 \\ \text{diverges} & \text{if } p \leq 1 \end{cases}$.

Proof: If $p = 1$, then $\sum_{k=1}^{\infty} \frac{1}{k^p} = \sum_{k=1}^{\infty} \frac{1}{k}$, the harmonic series, which we already know diverges (by Section 10.3 or, using the Integral Test, since $\int_{1}^{\infty} \frac{1}{x} dx$ diverges to infinity.)

The proof for $p \neq 1$ is a straightforward application of the Integral Test on $f(x) = 1/x^p$.

If $p \neq 1$, then
$$\int_{1}^{\infty} \frac{1}{x^p} dx = \int_{1}^{\infty} x^{-p} dx = \lim_{A \to \infty} \int_{1}^{A} x^{-p} dx$$
$$= \lim_{A \to \infty} \left(\frac{1}{1-p}\right) \cdot x^{1-p} \Big|_{1}^{A}$$
$$= \lim_{A \to \infty} \left(\frac{1}{1-p}\right) \cdot A^{1-p} - \left(\frac{1}{1-p}\right) \cdot 1$$

As we examine the limit of A^{1-p}, there are two cases to consider: $p < 1$ and $p > 1$.

If **p < 1**, then $1 - p > 0$ so A^{1-p} approaches infinity as A approaches infinity. Then $\int_{1}^{\infty} \frac{1}{x^p} dx$ diverges, so, by the Integral Test, $\sum_{k=1}^{\infty} \frac{1}{k^p}$ diverges.

If **p > 1**, then $p - 1 > 0$ and $A^{1-p} = \frac{1}{A^{p-1}}$ approaches 0 as A approaches infinity. Then $\int_{1}^{\infty} \frac{1}{x^p} dx$ converges, so, by the Integral Test, $\sum_{k=1}^{\infty} \frac{1}{k^p}$ converges.

Example 3: Use the P–Test to determine whether (a) $\sum_{k=1}^{\infty} \frac{1}{k^2}$ and (b) $\sum_{k=4}^{\infty} \frac{1}{\sqrt{k}}$ converge.

Solution: The convergence of both series have already been determined using the Integral Test, but the P–Test is much easier to apply.

(a) $p = 2 > 1$ so $\sum_{k=1}^{\infty} \frac{1}{k^2}$ converges. (b) $p = 1/2 < 1$ so $\sum_{k=4}^{\infty} \frac{1}{\sqrt{k}}$ diverges.

The P–Test is very easy to use (Is the exponent $p > 1$ or is $p \leq 1$?), and it is also very useful. In the next section we will compare new series with series whose convergence we already know, and most often this comparison is with some P–series whose convergence we know about from the P–Test.

Note: **The P–Test does not give the value of the sum, it only answers the question of whether the series converges of diverges.**

PROBLEMS

In problems 1 – 15 show that the function determined by the terms of the given series satisfies the hypotheses of the Integral Test, and then use the Integral Test to determine whether the series converges or diverges.

1. $\sum_{k=1}^{\infty} \dfrac{1}{2k+5}$

2. $\sum_{k=1}^{\infty} \dfrac{1}{(2k+5)^2}$

3. $\sum_{k=1}^{\infty} \dfrac{1}{(2k+5)^{3/2}}$

4. $\sum_{k=1}^{\infty} \dfrac{\ln(k)}{k}$

5. $\sum_{k=2}^{\infty} \dfrac{1}{k \cdot (\ln(k))^2}$

6. $\sum_{k=1}^{\infty} \dfrac{1}{k^2} \cdot \sin\left(\dfrac{1}{k}\right)$

7. $\sum_{k=1}^{\infty} \dfrac{1}{k^2+1}$

8. $\sum_{k=1}^{\infty} \dfrac{1}{k^2+100}$

9. $\sum_{k=1}^{\infty} \left\{\dfrac{1}{k} - \dfrac{1}{k+3}\right\}$

10. $\sum_{k=1}^{\infty} \left\{\dfrac{1}{k} - \dfrac{1}{k+1}\right\}$

11. $\sum_{k=1}^{\infty} \dfrac{1}{k(k+5)}$

12. $\sum_{k=2}^{\infty} \dfrac{1}{k^2-1}$

13. $\sum_{k=1}^{\infty} k \cdot e^{-(k^2)}$

14. $\sum_{k=1}^{\infty} k^2 \cdot e^{-(k^3)}$

15. $\sum_{k=1}^{\infty} \dfrac{1}{\sqrt{6k+10}}$

For problems 16 – 20, (a) use the P–Test to determine whether the given series converges, and then (b) use the Integral Test to verify your convergence conclusion of part (a).

16. $\sum_{k=1}^{\infty} \dfrac{1}{k^2}$

17. $\sum_{k=1}^{\infty} \dfrac{1}{k^3}$

18. $\sum_{k=2}^{\infty} \dfrac{1}{k}$

19. $\sum_{k=2}^{\infty} \dfrac{1}{\sqrt{k}}$

20. $\sum_{k=3}^{\infty} \dfrac{1}{k^{2/3}}$

21. $\sum_{k=3}^{\infty} \dfrac{1}{k^{3/2}}$

In the proof of the Integral Test, we derived an inequality bounding the values of the partial sums $s_n = \sum_{k=1}^{n} a_k$ between the values of two integrals: $\int_{1}^{n+1} f(x)\,dx \leq \sum_{k=1}^{n} a_k \leq a_1 + \int_{1}^{n} f(x)\,dx$. For problems 22 – 27, use this inequality to determine bounds on the values of s_{10}, s_{100}, and $s_{1,000,000}$ for the given series.

22. $\sum_{k=1}^{\infty} \dfrac{1}{k^2}$ (Note: The exact value of $\sum_{k=1}^{\infty} \dfrac{1}{k^2}$ is $\dfrac{\pi^2}{6}$ but it beyond our means to prove that in this course.)

23. $\sum_{k=1}^{\infty} \frac{1}{k^3}$

24. $\sum_{k=1}^{\infty} \frac{1}{k}$

25. $\sum_{k=1}^{\infty} \frac{1}{k+1000}$

26. $\sum_{k=1}^{\infty} \frac{1}{k^2+1}$

27. $\sum_{k=1}^{\infty} \frac{1}{k^2+100}$

28. **Euler's Constant:** Define $g_1 = 1 - \ln(1) = 1$,

    $g_2 = (1 + \frac{1}{2}) - \ln(2) \approx 0.806853$,

    $g_3 = (1 + \frac{1}{2} + \frac{1}{3}) - \ln(3) \approx 0.734721$, and, in general,

    $g_n = (1 + \frac{1}{2} + \frac{1}{3} + \frac{1}{4} + \ldots + \frac{1}{n}) - \ln(n)$.

    (a) Make several copies of Fig. 7, and shade the regions represented by g_2, g_3, g_4, and g_n.

    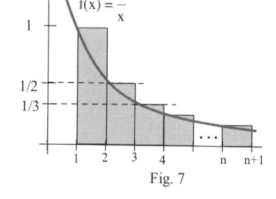

    Fig. 7

    (b) Provide a geometric argument that $g_n > 0$ for all $n \geq 1$.

    (c) Provide a geometric argument that $\{ g_n \}$ is monotonically decreasing: $g_{n+1} < g_n$ for all $n \geq 1$.

    (d) Conclude from parts (b) and (c) and the Monotone Convergence Theorem (Section 10.1) that $\{ g_n \}$ converges.

    (Note: The value to which $\{ g_n \}$ converges is denoted by "γ," the lower case Greek letter gamma, and is called Euler's constant. It is not even known if γ is a rational number. $\gamma \approx 0.5772157 \cdots$.)

29. (a) Show that the integral $\int_{2}^{\infty} \frac{1}{x \cdot (\ln x)^q} \, dx$ converges for $q > 1$ and diverges for $q \leq 1$.

    (b) Use the result of part (a) to state a "Q test" for $\sum_{k=2}^{\infty} \frac{1}{k \cdot (\ln k)^q}$.

In problems 30 – 33, use the result of Problem 29 to determine whether the given series converge.

30. $\sum_{k=2}^{\infty} \frac{1}{k \cdot \ln(k)}$

31. $\sum_{k=2}^{\infty} \frac{1}{k \cdot (\ln k)^3}$

32. $\sum_{k=2}^{\infty} \frac{1}{k \cdot \sqrt{\ln k}}$

33. $\sum_{k=2}^{\infty} \frac{1}{k \cdot \ln(k^3)}$

Practice Answers

Practice 1: { total area of rectangular pieces } > { area under the curve in Fig. 3a } so if the shaded area in Fig. 3a is infinite, then the shaded area in Fig. 3b is also infinite.

Practice 2: (a) Let $f(x) = \dfrac{1}{\sqrt{x}}$. Then $a_k = f(k)$ and f is continuous, positive and decreasing on $[1, \infty)$.

Then $\displaystyle\int_4^{\infty} \dfrac{4}{\sqrt{x}}\, dx = \lim_{n \to \infty} \int_1^n \dfrac{1}{\sqrt{x}}\, dx = \lim_{n \to \infty} 2x^{1/2} \Big|_4^n$

$= \lim_{n \to \infty} 2\sqrt{n} - 2\sqrt{4} = \infty$.

The integral $\displaystyle\int_4^{\infty} \dfrac{1}{\sqrt{x}}\, dx$ diverges so the series $\displaystyle\sum_{k=4}^{\infty} \dfrac{1}{\sqrt{k}}$ diverges.

(Note: It will be easier to determine that this series diverges by using the P–Test which occurs right after this Practice problem in the text.)

(b) Let $f(x) = e^{-x}$. Then $e^{-k} = a_k = f(k)$ and f is continuous, positive and decreasing on $[1, \infty)$.

Then $\displaystyle\int_1^{\infty} e^{-x}\, dx = \lim_{n \to \infty} \int_1^n e^{-x}\, dx = \lim_{n \to \infty} -e^{-x} \Big|_1^n = \lim_{n \to \infty} \left(-e^{-n}\right) - \left(-e^{-1}\right)$

$= \displaystyle\lim_{n \to \infty} \left(-\dfrac{1}{e^n}\right) - \left(-\dfrac{1}{e}\right) = \dfrac{1}{e} \approx 0.368$.

The integral $\displaystyle\int_1^{\infty} e^{-x}\, dx$ converges so the series $\displaystyle\sum_{k=1}^{\infty} e^{-k}$ converges.

10.5 POSITIVE TERM SERIES: COMPARISON TESTS

This section discusses how to determine whether some series converge or diverge by comparing them with other series which we already know converge or diverge. In the basic Comparison Test we compare the two series term by term. In the more powerful Limit Comparison Test, we compare limits of ratios of the terms of the two series. Finally, we focus on the parts of the terms of a series that determine whether the series converges or diverges.

Comparison Test

Informally, if the individual terms of our series are smaller than the corresponding terms of a known convergent series, then our series converges. If our series is larger, term by term, than a known divergent series then our series diverges. If the individual terms of our series are larger than the corresponding terms of a convergent series or smaller than the corresponding terms of a divergent series, then our series may converge or diverge — the Comparison Test does not tell us.

Comparison Test

Suppose we want to determine whether $\sum_{k=1}^{\infty} a_k$ converges or diverges.

(a) If there is a convergent series $\sum_{k=1}^{\infty} c_k$ with $0 < a_k \leq c_k$ for all k, then $\sum_{k=1}^{\infty} a_k$ converges.

(b) If there is a divergent series $\sum_{k=1}^{\infty} d_k$ with $a_k \geq d_k > 0$ for all k, then $\sum_{k=1}^{\infty} a_k$ diverges.

Proof: Since all of the terms of the $a_k, c_k,$ and d_k series are positive, their sequences of partial sums are all monotonic increasing. The proof compares the partial sums of the various series.

(a) Suppose that $0 < a_k \leq c_k$ for all k and that $\sum_{k=1}^{\infty} c_k$ converges. Since $\sum_{k=1}^{\infty} c_k$ converges,

then the partial sums $t_n = \sum_{k=1}^{n} c_k$ approach a finite limit: $\lim_{n \to \infty} t_n = L$.

For each n, $s_n \leq t_n$ (why?) so $\lim_{n \to \infty} s_n \leq \lim_{n \to \infty} t_n = L$, and the sequence $\{s_n\}$ is bounded by L. Finally, by the Monotone Convergence Theorem, we can conclude that $\{s_n\}$ converges and that the series $\sum_{k=1}^{\infty} a_k$ converges.

(b) Suppose that $a_k \geq d_k > 0$ for all k and that $\sum_{k=1}^{\infty} d_k$ diverges. Since $\sum_{k=1}^{\infty} d_k$ diverges, then the partial sums $u_n = \sum_{k=1}^{n} d_k$ approach infinity: $\lim_{n \to \infty} u_n = \infty$. Then

$$\lim_{n \to \infty} s_n \geq \lim_{n \to \infty} u_n = \infty$$

so the sequence of partial sums $\{s_n\}$ diverges and the series $\sum_{k=1}^{\infty} a_k$ diverges.

The Comparison Test requires that we select and compare our series against a series whose convergence or divergence is known, and that choice requires that we know a collection of series that converge and some that diverge. Typically, we pick a p–series or a geometric series to compare with our series, but this choice requires some experience and practice.

Example 1: Use the Comparison Test to determine the convergence or divergence of

(a) $\sum_{k=1}^{\infty} \frac{1}{k^2 + 3}$ and (b) $\sum_{k=1}^{\infty} \frac{k+1}{k^2}$.

Solution: For these two series it is useful to compare with p–series for appropriate values of p.

(a) For all k, $\frac{1}{k^2 + 3} < \frac{1}{k^2}$, and $\sum_{k=1}^{\infty} \frac{1}{k^2}$ converges (P–Test, p = 2)

so $\sum_{k=1}^{\infty} \frac{1}{k^2 + 3}$ converges.

(b) For all k, $\frac{k+1}{k^2} = \frac{1}{k} + \frac{1}{k^2} > \frac{1}{k}$.

Since $\sum_{k=1}^{\infty} \frac{1}{k}$ diverges (P–Test, p = 1), we conclude that $\sum_{k=1}^{\infty} \frac{k+1}{k^2}$ diverges.

Practice 1: Use the Comparison Test to determine the convergence or divergence of

(a) $\sum_{k=3}^{\infty} \frac{1}{\sqrt{k-2}}$ and (b) $\sum_{k=1}^{\infty} \frac{1}{2^k + 7}$.

Example 2: A student has shown algebraically that $\frac{1}{k^2} < \frac{1}{k^2-1} < \frac{1}{k}$ for all $k \geq 2$. From this information and the Comparison Test, what can the student conclude about the convergence of the series $\sum_{k=2}^{\infty} \frac{1}{k^2-1}$?

Solution: Nothing. The Comparison Test only gives a definitive answer if our series is smaller than a convergent series or if our series is larger than a divergent series. In this example, our series is larger than a convergent series, $\sum \frac{1}{k^2}$, and is smaller than a divergent series, $\sum \frac{1}{k}$, so we can not conclude anything about the convergence of $\sum_{k=2}^{\infty} \frac{1}{k^2-1}$.

However, we can show that if $k \geq 2$ then $\frac{1}{k^2-1} < \frac{2}{k^2}$. Since $\sum_{k=2}^{\infty} \frac{2}{k^2}$ converges (P–Test), we can conclude that $\sum_{k=2}^{\infty} \frac{1}{k^2-1}$ converges. Next in this section we present a variation on the Comparison Test that allows us to quickly conclude that $\sum_{k=2}^{\infty} \frac{1}{k^2-1}$ converges.

Limit Comparison Test

The exact value of the sum of a series depends on every part of the terms of the series, but if we are only asking about convergence or divergence, some parts of the terms can be safely ignored. For example, the three series with terms $1/k^2$, $1/(k^2+1)$, and $1/(k^2-1)$ converge to different values,

$$\sum_{k=2}^{\infty} \frac{1}{k^2} \approx 0.645, \quad \sum_{k=2}^{\infty} \frac{1}{k^2+1} \approx 0.577, \quad \sum_{k=2}^{\infty} \frac{1}{k^2-1} = 0.750,$$

but they all do converge. The "+ 1" and "– 1" in the denominators affect the value of the final sum, but they do not affect whether that sum is finite or infinite. When k is a large number, the values of $1/(k^2+1)$ and $1/(k^2-1)$ are both very close to the value of $1/k^2$, and the convergence or divergence of the series $\sum_{k=2}^{\infty} \frac{1}{k^2+1}$ and $\sum_{k=2}^{\infty} \frac{1}{k^2-1}$ can be predicted from the convergence or divergence of the series $\sum_{k=2}^{\infty} \frac{1}{k^2}$.

The Limit Comparison Test states these ideas precisely.

> **Limit Comparison Test**
>
> Suppose $a_k > 0$ for all k, and we want to determine whether $\sum_{k=1}^{\infty} a_k$ converges or diverges.
>
> If there is a series $\sum_{k=1}^{\infty} b_k$ so that $\lim_{k \to \infty} \frac{a_k}{b_k} = L$, a **positive, finite** value,
>
> then $\sum_{k=1}^{\infty} a_k$ and $\sum_{k=1}^{\infty} b_k$ both converge or both diverge.

Idea for a proof: The key idea is that if $\lim_{k \to \infty} \frac{a_k}{b_k} = L$ is a **positive, finite** value, then, when n is very large, $\frac{a_k}{b_k} \approx L$ so $a_k \approx L \cdot b_k$ and $\sum_{k=N}^{\infty} a_k \approx L \cdot \sum_{k=N}^{\infty} b_k$. If one of these series converges, then so does the other. If one of these series diverges, then so does the other. When n is a relatively small number, the a_k and b_k values may not have a ratio close to L, but the first "few" values of a series do not affect the convergence or divergence of the series. A proof of the Limit Comparison Test is given in an Appendix after the Practice Answers.

Example 3: Put $a_k = \frac{1}{k^2}$, $b_k = \frac{1}{k^2 + 1}$, and $c_k = \frac{1}{k^2 - 1}$ and show that $\lim_{k \to \infty} \frac{a_k}{b_k} = 1$

and $\lim_{k \to \infty} \frac{a_k}{c_k} = 1$. Since $\sum_{k=2}^{\infty} \frac{1}{k^2}$ converges (P–Test, p = 2) we can conclude

that $\sum_{k=2}^{\infty} \frac{1}{k^2 + 1}$ and $\sum_{k=2}^{\infty} \frac{1}{k^2 - 1}$ both converge too.

Solution: $\frac{a_k}{b_k} = \frac{1/k^2}{1/(k^2 + 1)} = \frac{k^2 + 1}{k^2} = 1 + \frac{1}{k^2} \longrightarrow 1$ so L = 1 is positive and finite.

Similarly, $\frac{a_k}{c_k} = \frac{1/k^2}{1/(k^2 - 1)} = \frac{k^2 - 1}{k^2} = 1 - \frac{1}{k^2} \longrightarrow 1$ so L = 1 is positive and finite.

Practice 2: (a) Find a p–series to "limit–compare" with $\sum_{k=2}^{\infty} \frac{k^2 + 5k}{k^3 + k^2 + 7}$.

(Suggestion: put $a_k = \frac{k^2 + 5k}{k^3 + k^2 + 7}$ and find a value of p so that $b_k = \frac{1}{k^p}$ and $\lim_{k \to \infty} \frac{a_k}{b_k} = L$, a positive, finite value.) Does $\sum_{k=2}^{\infty} \frac{k^2 + 5k}{k^3 + k^2 + 7}$ converge?

(b) Find a p–series to compare with $\sum_{k=3}^{\infty} \frac{5}{\sqrt{k^4 - 11}}$. Does $\sum_{k=3}^{\infty} \frac{5}{\sqrt{k^4 - 11}}$ converge?

The Limit Comparison Test is particularly useful because it allows us to ignore some parts of the terms that cause algebraic difficulties but that have no effect on the convergence of the series.

Using "Dominant Terms"

To use the Limit Comparison Test we need to pick a new series to compare with our given series. One effective way to pick the new series is to form the new series using the largest power of the variable (dominant term) from the numerator and the largest power of the variable (dominant term) from the denominator. The "dominant term" series consists of $\frac{\text{dominant term in the numerator}}{\text{dominant term in the denominator}}$. Then the Limit Comparison Test allows us to conclude that the original series converges if and only if the "dominant term" series converges.

Example 4: For each of the given series, form a new series consisting of the dominant terms from the numerator and the denominator. Does the series of dominant terms converge?

(a) $\sum_{k=3}^{\infty} \frac{5k^2 - 3k + 2}{17 + 2k^4}$ (b) $\sum_{k=1}^{\infty} \frac{1 + 4k}{\sqrt{k^3 + 5k}}$ (c) $\sum_{k=1}^{\infty} \frac{k^{23} + 1}{5k^{10} + k^{26} + 3}$.

Solution: (a) The dominant terms of the numerator and denominator are $5k^2$ and $2k^4$, respectively, so the "dominant term" series is $\sum_{k=3}^{\infty} \frac{5k^2}{2k^4} = \frac{5}{2} \sum_{k=3}^{\infty} \frac{1}{k^2}$ which converges (P–Test, p = 2).

(b) The dominant terms of the numerator and denominator are $4k$ and $k^{3/2}$, respectively, so the "dominant term" series is $\sum_{k=1}^{\infty} \frac{4k}{k^{3/2}} = 4 \sum_{k=3}^{\infty} \frac{1}{k^{1/2}}$ which diverges (P–Test, p = 1/2).

(c) The dominant terms of the numerator and denominator are k^{23} and k^{26}, respectively, so the "dominant term" series is $\sum_{k=1}^{\infty} \frac{k^{23}}{k^{26}} = \sum_{k=1}^{\infty} \frac{1}{k^3}$ which converges (P–Test, p = 3).

Using the Limit Comparison Test to compare the given series with the "dominant term" series, we can conclude that the given series (a) and (c) converge and that the given series (b) diverges.

Practice 3: For each of the given series, form a new series consisting of the dominant terms from the numerator and the denominator. Does the series of dominant terms converge? Do the given series converge?

(a) $\displaystyle\sum_{k=1}^{\infty} \frac{3k^4 - 5k + 2}{1 + 17k^2 + 9k^5}$ (b) $\displaystyle\sum_{k=1}^{\infty} \frac{\sqrt{1+9k}}{k^2 + 5k - 2}$ (c) $\displaystyle\sum_{k=1}^{\infty} \frac{k^{25} + 1}{5k^{10} + k^{26} + 3}$.

Experienced users of series commonly use "dominant terms" to make quick and accurate judgments about the convergence or divergence of a series. With practice, so can you.

PROBLEMS

In problems 1 – 12 use the Comparison Test to determine whether the given series converge or diverge.

1. $\displaystyle\sum_{k=1}^{\infty} \frac{\cos^2(k)}{k^2}$
2. $\displaystyle\sum_{k=1}^{\infty} \frac{3}{k^3 + 7}$
3. $\displaystyle\sum_{n=3}^{\infty} \frac{5}{n-1}$

4. $\displaystyle\sum_{n=1}^{\infty} \frac{2+\sin(n)}{n^3}$
5. $\displaystyle\sum_{j=1}^{\infty} \frac{3+\cos(j)}{j}$
6. $\displaystyle\sum_{j=1}^{\infty} \frac{\arctan(j)}{j^{3/2}}$

7. $\displaystyle\sum_{k=1}^{\infty} \frac{\ln(k)}{k}$
8. $\displaystyle\sum_{k=1}^{\infty} \frac{k-1}{k \cdot 1.5^k}$
9. $\displaystyle\sum_{k=1}^{\infty} \frac{k+9}{k \cdot 2^k}$

10. $\displaystyle\sum_{n=1}^{\infty} \frac{n^3 + 7}{n^4 - 1}$
11. $\displaystyle\sum_{n=1}^{\infty} \frac{1}{1+2+3+\ldots+(n-1)+n}$
12. $\displaystyle\sum_{k=1}^{\infty} \frac{1}{k!} = \sum_{k=1}^{\infty} \frac{1}{1\cdot 2\cdot 3\cdot \ldots \cdot (k-1)\cdot k}$

In problems 13 – 21 use the Limit Comparison Test (or the Nth Term Test) to determine whether the given series converge or diverge.

13. $\displaystyle\sum_{k=3}^{\infty} \frac{k+1}{k^2 + 4}$
14. $\displaystyle\sum_{j=1}^{\infty} \frac{7}{\sqrt{j^3 + 3}}$
15. $\displaystyle\sum_{w=1}^{\infty} \frac{5}{w+1}$

16. $\displaystyle\sum_{n=1}^{\infty} \frac{7n^3 - 4n + 3}{3n^4 + 7n^3 + 9}$
17. $\displaystyle\sum_{k=1}^{\infty} \frac{k^3}{(1+k^2)^3}$
18. $\displaystyle\sum_{k=1}^{\infty} \left(\frac{\arctan(k)}{k}\right)^2$

19. $\displaystyle\sum_{n=1}^{\infty} \left(\frac{5 - \frac{1}{n}}{n}\right)^3$
20. $\displaystyle\sum_{w=1}^{\infty} \left(1 + \frac{1}{w}\right)^w$
21. $\displaystyle\sum_{j=1}^{\infty} \left(1 - \frac{1}{j}\right)^j$

In problems 22 – 30 use "dominant term" series to determine whether the given series converge or diverge.

22. $\sum_{n=3}^{\infty} \dfrac{n+100}{n^2-4}$

23. $\sum_{k=1}^{\infty} \dfrac{7k}{\sqrt{k^3+5}}$

24. $\sum_{k=1}^{\infty} \dfrac{5}{k+1}$

25. $\sum_{j=1}^{\infty} \dfrac{j^3-4j+3}{2j^4+7j^6+9}$

26. $\sum_{n=1}^{\infty} \dfrac{5n^3+7n^2+9}{(1+n^3)^2}$

27. $\sum_{n=1}^{\infty} \left(\dfrac{\arctan(3n)}{2n}\right)^2$

28. $\sum_{k=1}^{\infty} \left(\dfrac{3-\frac{1}{k}}{k}\right)^2$

29. $\sum_{j=1}^{\infty} \dfrac{\sqrt{j^3+4j^2}}{j^2+3j-2}$

30. $\sum_{k=1}^{\infty} \dfrac{k+9}{k \cdot 2^k}$

Putting it all together

In problems 31 – 51 use any of the methods from this or previous sections to determine whether the given series converge or diverge. Give reasons for your answers.

31. $\sum_{n=2}^{\infty} \dfrac{n^2+10}{n^3-2}$

32. $\sum_{k=1}^{\infty} \dfrac{3k}{\sqrt{k^5+7}}$

33. $\sum_{k=1}^{\infty} \dfrac{3}{2k+1}$

34. $\sum_{j=1}^{\infty} \dfrac{j^2-j+1}{3j^4+2j^2+1}$

35. $\sum_{n=1}^{\infty} \dfrac{2n^3+n^2+6}{(3+n^2)^2}$

36. $\sum_{n=1}^{\infty} \left(\dfrac{\arctan(2n)}{3n}\right)^3$

37. $\sum_{k=1}^{\infty} \left(\dfrac{1-\frac{2}{k}}{k}\right)^3$

38. $\sum_{j=1}^{\infty} \dfrac{\sqrt{j^2+4j}}{j^3-2}$

39. $\sum_{k=1}^{\infty} \dfrac{k+5}{k \cdot 3^k}$

40. $\sum_{n=1}^{\infty} \dfrac{1+\sin(n)}{n^2+4}$

41. $\sum_{k=1}^{\infty} \dfrac{k+2}{\sqrt{k^2+1}}$

42. $\sum_{k=1}^{\infty} \dfrac{\sin(k\pi)}{k+1}$

43. $\sum_{j=1}^{\infty} \dfrac{3}{e^j+j}$

44. $\sum_{n=1}^{\infty} \dfrac{(2+3n)^2+9}{(1+n^3)^2}$

45. $\sum_{n=1}^{\infty} \left(\dfrac{\tan(3)}{2+n}\right)^2$

46. $\sum_{n=1}^{\infty} \sin^2\left(\dfrac{1}{n}\right)$

47. $\sum_{n=1}^{\infty} \sin^3\left(\dfrac{1}{n}\right)$

48. $\sum_{n=1}^{\infty} \cos^2\left(\dfrac{1}{n}\right)$

49. $\sum_{j=1}^{\infty} \cos^3\left(\dfrac{1}{j}\right)$

50. $\sum_{n=1}^{\infty} \tan^2\left(\dfrac{1}{n}\right)$

51. $\sum_{n=1}^{\infty} \left(1-\dfrac{2}{n}\right)^n$

Review for Positive Term Series: Converge or Diverge

State whether the given series converge or diverge and give reasons for your answer. You may need any of the methods discussed so far as well as some ingenuity.

R1. $\sum_{n=3}^{\infty} \dfrac{5}{3^n}$

R2. $\sum_{j=3}^{\infty} \dfrac{5+\cos(j^3)}{j^2}$

R3. $\sum_{w=1}^{\infty} \dfrac{2}{3+\sin(w^3)}$

R4. $\sum_{n=1}^{\infty} \dfrac{5}{(1/3)^n}$

R5. $\sum_{k=1}^{\infty} e^{-k}$

R6. $\sum_{w=1}^{\infty} \sin^2\left(\dfrac{1}{w}\right)$ (Hint: for $0 \le x \le 1$, $\sin(x) \le x$)

R7. $\sum_{k=1}^{\infty} \sin\left(\dfrac{1}{k^3}\right)$ (see the R6 hint)

R8. $\sum_{j=1}^{\infty} \cos^2\left(\dfrac{1}{j}\right)$

R9. $\sum_{n=3}^{\infty} \dfrac{5+\cos(n^2)}{n^3}$

R10. $\sum_{k=1}^{\infty} \dfrac{1}{k \cdot (3+\ln(k))}$

R11. $\sum_{j=1}^{\infty} \dfrac{1}{j \cdot (3+\ln(j))^2}$

R12. $\sum_{n=1}^{\infty} \dfrac{4}{n \cdot \arctan(n)}$

R13. $\sum_{n=3}^{\infty} \dfrac{4 \cdot \arctan(n)}{n}$

R14. $\sum_{k=1}^{\infty} \dfrac{\ln(k)}{k^3}$

R15. $\sum_{k=1}^{\infty} \dfrac{\ln(k)}{k^2}$

R16. $\sum_{j=1}^{\infty} \left(\dfrac{j}{2j+3}\right)^j$

R17. $\sum_{n=1}^{\infty} \dfrac{1+n}{1+n^2}$

R18. $\sum_{n=1}^{\infty} \left[\sin(n) - \sin(n+1)\right]$

R19. $\sum_{k=1}^{\infty} \sqrt{\dfrac{k^3+5}{k^5+3}}$

R20. $\sum_{k=1}^{\infty} \dfrac{1}{k^k}$

R21. $\sum_{n=1}^{\infty} n^{1/n}$

Practice Answers

Practice 1: (a) For $k > 3$, $0 < k-2 < k$ so $0 < \sqrt{k-2} < \sqrt{k}$ and $\dfrac{1}{\sqrt{k-2}} > \dfrac{1}{\sqrt{k}} = \dfrac{1}{k^{1/2}}$.

$\sum_{k=3}^{\infty} \dfrac{1}{k^{1/2}}$ diverges (P–test, p = 1/2) so $\sum_{k=3}^{\infty} \dfrac{1}{\sqrt{k-2}}$ diverges.

(b) For $k > 1$, $2^k + 7 > 2^k > 0$ so $\dfrac{1}{2^k + 7} < \dfrac{1}{2^k}$.

$\sum_{k=3}^{\infty} \dfrac{1}{2^k} = \sum_{k=3}^{\infty} \left(\dfrac{1}{2}\right)^k$ which is a convergent geometric series (r = 1/2) so $\sum_{k=3}^{\infty} \dfrac{1}{2^k + 7}$ converges.

Practice 2: (a) $a_k = \dfrac{k^2 + 5k}{k^3 + k^2 + 7}$. Put $b_k = \dfrac{1}{k^1}$. Then

$$\dfrac{a_k}{b_k} = \dfrac{\frac{k^2 + 5k}{k^3 + k^2 + 7}}{\frac{1}{k^1}} = \left(\dfrac{k^1}{1}\right)\dfrac{k^2 + 5k}{k^3 + k^2 + 7} = \dfrac{k^3 + 5k^2}{k^3 + k^2 + 7} = \dfrac{k^3\left(1 + \frac{5}{k}\right)}{k^3\left(1 + \frac{1}{k} + \frac{7}{k^3}\right)} \longrightarrow 1$$

so $L = 1$ is positive and finite, and $\sum_{k=1}^{\infty} a_k$ and $\sum_{k=1}^{\infty} b_k$ both converge or both diverge.

Since we know $\sum_{k=1}^{\infty} b_k = \sum_{k=1}^{\infty} \dfrac{1}{k}$ diverges (P–test, p=1 or as the Harmonic series),

we can conclude that $\sum_{k=1}^{\infty} a_k = \sum_{k=1}^{\infty} \dfrac{k^2 + 5k}{k^3 + k^2 + 7}$ diverges

(b) $a_k = \dfrac{5}{\sqrt{k^4 - 11}}$. Put $b_k = \dfrac{1}{\sqrt{k^4}} = \dfrac{1}{k^2}$. Then

$$\dfrac{a_k}{b_k} = \dfrac{\frac{5}{\sqrt{k^4 - 11}}}{\frac{1}{k^2}} = \dfrac{k^2}{1} \cdot \dfrac{5}{\sqrt{k^4 - 11}} = \dfrac{5}{1}\sqrt{\dfrac{k^4}{k^4 - 11}} \longrightarrow \dfrac{5}{1}\sqrt{1} = 5$$

so $L = 5$ is positive and finite, and $\sum_{k=1}^{\infty} a_k$ and $\sum_{k=1}^{\infty} b_k$ both converge or both diverge.

Since we know $\sum_{k=1}^{\infty} b_k = \sum_{k=1}^{\infty} \dfrac{1}{k^2}$ converges (P–test, p=2), we can conclude that

$\sum_{k=1}^{\infty} a_k = \sum_{k=1}^{\infty} \dfrac{5}{\sqrt{k^4 - 11}}$ converges.

Practice 3: (a) $\sum_{k=1}^{\infty} \dfrac{3k^4}{9k^5} = \dfrac{1}{3}\sum_{k=1}^{\infty} \dfrac{1}{k}$ which diverges (P–test, p = 1) so $\sum_{k=1}^{\infty} \dfrac{3k^4 - 5k + 2}{1 + 17k^2 + 9k^5}$ diverges.

(b) $\sum_{k=1}^{\infty} \dfrac{\sqrt{9k}}{k^2} = 3\sum_{k=1}^{\infty} \dfrac{k^{1/2}}{k^2} = 3\sum_{k=1}^{\infty} \dfrac{1}{k^{3/2}}$ which converges (P–test, p = 3/2)

so $\sum_{k=1}^{\infty} \dfrac{\sqrt{1 + 9k}}{k^2 + 5k - 2}$ converges.

(c) $\sum_{k=1}^{\infty} \frac{k^{25}}{k^{26}} = \sum_{k=1}^{\infty} \frac{1}{k}$ which diverges (P–test, p = 1) so $\sum_{k=1}^{\infty} \frac{k^{25}+1}{5k^{10}+k^{26}+3}$ diverges.

Appendix: Proof of the Limit Comparison Test

(a) Suppose $\sum_{k=1}^{\infty} b_k$ converges and that $\lim_{k \to \infty} \frac{a_k}{b_k} = L$, a **positive, finite** value.

Since $\lim_{k \to \infty} \frac{a_k}{b_k} = L$, there is a value N so that $\frac{a_k}{b_k} < L+1$ when $k \geq N$. Then

$a_k < b_k \cdot (L+1)$ when $k \geq N$, and $\sum_{k=N}^{\infty} a_k < (L+1) \cdot \sum_{k=N}^{\infty} b_k$. Since $\sum_{k=N}^{\infty} b_k$

converges, we can conclude that $\sum_{k=N}^{\infty} a_k$ converges so $\sum_{k=1}^{\infty} a_k$ also converges.

(b) Suppose $\sum_{k=1}^{\infty} b_k$ diverges and that $\lim_{k \to \infty} \frac{a_k}{b_k} = L$, a **positive, finite** value.

Since $\lim_{k \to \infty} \frac{a_k}{b_k} = L$, there is a value N so that $\frac{a_k}{b_k} > L/2 > 0$ when $k \geq N$. Then

$a_k > \frac{L}{2} \cdot b_k$ when $k \geq N$, and $\sum_{k=N}^{\infty} a_k \geq \frac{L}{2} \cdot \sum_{k=N}^{\infty} b_k$. Since $\sum_{k=N}^{\infty} b_k$ diverges, we

can conclude that $\sum_{k=N}^{\infty} a_k$ diverges so $\sum_{k=1}^{\infty} a_k$ also diverges.

10.6 ALTERNATING SERIES

In the last two sections we considered tests for the convergence of series whose terms were all positive. In this section we examine series whose terms change signs in a special way: they alternate between positive and negative. And we present a very easy–to–use test to determine if these alternating series converge.

An **alternating series** is a series whose terms alternate between positive and negative. For example, the following are alternating series:

(1) $\quad 1 - \frac{1}{2} + \frac{1}{3} - \frac{1}{4} + \frac{1}{5} - \ldots + (-1)^{k+1} \frac{1}{k} + \ldots = \sum_{k=1}^{\infty} (-1)^{k+1} \frac{1}{k} \quad$ (alternating harmonic series)

(2) $\quad -\frac{1}{3} + \frac{2}{4} - \frac{3}{5} + \frac{4}{6} - \frac{5}{7} + \ldots + (-1)^{k} \frac{k}{k+2} + \ldots = \sum_{k=1}^{\infty} (-1)^{k} \frac{k}{k+2}$

(3) $\quad -\frac{1}{\sqrt{3}} + \frac{1}{\sqrt{5}} - \frac{1}{\sqrt{7}} + \frac{1}{\sqrt{9}} + \frac{1}{\sqrt{11}} - \ldots + (-1)^{k} \frac{1}{\sqrt{2k+1}} + \ldots = \sum_{k=1}^{\infty} (-1)^{k} \frac{1}{\sqrt{2k+1}}$.

Figures 1, 2 and 3 show graphs and tables of values of several partial sums s_n for each of these series. As we move from left to right in each graph (as n increases), the partial sums alternately get larger and smaller, a common pattern for the partial sums of alternating series. The same pattern appears in the tables.

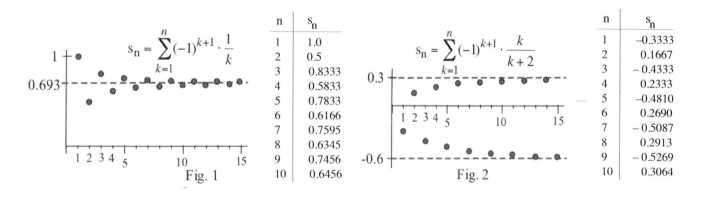

Fig. 1

Fig. 2

Fig. 3

Alternating Series Test

The following result provides a very easy way to determine that some alternating series converge. It says that if the absolute values of the terms decrease monotonically to 0 then the series converges. This is the main result for alternating series.

Alternating Series Test

If the numbers a_n satisfy the three conditions

(i) $a_n > 0$ for all n (each a_n is positive)

(ii) $a_n > a_{n+1}$ (the terms a_n are monotonically decreasing)

(iii) $\lim_{n \to \infty} a_n = 0$

then the **alternating series** $\sum_{n=1}^{\infty} (-1)^{n+1} a_n$ converges.

Proof: In order to show that the alternating series converges, we need to show that the sequence of partial sums approaches a finite limit, and we do so in this case by showing that the sequence of even partial sums $\{s_2, s_4, s_6, \ldots\}$ and the sequence of odd partial sums $\{s_1, s_3, s_5, \ldots\}$ each approach the same value.

Even partial sums:

$s_2 = a_1 - a_2 > 0$ since $a_1 > a_2$

$s_4 = a_1 - a_2 + a_3 - a_4 = s_2 + (a_3 - a_4) > s_2$ since $a_3 > a_4$

$s_6 = a_1 - a_2 + a_3 - a_4 + a_5 - a_6 = s_4 + (a_5 - a_6) > s_4$ since $a_5 > a_6$.

In general, the sequence of even partial sums is positive and increasing

$s_{2n+2} = s_{2n} + (a_{2n+1} - a_{2n+2}) > s_{2n} > 0$ since $a_{2n+1} > a_{2n+2}$.

Also,

$s_{2n} = a_1 - a_2 + a_3 - a_4 + a_5 - \ldots - a_{2n-2} + a_{2n-1} - a_{2n}$

$\quad = a_1 - (a_2 - a_3) - (a_4 - a_5) \ldots - (a_{2n-2} - a_{2n-1}) - a_{2n} < a_1$.

so the sequence of even partial sums is bounded above by a_1.

Since the sequence $\{s_2, s_4, s_6, \ldots\}$ of even partial sums is increasing and bounded, the Monotone Convergence Theorem of Section 10.1 tells us that sequence of even partial sums converges to some finite limit: $\lim_{n \to \infty} s_{2n} = L$.

Odd partial sums:

$$s_{2n+1} = s_{2n} + a_{2n+1} \quad \text{so} \quad \lim_{n \to \infty} s_{2n+1} = \lim_{n \to \infty} s_{2n} + \lim_{n \to \infty} a_{2n+1} = L + 0 = L.$$

Since the sequence of even partial sums and the sequence of odd partial sums both approach the same limit L, we can conclude that the limit of the sequence of partial sums is L and that the series $\sum_{n=1}^{\infty} (-1)^{n+1} a_n$ converges to L: $\sum_{n=1}^{\infty} (-1)^{n+1} a_n = L$.

Example 1: Show that each of the three alternating series satisfies the three conditions in the hypothesis of the Alternating Series Test. Then we can conclude that each of them converges.

(a) $1 - \frac{1}{2} + \frac{1}{3} - \frac{1}{4} + \ldots = \sum_{n=1}^{\infty} (-1)^{n+1} \frac{1}{n}$

(b) $\frac{3}{1} - \frac{3}{\sqrt{2}} + \frac{3}{\sqrt{3}} - \frac{3}{\sqrt{4}} + \ldots = \sum_{n=1}^{\infty} (-1)^{n+1} \frac{3}{\sqrt{n}}$

(c) $\frac{7}{2 \cdot \ln(2)} - \frac{7}{3 \cdot \ln(3)} + \frac{7}{4 \cdot \ln(4)} - \ldots = \sum_{n=2}^{\infty} (-1)^{n} \frac{7}{n \cdot \ln(n)}$.

Solution: (a) This series is called the **alternating harmonic series**. (i) $a_n = \frac{1}{n} > 0$ for all positive n. (ii) Since $n < n+1$, then $\frac{1}{n} > \frac{1}{n+1}$ and $a_n > a_{n+1}$.

(iii) $\lim_{n \to \infty} a_n = \lim_{n \to \infty} \frac{1}{n} = 0$. Therefore, the alternating harmonic series converges.

Fig. 1 shows some partial sums for the alternating harmonic series.

(b) $a_n = \frac{3}{\sqrt{n}} > 0$. Since $n < n+1$, we have $\sqrt{n} < \sqrt{n+1}$ and $\frac{3}{\sqrt{n}} > \frac{3}{\sqrt{n+1}}$.

$\lim_{n \to \infty} a_n = \lim_{n \to \infty} \frac{3}{\sqrt{n}} = 0$. The series $\frac{3}{1} + \frac{3}{\sqrt{2}} + \frac{3}{\sqrt{3}} + \frac{3}{\sqrt{4}} + \ldots$ converges.

(c) $a_n = \frac{7}{n \cdot \ln(n)} > 0$ for $n \geq 2$. Since $n < n+1$ and $\ln(n) < \ln(n+1)$, we have

$n \cdot \ln(n) < (n+1) \cdot \ln(n+1)$ and $\frac{7}{n \cdot \ln(n)} > \frac{7}{(n+1) \cdot \ln(n+1)}$.

$\lim_{n \to \infty} a_n = \lim_{n \to \infty} \frac{7}{n \cdot \ln(n)} = 0$. The series $\sum_{n=2}^{\infty} (-1)^n \frac{7}{n \cdot \ln(n)}$ converges.

Practice 1: Show that these two alternating series satisfy the three conditions in the hypothesis of the Alternating Series Test. Then we can conclude that each of them converges.

(a) $\quad 1 - \frac{1}{4} + \frac{1}{9} - \frac{1}{16} + \ldots = \sum_{n=1}^{\infty} (-1)^{n+1} \frac{1}{n^2}$

(b) $\quad \frac{3}{\ln(2)} - \frac{3}{\ln(3)} + \frac{3}{\ln(4)} - \frac{3}{\ln(5)} + \ldots = \sum_{n=2}^{\infty} (-1)^n \frac{3}{\ln(n)}$

Example 2: Does $\sum_{n=1}^{\infty} (-1)^{n+1} \frac{n}{n+2}$ converge?

Solution: $a_n = \frac{n}{n+2} > 0$, but $\lim_{n \to \infty} a_n = \lim_{n \to \infty} \frac{n}{n+2} = 1 \neq 0$. Since the terms do not approach 0, we can conclude from the Nth Term Test For Divergence (Section 10.2) that the series diverges.

Fig. 2 shows some of the partial sums for this series. You should notice that the even and the odd partial sums in Fig. 2 are approaching two different values.

Practice 2: (a) Does $\sum_{n=1}^{\infty} (-1)^{n+1} n$ converge? (b) Does $\sum_{n=1}^{\infty} (-1)^{n+1} \frac{1}{\sqrt{2n+1}}$ converge?

Example Of A Divergent Alternating Series

If the terms of a series, any series, do not approach 0, then the series must diverge (Nth Term Test For Divergence). If the terms do approach 0 the series may converge or may diverge. There are divergent alternating series whose terms approach 0 (but the approach to 0 is not monotonic). For example, $\frac{3}{2} - \frac{1}{2} + \frac{3}{4} - \frac{1}{4} + \frac{3}{6} - \frac{1}{6} + \frac{3}{8} - \frac{1}{8} + \ldots$ is an alternating series whose terms approach 0, but the series diverges.

The even partial sums of our new series are

$s_2 = \frac{3}{2} - \frac{1}{2} = 1$,

$s_4 = \frac{3}{2} - \frac{1}{2} + \frac{3}{4} - \frac{1}{4} = (\frac{3}{2} - \frac{1}{2}) + (\frac{3}{4} - \frac{1}{4}) = 1 + \frac{1}{2}$,

$s_6 = \frac{3}{2} - \frac{1}{2} + \frac{3}{4} - \frac{1}{4} + \frac{3}{6} - \frac{1}{6} = (\frac{3}{2} - \frac{1}{2}) + (\frac{3}{4} - \frac{1}{4}) + (\frac{3}{6} - \frac{1}{6}) = 1 + \frac{1}{2} + \frac{1}{3}$,

and $s_{2n} = 1 + \frac{1}{2} + \frac{1}{3} + \ldots + \frac{1}{n}$.

You should recognize that these partial sums are the partial sums of the harmonic series, a divergent series, so the partial sums of our new series diverge and our new series is divergent. If the terms of an alternating series approach 0, but not monotonically, then the Alternating Series Test does not apply, and the series may converge or it may diverge.

Approximating the Sum of an Alternating Series

If we know that a series converges and if we add the first "many" terms together, then we expect that the resulting partial sum is close to the value S obtained by adding all of the terms together. Generally, however, we do not know how close the partial sum is to S. The situation with many alternating series is much nicer. The next result says that for some alternating series (those that satisfy the three conditions in the next box), the difference between S and the nth partial sum of the alternating series, $|S - s_n|$, is less than the magnitude of the next term in the series, a_{n+1}.

Estimation Bound for Alternating Series

If $\quad$ S is the sum of an alternating series $\sum_{n=1}^{\infty} (-1)^{n+1} a_n$

that satisfies the three conditions
(i) $\quad a_n > 0$ for all n $\quad$ (each a_n is positive),
(ii) $\quad a_n > a_{n+1}$ $\quad$ (the terms are monotonically decreasing),
and (iii) $\quad \lim_{n \to \infty} a_n = 0$ $\quad$ (the terms approach 0),

then the nth partial sum s_n is within a_{n+1} of the sum S: $\quad s_n - a_{n+1} < S < s_n + a_{n+1}$
and $|$ approximation "error" using s_n as an estimate of S $| = |S - s_n| < a_{n+1}$.

Note: This Estimation Bound only applies to alternating series. It is tempting, but wrong, to use it with other types of series.

Geometric idea behind the Estimation Bound:

If we have an alternating series that satisfies the hypothesis of the Estimation Bound, then the graph of the sequence $\{s_n\}$ of partial sums is "trumpet–shaped" or "funnel–shaped" (Fig. 4). The partial sums are alternately above and below the value S, and they "squeeze" in on the value S. Since the distance from s_n to S is less than the distance between the successive terms s_n and s_{n+1} (Fig. 5), then
$|S - s_n| < |s_n - s_{n+1}| = a_{n+1}$.

Proof of the Estimation Bound for Alternating Series:

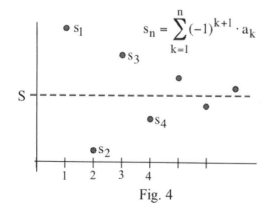

Fig. 4

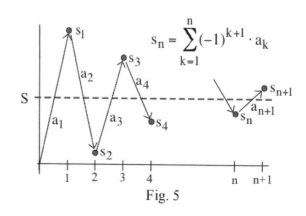

Fig. 5

10.6 Alternating Series

$$S - s_n = (a_1 - a_2 + a_3 - a_4 + \ldots + (-1)^{n+1} a_n + (-1)^{n+2} a_{n+1} \ldots) - (a_1 - a_2 + a_3 - a_4 + \ldots + (-1)^{n+1} a_n)$$
$$= (-1)^{n+2} a_{n+1} + (-1)^{n+3} a_{n+2} + (-1)^{n+4} a_{n+3} + \ldots$$
$$= (-1)^{n+2} (a_{n+1} - a_{n+2} + a_{n+3} - \ldots).$$

Then $0 \leq |S - s_n|\ = |(-1)^{n+2}(a_{n+1} - a_{n+2} + a_{n+3} - a_{n+4} + a_{n+5} \ldots)|$
$= a_{n+1} - a_{n+2} + a_{n+3} - a_{n+4} + a_{n+5} \ldots$ since $|(-1)^{n+2}| = 1$ and the rest is positive
$= a_{n+1} - (a_{n+2} - a_{n+3}) - (a_{n+4} - a_{n+5}) - \ldots < a_{n+1}$.

The Estimation Bound is typically used in two different ways. Sometimes we know the value of n, and we want to know how close s_n is to S. Sometimes we know how close we want s_n to be to S, and we want to find a value of n to ensure that level of closeness. The next two Examples illustrate these two different uses of the Estimation Bound.

Example 3: How close is $\sum_{n=1}^{4} (-1)^{n+1} \frac{1}{n^2} = 1 - \frac{1}{4} + \frac{1}{9} - \frac{1}{16} \approx 0.79861$ to the sum $\sum_{n=1}^{\infty} (-1)^{n+1} \frac{1}{n^2}$?

Solution: $a_n = \frac{1}{n^2}$ and $s_4 = \sum_{n=1}^{4} (-1)^{n+1} \frac{1}{n^2} = 1 - \frac{1}{4} + \frac{1}{9} - \frac{1}{16} \approx 0.79861$ so, by the Estimation Bound, we

can conclude that $|S - s_4| < a_5 = \frac{1}{25} = 0.04$: 0.79861 is within 0.04 of the exact value S. Then

$0.79861 - 0.04 < S < 0.79861 + 0.04$ and $0.75861 < S < 0.83861$.

Similarly, $s_9 = 1 - \frac{1}{4} + \frac{1}{9} - \frac{1}{16} + \ldots + \frac{1}{81} \approx 0.82796$ is within $a_{10} = \frac{1}{100} = 0.01$ of the exact

value of S, and $s_{99} \approx 0.822517$ is within $a_{100} = \frac{1}{100^2} = 0.0001$ of the exact value of S.

Then $0.822517 - 0.0001 < S < 0.822517 + 0.0001$ and $0.822417 < S < 0.822617$.

Practice 3: Evaluate s_4 and s_9 for the alternating series $\sum_{n=1}^{\infty} (-1)^{n+1} \frac{1}{n^3}$ and determine bounds

for $|S - s_4|$ and $|S - s_9|$.

Example 4: Find the number of terms needed so that s_n is within 0.001 of the exact

value of $\sum_{n=1}^{\infty} (-1)^{n+1} \frac{1}{n!}$ and evaluate s_n.

Solution: We know $|S - s_n| < a_{n+1}$ so we want to find n so that $a_{n+1} \leq 0.001 = \frac{1}{1000}$. With a little

numerical experimentation on a calculator, we see that $6! = 720$ and $1/720$ is not small enough, but

$7! = 5{,}040 > 1{,}000$ so $\frac{1}{7!} = \frac{1}{5040} \approx 0.000198 < 0.001$. Since $n+1 = 7$, $s_6 \approx 0.631944$ is the first partial sum guaranteed to be within 0.001 of S. In fact, s_6 is within $\frac{1}{5040} \approx 0.000198$ of S, so $0.631746 < S < 0.632142$.

Practice 4: Find the number of terms needed so s_n is within 0.001 of the value of $\sum_{n=1}^{\infty} (-1)^{n+1} \frac{1}{n^3 + 5}$ and evaluate s_n.

The Estimation Bound guarantees that s_n is **within** a_{n+1} of S. In fact, s_n is often much closer than a_{n+1} to S. The value a_{n+1} is an **upper bound** on how far s_n can be from S.

Note 1: The first **finite** number of terms do not affect the convergence or divergence of a series (they do effect the sum S) so we can use the Alternating Series Test and the Estimation Bound if the terms of a series "eventually" satisfy the conditions of the hypotheses of these results. By "eventually" we mean there is a value M so that for $n > M$ the series is an alternating series.

Note 2: If a series has some positive terms and some negative terms and if those terms do NOT "eventually" alternate signs, then we can NOT use the Alternating Series Test – it simply does not apply to such series.

PROBLEMS

In problems 1 – 6 you are given the values of the first four **terms** of a series. (a) Calculate and graph the first four partial sums for each series. (b) Which of the series are not alternating series?

1. $1, -0.8, 0.6, -0.4$
2. $-1, 1.5, -0.7, 1$
3. $-1, 2, -3, 4$
4. $2, -1, -0.5, 0.3$
5. $-1, -0.6, 0.4, 0.2$
6. $2, -1, 0.5, -0.3$

In problems 7 – 12 you are given the values of the first five **partial sums** of a series. Which of the series are not alternating series. Why?

7. $2, 1, 3, 2, 4$
8. $2, 1, 1.8, 1.4, 1.6$
9. $2, 3, 2.1, 2.9, 2.8$
10. $-3, -1, -2.5, -1.5, -2$
11. $-1, 1, -0.8, -0.6, -0.4$
12. $-2.3, -1.6, -1.4, -1.8, -1.7$

13. Fig. 6 shows the graphs of the partial sums of three series. Which is/are not the partial sums of alternating series? Why?

14. Fig. 7 shows the graphs of the partial sums of three series. Which is/are not the partial sums of alternating series? Why?

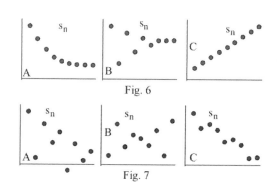

Fig. 6

Fig. 7

15. Fig. 8 shows the graphs of the partial sums of three series. Which is/are not the partial sums of alternating series? Why?

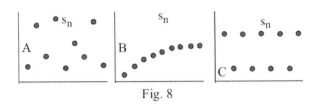
Fig. 8

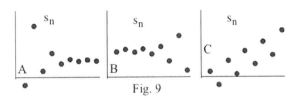
Fig. 9

16. Fig. 9 shows the graphs of the partial sums of three series. Which is/are not the partial sums of alternating series? Why?

In problems 17 – 31 determine whether the given series converge or diverge.

17. $\sum_{n=1}^{\infty} (-1)^{n+1} \frac{1}{n+5}$

18. $\sum_{n=2}^{\infty} (-1)^{n} \frac{1}{\ln(n)}$

19. $\sum_{n=1}^{\infty} (-1)^{n+1} \frac{n}{n^2+3}$

20. $\sum_{n=5}^{\infty} (-0.99)^{n+1}$

21. $\sum_{n=1}^{\infty} (-1)^{n} \frac{|n+3|}{|n+7|}$

22. $\sum_{n=1}^{\infty} (-1)^{n+1} \sin(\frac{1}{n})$

23. $\sum_{n=4}^{\infty} \frac{\cos(n\pi)}{n}$

24. $\sum_{n=3}^{\infty} \frac{\sin(n\pi)}{n}$

25. $\sum_{n=1}^{\infty} (-1)^{n+1} \frac{\ln(n)}{n}$

26. $\sum_{n=3}^{\infty} (-1)^{n+1} \frac{\ln(n)}{\ln(n^3)}$

27. $\sum_{n=2}^{\infty} (-1)^{n+1} \frac{\ln(n^3)}{\ln(n^{10})}$

28. $\sum_{n=1}^{\infty} (-1)^{n+1} \frac{5}{\sqrt{n+7}}$

29. $\sum_{n=1}^{\infty} \frac{(-2)^{n+1}}{1+(-3)^{n}}$

30. $\sum_{n=3}^{\infty} (-1)^{n+1} \frac{(-2)^{n+1}}{1+3^{n}}$

31. $\sum_{n=1}^{\infty} \cos(n\pi) \cdot \sin(\pi/n)$

In problems 32 – 40, (a) calculate s_4 for each series and determine an upper bound for how far s_4 is from the exact value S of the infinite series. Then (b) use s_4 find lower and upper bounds on the value of S so that {lower bound} < S < {upper bound}.

32. $\sum_{n=1}^{\infty} (-1)^{n+1} \frac{1}{n+6}$

33. $\sum_{n=1}^{\infty} (-1)^{n} \frac{1}{\ln(n+1)}$

34. $\sum_{n=1}^{\infty} (-1)^{n+1} \frac{2}{\sqrt{n+21}}$

35. $\sum_{n=1}^{\infty} (-0.8)^{n+1}$

36. $\sum_{n=1}^{\infty} (-\frac{1}{3})^{n}$

37. $\sum_{n=1}^{\infty} (-1)^{n+1} \sin(\frac{1}{n})$

38. $\sum_{n=1}^{\infty} (-1)^{n+1} \frac{1}{n^2}$

39. $\sum_{n=1}^{\infty} (-1)^{n} \frac{1}{n^3}$

40. $\sum_{n=1}^{\infty} \frac{\cos(n\pi)}{n+\ln(n)}$

10.6 Alternating Series

In problems 41 – 50 find the number of terms needed to guarantee that s_n is within the specified distance D of the exact value S of the sum of the given series.

41. $\sum_{n=1}^{\infty} \frac{(-1)^{n+1}}{n+6}$, D = 0.01

42. $\sum_{n=1}^{\infty} \frac{(-1)^n}{\ln(n+1)}$, D = 0.01

43. $\sum_{n=1}^{\infty} \frac{(-1)^n}{\sqrt{n}}$, D = 0.01

44. $\sum_{n=1}^{\infty} (-0.8)^{n+1}$, D = 0.003

45. $\sum_{n=1}^{\infty} (-\frac{1}{3})^n$, D = 0.002

46. $\sum_{n=1}^{\infty} (-1)^{n+1} \sin(\frac{1}{n})$, D = 0.06

47. $\sum_{n=1}^{\infty} (-1)^{n+1} \frac{1}{n^2}$, D = 0.001

48. $\sum_{n=1}^{\infty} (-1)^n \frac{1}{n^3}$, D = 0.0001

49. $\sum_{n=1}^{\infty} \frac{\cos(n\pi)}{n + \ln(n)}$, D = 0.04

Problems 50 – 60 ask you to use the series S(x), C(x), and E(x) given below. For each problem, (a) substitute the given value for x in the series, (b) evaluate s_3, the sum of the first three terms of the series, (c) determine an upper bound on the error $|\text{actual sum} - s_3| = |S - s_3|$.
(The partial sums of S(x), C(x), and E(x) approximate sin(x), cos(x), and e^x, respectively.)

$$S(x) = x - \frac{x^3}{2\cdot 3} + \frac{x^5}{2\cdot 3\cdot 4\cdot 5} - \frac{x^7}{2\cdot 3\cdot 4\cdot 5\cdot 6\cdot 7} + \ldots + (-1)^n \frac{x^{2n+1}}{(2n+1)!} + \ldots$$

$$C(x) = 1 - \frac{x^2}{2} + \frac{x^4}{2\cdot 3\cdot 4} - \frac{x^6}{2\cdot 3\cdot 4\cdot 5\cdot 6} + \ldots + (-1)^n \frac{x^{2n}}{(2n)!} + \ldots$$

$$E(x) = 1 + x + \frac{x^2}{2} + \frac{x^3}{2\cdot 3} + \frac{x^4}{2\cdot 3\cdot 4} + \ldots + \frac{x^n}{n!} + \ldots$$

50. x = 0.5 in S(x)
51. x = 0.3 in S(x)
52. x = 1 in S(x)
53. x = 0.1 in S(x)
54. x = –0.3 in S(x)
55. x = 1 in C(x)
56. x = 0.5 in C(x)
57. x = –0.2 in C(x)
58. x = –0.3 in C(x)
59. x = –1 in E(x)
60. x = –0.5 in E(x)
61. x = –0.2 in E(x)

Practice Answers

Practice 1: (a) (i) $a_n = \frac{1}{n^2} > 0$ for all n. (ii) $n^2 < (n+1)^2$ so $\frac{1}{n^2} > \frac{1}{(n+1)^2}$ and $a_n > a_{n+1}$.

(iii) $\lim_{n\to\infty} a_n = \lim_{n\to\infty} \frac{1}{n^2} = 0$. The series $\sum_{n=1}^{\infty} (-1)^{n+1} \frac{1}{n^2}$ converges.

(b) (i) $a_n = \frac{3}{\ln(n)} > 0$ for all n≥2. (ii) $\ln(n) < \ln(n+1)$ so $\frac{3}{\ln(n)} > \frac{3}{\ln(n+1)}$ and $a_n > a_{n+1}$.

(iii) $\lim_{n\to\infty} a_n = \lim_{n\to\infty} \frac{3}{\ln(n)} = 0$. The series $\sum_{n=1}^{\infty} (-1)^n \frac{3}{\ln(n)}$ converges.

Practice 2: (a) $|(-1)^{n+1} n| = |n| \longrightarrow \infty \neq 0$ so the terms $(-1)^{n+1} n$ do NOT approach 0, and the **series diverges** by the nth Term Test for Divergence (Section 10.2).

(b) (i) $a_n = \dfrac{1}{\sqrt{2n+1}} > 0$ for all $n \geq 1$.

(ii) $2n+1 < 2(n+1)+1$ so $\sqrt{2n+1} < \sqrt{2(n+1)+1}$ and $\dfrac{1}{\sqrt{2n+1}} > \dfrac{1}{\sqrt{2(n+1)+1}}$ so $a_n > a_{n+1}$.

(iii) $\lim\limits_{n \to \infty} a_n = \lim\limits_{n \to \infty} \dfrac{1}{\sqrt{2n+1}} = 0$. Therefore, by the Alternating Series Test,

the series $\sum\limits_{n=1}^{\infty} (-1)^{n+1} \dfrac{1}{\sqrt{2n+1}}$ converges.

Practice 3: $\sum\limits_{n=1}^{\infty} (-1)^{n+1} \dfrac{1}{n^3}$: $s_4 = 1 - \dfrac{1}{8} + \dfrac{1}{27} - \dfrac{1}{64} \approx 0.896412$.

$|S - s_4| < |a_5| = |(-1)^6 \dfrac{1}{5^3}| = \dfrac{1}{125} = 0.008$ so $s_4 - 0.008 < S < s_4 + 0.008$

and $0.888412 < S < 0.904412$.

$s_9 \approx 0.9021164$. $|S - s_9| < |a_{10}| = |(-1)^{11} \dfrac{1}{10^3}| = \dfrac{1}{1000} = 0.001$ so

$s_9 - 0.001 < S < s_9 + 0.001$ and $0.901116 < S < 0.903116$.

Practice 4: We need to find a value for n so $|a_{n+1}| < 0.001$.

$|a_{n+1}| = \left|(-1)^{n+2} \dfrac{1}{(n+1)^3 + 5}\right| = \dfrac{1}{(n+1)^3 + 5}$

n	$|a_{n+1}|$
1	$\dfrac{1}{2^3+5} = \dfrac{1}{13} \approx 0.0769$
2	$\dfrac{1}{3^3+5} = \dfrac{1}{32} \approx 0.03125$
8	$\dfrac{1}{9^3+5} = \dfrac{1}{734} \approx 0.00136$
9	$\dfrac{1}{10^3+5} = \dfrac{1}{1005} \approx 0.000995 < 0.001$

Since $|a_{10}| < 0.001$ we can be sure that s_9 is within 0.001 of S.

$s_9 = \sum\limits_{n=1}^{9} (-1)^{n+1} \dfrac{1}{n^3+5} = \dfrac{1}{6} - \dfrac{1}{13} + \dfrac{1}{32} - \ldots + \dfrac{1}{1005} \approx 0.112156571145$ so

$0.111156571145 < S < 0.113156571145$.

10.7 ABSOLUTE CONVERGENCE and the RATIO TEST

The series we examined in the previous sections all behaved very regularly with regard to the signs of the terms: the signs of the terms were either all the same or they alternated between + and −. However, there are series whose signs do not behave in such regular ways, and in this section we examine some techniques for determining whether those series converge or diverge.

Two Examples

Suppose we have the two series whose terms have magnitudes $\frac{1}{n}$ and $\frac{1}{n^2}$, but we don't know whether each term is positive or negative:

(a) $\sum_{n=1}^{\infty} *\frac{1}{n} = (*1) + (*\frac{1}{2}) + (*\frac{1}{3}) + (*\frac{1}{4}) + \ldots + (*\frac{1}{n}) + \ldots$ where each $*$ is either $+$ or $-$.

(b) $\sum_{n=1}^{\infty} *\frac{1}{n^2} = (*1) + (*\frac{1}{4}) + (*\frac{1}{9}) + (*\frac{1}{16}) + \ldots + (*\frac{1}{n^2}) + \ldots$ where each $*$ is either $+$ or $-$.

These two series behave very differently depending on how we replace the each "*" with + or − signs.

Series (a): $\sum_{n=1}^{\infty} *\frac{1}{n}$

If we **always** replace "*" with a +, then we have the harmonic series which diverges.

If we **always** replace "*" with a −, then we have the −1 times the harmonic series which diverges.

If we **alternate** replacing "*" with + and −, then we have the alternating harmonic series which converges.

The answer we have to give to the question "Does series (a) converge?" is "It depends on how the signs of the terms are chosen."

Series (b): $\sum_{n=1}^{\infty} *\frac{1}{n^2}$

If we always replace "*" with a +, then we have a series which converges (by the P–Test with p = 2). This is the largest value series (b) can have.

If we always replace "*" with a −, then we have −1 times a convergent series so the series converges. This is the smallest value series (b) can have.

If we replace the "*" with + or − in some other way, then the result is some number between the largest and smallest possible values (both of which are finite numbers), and the result must be some finite number.

The answer to the question "Does series (b) converge?" is "Yes, no matter how the $*$ are replaced with + or −." The value of the sum is affected by how the signs are chosen, but the series does converge.

Series (a) and (b) illustrate the distinction we want to examine in this section. Series (a) is an example of a **conditionally convergent** series since the convergence depends on how the "∗" are replaced. Series (b) is an example of an **absolutely convergent** series since it does not matter how the "∗" are replaced. Series (b) is convergent even when we are adding all positive terms or all negative terms.

The following definitions make the distinctions precise.

Definitions

> **Definition:** A series $\sum_{n=1}^{\infty} a_n$ is **absolutely convergent** if $\sum_{n=1}^{\infty} |a_n|$ converges.

Series (b) is absolutely convergent.

> **Definition:** A series $\sum_{n=1}^{\infty} a_n$ is **conditionally convergent** if it is convergent but not absolutely convergent (i.e., if $\sum_{n=1}^{\infty} a_n$ converges, but $\sum_{n=1}^{\infty} |a_n|$ diverges).

The alternating harmonic series $\sum_{n=1}^{\infty} (-1)^{n+1} \frac{1}{n}$ is conditionally convergent because

$\sum_{n=1}^{\infty} (-1)^{n+1} \frac{1}{n}$ converges but $\sum_{n=1}^{\infty} \left| (-1)^{n+1} \frac{1}{n} \right| = \sum_{n=1}^{\infty} \frac{1}{n}$ diverges.

Example 1: Determine whether these series are absolutely convergent, conditionally convergent, or divergent.

(a) $\sum_{n=1}^{\infty} (-1)^{n+1} \frac{1}{\sqrt{n}}$ (b) $\sum_{k=1}^{\infty} \frac{\sin(k)}{k^2}$ (c) $\sum_{j=1}^{\infty} (-1)^{j+1} \frac{j^2}{j+1}$

Solution: (a) $\sum_{n=1}^{\infty} \left| (-1)^{n+1} \frac{1}{\sqrt{n}} \right| = \sum_{n=1}^{\infty} \frac{1}{\sqrt{n}}$ which diverges by the P–Test with $p = 1/2 < 1$

so the series is not absolutely convergent. $\frac{1}{\sqrt{n}} \longrightarrow 0$ monotonically so the

alternating series $\sum_{n=1}^{\infty} (-1)^{n+1} \frac{1}{\sqrt{n}}$ converges by the Alternating Series Test.

The series $\sum_{n=1}^{\infty} (-1)^{n+1} \frac{1}{\sqrt{n}}$ is conditionally convergent.

(b) $\sum_{k=1}^{\infty} \left| \frac{\sin(k)}{k^2} \right| \leq \sum_{k=1}^{\infty} \frac{1}{k^2}$ which converges by the P–Test,

so $\sum_{k=1}^{\infty} \frac{\sin(k)}{k^2}$ is absolutely convergent.

(c) $\left| (-1)^{j+1} \frac{j^2}{j+1} \right| = j - 1 + \frac{1}{j+1} \longrightarrow \infty \neq 0$ so $\sum_{j=1}^{\infty} (-1)^{j+1} \frac{j^2}{j+1}$ diverges.

Practice 1: Determine whether these series are absolutely convergent, conditionally convergent, or divergent.

(a) $\sum_{n=2}^{\infty} (-1)^n \frac{5}{\ln(n)}$ (b) $\sum_{k=1}^{\infty} \frac{\cos(\pi k)}{k^2}$

An important result about absolutely convergent series is that they are also convergent.

Absolute Convergence Theorem

Every absolutely convergent series is convergent: if $\sum_{n=1}^{\infty} |a_n|$ converges, then $\sum_{n=1}^{\infty} a_n$ converges.

Since the series $\sum_{n=1}^{\infty} \left| * \frac{1}{n^2} \right| = \sum_{n=1}^{\infty} \frac{1}{n^2}$ is absolutely convergent, the series $\sum_{n=1}^{\infty} * \frac{1}{n^2}$ converges no matter how the "*" is replaced with + and − signs.

Proof of the Absolute Convergence Theorem:

If $a_n \geq 0$ then $a_n = |a_n|$, and if $a_n < 0$ then $a_n = -|a_n|$. In either case we have $-|a_n| \leq a_n \leq |a_n|$.

Adding $|a_n|$ to each piece of $-|a_n| \leq a_n \leq |a_n|$, we have $0 \leq |a_n| + a_n \leq 2|a_n|$ for all n.

Let $b_n = |a_n| + a_n \geq 0$ for all n.

Since $b_n \geq 0$ and $b_n \leq 2|a_n|$, the terms of a convergent series, then, by the Comparison Test we know that $\sum_{n=1}^{\infty} b_n$ converges.

Finally, $a_n = (|a_n| + a_n) - |a_n| = b_n - |a_n|$ so the series $\sum_{n=1}^{\infty} a_n = \sum_{n=1}^{\infty} b_n - \sum_{n=1}^{\infty} |a_n|$

is the difference of two convergent series, and we can conclude that the series $\sum_{n=1}^{\infty} a_n$ converges.

The following corollary (the contrapositive form of the Absolute Convergence Theorem) is sometimes useful for showing that a series is not absolutely convergent.

> **Corollary**:
>
> If a series is not convergent, then it is not absolutely convergent:
>
> $$\text{if } \sum_{n=1}^{\infty} a_n \text{ diverges, then } \sum_{n=1}^{\infty} |a_n| \text{ diverges.}$$

The Ratio Test

The following test is useful for determining whether a given series is absolutely convergent, and it will be used often in Section 10.8 when we want to determine where a power series converges. It says that we can be certain that a given series is absolutely convergent, if the limit of the ratios of successive terms has a value less than 1. The Ratio Test is very important and will be used very often in the next sections on power series.

> **The Ratio Test**
>
> Suppose $\lim_{n \to \infty} \left| \dfrac{a_{n+1}}{a_n} \right| = L$.
>
> (a) If $L < 1$, then the series $\sum_{n=1}^{\infty} a_n$ is absolutely convergent (and also convergent).
>
> (b) If $L > 1$, then the series $\sum_{n=1}^{\infty} a_n$ is divergent.
>
> (c) If $L = 1$, then the series $\sum_{n=1}^{\infty} a_n$ may converge or may diverge (the Ratio Test does not help).

A proof of the Ratio Test is rather long and is included in an Appendix after the Practice Answers for this section. Part (a) is proved by showing that if $L < 1$, then the series is, term–by–term, less than a convergent geometric series. Part (b) is proved by showing that if $L > 1$, then the terms of the series do not approach 0 so the series diverges. Part (c) is proved by giving two series, one convergent and one divergent, that both have $L = 1$.

One powerful aspect of the Ratio Test is that it is very "mechanical" — we simply calculate a particular limit and then we (often) have a conclusion about the convergence or divergence of the series.

Example 2: Use the Ratio Test to determine if these series are absolutely convergent:

(a) $\sum_{n=1}^{\infty} \dfrac{2^n n}{5^n}$
(b) $\sum_{n=1}^{\infty} \dfrac{n^2}{n!}$.

Solution: (a) $a_n = \dfrac{2^n n}{5^n}$ so $a_{n+1} = \dfrac{2^{n+1}(n+1)}{5^{n+1}}$. Then

$$\left|\dfrac{a_{n+1}}{a_n}\right| = \left|\dfrac{\dfrac{2^{n+1}(n+1)}{5^{n+1}}}{\dfrac{2^n n}{5^n}}\right| = \left|\dfrac{2^{n+1}}{2^n}\dfrac{5^n}{5^{n+1}}\dfrac{n+1}{n}\right| = \left|\dfrac{2}{5}\dfrac{n+1}{n}\right| \longrightarrow \dfrac{2}{5} < 1$$

so $\sum\limits_{n=1}^{\infty} \dfrac{2^n n}{5^n}$ is absolutely convergent.

(b) $a_n = \dfrac{n^2}{n!}$ so $a_{n+1} = \dfrac{(n+1)^2}{(n+1)!}$. Then

$$\left|\dfrac{a_{n+1}}{a_n}\right| = \left|\dfrac{\dfrac{(n+1)^2}{(n+1)!}}{\dfrac{n^2}{n!}}\right| = \left|\dfrac{n!}{(n+1)!}\dfrac{(n+1)^2}{n^2}\right| = \left|\dfrac{1}{n+1}\left(\dfrac{n+1}{n}\right)^2\right| \longrightarrow 0 < 1$$

so $\sum\limits_{n=1}^{\infty} \dfrac{n^2}{n!}$ is absolutely convergent.

Practice 2: Use the Ratio Test to determine if these series are absolutely convergent:

(a) $\sum\limits_{n=1}^{\infty} (-1)^{n+1}\dfrac{e^n}{n!}$ (b) $\sum\limits_{n=1}^{\infty} \dfrac{n^5}{3^n}$.

The Ratio Test is very useful for determining values of a variable which guarantee the absolute convergence (and convergence) of a series. This is a method we will use often in the rest of this chapter.

Example 3: For which values of x is the series $\sum\limits_{n=1}^{\infty} \dfrac{(x-3)^n}{n}$ absolutely convergent?

Solution: $a_n = \dfrac{(x-3)^n}{n}$ so $a_{n+1} = \dfrac{(x-3)^{n+1}}{n+1}$. Then

$$\left|\dfrac{a_{n+1}}{a_n}\right| = \left|\dfrac{\dfrac{(x-3)^{n+1}}{n+1}}{\dfrac{(x-3)^n}{n}}\right| = \left|\dfrac{(x-3)^{n+1}}{(x-3)^n}\dfrac{n}{n+1}\right| \to |x-3| = L.$$

Now we simply need to solve the inequality $|x-3| < 1$ ($L < 1$) for x.

If $|x-3| < 1$, then $-1 < x-3 < 1$ so **$2 < x < 4$**.

If $2 < x < 4$, then $L = |x-3| < 1$ so $\sum\limits_{n=1}^{\infty} \dfrac{(x-3)^n}{n}$ is absolutely convergent (and convergent).

If $x < 2$ or $x > 4$ then $L > 1$ so the series diverges. (What happens at the "endpoints," x=2 and x=4, where L=1?)

Practice 3: For which values of x is the series $\sum\limits_{n=1}^{\infty} \dfrac{(x-5)^n}{n^2}$ absolutely convergent?

Note: If the terms of a series contain **factorials** or things raised to the **nth power**, it is usually a good idea to use the Ratio Test as the first test you try.

Rearrangements

Absolutely convergent series share an important property with finite sums — no matter what order we add the numbers, the sum is always the same. Stated another way, the sum of an absolutely convergent series is always the same value, even if the terms are rearranged.

Conditionally convergent series do not have this property — the order in which we add the terms does matter. If the terms of a conditionally convergent series are rearranged or reordered, the sum after the rearrangement may be different than the sum before the rearrangement. A rather amazing fact is that the terms of a conditionally convergent series can be rearranged to obtain any sum we want.

We illustrate this strange result by showing that we can rearrange the alternating harmonic series, a series conditionally convergent to approximately 0.69, so that the sum of the rearranged series is 2 rather than 0.69..

$$\sum_{n=1}^{\infty} (-1)^{n+1} \frac{1}{n} = 1 - \frac{1}{2} + \frac{1}{3} - \frac{1}{4} + \frac{1}{5} - \frac{1}{6} + \ldots$$ is conditionally convergent to approximately 0.69.

First we note that the sum of the positive terms alone is a divergent series:

$$1 + \frac{1}{3} + \frac{1}{5} + \frac{1}{7} + \ldots = \sum_{n=1}^{\infty} \frac{1}{2n-1}$$ which diverges by the Limit Comparison Test,

so the partial sums of the positive terms eventually exceed any number we pick.

Similarly, the sum of the negative terms alone is also a divergent series:

$$-\frac{1}{2} - \frac{1}{4} - \frac{1}{6} - \frac{1}{8} + \ldots = -\frac{1}{2} \sum_{n=1}^{\infty} \frac{1}{n}$$ which diverges,

so the partial sums of the negative terms eventually become as large negatively as we want.

Finally, we pick a target number we want for the sum (in this illustration, we picked a target of 2).

Then the following clever strategy tells us how to chose the order of the terms, the rearrangement, so the sum of the rearranged series is the target number, 2 :

(1) select the positive terms, in the order they appear in the original series, until the partial sum exceeds our target number, then

(2) select the negative terms, in the order they appear in the original series, until the partial sum falls below our target number, and

(3) keep repeating steps (1) and (2) with the previously unused terms of the original series.

In order to rearrange the terms of the alternating harmonic series so the sum is 2, we pick positive terms, in order, until the partial sum **is larger than** 2. This requires the first 8 positive terms:

$s_1 = 1$,

$s_2 = 1 + \frac{1}{3} \approx 1.3333, \ldots$

$s_7 = 1 + \frac{1}{3} + \frac{1}{5} + \frac{1}{7} + \frac{1}{9} + \frac{1}{11} + \frac{1}{13} \approx 1.955133755$

$s_8 = 1 + \frac{1}{3} + \frac{1}{5} + \frac{1}{7} + \frac{1}{9} + \frac{1}{11} + \frac{1}{13} + \frac{1}{15} \approx 2.021800422$.

Then we pick negative terms, in order, until the partial sum **is less than** two. This only requires 1 negative term:

$s_9 = 1 + \frac{1}{3} + \frac{1}{5} + \frac{1}{7} + \frac{1}{9} + \frac{1}{11} + \frac{1}{13} + \frac{1}{15} - \frac{\mathbf{1}}{\mathbf{2}} \approx 1.521800422$.

Then we pick more unused positive terms until the partial sum exceeds 2. This requires many more positive terms:

$s_{10} = 1 + \frac{1}{3} + \frac{1}{5} + \frac{1}{7} + \frac{1}{9} + \frac{1}{11} + \frac{1}{13} + \frac{1}{15} - \frac{1}{2} + \frac{1}{17} \approx 1.580623951$

$s_{11} = 1 + \frac{1}{3} + \frac{1}{5} + \frac{1}{7} + \frac{1}{9} + \frac{1}{11} + \frac{1}{13} + \frac{1}{15} - \frac{1}{2} + \frac{1}{17} + \frac{1}{19} \approx 1.63325553$

$s_{12} = s_{11} + \frac{1}{21} \approx 1.680874578$, $s_{13} = s_{12} + \frac{1}{23} \approx 1.724352839$

$s_{14} \approx 1.764352839$, $s_{15} \approx 1.801389876$, $s_{16} \approx 1.835872634$, $s_{17} \approx 1.868130699$

$s_{18} \approx 1.898433729$, $s_{19} \approx 1.927005158$, $s_{20} \approx 1.954032185$, $s_{21} \approx 1.97967321$,

$s_{22} \approx s_{21} + \frac{1}{41} \approx 2.004063454$.

Then we pick more previously unused negative terms, in order, until the partial sum is less than two. Again only 1 negative term is required:

$s_{23} = s_{22} - \frac{\mathbf{1}}{\mathbf{4}} \approx 1.754063454$.

As we continue to repeat this process, we "eventually" use all of the terms of the original conditionally convergent series and the partial sums of the new "rearranged" series get, and stay, arbitrarily close to the target number, 2. The same method can be used to rearrange the terms of the alternating harmonic series (or any conditionally convergent series) to sum to 0.3, 3, 30 or any positive target number we want. How do you think the strategy needs to be changed to rearrange a conditionally convergent series to sum to a **negative** target number?

PROBLEMS

In problems 1 – 30 determine whether the given series Converge Absolutely, Converge Conditionally, or Diverge and give reasons for your conclusions.

1. $\sum_{n=1}^{\infty} (-1)^n \frac{1}{n+2}$

2. $\sum_{n=1}^{\infty} (-1)^{n+1} \frac{1}{\sqrt{n}}$

3. $\sum_{n=1}^{\infty} (-1)^{n+1} \frac{5}{n^3}$

4. $\sum_{n=1}^{\infty} (-1)^n \frac{1}{2 + \ln(n)}$

5. $\sum_{n=1}^{\infty} (-0.5)^n$

6. $\sum_{n=1}^{\infty} (-0.5)^{-n}$

7. $\sum_{n=1}^{\infty} (-1)^n \frac{1}{n^2}$

8. $\sum_{n=1}^{\infty} (-1)^{n+1} \frac{1}{3+n^2}$

9. $\sum_{n=1}^{\infty} (-1)^{n+1} \frac{\ln(n)}{n}$

10. $\sum_{n=1}^{\infty} (-1)^{n+1} \frac{\ln(n)}{n^2}$

11. $\sum_{n=1}^{\infty} (-1)^n \frac{1}{n + \ln(n)}$

12. $\sum_{n=1}^{\infty} (-1)^{n+1} \frac{5}{\sqrt{n+7}}$

13. $\sum_{n=1}^{\infty} (-1)^n \sin(\frac{1}{n})$ (recall that if $0 < x < 1$, then $0 < \sin(x) < x$)

14. $\sum_{n=1}^{\infty} (-1)^n \sin(\frac{1}{n^2})$

15. $\sum_{n=1}^{\infty} (-1)^n \sqrt{n} \sin(\frac{1}{n^2})$

16. $\sum_{n=1}^{\infty} \frac{\cos(\pi n)}{n}$

17. $\sum_{n=1}^{\infty} (-1)^{n+1} \frac{\ln(n)}{\ln(n^3)}$

18. $\sum_{n=1}^{\infty} (-1)^{n+1} \frac{n^2+7}{n^2+10}$

19. $\sum_{n=1}^{\infty} (-1)^{n+1} \frac{n^2+7}{n^3+10}$

20. $\sum_{n=1}^{\infty} (-1)^{n+1} \frac{n^2+7}{n^4+10}$

21. $\sum_{n=1}^{\infty} (-1)^{n+1} \frac{(n^2+7)^2}{n^2+10}$

22. $\sum_{n=1}^{\infty} \frac{\sin(n)}{n^2}$

23. $\sum_{n=1}^{\infty} \frac{\sin(\pi n)}{n}$

24. $\sum_{n=1}^{\infty} (-n)^{-n}$

25. $\sum_{n=1}^{\infty} (-1)^{n+1} \frac{1+\sqrt{3n}}{n+2}$

26. $\sum_{n=1}^{\infty} \frac{(-2)^n}{n^2}$

27. $\sum_{n=1}^{\infty} (-1)^{n+1} \frac{(-3)^n}{n^3}$

28. $\sum_{n=2}^{\infty} (-1)^{n+1} \left(\frac{\ln(n)}{\ln(n^5)} \right)^2$

29. $\sum_{n=1}^{\infty} \frac{(-2)^n}{n \cdot 3^n}$

30. $\sum_{n=1}^{\infty} \frac{3^n}{n^3}$

The Ratio Test is commonly used with series that contain factorials, and factorials are also going to become more common in the next few sections. Problems 31 – 40 ask you to simplify factorial expressions in order to get ready for that usage.

(Definitions: $0! = 1$, $1! = 1$, $2! = 1 \cdot 2 = 2$, $3! = 1 \cdot 2 \cdot 3 = 6$, $4! = 24$, and, in general, $n! = 1 \cdot 2 \cdot 3 \cdot \ldots \cdot (n-1) \cdot n$)

31. Show that $\frac{n!}{(n+1)!} = \frac{1}{n+1}$

32. Show that $\frac{n!}{(n+2)!} = \frac{1}{(n+1)(n+2)}$

33. Show that $\frac{n!}{(n+3)!} = \frac{1}{(n+1)(n+2)(n+3)}$

34. Show that $\frac{(n+1)!}{(n+2)!} = \frac{1}{n+2}$

35. Show that $\frac{(n-1)!}{(n+1)!} = \frac{1}{(n)(n+1)}$

36. Show that $\frac{2(n!)}{(2n)!} = \frac{2}{(n+1) \cdot (n+2) \cdot \ldots \cdot (2n)}$

37. Show that $\frac{(2n)!}{(2n+1)!} = \frac{1}{2n+1}$

38. Show that $\frac{(2n)!}{(2(n+1))!} = \frac{1}{(2n+1)(2n+2)}$

39. Show that $\dfrac{n^n}{n!} = \dfrac{n}{1} \cdot \dfrac{n}{2} \cdot \dfrac{n}{3} \cdot \ldots \cdot \dfrac{n}{n-1} \cdot \dfrac{n}{n}$

40. For $n > 0$, which is larger, $7!$ or $\dfrac{(n+7)!}{n!}$?

In problems 41 – 56, (a) determine the value of $L = \lim\limits_{n \to \infty} \left| \dfrac{a_{n+1}}{a_n} \right|$ for the given series, and (b) state the conclusion of the Ratio Test as it applies to the series. (c) If the Ratio Test is inconclusive, use some other method to determine if the given series converges or diverges.

41. $\sum\limits_{n=1}^{\infty} \dfrac{1}{n}$

42. $\sum\limits_{n=1}^{\infty} \dfrac{1}{n^2}$

43. $\sum\limits_{n=2}^{\infty} \dfrac{1}{n^3}$

44. $\sum\limits_{n=2}^{\infty} \dfrac{1}{\sqrt{n}}$

45. $\sum\limits_{n=1}^{\infty} \left(\dfrac{1}{2}\right)^n$

46. $\sum\limits_{n=1}^{\infty} \left(\dfrac{1}{3}\right)^n$

47. $\sum\limits_{n=5}^{\infty} 1^n$

48. $\sum\limits_{n=1}^{\infty} (-2)^n$

49. $\sum\limits_{n=1}^{\infty} \dfrac{1}{n!}$

50. $\sum\limits_{n=1}^{\infty} \dfrac{5}{n!}$

51. $\sum\limits_{n=1}^{\infty} \dfrac{2^n}{n!}$

52. $\sum\limits_{n=1}^{\infty} \dfrac{5^n}{n!}$

53. $\sum\limits_{n=1}^{\infty} \left(\dfrac{1}{2}\right)^{3n}$

54. $\sum\limits_{n=1}^{\infty} \left(\dfrac{1}{3}\right)^{2n}$

55. $\sum\limits_{n=5}^{\infty} (0.9)^{2n+1}$

56. $\sum\limits_{n=5}^{\infty} (-0.8)^{2n+1}$

For each series in problems 57 – 74 find the values of x for which the value of $L = \lim\limits_{n \to \infty} \left| \dfrac{a_{n+1}}{a_n} \right| < 1$. From the Ratio Test we can conclude that each series converges absolutely (and thus converges) for those values of x.

57. $\sum\limits_{n=5}^{\infty} (x-5)^n$

58. $\sum\limits_{n=1}^{\infty} \dfrac{(x-5)^n}{n}$

59. $\sum\limits_{n=1}^{\infty} \dfrac{(x-5)^n}{n^2}$

60. $\sum\limits_{n=1}^{\infty} \dfrac{(x-2)^n}{n^2}$

61. $\sum\limits_{n=1}^{\infty} \dfrac{(x-2)^n}{n!}$

62. $\sum\limits_{n=1}^{\infty} \dfrac{(x-10)^n}{n!}$

63. $\sum\limits_{n=1}^{\infty} \dfrac{(2x-12)^n}{n^2}$

64. $\sum\limits_{n=1}^{\infty} \dfrac{(4x-12)^n}{n^2}$

65. $\sum\limits_{n=1}^{\infty} \dfrac{(6x-12)^n}{n!}$

66. $\sum\limits_{n=1}^{\infty} (x-3)^{2n}$

67. $\sum\limits_{n=2}^{\infty} \dfrac{(x+1)^{2n}}{n}$

68. $\sum\limits_{n=2}^{\infty} \dfrac{(x+2)^{2n+1}}{n^2}$

69. $\sum_{n=1}^{\infty} \frac{(x-5)^{3n+1}}{n^2}$
70. $\sum_{n=1}^{\infty} \frac{(x+4)^{2n+1}}{n!}$
71. $\sum_{n=1}^{\infty} \frac{(x+3)^{2n-1}}{(n+1)!}$

72. $S(x) = \sum_{n=0}^{\infty} (-1)^n \frac{x^{2n+1}}{(2n+1)!}$
73. $C(x) = \sum_{n=0}^{\infty} (-1)^n \frac{x^{2n}}{(2n)!}$
74. $E(x) = \sum_{n=0}^{\infty} \frac{x^n}{n!}$

Rearrangements

In problems 75 – 80 use the strategy in the illustrative example to find the first 15 terms of a rearrangement of the given conditionally convergent series so that the rearranged series converges to the given target number.

75. $\sum_{n=1}^{\infty} (-1)^n \frac{1}{n}$, target number = 0.3 .
76. $\sum_{n=1}^{\infty} (-1)^{n+1} \frac{1}{n}$, target number = 0.7 .

77. $\sum_{n=1}^{\infty} (-1)^{n+1} \frac{1}{n}$, target number = 1 .
78. $\sum_{n=1}^{\infty} (-1)^{n+1} \frac{1}{\sqrt{n}}$, target number = 0.4 .

79. $\sum_{n=1}^{\infty} (-1)^{n+1} \frac{1}{\sqrt{n}}$, target number = 1 .
80. $\sum_{n=1}^{\infty} (-1)^{n+1} \frac{1}{n}$, target number = –1 .

Practice Answers

Practice 1: (a) $\sum_{n=2}^{\infty} (-1)^n \frac{5}{\ln(n)}$ converges by the Alternating Series Test

$\sum_{n=2}^{\infty} \left| (-1)^n \frac{5}{\ln(n)} \right| = \sum_{n=2}^{\infty} \frac{5}{\ln(n)}$ which diverges by comparison with the divergent series $\sum_{n=2}^{\infty} \frac{5}{n}$.

Therefore, $\sum_{n=2}^{\infty} (-1)^n \frac{5}{\ln(n)}$ is **conditionally convergent** .

(b) $\sum_{k=1}^{\infty} \frac{\cos(\pi k)}{k^2} = \frac{-1}{1^2} + \frac{1}{2^2} + \frac{-1}{3^2} + \ldots$

$\sum_{k=1}^{\infty} \left| \frac{\cos(\pi k)}{k^2} \right| = \sum_{k=1}^{\infty} \frac{1}{k^2}$ which is convergent (P–test, p = 2) so

$\sum_{k=1}^{\infty} \frac{\cos(\pi k)}{k^2}$ is **absolutely convergent** (and convergent).

Practice 2: (a) $a_n = (-1)^{n+1} \dfrac{e^n}{n!}$ so $a_{n+1} = (-1)^{n+2} \dfrac{e^{n+1}}{(n+1)!}$. Then

$$\left| \dfrac{a_{n+1}}{a_n} \right| = \left| \dfrac{(-1)^{n+2} \dfrac{e^{n+1}}{(n+1)!}}{(-1)^{n+1} \dfrac{e^n}{n!}} \right| = \left| \dfrac{e^{n+1}}{e^n} \dfrac{n!}{(n+1)!} \right| = \left| \dfrac{e}{n+1} \right| \longrightarrow 0 < 1$$

so $\sum\limits_{n=1}^{\infty} (-1)^{n+1} \dfrac{e^n}{n!}$ is absolutely convergent.

(b) $a_n = \dfrac{n^5}{3^n}$ so $a_{n+1} = \dfrac{(n+1)^5}{3^{n+1}}$. Then

$$\left| \dfrac{a_{n+1}}{a_n} \right| = \left| \dfrac{\dfrac{(n+1)^5}{3^{n+1}}}{\dfrac{n^5}{3^n}} \right| = \left| \dfrac{3^n}{3^{n+1}} \dfrac{(n+1)^5}{n^5} \right| = \left| \dfrac{1}{3}\left(\dfrac{n+1}{n}\right)^5 \right| \longrightarrow \dfrac{1}{3} < 1$$

so $\sum\limits_{n=1}^{\infty} \dfrac{n^5}{3^n}$ is absolutely convergent.

Practice 3: $a_n = \dfrac{(x-5)^n}{n^2}$ so $a_{n+1} = \dfrac{(x-5)^{n+1}}{(n+1)^2}$. Then

$$\left| \dfrac{a_{n+1}}{a_n} \right| = \left| \dfrac{\dfrac{(x-5)^{n+1}}{(n+1)^2}}{\dfrac{(x-5)^n}{n^2}} \right| = \left| \dfrac{(x-5)^{n+1}}{(x-5)^n} \dfrac{n^2}{(n+1)^2} \right| \to |x-5| = L.$$

Now we simply need to solve the inequality $|x-5| < 1$ ($L < 1$) for x.

If $|x-5| < 1$, then $-1 < x - 5 < 1$ so **$4 < x < 6$**.

If $4 < x < 6$, then $L = |x-5| < 1$ so $\sum\limits_{n=1}^{\infty} \dfrac{(x-5)^n}{n^2}$ is absolutely convergent (and convergent).

Also, if $x < 4$ or $x > 6$ then $L > 1$ so $\sum\limits_{n=1}^{\infty} \dfrac{(x-5)^n}{n^2}$ diverges.

"Endpoints": If $x = 4$, then $\sum\limits_{n=1}^{\infty} \dfrac{(x-5)^n}{n^2} = \sum\limits_{n=1}^{\infty} \dfrac{(-1)^n}{n^2}$ which converges.

If $x = 6$, then $\sum\limits_{n=1}^{\infty} \dfrac{(x-5)^n}{n^2} = \sum\limits_{n=1}^{\infty} \dfrac{1}{n^2}$ which converges.

Appendix: A Proof of the Ratio Test

The Ratio Test has three parts, (a), (b), and (c), and each part requires a separate proof.

(a) $L<1 \Rightarrow$ the series is absolutely convergent. The basic pattern of the proof for this part is to show that the given series is, term–by–term, less than a convergent geometric series. Then we conclude by the Comparison Test that the given series converges.

Suppose $\lim_{n \to \infty} \left| \frac{a_{n+1}}{a_n} \right| = L < 1$. Then there is a number r between L and 1 so that the ratios $\left| \frac{a_{n+1}}{a_n} \right|$ are eventually (for all n > some N) less than r:

for all $n > N$, $\left| \frac{a_{n+1}}{a_n} \right| < r < 1$.

Then $|a_{N+1}| < r|a_N|$, $|a_{N+2}| < r|a_{N+1}| < r^2|a_N|$, $|a_{N+3}| < r|a_{N+2}| < r^3|a_N|$,
and, in general, $|a_{N+k}| < r^k|a_N|$.

So $|a_N| + |a_{N+1}| + |a_{N+2}| + |a_{N+3}| + |a_{N+4}| + \ldots$

$< |a_N| + r|a_N| + r^2|a_N| + r^3|a_N| + r^4|a_N| + \ldots$

$= |a_N| \cdot \{ 1 + r + r^2 + r^3 + r^4 + \ldots \}$

$= |a_N| \cdot \frac{1}{1-r}$, since the powers of r form a convergent geometric series,

and the series $\sum_{n=N}^{\infty} |a_n|$ is convergent by the Comparison Test.

Finally, we can conclude that the series $\sum_{n=1}^{\infty} |a_n| = \sum_{n=1}^{N-1} |a_n| + \sum_{n=N}^{\infty} |a_n|$ is convergent, since it is the sum of two convergent series. $\sum_{n=1}^{\infty} a_n$ is absolutely convergent.

(b) $L>1 \Rightarrow$ the series is divergent. The basic idea in this part is to show that the terms of the given series do not approach 0. Then we can conclude by the Nth Term Test that the given series diverges.

Suppose $\lim_{n \to \infty} \left| \frac{a_{n+1}}{a_n} \right| = L > 1$. Then the ratios $\left| \frac{a_{n+1}}{a_n} \right|$ are eventually (for all n > some N) larger than 1 so $|a_{N+1}| > |a_N|$, $|a_{N+2}| > |a_{N+1}| > |a_N|$, $|a_{N+3}| > |a_{N+2}| > |a_N|$, and, for all $k > N$, $|a_k| > |a_N|$.

Thus $\lim_{n \to \infty} |a_n| \geq |a_N| \neq 0$ and $\lim_{n \to \infty} a_n \neq 0$, so by the Nth Term Test for Divergence (Section 10.2) we can conclude that the series $\sum_{n=1}^{\infty} a_n$ is divergent.

(c) $L=1 \Rightarrow$ nothing. Part (c) can be verified by giving two series, one convergent and one divergent, for which $L = 1$.

If $a_n = \dfrac{1}{n^2}$, then $\displaystyle\sum_{n=1}^{\infty} a_n = \sum_{n=1}^{\infty} \dfrac{1}{n^2}$ is convergent by the P–Test with p=2, and

$$\lim_{n\to\infty} \left| \dfrac{a_{n+1}}{a_n} \right| = \lim_{n\to\infty} \left| \dfrac{1/(n+1)^2}{1/n^2} \right| = \lim_{n\to\infty} \left(\dfrac{n}{n+1} \right)^2 = 1.$$

If $a_n = \dfrac{1}{n}$, then $\displaystyle\sum_{n=1}^{\infty} a_n = \sum_{n=1}^{\infty} \dfrac{1}{n}$ is the divergent harmonic series, and

$$\lim_{n\to\infty} \left| \dfrac{a_{n+1}}{a_n} \right| = \lim_{n\to\infty} \left| \dfrac{1/(n+1)}{1/n} \right| = \lim_{n\to\infty} \dfrac{n}{n+1} = 1.$$

Since $L = 1$ for each of these series, one convergent and one divergent, knowing that $L = 1$ for a series does not let us conclude that the series converges or that it diverges.

10.8 POWER SERIES: $\sum_{n=0}^{\infty} a_n x^n$ and $\sum_{n=0}^{\infty} a_n (x-c)^n$

So far most of the series we have examined have consisted of numbers (numerical series), but the most important series contain powers of a variable, and they define functions of that variable.

Definition of Power Series

A **power series** is an expression of the form $\sum_{n=0}^{\infty} a_n x^n = a_0 + a_1 x + a_2 x^2 + a_3 x^3 + \ldots + a_n x^n + \ldots$

where $a_0, a_1, a_2, a_3, \ldots$ are constants, called the coefficients of the series, and x is a variable.

(Note: For $n = 0$ we use the convention for power series that $x^0 = 1$ even when $x = 0$. This convention simply makes it easier for us to represent the series using the summation notation.)

For each value of the variable, the power series is simply a numerical series that may converge or diverge. If the power series does converge, the value of the function is the sum of the series, and the domain of the function is the set of x values for which the series converges. Power series are particularly important in mathematics and applications because many important functions such as $\sin(x)$, $\cos(x)$, e^x, and $\ln(x)$ can be represented and approximated by power series.

The following are examples of power series:

$$f(x) = 1 + x + x^2 + x^3 + x^4 + \ldots = \sum_{n=0}^{\infty} x^n \quad (a_n = 1 \text{ for all } n)$$

$$g(x) = 1 + x + \frac{x^2}{2!} + \frac{x^3}{3!} + \frac{x^4}{4!} + \ldots = \sum_{n=0}^{\infty} \frac{x^n}{n!} \quad (a_n = \frac{1}{n!} \text{ for all n and with the definition that } 0! = 1)$$

$$h(x) = x - \frac{x^3}{3!} + \frac{x^5}{5!} - \frac{x^7}{7!} + \ldots = \sum_{n=0}^{\infty} (-1)^n \cdot \frac{x^{2n+1}}{(2n+1)!} \quad (a_n = \frac{(-1)^n}{(2n+1)!} \text{ for all n.)}$$

Power series look like long (very long) polynomials, and in many ways they behave like polynomials.

This section focuses on what a power series is and on determining where a given power series converges. Section 10.9 looks at the arithmetic (sums, differences, products) and calculus (derivative and integrals) of power series. Section 10.10 examines how to represent particular functions such as $\sin(x)$ and e^x and others as power series.

Finding Where a Power Series Converges

The power series $f(x) = \sum_{n=0}^{\infty} a_n x^n$ always converges at $x = 0$: $f(0) = \sum_{n=0}^{\infty} a_n (0)^n = a_0$. To find which other values of x make a power series converge, we could try x values one–by–one, but that is very inefficient and time consuming. Instead, the Ratio Test allows us to get answers for lots of x values all at once. (Since we are using a_n to represent the coefficient of the nth term $a_n x^n$, we let $c_n = a_n x^n$ represent the nth term and we use the ratio $\dfrac{c_{n+1}}{c_n}$.)

Example 1: Find all of the values of x for which the power series $\sum_{n=0}^{\infty} (2n+1) \cdot x^n$ converges.

Solution: $c_n = (2n+1) \cdot x^n$ so $c_{n+1} = (2(n+1)+1) \cdot x^{n+1} = (2n+3) \cdot x^{n+1}$. Then, using the Ratio Test,

$$\left| \frac{c_{n+1}}{c_n} \right| = \left| \frac{(2n+3) \cdot x^{n+1}}{(2n+1) \cdot x^n} \right| = \left| \frac{2n+3}{2n+1} \cdot \frac{x^{n+1}}{x^n} \right| = \left| \frac{2n+3}{2n+1} \cdot x \right| \to |x| = L.$$

From the Ratio Test we know the series converges if $L < 1$: if $|x| < 1$ or, in other words, $-1 < x < 1$. We also know the series diverges if $L > 1$ so the series diverges if $|x| > 1$: if $x > 1$ or $x < -1$. Finally, we need to check the two remaining values of x: the **endpoints** $x = -1$ and $x = 1$.

When $x = 1$, $\sum_{n=0}^{\infty} (2n+1) \cdot x^n = \sum_{n=0}^{\infty} (2n+1) \cdot 1^n = \sum_{n=0}^{\infty} (2n+1)$ which diverges since the terms do not approach 0. When $x = -1$, $\sum_{n=0}^{\infty} (2n+1) \cdot x^n = \sum_{n=0}^{\infty} (2n+1) \cdot (-1)^n$ which also diverges because the terms do not approach 0.

The power series $\sum_{n=0}^{\infty} (2n+1) \cdot x^n$ converges if and only if $-1 < x < 1$. In other words, the series converges when x is in the interval $(-1, 1)$ and it diverges when x is outside the interval $(-1, 1)$.

Example 2: Find all of the values of x for which the power series $\sum_{n=1}^{\infty} \dfrac{x^n}{n \cdot 3^n}$ converges.

Solution: $c_n = \dfrac{x^n}{n \cdot 3^n}$ so $c_{n+1} = \dfrac{x^{n+1}}{(n+1) \cdot 3^{n+1}}$. Using the Ratio Test,

$$\left| \frac{c_{n+1}}{c_n} \right| = \left| \frac{\frac{x^{n+1}}{(n+1) \cdot 3^{n+1}}}{\frac{x^n}{n \cdot 3^n}} \right| = \left| \frac{n}{n+1} \cdot \frac{3^n}{3^{n+1}} \cdot \frac{x^{n+1}}{x^n} \right| = \left| \frac{n}{n+1} \cdot \frac{1}{3} \cdot x \right| \to \left| \frac{x}{3} \right| = L.$$

Solving $|\frac{x}{3}| < 1$, we have $-1 < \frac{x}{3} < 1$ or $-3 < x < 3$. The series converges for $-3 < x < 3$ and it diverges for $x < -3$ and for $x > 3$.

Finally, we need to check the two remaining values of x: the **endpoints** $x = -3$ and $x = 3$.

When $x = -3$, $\sum_{n=1}^{\infty} \frac{x^n}{n \cdot 3^n} = \sum_{n=1}^{\infty} \frac{(-3)^n}{n \cdot 3^n} = \sum_{n=1}^{\infty} \frac{(-1)^n}{n}$ which converges by the Alternating Series Test.

When $x = 3$, $\sum_{n=1}^{\infty} \frac{x^n}{n \cdot 3^n} = \sum_{n=1}^{\infty} \frac{(3)^n}{n \cdot 3^n} = \sum_{n=1}^{\infty} \frac{1}{n}$, the harmonic series, which diverges.

In summary, the power series $\sum_{n=1}^{\infty} \frac{x^n}{n \cdot 3^n}$ converges if $-3 \leq x < 3$, if x is in the interval $[-3, 3)$.

Note: The Ratio Test is very powerful for determining where a power series converges: put $c_n = a_n x^n$, calculate the limit of the ratio $\left|\frac{c_{n+1}}{c_n}\right|$, and then solve the resulting absolute value inequality for x.

Typically, we also need to check the endpoints of the interval by replacing x with the two endpoint values and then determining if the resulting numerical series converge or diverge at these endpoints. The Ratio Test does not help with the endpoints.

Practice 1: Find all of the values of x for which the series $\sum_{n=1}^{\infty} \frac{5^n \cdot x^n}{n}$ converges.

Interval of Convergence, Radius of Convergence

In each of the previous examples, the values of x for which the power series converge form an interval. The next theorem says that always happens.

Interval of Convergence Theorem for Power Series

The values of x for which the power series $\sum_{n=0}^{\infty} a_n x^n$ converges form an interval.

(i) If this power series converges for $x = c$, then the series converges for all satisfying $|x| < |c|$.

(ii) If this power series diverges for $x = d$, then the series diverges for all satisfying $|x| > |d|$.

A proof of the Interval of Convergence Theorem is given after the Practice Answers.

Meaning of the Interval of Convergence Theorem: If a power series $\sum_{n=0}^{\infty} a_n x^n$ converges for a value $x = c$, then the series also converges for all values of x closer to the origin than c. If the power series diverges for a value $x = d$, then the power series diverges for all values of x farther from the origin than d.

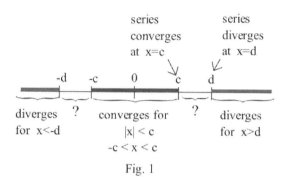

Fig. 1

This guarantees that the values where the power series converges form an interval, from $-|c|$ to $|c|$. This Theorem does **not** tell us about the convergence of the power series at the endpoints of the interval — we need to check those two points individually. This Theorem also does **not** tell us about the convergence of the power series for values of x with $|c| < |x| < |d|$. See Fig. 1.

Example 3: Suppose we know that a power series $\sum_{n=0}^{\infty} a_n x^n$ converges at $x = 4$ and diverges at $x = 9$.

What can we conclude (converge or diverge or not enough information) about the series when $x = 2, -3, -4, 5, -6, 8, -9, 10, -11$?

Solution: We know the power series converges at $x = 4$ ($c = 4$) so we can conclude that the series converges for $x = 2$ and $x = -3$ since $|2|<|4|$ and $|-3|<|4|$.

We know the power series diverges at $x = 9$ ($d = 9$) so we can conclude that the series diverges for $x = 10$ and $x = -11$ since $|10|>|9|$ and $|-11|>|9|$.

The remaining values of x ($-4, 5, -6, 8,$ and -9) do not satisfy $|x| < 4$ or $|x| > 9$ so the series may converge or may diverge — we don't have enough information.

Fig. 2 shows the regions where convergence of this power series is guaranteed, where divergence is guaranteed, and where we don't have enough information.

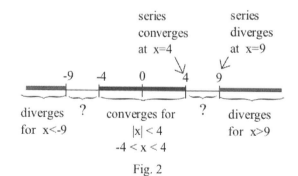

Fig. 2

Practice 2: Suppose we know that a power series $\sum_{n=0}^{\infty} a_n x^n$ converges at $x = 3$ and diverges at $x = -7$.

What can we say about the convergence of the series for $x = -1, 2, -3, 4, -6, 7, -8$, and 17? Sketch the regions of known convergence and known divergence.

Definition: The **interval of convergence** of a power series $\sum_{n=0}^{\infty} a_n x^n$ is the interval of values of x for which the series converges.

From Example 1 we know the interval of convergence of $\sum_{n=0}^{\infty} (2n+1) \cdot x^n$ is $(-1, 1)$.

From Example 2 we know the interval of convergence of $\sum_{n=1}^{\infty} \dfrac{x^n}{n \cdot 3^n}$ is $[-3, 3)$.

Definition: For each power series $\sum_{n=0}^{\infty} a_n x^n$ there is a number R, called the **radius of convergence**, so that the series converges for $|x| < R$ and diverges when $|x| > R$. (The series may converge or may diverge at $|x| = R$.) (Fig. 3)

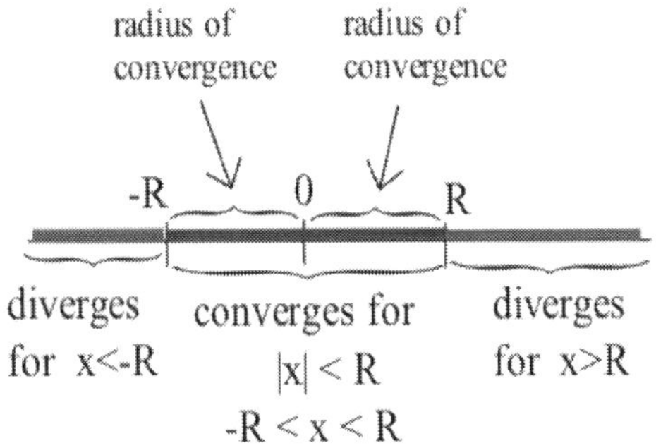

Fig. 3

The radius of convergence is half of the length of the interval of convergence.

Example 4: What is the radius of convergence of each series in Examples 1 and 2?

Solution: The power series $\sum_{n=0}^{\infty} (2n+1) \cdot x^n$ converges if $-1 < x < 1$ so $R = 1$.

The power series $\sum_{n=1}^{\infty} \dfrac{x^n}{n \cdot 3^n}$ converges if $-3 \leq x < 3$, so $R = 3$.

The convergence or divergence of a power series at the endpoints of the interval of convergence does not affect the value of the radius of convergence R, and the value of R does not tell us about the convergence of the power series at the endpoints of the interval of convergence, at $x = R$ and $x = -R$.

Practice 3: What is the radius of convergence of the series in Practice 1?

> **Summary**
>
> For a power series $\sum_{n=0}^{\infty} a_n x^n$ exactly one of these three situations occurs:
>
> (i) the series converges only for $x = 0$. (Then we say the radius of convergence is 0.)
>
> (ii) the series converges for all x with $|x| < R$ and diverges for all x with $|x| > R$. (Then we say the radius of convergence is R.)
>
> (iii) the series converges for all values of x. (Then we say the radius of convergence is infinite.)

The following list shows the intervals and radii of convergence for several power series. Four of the series in the list have the same radius of convergence, $R = 1$, but slightly different intervals of convergence. (The $\sum$ below simply means a series whose starting index is some finite value of n and whose upper index is ∞.)

Series	Radius of Convergence	Interval of Convergence	Series converges for x Values in the Shaded Intervals
$\sum n! \cdot x^n$	$R = 0$	$\{0\}$, a single point	
$\sum x^n$	$R = 1$	$(-1, 1)$	
$\sum \dfrac{x^n}{n}$	$R = 1$	$[-1, 1)$	
$\sum (-1)^n \dfrac{x^n}{n}$	$R = 1$	$(-1, 1]$	
$\sum \dfrac{x^n}{n^2}$	$R = 1$	$[-1, 1]$	
$\sum \dfrac{x^n}{2^n}$	$R = 2$	$(-2, 2)$	
$\sum \dfrac{x^n}{n!}$	$R = \infty$	$(-\infty, \infty)$	

Power Series Centered at c

Sometimes it is useful to "shift" a power series. These shifted power series contain powers of "x – c" instead of powers of "x," but many of the properties we have examined still hold.

10.8 Power Series

Definition: A **power series centered at c** is a series of the form

$$\sum_{n=0}^{\infty} a_n(x-c)^n = a_0 + a_1(x-c) + a_2(x-c)^2 + a_3(x-c)^3 + \ldots + a_n(x-c)^n + \ldots$$

where $a_0, a_1, a_2, a_3, \ldots$ are constants, called the coefficients of the series, and x is a variable.

Note: As usual for power series, for $n = 0$ we use the convention that $(x-c)^0 = 1$, even when $x = c$.

A power series centered at c always converges for $x = c$, and the interval of convergence is an interval **centered at c**. The radius of convergence is half the length of the interval of convergence (Fig. 4).

If a power series centered at c converges for a value of x, then the series converges for all values closer to c. If a power series centered at c diverges for a value of x, then the series diverges for all values farther from c.

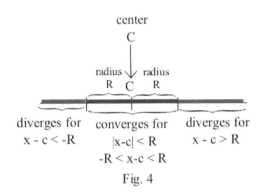

Fig. 4

Example 5: Suppose we know that a power series $\sum_{n=0}^{\infty} a_n(x-4)^n$ converges at $x = 6$ and diverges at $x = 0$. What can we conclude (converge or diverge or not enough information) about the series when $x = 3, 9, -1, 2,$ and 7?

Solution: We know the power series converges at $x = 6$ so we can conclude that the series converges for values of x **closer to 4** than $|6 - 4| = 2$ units: the series converges at $x = 3$.

We know the power series diverges at $x = 0$ so we can conclude that the series diverges for values of x **farther from 4** than $|0 - 4| = 4$ units: the series diverges at $x = 9$.

The remaining values of x $(-1, 2$ and $7)$ do not satisfy $|x - 4| < 2$ or $|x - 4| > 4$ so the series may converge or may diverge.

Fig. 5 shows the regions where convergence of this power series is guaranteed, where divergence is guaranteed, and where we don't have enough information.

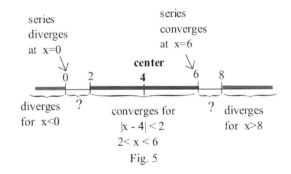

Fig. 5

Practice 4: Suppose we know that a power series $\sum_{n=0}^{\infty} a_n(x+5)^n$ converges at $x=-1$ and diverges at $x=1$. What can we conclude about the series when $x=-2, -9, 0, -11,$ and 3?

The Ratio Test is still our primary tool for finding an interval of convergence of a power series, even if the power series is centered at c rather than at 0.

Example 6: Find the interval and radius of convergence of $\sum_{n=1}^{\infty} \frac{(x-5)^n}{n \cdot 2^n}$.

Solution: $c_n = \frac{(x-5)^n}{n \cdot 2^n}$ so $c_{n+1} = \frac{(x-5)^{n+1}}{(n+1) \cdot 2^{n+1}}$. The ratio $\left| \frac{c_{n+1}}{c_n} \right|$ for the Ratio Test appears to be messy, but if we group the similar pieces algebraically, the ratio simplifies nicely:

$$\left| \frac{c_{n+1}}{c_n} \right| = \left| \frac{\frac{(x-5)^{n+1}}{(n+1) \cdot 2^{n+1}}}{\frac{(x-5)^n}{n \cdot 2^n}} \right| = \left| \frac{n}{n+1} \cdot \frac{2^n}{2^{n+1}} \cdot \frac{(x-5)^{n+1}}{(x-5)^n} \right| = \left| \frac{n}{n+1} \cdot \frac{1}{2} \cdot (x-5) \right| \to \left| \frac{(x-5)}{2} \right| = L.$$

Solving $\left| \frac{(x-5)}{2} \right| < 1$, we have $-1 < \frac{(x-5)}{2} < 1$ so $-2 < x-5 < 2$ and $3 < x < 7$. The series converges for $3 < x < 7$, and it diverges for $x < 3$ and for $x > 7$.

Finally, we need to check the two remaining values of x: the **endpoints** $x = 3$ and $x = 7$.

When $x = 3$, $\sum_{n=1}^{\infty} \frac{(x-5)^n}{n \cdot 2^n} = \sum_{n=1}^{\infty} \frac{(-2)^n}{n \cdot 2^n} = \sum_{n=1}^{\infty} \frac{(-1)^n}{n}$ which converges by the Alternating Series Test.

When $x = 7$, $\sum_{n=1}^{\infty} \frac{(x-5)^n}{n \cdot 2^n} = \sum_{n=1}^{\infty} \frac{(2)^n}{n \cdot 2^n} = \sum_{n=1}^{\infty} \frac{1}{n}$, the harmonic series, which diverges.

The power series $\sum_{n=1}^{\infty} \frac{(x-5)^n}{n \cdot 2^n}$ converges if $3 \leq x < 7$. The interval of convergence is $[3, 7)$, and the radius of convergence is $R = \frac{1}{2}$ (length of the interval of convergence) $= \frac{1}{2}(7-3) = 2$.

Practice 5: Find the interval and radius of convergence of $\sum_{n=0}^{\infty} \frac{n \cdot (x-3)^n}{5^n}$.

10.8 Power Series

A power series looks like a very long polynomial. However, a regular polynomial with a finite number of terms is defined at every value of x, but a power series may diverge for many values of the variable x. As we continue to work with power series we need to be alert to where the power series converges (and behaves like a finite polynomial) and where the power series diverges. We need to know the interval of convergence of the power series, and, typically, we use the Ratio Test to find that interval.

PROBLEMS

In problems 1 – 24, (a) find all values of x for which each given power series converges, and (b) graph the interval of convergence for the series on a number line.

1. $\sum_{n=1}^{\infty} x^n$
2. $\sum_{n=1}^{\infty} (x-3)^n$
3. $\sum_{n=1}^{\infty} (x+2)^n$
4. $\sum_{n=1}^{\infty} (x+5)^n$

5. $\sum_{n=3}^{\infty} \frac{x^n}{n}$
6. $\sum_{n=3}^{\infty} \frac{(x-2)^n}{n}$
7. $\sum_{n=1}^{\infty} \frac{(x+3)^n}{n}$
8. $\sum_{n=1}^{\infty} \frac{(x-5)^n}{n^2}$

9. $\sum_{n=1}^{\infty} \frac{(x-7)^{2n+1}}{n^2}$
10. $\sum_{n=1}^{\infty} \frac{(x+1)^{2n}}{n^3}$
11. $\sum_{n=2}^{\infty} (2x)^n$
12. $\sum_{n=2}^{\infty} (5x)^n$

13. $\sum_{n=1}^{\infty} (\frac{x}{3})^{2n+1}$
14. $\sum_{n=1}^{\infty} n(\frac{x}{4})^{2n+1}$
15. $\sum_{n=1}^{\infty} (2x-6)^n$
16. $\sum_{n=1}^{\infty} (3x+1)^n$

17. $\sum_{n=0}^{\infty} \frac{x^n}{n!}$
18. $\sum_{n=0}^{\infty} \frac{(x-5)^n}{n!}$
19. $\sum_{n=1}^{\infty} \frac{n^2 \cdot x^n}{3^n}$
20. $\sum_{n=1}^{\infty} \frac{n^5 \cdot x^n}{3^n}$

21. $\sum_{n=1}^{\infty} n! \cdot x^n$
22. $\sum_{n=1}^{\infty} n! \cdot (x+2)^n$
23. $\sum_{n=3}^{\infty} n! \cdot (x-7)^n$
24. $\sum_{n=0}^{\infty} \frac{3^n \cdot x^n}{n!}$

In problems 25 – 34, the letters "a" and "b" represent positive constants. Find all values of x for which each given power series converges.

25. $\sum_{n=1}^{\infty} (x-a)^n$
26. $\sum_{n=1}^{\infty} (x+b)^n$
27. $\sum_{n=1}^{\infty} \frac{(x-a)^n}{n}$
28. $\sum_{n=1}^{\infty} \frac{(x-a)^n}{n^2}$

29. $\sum_{n=1}^{\infty} (ax)^n$
30. $\sum_{n=1}^{\infty} (\frac{x}{a})^n$
31. $\sum_{n=1}^{\infty} (ax-b)^n$
32. $\sum_{n=1}^{\infty} (ax+b)^n$

33. A friend claimed that the interval of convergence for a power series of the form $\sum_{n=1}^{\infty} a_n \cdot (x-4)^n$ is the interval $(1, 9)$. Without checking your friend's work, how can you be certain that your friend is wrong?

34. Which of the following intervals are possible intervals of convergence for the power series in problem 33?
    $(2, 6), (0, 4), x = 0, [1, 7], (-1, 9], x = 4, [3, 5), [-4, 4), x = 3$.

35. Which of the following intervals are possible intervals of convergence for $\sum_{n=1}^{\infty} a_n \cdot (x-7)^n$?
    $(3, 10), (5, 9), x = 0, [1, 13], (-1, 15], x = 4, [3, 11), [0, 14), x = 7$.

36. Fill in each blank with a number so the resulting interval could be the interval of convergence for the power series $\sum_{n=1}^{\infty} a_n \cdot (x-3)^n$: $(0, \_\_\_), (\_\_\_, 7), [1, \_\_\_], (\_\_\_, 15], [\_\_\_, 11), [0, \_\_\_), x = \_\_\_$.

37. Fill in each blank with a number so the resulting interval could be the interval of convergence for the power series $\sum_{n=1}^{\infty} a_n \cdot (x-1)^n$: $(0, \_\_\_), (\_\_\_, 7), [1, \_\_\_], (\_\_\_, 5], [\_\_\_, 11), [0, \_\_\_), x = \_\_\_$.

In problems 38 – 45, use the patterns you noticed in the earlier problems and examples to build a power series with the given intervals of convergence. (There are many possible correct answers — find one.)

38. $(-5, 5)$ 39. $[-3, 3)$ 40. $[-2, 2]$ 41. $(0, 6)$

42. $[0, 8)$ 43. $[2, 8)$ 44. $[3, 7]$ 45. $x = 3$

In problems 46 – 59, find the interval of convergence for each series. For x in the interval of convergence, find the sum of the series as a function of x. (Hint: You know how to find the sum of a geometric series.)

46. $\sum_{n=0}^{\infty} x^n$ 47. $\sum_{n=0}^{\infty} (x-3)^n$ 48. $\sum_{n=0}^{\infty} (2x)^n$ 49. $\sum_{n=0}^{\infty} (3x)^n$

50. $\sum_{n=0}^{\infty} x^{2n}$ 51. $\sum_{n=0}^{\infty} x^{3n}$ 52. $\sum_{n=0}^{\infty} (\frac{x-6}{2})^n$ 53. $\sum_{n=0}^{\infty} (\frac{x-6}{5})^n$

54. $\sum_{n=0}^{\infty} (\frac{x}{2})^n$ 55. $\sum_{n=0}^{\infty} (\frac{x}{5})^n$ 56. $\sum_{n=0}^{\infty} (\frac{x}{3})^{2n}$ 57. $\sum_{n=0}^{\infty} (\frac{x}{2})^{3n}$

58. $\sum_{n=0}^{\infty} (\frac{1}{2}\sin(x))^n$ 59. $\sum_{n=0}^{\infty} (\frac{1}{3}\cos(x))^n$

Practice Answers

Practice 1: $c_n = \dfrac{5^n \cdot x^n}{n}$ so $c_{n+1} = \dfrac{5^{n+1} \cdot x^{n+1}}{n+1}$. Then, using the Ratio Test,

$$\left| \frac{c_{n+1}}{c_n} \right| = \left| \frac{\dfrac{5^{n+1} x^{n+1}}{n+1}}{\dfrac{5^n x^n}{n}} \right| = \left| \frac{n}{n+1} \cdot \frac{5^{n+1}}{5^n} \cdot \frac{x^{n+1}}{x^n} \right| = \left| \frac{n}{n+1} \cdot 5 \cdot x \right| \to |5x| = L.$$

Solving $|5x| < 1$, we have $-1 < 5x < 1$ or $-1/5 < x < 1/5$. The series converges for $-1/5 < x < 1/5$ and it diverges for $x < -1/5$ and for $x > 1/5$.

Finally, we need to check the two remaining values of x: the **endpoints** $x = -1/5$ and $x = 1/5$.

When $x = -1/5$, $\displaystyle\sum_{n=1}^{\infty} \frac{5^n x^n}{n} = \sum_{n=1}^{\infty} \frac{5^n (-1/5)^n}{n} = \sum_{k=1}^{\infty} \frac{(-1)^n}{n}$ which converges by the Alternating Series Test.

When $x = 1/5$, $\displaystyle\sum_{n=1}^{\infty} \frac{5^n x^n}{n} = \sum_{n=1}^{\infty} \frac{5^n (1/5)^n}{n} = \sum_{k=1}^{\infty} \frac{1}{n}$, the harmonic series, which diverges.

In summary, the power series $\displaystyle\sum_{n=1}^{\infty} \frac{5^n x^n}{n}$ converges if $-1/5 \leq x < 1/5$, if x is in the interval $[-1/5, 1/5)$.

Practice 2: The series converges at $x = -1$ and $x = 2$.

The convergence is unknown at

$x = -3$ (endpoint?), 4, -6, and 7 (endpoint?)

The series diverges at $x = -8$ and $x = 17$.

The regions of known convergence and known divergence are shown in Fig. 6.

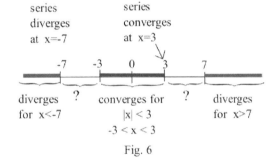

Fig. 6

Practice 3: The radius of convergence is $1/5$.

Practice 4: The series converges at $x = -2$.

The convergence is unknown at

$x = -9$ (endpoint?), 0, and -11 (endpoint?)

The series diverges at $x = 3$.

The regions of known convergence and known divergence are shown in Fig. 7.

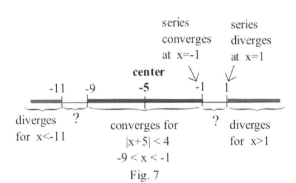

Fig. 7

Practice 5: $c_n = \dfrac{n \cdot (x-3)^n}{5^n}$ so $c_{n+1} = \dfrac{(n+1) \cdot (x-3)^{n+1}}{5^{n+1}}$. Then, using the Ratio Test,

$$\left| \frac{c_{n+1}}{c_n} \right| = \left| \frac{\dfrac{(n+1) \cdot (x-3)^{n+1}}{5^{n+1}}}{\dfrac{n \cdot (x-3)^n}{5^n}} \right| = \left| \frac{n+1}{n} \cdot \frac{5^n}{5^{n+1}} \cdot \frac{(x-3)^{n+1}}{(x-3)^n} \right| = \left| \frac{n+1}{n} \cdot \frac{1}{5} \cdot (x-3) \right| \to \left| \frac{x-3}{5} \right| = L.$$

Solving $\left|\dfrac{x-3}{5}\right| < 1$, we have $-1 < \dfrac{x-3}{5} < 1$ so $-5 < x-3 < 8$. The series converges for $-2 < x < 8$ and it diverges for $x < -2$ and for $x > 8$. The radius of convergence is 5.

To find the interval of convergence we still need to check the endpoints $x = -2$ and $x = 8$:

When $x = -2$, $\displaystyle\sum_{n=0}^{\infty} \dfrac{n \cdot (x-3)^n}{5^n} = \sum_{n=1}^{\infty} \dfrac{n \cdot (-5)^n}{5^n} = \sum_{n=1}^{\infty} n \cdot (-1)^n$ which diverges by the nth Term Test.

When $x = 8$, $\displaystyle\sum_{n=0}^{\infty} \dfrac{n \cdot (x-3)^n}{5^n} = \sum_{n=1}^{\infty} \dfrac{n \cdot (5)^n}{5^n} = \sum_{n=1}^{\infty} n$ which diverges by the nth Term Test.

The interval of convergence is $-2 < x < 8$ or $(-2, 8)$.

Appendix: Proof of the Interval of Convergence Theorem

(i) Suppose the power series $\displaystyle\sum_{n=0}^{\infty} a_n x^n$ converges at $x = c$: $\displaystyle\sum_{n=0}^{\infty} a_n c^n$ converges.

Then the terms of the series $a_n c^n$ must approach 0, $\displaystyle\lim_{n \to \infty} a_n c^n = 0$, so there is a number N so that if $n \geq N$, then $|a_n c^n| < 1$ and $|a_n| < \dfrac{1}{|c^n|}$ for all $n > N$.

If $|x| < |c|$, then $\displaystyle\sum_{n=0}^{\infty} |a_n x^n| = \{|a_0| + |a_1 x| + \ldots + |a_{N-1} x^{N-1}|\} + \{|a_N x^N| + |a_{N+1} x^{N+1}| + \ldots\}$.

The first piece, $|a_0| + |a_1 x| + \ldots + |a_{N-1} x^{N-1}|$, consists of a finite number of terms so it is a finite number.

The second piece, $|a_N x^N| + |a_{N+1} x^{N+1}| + \ldots$ is less than a convergent geometric series:

$$|a_N x^N| + |a_{N+1} x^{N+1}| + |a_{N+2} x^{N+2}| + |a_{N+3} x^{N+3}| + \ldots$$

$$< \frac{1}{|c^N|}|x^N| + \frac{1}{|c^{N+1}|}|x^{N+1}| + \frac{1}{|c^{N+2}|}|x^{N+2}| + \frac{1}{|c^{N+3}|}|x^{N+3}| + \ldots$$

$$= \left|\frac{x}{c}\right|^N + \left|\frac{x}{c}\right|^{N+1} + \left|\frac{x}{c}\right|^{N+2} + \left|\frac{x}{c}\right|^{N+3} + \ldots \text{ which converges since } |x| < |c| \text{ and } \left|\frac{x}{c}\right| < 1.$$

If $|x| < |c|$, then $\displaystyle\sum_{n=0}^{\infty} |a_n x^n|$ converges so $\displaystyle\sum_{n=0}^{\infty} a_n x^n$ converges.

(ii) This part follows from part (i) of the Theorem. Suppose that the power series $\sum_{n=0}^{\infty} a_n x^n$ diverges at $x = d$: $\sum_{n=0}^{\infty} a_n d^n$ diverges. If the series converges for some x_0 with $|x_0| > |d|$, we can put $c = x_0$ and conclude from part (i) that the series must converge at $x = d$ because $|c| = |x_0| > |d|$. This contradicts the fact that the series diverges at $x = d$, so the series cannot converge for any x_0 with $|x_0| > |d|$.

If $\sum_{n=0}^{\infty} a_n d^n$ diverges and $|x| > |d|$, then the power series $\sum_{n=0}^{\infty} a_n x^n$ diverges.

10.9 REPRESENTING FUNCTIONS AS POWER SERIES

Power series define functions, but how are these power series functions related to functions we know about such as $\sin(x)$, $\cos(x)$, e^x, and $\ln(x)$? How can we represent common functions as power series, and why would we want to do so? The next two sections provide partial answers to these questions. In this section we start with a function defined by a geometric series and show how we can obtain power series representations for several related functions. And we look at a few ways in which power series representations of functions are used. The next section examines a more general method for obtaining power series representations for functions.

The foundation for the examples in this section is a power series whose sum we know. The power series $\sum_{n=0}^{\infty} x^n$ is also a geometric series, with the common ratio $r = x$, and, for $|x| < 1$, we know the sum of the series is $\frac{1}{1-x}$.

Geometric Series Formula

For $|x| < 1$, $\sum_{n=0}^{\infty} x^n = 1 + x + x^2 + x^3 + x^4 + \ldots = \frac{1}{1-x}$.

One simple but powerful method of obtaining power series for related functions is to replace each "x" with a function of x.

Substitution in Power Series

Suppose $f(x)$ is defined by a power series

$$f(x) = \sum_{n=0}^{\infty} a_n x^n = a_0 + a_1 x + a_2 x^2 + a_3 x^3 + \ldots + a_n x^n + \ldots$$

that converges for $-R < x < R$.

Then $f(x^p) = \sum_{n=0}^{\infty} a_n \{x^p\}^n = a_0 + a_1 x^p + a_2 \{x^p\}^2 + a_3 \{x^p\}^3 + \ldots + a_n \{x^p\}^n + \ldots$

$= a_0 + a_1 x^p + a_2 x^{2p} + a_3 x^{3p} + \ldots + a_n x^{np} + \ldots$

converges for $-R < x^p < R$, and

$f(x-c) = \sum_{n=0}^{\infty} a_n (x-c)^n = a_0 + a_1(x-c) + a_2(x-c)^2 + a_3(x-c)^3 + \ldots + a_n(x-c)^n + \ldots$

converges for $-R < x - c < R$.

We can use this substitution method to obtain power series for some functions related to the Geometric Series Formula $\frac{1}{1-x}$.

Example 1: Find power series for (a) $\frac{1}{1-x^2}$, (b) $\frac{1}{1+x}$, and (c) $\frac{x}{1-x}$.

Solution:

(a) Substituting "x^2" for "x" in the Geometric Series Formula, we get

$$\frac{1}{1-x^2} = 1 + \{x^2\} + \{x^2\}^2 + \{x^2\}^3 + \{x^2\}^4 + \ldots = \sum_{n=0}^{\infty} (x^2)^n$$

$$= 1 + x^2 + x^4 + x^6 + x^8 + \ldots = \sum_{n=0}^{\infty} x^{2n} \quad \text{for } -1 < x < 1.$$

(b) Substituting "$-x$" for "x" in the Geometric Series Formula,

$$\frac{1}{1+x} = \frac{1}{1-(-x)} = 1 + \{-x\} + \{-x\}^2 + \{-x\}^3 + \{-x\}^4 + \ldots = \sum_{n=0}^{\infty} (-x)^n$$

$$= 1 - x + x^2 - x^3 + x^4 + \ldots = \sum_{n=0}^{\infty} (-1)^n x^n \quad \text{for } -1 < x < 1.$$

(c) We need to recognize that $\frac{x}{1-x}$ is a product, $\frac{x}{1-x} = x \cdot \frac{1}{1-x}$. Then

$$x \cdot \frac{1}{1-x} = x \{ 1 + x + x^2 + x^3 + x^4 + \ldots \}$$

$$= x + x^2 + x^3 + x^4 + x^5 + \ldots = \sum_{n=0}^{\infty} x^{n+1} \quad \text{or, equivalently,} \quad \sum_{n=1}^{\infty} x^n \quad \text{for } -1 < x < 1.$$

Practice 1: Find power series for (a) $\frac{1}{1-x^3}$, (b) $\frac{1}{1+x^2}$, and (c) $\frac{5x}{1+x}$.

One of the features of polynomials that makes them very easy to differentiate and integrate is that we can differentiate and integrate them term–by–term. The same result is true for power series.

> **Term–by–Term Differentiation and Integration of Power Series**
>
> Suppose $f(x)$ is defined by a power series
>
> $$f(x) = \sum_{n=0}^{\infty} a_n x^n = a_0 + a_1 x + a_2 x^2 + a_3 x^3 + \ldots + a_n x^n + \ldots$$
>
> that converges for $-R < x < R$.
>
> Then,
>
> (a) the derivative of f is given by the power series obtained by term–by–term differentiation of f:
>
> $$f'(x) = \sum_{n=1}^{\infty} n \cdot a_n x^{n-1} = a_1 + 2a_2 x + 3a_3 x^2 + 4a_4 x^3 + \ldots + n \cdot a_n x^{n-1} + \ldots$$
>
> (b) an antiderivative of f is given by the power series obtained by term–by–term integration of f:
>
> $$\int f(x)\, dx = C + \sum_{n=0}^{\infty} a_n \frac{x^{n+1}}{n+1} = C + a_0 x + a_1 \frac{x^2}{2} + a_2 \frac{x^3}{3} + a_3 \frac{x^4}{4} + \ldots + a_n \frac{x^{n+1}}{n+1} + \ldots$$
>
> The power series for the derivative and antiderivative of f converge for $-R < x < R$. (The power series for f, f' and the antiderivative of f may differ in whether or not they converge at the **endpoints** of the interval of convergence, but they all converge for $-R < x < R$.)

The proof of this result is rather long and technical and is omitted.

Like the previous substitution method, term–by–term differentiation and integration can be used to obtain power series for some functions related to the Geometric Series Formula $\frac{1}{1-x}$.

Example 2: Find power series for (a) $\ln(1-x)$, and (b) $\arctan(x)$.

Solution: These two are more challenging than the previous examples because we need to recognize that these two functions are integrals of functions whose power series we already know.

(a) $\ln(1-x) = \int \frac{-1}{1-x}\, dx = -\int \{ 1 + x + x^2 + x^3 + x^4 + \ldots \}\, dx$

$$= -\{ x + \frac{x^2}{2} + \frac{x^3}{3} + \frac{x^4}{4} + \ldots \} + C = C - \sum_{n=1}^{\infty} \frac{x^n}{n}.$$

We can find the value of C by using the fact that $\ln(1) = 0$:

for $x = 0$, $0 = \ln(1-0) = C - \sum_{n=1}^{\infty} \frac{0^n}{n} = C$ so $C = 0$ and

$$\ln(1-x) = -\{ x + \frac{x^2}{2} + \frac{x^3}{3} + \frac{x^4}{4} + \ldots \} = -\sum_{n=1}^{\infty} \frac{x^n}{n} \quad \text{for } -1 \leq x < 1.$$

(b) $\arctan(x) = \int \frac{1}{1+x^2} dx = \int \{ 1 - x^2 + x^4 - x^6 + x^8 - \ldots \} dx$

$$= C + \{ x - \frac{x^3}{3} + \frac{x^5}{5} - \frac{x^7}{7} + \frac{x^9}{9} - \ldots \} = C + \sum_{n=0}^{\infty} (-1)^n \frac{x^{2n+1}}{2n+1}.$$

We can find the value of C by using the fact that $\arctan(0) = 0$:

for $x = 0$, $0 = \arctan(0) = C + \sum_{n=0}^{\infty} (-1)^n \frac{0^{2n+1}}{2n+1} = C$ so $C = 0$ and

$$\arctan(x) = x - \frac{x^3}{3} + \frac{x^5}{5} - \frac{x^7}{7} + \frac{x^9}{9} - \ldots = \sum_{n=0}^{\infty} (-1)^n \frac{x^{2n+1}}{2n+1} \quad \text{for } -1 \leq x \leq 1.$$

Practice 2: Find a power series for $\ln(1+x)$.

Power series can also be used to help us evaluate definite integrals.

Example 3: Use the power series for $\arctan(x)$ to represent the definite integral $\int_0^{0.5} \arctan(x)\, dx$ as a numerical series. Then approximate the value of the integral by calculating the sum of the first four terms of the numerical series.

Solution: $\int_0^{0.5} \arctan(x)\, dx = \int_0^{0.5} \{ x - \frac{x^3}{3} + \frac{x^5}{5} - \frac{x^7}{7} + \frac{x^9}{9} - \ldots \} dx$

$$= \frac{x^2}{2} - \frac{x^4}{4 \cdot 3} + \frac{x^6}{6 \cdot 5} - \frac{x^8}{8 \cdot 7} + \frac{x^{10}}{10 \cdot 9} - \ldots \Big|_0^{0.5}$$

$$= \{ \frac{1}{2}(0.5)^2 - \frac{1}{12}(0.5)^4 + \frac{1}{30}(0.5)^6 - \frac{1}{56}(0.5)^8 + \frac{1}{90}(0.5)^{10} - \ldots \} - \{ 0 \}.$$

The sum of the first four terms is approximately 0.120243. Since the numerical series is an alternating series, we know that the fourth partial sum, 0.120243, is within the value of the next term, $\frac{1}{90}(0.5)^{10} \approx 0.000011$, of the exact value of the sum. The exact value of the definite integral is between $0.120243 - 0.000011 = 0.120232$ and $0.120243 + 0.000011 = 0.120254$.

Practice 3: Use the power series for $x^2 \cdot \ln(1+x)$ to represent the definite integral $\int_0^{0.2} x^2 \cdot \ln(1+x)\, dx$ as a numerical series. Then approximate the value of the integral by calculating the sum of the first three terms of the numerical series.

All of the power series used in this section have followed from the Geometric Series Formula, and their main purpose here was to illustrate some uses of substitution and term–by–term differentiation and integration to obtain power series for related functions. Many functions, however, are not related to a geometric series, and the next section discusses a method for representing them using power series. Once we can represent these new functions using power series, we can then use substitution and term–by–term differentiation and integration to obtain power series for functions related to them.

The following table collects some of the power series representations we have obtained in this section.

Table of Series Based on $\dfrac{1}{1-x} = \sum\limits_{n=0}^{\infty} x^n$

$\dfrac{1}{1-x} = 1 + x + x^2 + x^3 + x^4 + \ldots = \sum\limits_{n=0}^{\infty} x^n$ with interval of convergence $(-1, 1)$.

$\dfrac{1}{1+x} = 1 - x + x^2 - x^3 + x^4 + \ldots = \sum\limits_{n=0}^{\infty} (-1)^n x^n = \dfrac{1}{1-(-x)}$

$\dfrac{1}{1-x^2} = 1 + x^2 + x^4 + x^6 + x^8 + \ldots = \sum\limits_{n=0}^{\infty} x^{2n} = \dfrac{1}{1-(x^2)}$

$\dfrac{1}{1+x^2} = 1 - x^2 + x^4 - x^6 + x^8 - \ldots = \sum\limits_{n=0}^{\infty} (-1)^n x^{2n} = \dfrac{1}{1-(-x^2)}$

$\dfrac{1}{1-x^3} = 1 + x^3 + x^6 + x^9 + x^{12} + \ldots = \sum\limits_{n=0}^{\infty} x^{3n} = \dfrac{1}{1-(x^3)}$

$\ln(1-x) = -\left\{ x + \dfrac{x^2}{2} + \dfrac{x^3}{3} + \dfrac{x^4}{4} + \ldots \right\} = -\sum\limits_{n=1}^{\infty} \dfrac{x^n}{n} = \int \dfrac{-1}{1-x}\, dx$

$\ln(1+x) = x - \dfrac{x^2}{2} + \dfrac{x^3}{3} - \dfrac{x^4}{4} + \ldots = \sum\limits_{n=1}^{\infty} (-1)^{n+1} \dfrac{x^n}{n} = \int \dfrac{1}{1+x}\, dx$

$$\arctan(x) = x - \frac{x^3}{3} + \frac{x^5}{5} - \frac{x^7}{7} + \frac{x^9}{9} - \ldots = \sum_{n=0}^{\infty} (-1)^n \frac{x^{2n+1}}{2n+1} = \int \frac{1}{1+x^2} dx$$

$$\frac{1}{(1-x)^2} = 1 + 2x + 3x^2 + 4x^3 + 5x^4 + \ldots = \sum_{n=1}^{\infty} n \cdot x^{n-1} = D\left(\frac{1}{1-x}\right)$$

$$\frac{1}{(1+x)^2} = 1 - 2x + 3x^2 - 4x^3 + 5x^4 - \ldots = \sum_{n=1}^{\infty} (-1)^n n \cdot x^{n-1} = D\left(\frac{-1}{1+x}\right)$$

Fig. 1 shows the graphs of arctan(x) and the first few polynomials $x, x - \frac{x^3}{3}, x - \frac{x^3}{3} + \frac{x^5}{5}$ that approximate arctan(x). This type of approximation will be discussed a little in Section 10.10 and a lot in Section 10.11.

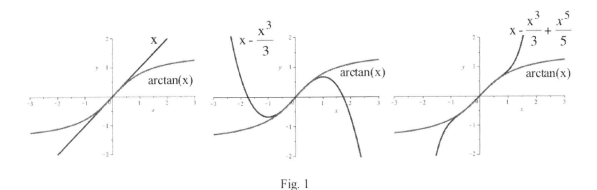

Fig. 1

PROBLEMS

In problems 1 – 14, use the substitution method and a known power series to find power series for the given functions.

1. $\dfrac{1}{1-x^4}$

2. $\dfrac{1}{1-x^5}$

3. $\dfrac{1}{1+x^4}$

4. $\dfrac{1}{1+x^5}$

5. $\dfrac{1}{5+x} = \dfrac{1}{5} \cdot \dfrac{1}{1+(x/5)}$

6. $\dfrac{1}{3-x} = \dfrac{1}{3} \cdot \dfrac{1}{1-(x/3)}$

7. $\dfrac{x^2}{1+x^3}$

8. $\dfrac{x}{1+x^4}$

9. $\ln(1+x^2)$

10. $\ln(1+x^3)$

11. $\arctan(x^2)$

12. $\arctan(x^3)$

13. $\dfrac{1}{(1-x^2)^2}$

14. $\dfrac{1}{(1+x^2)^2}$

In problems 15–21, represent each definite integral as a numerical series. Calculate the sum of the first three terms for each series.

15. $\displaystyle\int_0^{0.5} \frac{1}{1-x^3}\, dx$

16. $\displaystyle\int_0^{0.5} \frac{1}{1+x^3}\, dx$

17. $\displaystyle\int_0^{0.6} \ln(1+x)\, dx$

18. $\displaystyle\int_0^{0.5} x^2 \cdot \arctan(x)\, dx$

19. $\displaystyle\int_0^{0.3} \frac{1}{(1-x)^2}\, dx$

20. $\displaystyle\int_0^{0.7} \frac{x^3}{(1-x)^2}\, dx$

In problems 21 – 27, use the Table of Series to help represent each function as a power series. Then calculate each limit.

21. $\displaystyle\lim_{x\to 0} \frac{\arctan(x)}{x}$

22. $\displaystyle\lim_{x\to 0} \frac{\ln(1-x)}{2x}$

23. $\displaystyle\lim_{x\to 0} \frac{\ln(1+x)}{2x}$

24. $\displaystyle\lim_{x\to 0} \frac{\arctan(x^2)}{x}$

25. $\displaystyle\lim_{x\to 0} \frac{\ln(1-x^2)}{3x}$

26. $\displaystyle\lim_{x\to 0} \frac{\ln(1+x^2)}{3x}$

In problems 27 – 32, use the Table of Series or the substitution method to determine a power series for each function and then determine the interval of convergence of each power series.

27. $\dfrac{1}{1+x}$

28. $\dfrac{1}{1-x^2}$

29. $\ln(1-x)$

30. $\ln(1+x)$

31. $\arctan(x)$

32. $\arctan(x^2)$

The series given below for $\sin(x)$ and e^x are derived in Section 10.10. Use these given series and the methods of this section to answer problems 33 – 42.

$$\sin(x) = x - \frac{x^3}{3!} + \frac{x^5}{5!} - \frac{x^7}{7!} + \ldots = \sum_{k=0}^{\infty}(-1)^k \frac{x^{2k+1}}{(2k+1)!}, \quad e^x = 1 + x + \frac{x^2}{2!} + \frac{x^3}{3!} + \frac{x^4}{4!} + \ldots = \sum_{k=0}^{\infty} \frac{x^k}{k!}$$

33. Find a power series for $\sin(x^2)$

34. Find a power series for $\sin(2x)$

35. Find a power series for $e^{(-x^2)}$

36. Find a power series for $e^{(2x)}$

37. Find a power series for $\cos(x)$

38. Find a power series for $\cos(x^2)$

39. Represent the integral as a numerical series: $\displaystyle\int_0^1 \sin(x^2)\, dx$

40. Represent the integral as a numerical series: $\displaystyle\int_0^1 e^{(-x^2)}\, dx$

41. $\displaystyle\lim_{x\to 0} \frac{\sin(x)}{x}$

42. $\displaystyle\lim_{x\to 0} \frac{\sin(x)-x}{x^3}$

Practice Answers

Practice 1: Geometric Series Formula: $\dfrac{1}{1-x} = \sum_{n=0}^{\infty} x^n$ for $|x| < 1$.

(a) Replacing "x" with "x^3" we have $\dfrac{1}{1-x^3} = \sum_{n=0}^{\infty} x^{3n}$ for $|x^3| < 1$ or $|x| < 1$.

(b) Replacing "x" with "$-x^2$" we have $\dfrac{1}{1-(-x^2)} = \dfrac{1}{1+x^2} = \sum_{n=0}^{\infty} (-x^2)^n = \sum_{n=0}^{\infty} (-1)^n x^{2n}$

for $|-x| < 1$ or $|x| < 1$.

(c) Using the result of part (b), $\dfrac{5x}{1+x} = 5x \dfrac{1}{1+x} = 5x \cdot \sum_{n=0}^{\infty} (-1)^n x^n = 5 \cdot \sum_{n=0}^{\infty} (-1)^n x^{n+1}$

for $|-x| < 1$ or $|x| < 1$.

Practice 2: $\ln(1+x) = \displaystyle\int \dfrac{1}{1+x} dx = \int \sum_{n=0}^{\infty} (-1)^n x^n \, dx = \int 1 - x + x^2 - x^3 + x^4 - \ldots \, dx$

$= x - \dfrac{x^2}{2} + \dfrac{x^3}{3} - \dfrac{x^4}{4} + \dfrac{x^5}{5} - \ldots + C$. Putting $x = 0$, we get $C = 0$. Then

$\ln(1+x) = \sum_{n=0}^{\infty} (-1)^n \dfrac{x^{n+1}}{n+1}$ or, equivalently, $\sum_{n=1}^{\infty} (-1)^{n+1} \dfrac{x^n}{n}$

Practice 3: Using the result of Practice 2,

$x^2 \cdot \ln(1+x) = x^2 \cdot \sum_{n=1}^{\infty} (-1)^{n+1} \dfrac{x^n}{n} = \sum_{n=1}^{\infty} (-1)^{n+1} \dfrac{x^{n+2}}{n}$. Then

$\displaystyle\int x^2 \cdot \ln(1+x) \, dx = \int \sum_{n=1}^{\infty} (-1)^{n+1} \dfrac{x^{n+2}}{n} \, dx = \sum_{n=1}^{\infty} \dfrac{(-1)^{n+1}}{n} \dfrac{x^{n+3}}{n+3} \Big|_0^{0.2}$

$= \left\{ \dfrac{1}{4}(0.2)^4 - \dfrac{1}{10}(0.2)^5 + \dfrac{1}{18}(0.2)^6 - \dfrac{1}{28}(0.2)^7 + \ldots \right\} - \{0\}$.

$s_3 = \dfrac{1}{4}(0.2)^4 - \dfrac{1}{10}(0.2)^5 + \dfrac{1}{18}(0.2)^6 \approx 0.0003716$ with an error less than $|a_4| = \dfrac{1}{28}(0.2)^7 \approx 4.57 \cdot 10^{-7}$

Appendix: Partial Sums of Power Series using MAPLE

MAPLE commands for $\sum_{n=1}^{100} \dfrac{(0.7)^n}{n}$, $\sum_{n=1}^{100} \dfrac{(0.7)^n}{n^2}$, $\sum_{n=0}^{100} \dfrac{(0.7)^n}{n!}$, $\sum_{n=0}^{100} (-1)^n \dfrac{(0.7)^n}{1+\sqrt{n}}$

> sum((.7)^n/n , n=1..100) ;   (then press ENTER key)

> sum((.7)^n/n^2 , n=1..100) ;

> sum((.7)^n/n! , n=0..100) ;

> sum((-1)^n*(.7)^n/(1+sqrt(n)) , n=0..100) ;

10.10 TAYLOR AND MACLAURIN SERIES

This section discusses a method for representing a variety of functions as power series, and power series representations are derived for $\sin(x)$, $\cos(x)$, e^x, and several functions related to them. These power series are used to evaluate the functions and limits and to approximate definite integrals.

We start with an examination of how to determine the formula for a polynomial from information about the polynomial when $x = 0$, and then this process is extended to determine a series representation for a function from information about the function when $x = 0$.

Polynomials

Polynomials are among the easiest functions to work with, and they have a variety of "nice" properties including the following:

The values of $P(x)$ and its derivatives at $x = 0$ completely determine the formula for $P(x)$.

If the values of $P(x)$ and all of its derivatives at $x = 0$ are known, then we can use those values to find a formula for $P(x)$.

Example 1: Suppose $P(x)$ is a cubic polynomial with $P(0) = 7$, $P'(0) = 5$, $P''(0) = 16$, and $P'''(0) = 18$. (Since $P(x)$ is a cubic, its higher derivatives are all 0.) Find a formula for $P(x)$.

Solution: Since $P(x)$ is a cubic polynomial, then $P(x) = a_0 + a_1 x + a_2 x^2 + a_3 x^3$ for some numbers a_0, a_1, a_2, and a_3. We want to find the values of those numbers, and we can do so by substituting 0 for x in the expressions for P, P', P'', and P''' and using the given information.

$7 = P(0) = a_0 + a_1 \cdot 0 + a_2 \cdot 0^2 + a_3 \cdot 0^3 = a_0$ so $a_0 = 7$.

$P'(x) = a_1 + 2a_2 x + 3a_3 x^2$. $5 = P'(0) = a_1 + 2a_2 \cdot 0 + 3a_3 \cdot 0^2 = a_1$ so $a_1 = 5$.

$P''(x) = 2a_2 + 6a_3 x$. $16 = P''(0) = 2a_2 + 6a_3 \cdot 0 = 2a_2$ so $a_2 = 16/2 = 8$.

$P'''(x) = 6a_3$. $18 = P'''(0) = 6a_3$ so $a_3 = 18/6 = 3$.

$P(x) = 7 + 5x + 8x^2 + 3x^3$. You should verify that this cubic polynomial and its derivatives have the values specified in the problem.

Practice 1: Suppose $P(x)$ is a 4th degree polynomial with $P(0) = -3$, $P'(0) = 4$, $P''(0) = 10$, $P'''(0) = 12$, and $P^{(4)}(0) = 24$. (Since $P(x)$ is a 4th degree polynomial, the higher derivatives are all 0.) Find a formula for $P(x)$.

For polynomials, the n_{th} derivative evaluated at $x = 0$ is $P^{(n)}(0) = (n)(n-1)(n-2)...(2)(1)\, a_n = n! \cdot a_n$, so the coefficient a_n of the nth term of the polynomial is $a_n = P^{(n)}(0)/n!$.

Series

In many important ways power series behave like polynomials, very big polynomials, and this is one of those ways. The next result says that if a function can be represented by a power series, then the coefficients of the power series just depend on the values of the derivatives of the function evaluated at 0.

Maclaurin Series for f(x)

If a function $f(x)$ has a power series representation $f(x) = \sum_{n=0}^{\infty} a_n x^n$ for $|x| < R$

then the coefficients are given by $a_n = \dfrac{f^{(n)}(0)}{n!}$.

The **Maclaurin Series** for $f(x)$ is

$$f(x) = \sum_{n=0}^{\infty} \frac{f^{(n)}(0)}{n!} x^n = f(0) + f'(0) \cdot x + \frac{f''(0)}{2!} \cdot x^2 + \frac{f'''(0)}{3!} \cdot x^3 + \ldots + \frac{f^{(n)}(0)}{n!} \cdot x^n + \ldots$$

Proof: Suppose $f(x) = \sum_{n=0}^{\infty} a_n x^n = a_0 + a_1 x + a_2 x^2 + a_3 x^3 + \ldots + a_n x^n + \ldots$ for $|x| < R$.

Then $f(0) = a_0 + a_1 \cdot 0 + a_2 \cdot 0^2 + a_3 \cdot 0^3 + \ldots + a_n \cdot 0^n + \ldots = a_0$ so $a_0 = f(0) = \dfrac{f^{(0)}(0)}{0!}$.

(We are using the conventions that $f^{(0)}(x) = f(x)$ and that $0! = 1$.)

$f'(x) = a_1 + 2a_2 x + 3a_3 x^2 + \ldots + na_n x^{n-1} + \ldots$

so $f'(0) = a_1 + 2a_2 \cdot 0 + 3a_3 \cdot 0^2 + \ldots + na_n \cdot 0^{n-1} + \ldots = a_1$ and $a_1 = \dfrac{f'(0)}{1!}$.

$f''(x) = 2a_2 + 2 \cdot 3 a_3 x + \ldots + (n-1) \cdot n \cdot a_n x^{n-2} + \ldots$

so $f''(0) = 2a_2 + 2 \cdot 3 a_3 \cdot 0 + \ldots + (n-1) \cdot n \cdot a_n \cdot 0^{n-2} + \ldots = 2a_2$ and $a_2 = \dfrac{f''(0)}{2} = \dfrac{f''(0)}{2!}$.

$f'''(x) = 2 \cdot 3 a_3 + \ldots + (n-2) \cdot (n-1) \cdot n \cdot a_n x^{n-3} + \ldots$

so $f'''(0) = 2 \cdot 3 a_3 + \ldots + (n-2) \cdot (n-1) \cdot n \cdot a_n \cdot 0^{n-3} + \ldots = 2 \cdot 3 a_3$ and $a_3 = \dfrac{f'''(0)}{2 \cdot 3} = \dfrac{f'''(0)}{3!}$.

In general, $f^{(n)}(x) = 1 \cdot 2 \cdot 3 \cdot \ldots \cdot (n-1) \cdot n\, a_n + \{$ terms still containing powers of $x\}$

so $f^{(n)}(0) = 1 \cdot 2 \cdot 3 \cdot \ldots \cdot (n-1) \cdot n\, a_n + \{0\} = n!\, a_n$ and $a_n = \dfrac{f^{(n)}(0)}{n!}$.

A similar result, and proof, is also true for a "shifted" power series, a power series centered at some value c. Such shifted series are called Taylor series.

Taylor Series for f(x) centered at c

If a function $f(x)$ has a power series representation $f(x) = \sum_{n=0}^{\infty} a_n (x-c)^n$ for $|x-c| < R$

then the coefficients are given by $a_n = \dfrac{f^{(n)}(c)}{n!}$.

The **Taylor Series** for $f(x)$ at c is

$$f(x) = \sum_{n=0}^{\infty} \frac{f^{(n)}(c)}{n!}(x-c)^n = f(c) + f'(c)\cdot(x-c) + \frac{f''(c)}{2!}\cdot(x-c)^2 + \frac{f'''(c)}{3!}\cdot(x-c)^3 + \ldots + \frac{f^{(n)}(c)}{n!}\cdot(x-c)^n + \ldots$$

The proof is very similar to the proof for Maclaurin series and is not included here.

A Maclaurin series is a Taylor series centered at $c = 0$, and Maclaurin series are a special case of Taylor series.

Note: These statements for Maclaurin series and Taylor series do not say that every function is or can be written as a power series. However, if a function is a power series, then its coefficients must follow the given pattern. Fortunately, most of the important functions such as $\sin(x), \cos(x), e^x$, and $\ln(x)$ can be written as power series.

You should notice that the first term of the Taylor series is simply the value of the function f at the point $x = c$: it provides the best constant function approximation of f near $x = c$. The sum of the first two terms of the Taylor series pattern for a function, $f(c) + f'(c)\cdot(x-c)$, is the formula for the tangent line to f at $x = c$ and is the linear approximation of $f(x)$ near $x = c$ that we first examined in Chapter 2. The Taylor series formula extends these approximations to higher degree polynomials, and the partial sums of the Taylor series provide higher degree polynomial approximations of f near $x = c$.

Taylor series and Maclaurin series were first discovered by the Scottish mathematician/astronomer James Gregory (1638–1675), but the results were not published until after his death. The English mathematician Brook Taylor (1685–1731) independently rediscovered these results and included them in a book in 1715. The Scottish mathematician/engineer Colin Maclaurin (1698–1746) quoted Taylor's work in his *Treatise on Fluxions* published in 1742. Maclaurin's book was widely read, and the Taylor series centered at $c = 0$ became known as Maclaurin series.

Example 2: Find the Maclaurin series for $f(x) = \sin(x)$ and the radius of convergence of the series.

Solution: $f(x) = \sin(x)$ so $f(0) = \sin(0)$ and $a_0 = f(0) = 0$.

$f'(x) = \cos(x)$ so $f'(0) = \cos(0) = 1$ and $a_1 = f'(0) = 1$.

$f''(x) = -\sin(x)$ so $f''(0) = -\sin(0) = 0$ and $a_2 = \dfrac{f''(0)}{2!} = 0$.

$f'''(x) = -\cos(x)$ so $f'''(0) = -\cos(0) = -1$ and $a_3 = \dfrac{f'''(0)}{3!} = \dfrac{-1}{3!}$.

$f^{(4)}(x) = \sin(x)$ and the pattern repeats:

$a_4 = 0, a_5 = \dfrac{1}{5!}, a_6 = 0, a_7 = \dfrac{-1}{7!}, a_8 = 0, a_9 = \dfrac{1}{9!}, \ldots$.

$$\sin(x) = x - \frac{x^3}{3!} + \frac{x^5}{5!} - \frac{x^7}{7!} + \frac{x^9}{9!} - \frac{x^{11}}{11!} + \ldots = \sum_{n=0}^{\infty} (-1)^n \frac{x^{2n+1}}{(2n+1)!}.$$

Notice that the Maclaurin series for $\sin(x)$ alternates and contains only odd powers of x.

We use the Ratio Test to find the radius of convergence. Let $c_n = (-1)^n \dfrac{x^{2n+1}}{(2n+1)!}$. Then

$$c_{n+1} = (-1)^{n+1} \frac{x^{2(n+1)+1}}{(2(n+1)+1)!} = (-1)^{n+1} \frac{x^{2n+3}}{(2n+3)!} \quad \text{so}$$

$$\left| \frac{c_{n+1}}{c_n} \right| = \left| \frac{(-1)^{n+1} \dfrac{x^{2n+3}}{(2n+3)!}}{(-1)^n \dfrac{x^{2n+1}}{(2n+1)!}} \right| = \left| \frac{x^{2n+3}}{x^{2n+1}} \frac{(2n+1)!}{(2n+3)!} \right| = \left| x^2 \frac{1}{(2n+2)(2n+3)} \right| \to 0 < 1$$

for every value of x.

The radius of convergence is $R = \infty$, and the interval of convergence is $(-\infty, \infty)$.

The Maclaurin series for $\sin(x)$ converges for every value of x.

Fig. 1 shows the graphs of $\sin(x)$ and the first few approximating polynomials x, $x - \dfrac{x^3}{3!}$, and $x - \dfrac{x^3}{3!} + \dfrac{x^5}{5!}$ for $-\pi \leq x \leq \pi$.

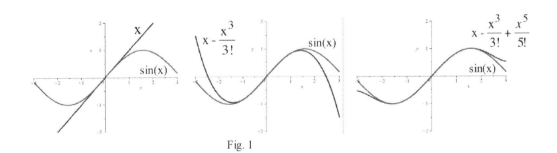

Fig. 1

By focusing our attention near $x = 0$, Fig. 1 shows the "goodness" of the Taylor polynomial fit to the function $f(x) = \sin(x)$. However, Fig. 2 shows that if x is not close to 0 then the values of the Taylor polynomials are far from the values of $f(x) = \sin(x)$. Typically the Taylor polynomials of a function are closest to the function when x is close to the number at which the series was centered, the value of c.

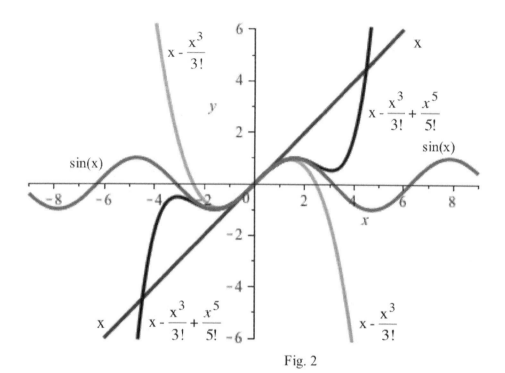

Fig. 2

Practice 2: Find the Maclaurin series for $f(x) = \cos(x)$ and the radius of convergence of the series. (Suggestion: Use the Term–by–Term Differentiation result of Section 10.9 .)

Once we have power series representations for $\sin(x)$ and $\cos(x)$, we can use the methods of Section 10.9 and the known series to approximate values of sine and cosine, determine power series representations of related functions, calculate limits, and approximate definite integrals.

Example 3: Use the series $\sin(x) = x - \dfrac{x^3}{3!} + \dfrac{x^5}{5!} - \dfrac{x^7}{7!} + \dfrac{x^9}{9!} - \dfrac{x^{11}}{11!} + \ldots$ to represent $\sin(0.5)$ as a numerical series. Approximate the value of $\sin(0.5)$ by calculating the partial sum of the first three non–zero terms and give a bound on the "error" between this approximation and the exact value of $\sin(0.5)$.

Solution: $\sin(0.5) = (0.5) - \dfrac{(0.5)^3}{3!} + \dfrac{(0.5)^5}{5!} - \dfrac{(0.5)^7}{7!} + \dfrac{(0.5)^9}{9!} - \dfrac{(0.5)^{11}}{11!} + \ldots$.

$\sin(0.5) \approx (0.5) - \dfrac{(0.5)^3}{3!} + \dfrac{(0.5)^5}{5!} = \dfrac{1}{2} - \dfrac{1}{48} + \dfrac{1}{3840} \approx 0.479427083333$.

Since this is an alternating series, the difference between the approximation and the exact value is less than the next term in the alternating series: "error" $< \dfrac{(0.5)^7}{7!} = \dfrac{1}{645120} \approx 0.00000155$.

If we use the sum of the first four nonzero terms to approximate the value of sin(0.5), then the "error" of the approximation is less than $\dfrac{(0.5)^9}{9!} = \dfrac{1}{185794560} \approx 5.4 \times 10^{-9}$.

We were able to obtain a bound for the error in the approximation of sin(0.5) because we were dealing with an alternating series, a type of series for which we have an error bound. However, many power series are not alternating series. In Section 10.11 we discuss a general error bound for Taylor series.

Practice 3: Use the sum of the first two nonzero terms of the Maclaurin series for cos(x) to approximate the value of cos(0.2). Give a bound on the "error" between this approximation and the exact value of cos(0.2).

Calculator Note: When you press the buttons on a calculator to evaluate sin(0.5) or cos(0.2), the calculator does not look up the answer in a table. Instead, the calculator is programmed with series representations for sine and cosine and other functions, and it calculates a partial sum of the appropriate series to obtain a numerical answer. It adds enough terms so that the 8 or 9 digits shown on the display are (usually) correct. In Section 10.11 we examine these methods in more detail and consider how to determine the number of terms needed in the partial sum to achieve the desired number of accurate digits in the answer.

Example 4: Represent $\sin(x^3)$ and $\int \sin(x^3)\,dx$ as power series. Then use the first three nonzero terms to approximate the value of $\int_0^1 \sin(x^3)\,dx$ and obtain a bound on the "error" of this approximation.

Solution: $\sin(x) = x - \dfrac{x^3}{3!} + \dfrac{x^5}{5!} - \dfrac{x^7}{7!} + \dfrac{x^9}{9!} - \dfrac{x^{11}}{11!} + \ldots$ so

$$\sin(x^3) = (x^3) - \dfrac{(x^3)^3}{3!} + \dfrac{(x^3)^5}{5!} - \dfrac{(x^3)^7}{7!} \ldots = x^3 - \dfrac{x^9}{3!} + \dfrac{x^{15}}{5!} - \dfrac{x^{21}}{7!} \ldots .$$

$$\int_0^1 \sin(x^3)\,dx = \int_0^1 \left\{ x^3 - \dfrac{x^9}{3!} + \dfrac{x^{15}}{5!} - \dfrac{x^{21}}{7!} \ldots \right\} dx = \dfrac{x^4}{4} - \dfrac{x^{10}}{10 \cdot 3!} + \dfrac{x^{16}}{16 \cdot 5!} - \dfrac{x^{22}}{22 \cdot 7!} + \ldots \Big|_0^1$$

$$= \left\{ \dfrac{1}{4} - \dfrac{1}{10 \cdot 3!} + \dfrac{1}{16 \cdot 5!} - \dfrac{1}{22 \cdot 7!} + \ldots \right\} - \{ 0 \} .$$

$$\frac{1}{4} - \frac{1}{10 \cdot 3!} + \frac{1}{16 \cdot 5!} \approx 0.2338542 \text{ and } \frac{1}{22 \cdot 7!} = \frac{1}{110880} \approx 0.0000090 \text{ so}$$

$$\int_0^1 \sin(x^3)\, dx \approx 0.2338542 \text{ and this approximation is within } 0.0000090 \text{ of the exact value.}$$

If we took just one more term, $\int_0^1 \sin(x^3)\, dx \approx \frac{1}{4} - \frac{1}{10 \cdot 3!} + \frac{1}{16 \cdot 5!} - \frac{1}{22 \cdot 7!} \approx 0.233845515$

is within $\frac{1}{28 \cdot 9!} \approx 0.000000098$ of the exact value of the integral.

Practice 4: Represent $x \cdot \cos(x^3)$ and $\int x \cdot \cos(x^3)\, dx$ as power series. Then use the first two nonzero terms to approximate the value of $\int_0^{1/2} x \cdot \cos(x^3)\, dx$ and obtain a bound on the "error" of this approximation.

Graphically

Each partial sum of the series $\sin(x) = x - \frac{x^3}{3!} + \frac{x^5}{5!} - \frac{x^7}{7!} + \frac{x^9}{9!} - \frac{x^{11}}{11!} + \ldots$ contains a finite number of terms and is simply a polynomial:

$$P_1(x) = x$$

$$P_3(x) = x - \frac{x^3}{3!} = x - \frac{1}{6}x^3$$

$$P_5(x) = x - \frac{x^3}{3!} + \frac{x^5}{5!} = x - \frac{1}{6}x^3 + \frac{1}{120}x^5, \ldots.$$

Fig. 1 showed the graphs of $\sin(x)$ and $P_1(x), P_3(x),$ and $P_5(x)$. As you saw, all of these polynomials are "good" approximations of $\sin(x)$ when x is very close to 0. The higher degree polynomials $P_n(x)$ provide "good" approximations of $\sin(x)$ over larger intervals.

Power Series for e^x

Example 5: Find the Maclaurin series for $f(x) = e^x$ and the radius of convergence of the series.

Solution: This is a very important series.

$f(x) = e^x$ so $f(0) = e^0 = 1$ and $a_0 = f(0) = 1$.

$f'(x) = e^x$ so $f'(0) = e^0 = 1$ and $a_1 = f'(0) = 1$.

$f''(x) = e^x$ so $f''(0) = e^0 = 1$ and $a_2 = \dfrac{f''(0)}{2!} = \dfrac{1}{2!}$.

For every value of n, $f^{(n)}(x) = e^x$ so $f^{(n)}(0) = e^0 = 1$ and $a_n = \dfrac{f''(0)}{n!} = \dfrac{1}{n!}$. Then

$$e^x = 1 + x + \frac{x^2}{2!} + \frac{x^3}{3!} + \frac{x^4}{4!} + \frac{x^5}{5!} + \ldots = \sum_{n=0}^{\infty} \frac{x^n}{n!} \ .$$

We can use the Ratio Test to find the radius of convergence. $c_n = \dfrac{x^n}{n!}$ so $c_{n+1} = \dfrac{x^{n+1}}{(n+1)!}$.

$$\left| \frac{c_{n+1}}{c_n} \right| = \left| \frac{\frac{x^{n+1}}{(n+1)!}}{\frac{x^n}{n!}} \right| = \left| \frac{x^{n+1}}{x^n} \cdot \frac{n!}{(n+1)!} \right| = \left| x \cdot \frac{1}{n+1} \right| \to 0 < 1 \text{ for every value of } x.$$

The radius of convergence is $R = \infty$, and the interval of convergence is $(-\infty, \infty)$. The Maclaurin series for e^x converges for every value of x.

Practice 5: Evaluate the partial sums of the first six terms of the numerical series for $e = e^1$ and $\dfrac{1}{\sqrt{e}} = e^{-1/2}$ and compare these partial sums with the values your calculator gives.

(Note: The numerical series for e^1 is not an alternating series so we do not have a bound for the approximation yet. We will in the next section.)

Fig. 3 shows the graphs of e^x and the approximating polynomials $1 + x$, $1 + x + \dfrac{x^2}{2!}$, and $1 + x + \dfrac{x^2}{2!} + \dfrac{x^3}{3!}$ for values of near 0.

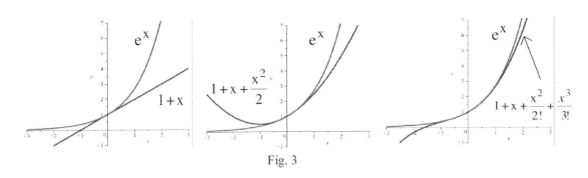

Fig. 3

> The following series converge for all values of x in the interval I:
>
> $$\sin(x) = x - \frac{x^3}{3!} + \frac{x^5}{5!} - \frac{x^7}{7!} + \frac{x^9}{9!} - \frac{x^{11}}{11!} + \ldots = \sum_{n=0}^{\infty} (-1)^n \frac{x^{2n+1}}{(2n+1)!} \qquad I = (-\infty, \infty).$$
>
> $$\cos(x) = 1 - \frac{x^2}{2!} + \frac{x^4}{4!} - \frac{x^6}{6!} + \frac{x^8}{8!} - \frac{x^{10}}{10!} + \ldots = \sum_{n=0}^{\infty} (-1)^n \frac{x^{2n}}{(2n)!} \qquad I = (-\infty, \infty).$$
>
> $$e^x = 1 + x + \frac{x^2}{2!} + \frac{x^3}{3!} + \frac{x^4}{4!} + \frac{x^5}{5!} + \ldots = \sum_{n=0}^{\infty} \frac{x^n}{n!} \qquad I = (-\infty, \infty).$$
>
> $$\ln(x) = (x-1) - \frac{1}{2}(x-1)^2 + \frac{1}{3}(x-1)^3 - \frac{1}{4}(x-1)^4 + \ldots = \sum_{n=1}^{\infty} (-1)^{n+1} \frac{(x-1)^n}{n} \qquad I = (0, 2].$$

Typically, these series converge very quickly to the value of the functions when x is close to 0 (or when x is close to 1 for ln(x)), but the convergence can be rather slow when x is far from 0. For example, the first 2 terms of the Taylor series for sine approximate sin(0.1) correctly to 6 decimal places, but 11 terms are needed to approximate sin(5) with the same accuracy.

Multiplying Power Series

We can add and subtract power series term–by–term, and we have already multiplied power series by single terms such as x and x^2, but occasionally it is useful to multiply a power series by another power series. The method for multiplying series is the same method we use to multiply a polynomial by another polynomial, but it becomes very tedious to get more than the first few terms of the resulting product.

Example 6: Find the first 5 nonzero terms of

$$\frac{1}{1-x} \cdot \sin(x) = (1 + x + x^2 + x^3 + \ldots) \cdot (x - \frac{x^3}{6} + \frac{x^5}{120} - \ldots).$$

Solution:

$$\begin{array}{r} 1 + x + x^2 + x^3 + x^4 + x^5 + \ldots \\ \text{times} \quad x \quad\quad - \dfrac{x^3}{6} \quad\quad + \dfrac{x^5}{120} - \ldots \\ \hline \end{array}$$

$\quad\quad\quad\quad x + x^2 + x^3 + x^4 + x^5 + \ldots$ (from multiplying by x)

$\quad\quad\quad\quad\quad\quad\quad\quad -\dfrac{1}{6}x^3 - \dfrac{1}{6}x^4 - \dfrac{1}{6}x^5 + \ldots$ (from multiplying by $-x^3/6$)

$\quad\quad\quad\quad\quad\quad\quad\quad\quad\quad\quad\quad\quad + \dfrac{1}{120}x^5 + \ldots$ (from multiplying by $x^5/120$)

$= \quad x + x^2 + \dfrac{5}{6}x^3 + \dfrac{5}{6}x^4 + \dfrac{101}{120}x^5 + \ldots$ (from adding previous terms)

Practice 6: Find the first 3 nonzero terms of

$$e^x \cdot \sin(x) = \left(1 + x + \dfrac{x^2}{2} + \dfrac{x^3}{6} + \ldots \right) \cdot \left(x - \dfrac{x^3}{6} + \dfrac{x^5}{120} - \ldots \right).$$

It is also possible to divide one power series by another power series using a procedure similar to "long division" of a polynomial by a polynomial, but we will not discuss that algorithm.

PROBLEMS

1. Find a formula for a polynomial P of degree 2 such that $P(0) = 7$, $P'(0) = -5$, and $P''(0) = 8$.

2. Find a formula for a polynomial P of degree 3 such that $P(0) = -1$, $P'(0) = 2$, $P''(0) = -5$, and $P'''(0) = 12$.

3. Find a formula for a polynomial P of degree 2 such that $P(3) = -2$, $P'(3) = 5$, and $P''(3) = 4$.

4. Find a formula for a polynomial P of degree 2 such that $P(1) = -2$, $P'(1) = 5$, and $P''(1) = 4$.

In problems 5 – 8, calculate the first several terms of the Maclaurin series for the given functions and compare with the series representations we found in Section 10.9 . (The series should be the same.)

5. $\ln(1+x)$ to the x^4 term
6. $\ln(1-x)$ to the x^4 term

7. $\arctan(x)$ to the x^3 term
8. $1/(1-x)$ to the x^4 term

In problems 9 – 12, calculate the first several terms of the Maclaurin series for the given functions.

9. $\cos(x)$ to the x^6 term
10. $\tan(x)$ to the x^5 term

11. $\sec(x)$ to the x^4 term
12. e^{3x} to the x^4 term

In problems 13 - 18, calculate the first several terms of the Taylor series for the given functions at the given point c.

13. $\ln(x)$ for $c = 1$

14. $\sin(x)$ for $c = \pi$

15. $\sin(x)$ for $c = \pi/2$

16. $\sqrt{x}$ for $c = 1$

17. $\sqrt{x}$ for $c = 9$

In problems 18 – 21, use the first three nonzero terms of the Maclaurin series for each function to approximate the numerical values. Then compare the Maclaurin series approximation with the value your calculator gives.

18. $\sin(0.1)$, $\sin(0.2)$, $\sin(0.5)$, $\sin(1)$, and $\sin(2)$

19. $\cos(0.1)$, $\cos(0.2)$, $\cos(0.5)$, $\cos(1)$, and $\cos(2)$

20. $\ln(1.1)$, $\ln(1.2)$, $\ln(1.3)$, $\ln(2)$, and $\ln(3)$

21. $\arctan(0.1)$, $\arctan(0.2)$, $\arctan(0.5)$, $\arctan(1)$, and $\arctan(2)$

In problems 22 – 27, calculate the first three nonzero terms of the power series for each of the integrals.

22. $\int \cos(x^2)\,dx$ and $\int \cos(x^3)\,dx$

23. $\int \sin(x^2)\,dx$ and $\int \sin(x^3)\,dx$

24. $\int e^{(x^2)}\,dx$ and $\int e^{(x^3)}\,dx$

25. $\int e^{(-x^2)}\,dx$ and $\int e^{(-x^3)}\,dx$

26. $\int \ln(x)\,dx$ and $\int x\ln(x)\,dx$

27. $\int x\sin(x)\,dx$ and $\int x^2\sin(x)\,dx$

In problems 28 – 35, use the series representation of these functions to calculate the limits.

28. $\lim\limits_{x \to 0} \dfrac{1 - \cos(x)}{x}$

29. $\lim\limits_{x \to 0} \dfrac{1 - \cos(x)}{x^2}$

30. $\lim\limits_{x \to 0} \dfrac{\ln(x)}{x - 1}$

31. $\lim\limits_{x \to 0} \dfrac{1 - e^x}{x}$

32. $\lim\limits_{x \to 0} \dfrac{1 + x - e^x}{x^2}$

33. $\lim\limits_{x \to 0} \dfrac{\sin(x)}{x}$

34. $\lim\limits_{x \to 0} \dfrac{x - \sin(x)}{x^3}$

35. $\lim\limits_{x \to 0} \dfrac{x - \dfrac{x^3}{6} - \sin(x)}{x^5}$

36. Use the series for e^x and e^{-x} to write a series representation for $\cosh(x) = \frac{1}{2}(e^x + e^{-x})$.

   (The function "cosh" is called the hyperbolic cosine function.)

37. Use the series for e^x and e^{-x} to write a series representation for $\sinh(x) = \frac{1}{2}(e^x - e^{-x})$.

   (The function "sinh" is called the hyperbolic sine function.)

38. Show that **D**(series for cosh(x) in problem 36) is the series for sinh(x) in problem 37.

39. Show that **D**(series for sinh(x) in problem 37) is the series for cosh(x) in problem 36.

Euler's Formula

So far we have only discussed series with real numbers, but sometimes it is useful to replace the variable with complex numbers. The next problems ask you to make such a substitution and then to derive and use one of the most famous formulas in mathematics, Euler's formula. (Recall that $i = \sqrt{-1}$ is called the complex unit and that its powers follow the pattern $i^2 = -1$, $i^3 = (i^2)(i) = -i$, $i^4 = (i^2)(i^2) = 1$, $i^5 = (i^4)(i) = i$, ...)

40. Start with the series $e^x = 1 + x + \frac{x^2}{2!} + \frac{x^3}{3!} + \frac{x^4}{4!} + \frac{x^5}{5!} + \frac{x^6}{6!} + \frac{x^7}{7!} + \frac{x^8}{8!} + ...$

    (a) Substitute "ix" for "x" and write a series for e^{ix}.

    (b) In part (a) simplify each power of i and write a simplified series for e^{ix}. (e.g., $(ix)^3/3!$ simplifies to $i^3 x^3/3! = -i \cdot x^3/3!$)

    (c) Sort the terms in the series in part (b) into those terms that do not contain i and those terms that do contain i. Then rewrite the series for e^{ix} in the form
    e^{ix} = { terms that did not contain i } + i·{ terms that did contain i }.

    (d) You should recognize the sum in each bracket in part (c) as the series for an elementary function (hint: think trigonometry). Rewrite the pattern in part (c) as
    e^{ix} = { function } + i·{ another function }.

41. The answer you should have gotten in problem 40d, $e^{ix} = \cos(x) + i \cdot \sin(x)$, is called Euler's formula. Use Euler's formula to calculate the values of $e^{i(\pi/2)}$ and $e^{\pi i}$.

42. Use Euler's formula to show that $e^{\pi i} + 1 = 0$. This is one of the most remarkable formulas in mathematics because it connects five of the most fundamental constants (the additive identity 0, the multiplicative identity 1, the complex unit i, and the two most commonly used irrational numbers π and e in a simple but non-obvious way.

Binomial Series

You have probably seen the pattern for expanding $(1+x)^n$ where n is a nonnegative integer:

$$(1+x)^0 = 1$$
$$(1+x)^1 = 1+x$$
$$(1+x)^2 = 1+2x+x^2$$
$$(1+x)^3 = 1+3x+3x^2+x^3$$
$$(1+x)^4 = 1+4x+6x^2+4x^3+x^4$$
$$(1+x)^5 = 1+5x+10x^2+10x^3+5x^4+x^5$$

```
Row        Pascal's Triangle
 0 ............ 1
 1 ..........  1  1
 2 .........  1  2  1
 3 ........ 1  3  3  1
 4 ......  1  4  6  4  1
 5 ..... 1  5  10  10  5  1
 6 .... 1  6  15  20  15  6  1
```

Each number in Pascal's Triangle is the sum of the two numbers closest to it in the row immediately above it.

Fig. 4

either using Pascal's triangle (Fig. 4) or using the binomial coefficients, written $\binom{n}{k}$ and defined as

$$\binom{n}{0} = 1 \text{ and } \binom{n}{k} = \frac{n(n-1)(n-2)\cdots(n-k+1)}{k!} = \frac{n!}{k!(n-k)!}.$$

43. Calculate the binomial coefficients $\binom{3}{0}, \binom{3}{1}, \binom{3}{2},$ and $\binom{3}{3}$ and verify that

   (i) they agree with the entries in the 3rd row of Pascal's triangle

   (ii) they agree with the coefficients of the terms of $(1+x)^3$.

44. Calculate the binomial coefficients $\binom{4}{0}, \binom{4}{1}, \binom{4}{2}, \binom{4}{3},$ and $\binom{4}{4}$ and verify that

   (i) they agree with the entries in the 4th row of Pascal's triangle

   (ii) they agree with the coefficients of the terms of $(1+x)^4$.

Using binomial coefficients, the pattern for nonnegative integer powers of $(1+x)$ can be described in a very compact way:

$$(1+x)^n = \sum_{k=0}^{n} \binom{n}{k} x^k.$$

When n is a positive integer, $(1+x)^n$ expands to be a polynomial of degree n.

But what happens when n is a negative integer or perhaps not even an integer? This was a question that Newton himself investigated, and it led him to a general pattern, called the Binomial Series Theorem, for $(1+x)^m$ when m is any real number. And now you can do it, too.

45. Let $f(x) = (1+x)^{5/2}$ and determine the first 5 terms of the Maclaurin series for $f(x)$.

46. Let $f(x) = (1+x)^{-3/2}$ and determine the first 5 terms of the Maclaurin series for $f(x)$.

47. Let $f(x) = (1+x)^m$ and determine the first 4 terms of the Maclaurin series for $f(x)$. This is the start of the derivation of the Binomial Series Theorem given below.

Binomial Series Theorem

If m is any real number and $|x| < 1$

then $(1+x)^m = 1 + mx + \dfrac{m(m-1)}{2}x^2 + \dfrac{m(m-1)(m-2)}{3!}x^3 + \ldots = \sum\limits_{k=0}^{\infty} \binom{m}{k} x^k$

where $\binom{m}{k} = \dfrac{m(m-1)(m-2)\cdots(m-k+1)}{k!}$ (for $k \geq 1$) and $\binom{m}{0} = 1$.

---

48. Use the Ratio Test to show that $\sum\limits_{k=0}^{\infty} \binom{m}{k} x^k$ converges for $|x| < 1$.

Practice Answers

Practice 1: $P(x) = a_0 + a_1 x + a_2 x^2 + a_3 x^3 + a_4 x^4$ with $P(0) = -3, P'(0) = 4, P''(0) = 10$,

$P'''(0) = 12$, and $P^{(4)}(0) = 24$.

$-3 = P(0) = a_0 + a_1 \cdot 0 + a_2 \cdot 0 + a_3 \cdot 0 + a_4 \cdot 0 = a_0$ so $a_0 = -3$

$P'(x) = a_1 + 2a_2 x + 3a_3 x^2 + 4a_4 x^3$

$4 = P'(0) = a_1 + 2a_2 \cdot 0 + 3a_3 \cdot 0 + 4a_4 \cdot 0 = a_1$ so $a_1 = 4$

$P''(x) = 2a_2 + 6a_3 x + 12a_4 x^2$

$10 = P''(0) = 2a_2 + 6a_3 \cdot 0 + 12a_4 \cdot 0 = 2a_2$ so $a_2 = 10/2 = 5$

$P'''(x) = 6a_3 + 24a_4 x$

$12 = P'''(0) = 6a_3 + 24a_4 \cdot 0 = 6a_3$ so $a_3 = 12/6 = 2$

$P^{(4)}(x) = 24a_4$

$24 = P^{(4)}(0) = 24a_4$ so $a_4 = 24/24 = 1$.

Then $P(x) = -3 + 4x + 5x^2 + 2x^3 + 1x^4$.

Practice 2: $\cos(x) = D(\sin(x)) = D\left(x - \dfrac{x^3}{3!} + \dfrac{x^5}{5!} - \dfrac{x^7}{7!} + \dfrac{x^9}{9!} - \dfrac{x^{11}}{11!} + \ldots\right)$

$$= 1 - 3 \cdot \dfrac{x^2}{3!} + 5 \cdot \dfrac{x^4}{5!} - 7 \cdot \dfrac{x^6}{7!} + 9 \cdot \dfrac{x^8}{9!} - 11 \cdot \dfrac{x^{10}}{11!} + \ldots$$

$$= 1 - \dfrac{x^2}{2!} + \dfrac{x^4}{4!} - \dfrac{x^6}{6!} + \dfrac{x^8}{8!} - \dfrac{x^{10}}{10!} + \ldots = \sum_{n=0}^{\infty} (-1)^n \dfrac{x^{2n}}{(2n)!}$$

Practice 3: $\cos(x) = 1 - \dfrac{x^2}{2!} + \dfrac{x^4}{4!} - \dfrac{x^6}{6!} + \dfrac{x^8}{8!} - \dfrac{x^{10}}{10!} + \ldots$

Using the first two nonzero terms, $\cos(0.2) \approx 1 - \dfrac{(0.2)^2}{2!} = 1 - \dfrac{0.04}{2} = 0.98$.

Since $\cos(0.2) = 1 - \dfrac{(0.2)^2}{2!} + \dfrac{(0.2)^4}{4!} - \dfrac{(0.2)^6}{6!} + \ldots$ is a convergent alternating series, the error is less than the absolute value of the next term.

Then $\cos(0.2) \approx 1 - \dfrac{(0.2)^2}{2!} = 0.98$ with an error less than $\left|\dfrac{(0.2)^4}{4!}\right| = \dfrac{0.0016}{24} \approx 0.000067$:

$|\cos(0.2) - 0.98| < 0.000067$. (In fact, $\cos(0.2) \approx 0.9800665778$.)

Practice 4: $\cos(x) = 1 - \dfrac{x^2}{2!} + \dfrac{x^4}{4!} - \dfrac{x^6}{6!} + \dfrac{x^8}{8!} - \dfrac{x^{10}}{10!} + \ldots$

$\cos(x^3) = 1 - \dfrac{(x^3)^2}{2!} + \dfrac{(x^3)^4}{4!} - \dfrac{(x^3)^6}{6!} + \dfrac{(x^3)^8}{8!} - \ldots = 1 - \dfrac{x^6}{2!} + \dfrac{x^{12}}{4!} - \dfrac{x^{18}}{6!} + \dfrac{x^{24}}{8!} - \dfrac{x^{30}}{10!} + \ldots$

$x \cdot \cos(x^3) = x - \dfrac{x^7}{2!} + \dfrac{x^{13}}{4!} - \dfrac{x^{19}}{6!} + \dfrac{x^{25}}{8!} - \dfrac{x^{31}}{10!} + \ldots$

$\displaystyle\int x \cdot \cos(x^3)\, dx = \dfrac{x^2}{2} - \dfrac{1}{8} \cdot \dfrac{x^8}{2!} + \dfrac{1}{14} \cdot \dfrac{x^{14}}{4!} - \dfrac{1}{20} \cdot \dfrac{x^{20}}{6!} + \dfrac{1}{26} \cdot \dfrac{x^{26}}{8!} - \dfrac{1}{32} \cdot \dfrac{x^{32}}{10!} + \ldots + C$

$\displaystyle\int_0^{1/2} x \cdot \cos(x^3)\, dx \approx \left.\dfrac{x^2}{2} - \dfrac{1}{8} \cdot \dfrac{x^8}{2!}\right|_0^{1/2} = \left(\dfrac{(0.5)^2}{2} - \dfrac{(0.5)^8}{2!\cdot 8}\right) - (0) \approx 0.124755859$

with $|\text{error}| \leq \dfrac{(0.5)^{14}}{4! \cdot 14} \approx 1.82 \cdot 10^{-7} = 0.000000182$.

Practice 5: $e^x = 1 + x + \dfrac{x^2}{2!} + \dfrac{x^3}{3!} + \dfrac{x^4}{4!} + \dfrac{x^5}{5!} + \ldots$. Using the first six terms,

$e^1 \approx 1 + 1 + \dfrac{1}{2!} + \dfrac{1}{3!} + \dfrac{1}{4!} + \dfrac{1}{5!} \approx 2.71666666666$ (My calculator gives $e^1 \approx 2.718281828$)

$e^{-1/2} \approx 1 + (-1/2) + \dfrac{(-1/2)^2}{2!} + \dfrac{(-1/2)^3}{3!} + \dfrac{(-1/2)^4}{4!} + \dfrac{(-1/2)^5}{5!} \approx 0.6065104167$

(My calculator gives $e^{-1/2} \approx 0.6065306597$).

Practice 6:

$$1 + x + \frac{x^2}{2} + \frac{x^3}{6} + \ldots \quad (= e^x)$$

times

$$x - \frac{x^3}{6} + \frac{x^5}{120} - \ldots \quad (= \sin(x))$$

$$x + x^2 + \frac{x^3}{2} + \frac{x^4}{6} + \ldots \quad \text{(from multiplying by } x\text{)}$$

$$-\frac{x^3}{6} - \frac{x^4}{6} - \frac{x^5}{12} - \ldots \quad \text{(from multiplying by } -x^3/6\text{)}$$

$$\frac{x^5}{120} + \ldots \quad \text{(from multiplying by } x^5/120\text{)}$$

---

product is $\quad x + x^2 + \frac{x^3}{3} + 0 - \frac{-9x^5}{120} + \ldots \quad$ (from adding the previous terms)

The sum of the first three nonzero terms is $e^x \cdot \sin(x) = x + x^2 + \frac{x^3}{3} + \ldots$.

MAPLE command to plot Fig. 2

plot({x, x-x^3/6, x-x^3/6+x^5/120, sin(x)}, x=-9..9, y=-6..6, color=[blue, green, black, red], thickness=3);

10.11 APPROXIMATION USING TAYLOR POLYNOMIALS

The previous two sections focused on obtaining power series representations for functions, finding their intervals of convergence, and using those power series to approximate values of functions, limits, and integrals. In the cases where the power series resulted in an alternating numerical series, we were also able to use the Estimation Bound for Alternating Series (Section 10.6) to get a bound on the "error:"

"error" = | {exact value} – {partial sum approximation} | < | next term in the series | .

If the power series did not result in an alternating numerical series, we did not have a bound on the size of the error of the approximation.

In this section we introduce Taylor Polynomials (partial sums of the Taylor Series) and obtain a bound on the approximation error, the value |{ exact value of f(x) } – { Taylor Polynomial approximation of f(x) }| . The bound we get is valid even if the Taylor series is not an alternating series, and the pattern for the error bound looks very much like **the next term in the series**, the first unused term in the partial sum of the Taylor series. In mathematics, this error bound is important for determining which functions are approximated by their Taylor series. In computer and calculator applications, the error bound is important to designers to ensure that their machines calculate enough digits of functions such as e^x and sin(x) for various values of x. In general, knowing this error bound can help us work efficiently by allowing us to use only the number of terms we really need.

We also examine graphically how well the Taylor Polynomials of f(x) approximate f(x)

Taylor Polynomials

If we add a finite number of terms of a power series, the result is a polynomial.

Definition

For a function f, the nth **degree Taylor Polynomial** (centered at c), written $P_n(x)$, is the partial sum of the terms up to the nth degree of the Taylor Series for f:

$$P_n(x) = \sum_{k=0}^{n} \frac{f^{(k)}(c)}{k!}(x-c)^k$$

$$= f(c) + f'(c)(x-c) + \frac{f''(c)}{2!}(x-c)^2 + \frac{f'''(c)}{3!}(x-c)^3 + \frac{f^{(4)}(c)}{4!}(x-c)^4 + \ldots + \frac{f^{(n)}(c)}{n!}(x-c)^n$$

Example 1: Write the first four Taylor Polynomials, $P_0(x)$ to $P_3(x)$, centered at 0 for e^x, and then graph them for $-1 < x < 1$.

Solution: The Maclaurin series for e^x is $e^x = 1 + x + \frac{x^2}{2!} + \frac{x^3}{3!} + \frac{x^4}{4!} + \frac{x^5}{5!} + \ldots = \sum_{n=0}^{\infty} \frac{x^n}{n!}$ so

$$P_0(x) = 1, \quad P_1(x) = 1 + x, \quad P_2(x) = 1 + x + \frac{x^2}{2}, \quad \text{and} \quad P_3(x) = 1 + x + \frac{x^2}{2} + \frac{x^3}{6}.$$

The graphs of e^x and $P_1(x), P_2(x)$, and $P_3(x)$ are shown in Fig. 1.

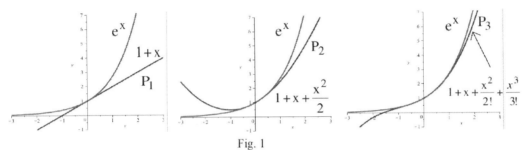

Fig. 1

Notice that $P_0(x)$ and e^x agree in value when $x = 0$,

$P_1(x)$, e^x, and their first derivatives agree in value when $x = 0$,

$P_2(x)$, e^x, their first derivatives, and their second derivatives agree in value when $x = 0$.

Practice 1: Write the Taylor Polynomials $P_0(x), P_2(x)$, and $P_4(x)$ centered at 0 for $\cos(x)$, and then graph them for $-\pi < x < \pi$. Write the Taylor Polynomials $P_1(x)$ and $P_3(x)$.

When we center the Taylor Polynomial at $x = c \neq 0$, the Taylor Polynomials approximate the function and its derivatives well for x close to c.

Example 2: Write the Taylor Polynomials $P_0(x), P_2(x)$, and $P_4(x)$ **centered at $3\pi/2$** for $\sin(x)$, and then graph them for $2 < x < 8$.

Solution: The Taylor series, centered at $3\pi/2$, for $\sin(x)$ is

$$\sin(x) = -1 + \frac{1}{2!}(x - 3\pi/2)^2 - \frac{1}{4!}(x - 3\pi/2)^4 + \frac{1}{6!}(x - 3\pi/2)^6 + \ldots = \sum_{n=0}^{\infty}(-1)^{n+1}\frac{1}{(2n)!}(x - 3\pi/2)^{2n}.$$

Then $P_0(x) = -1$, $P_2(x) = -1 + \frac{1}{2}(x - 3\pi/2)^2$, and $P_4(x) = -1 + \frac{1}{2}(x - 3\pi/2)^2 - \frac{1}{24}(x - 3\pi/2)^4$.

The graphs of $\sin(x), P_0(x), P_2(x)$, and $P_4(x)$ are shown in Fig. 2.

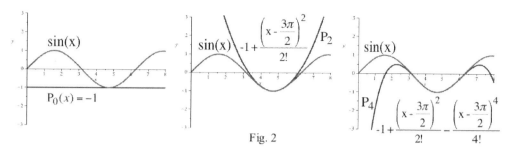

Fig. 2

Practice 2: Write the Taylor Polynomials $P_0(x), P_1(x),$ and $P_3(x)$ **centered at $\pi/2$** for $\cos(x)$, and then graph them for $-1 < x < 4$.

The Remainder

Approximation formulas such as the Taylor Polynomials are useful by themselves, but in many applied situations we want to know how good the approximation is or how many terms of a series are required to obtain a needed level of accuracy. If 2 terms of a series give you the needed level of accuracy for your application, it is a waste of time and money to use 100 terms. On the other hand, sometimes even 100 terms may not give the accuracy you need. Fortunately, it is possible to obtain a guarantee on how close a particular Taylor Polynomial approximation is to the exact value. Then we can work efficiently and use the number of terms that we need. The next theorem gives a pattern for the amount of "error" in our Taylor Polynomial approximation and can be used to obtain a bound on the size of the "error."

Taylor's Formula with Remainder

If f has n+1 derivatives in an interval I containing c, and x is in I,

then there is a number z, strictly <u>between</u> c and x, so that

$$f(x) = P_n(x) + R_n(x) \quad \text{where} \quad R_n(x) = \frac{f^{(n+1)}(z)}{(n+1)!}(x-c)^{n+1}.$$

This says that $f(x)$ is equal to the nth degree Taylor Polynomial plus a Remainder, and the Remainder $R_n(x)$ has the form given in the theorem. Notice that the pattern for R_n looks like the pattern for the $(n+1)^{st}$ term of the Taylor series for $f(x)$ except that it contains $f^{(n+1)}(z)$ instead of $f^{(n+1)}(c)$. This particular pattern for $R_n(x)$ is called the Lagrange form of the remainder, and is named for the French–Italian mathematician and astronomer Joseph Lagrange (1736–1813).

The main idea of the proof of the Taylor's Formula with Remainder is straightforward, but the technical details are rather complicated. The main idea and the technical details are given in the Appendix.

The pattern for the remainder, $\frac{f^{(n+1)}(z)}{(n+1)!}(x-c)^{n+1}$, contains three pieces, $(n+1)!$, $(x-c)^{n+1}$, and $f^{(n+1)}(z)$ for some z between x and c. The Taylor Remainder Formula is typically used in two ways:

In one type of use, the Taylor Polynomial is given so x, c, and n are known, and we can evaluate $(n+1)!$ and $(x-c)^{n+1}$ exactly. That leaves the piece $f^{(n+1)}(z)$ for some z between x and c. If we can find a bound for the value of $|f^{(n+1)}(z)|$ for all z between x and c, then we can put it together with the values of $(n+1)!$ and $(x-c)^{n+1}$ to obtain a bound for the remainder term $R_n(x)$.

In the other common usage, the amount of acceptable "error" is given, so x, c, and $R_n(x)$ are known, and we need to find a value of n that guarantees the required accuracy.

Corollary: A Bound for the Remainder $R_n(x)$

If   f has n+1 derivatives in an interval I containing c, and x is in I, and
$|f^{(n+1)}(z)| \leq M$ for **all** z between x and c,

then   "error" $= |f(x) - P_n(x)| = |R_n(x)| \leq M \cdot \frac{|x-c|^{n+1}}{(n+1)!}$.

Example 3: We plan to approximate the values of e^x with $P_3(x) = 1 + x + \frac{x^2}{2} + \frac{x^3}{6}$. Find a bound for the "error" of the approximation, $R_3(x)$, if x is in the interval

(a) $[-1, 1]$, (b) $[-3, 2]$ and (c) $[-0.2, 0.3]$.

Solution: $f(x) = e^x$, $c = 0$ (a Maclaurin series), $n = 3$, and $f^{(n+1)}(x) = f^{(4)}(x) = e^x$.

(a) For x in the interval $[-1, 1]$:

$|(x-c)^{n+1}| = |x^4| \leq |1^4| = 1$. $(n+1)! = 4! = 24$.

For x in $[-1, 1]$, $|f^{(n+1)}(x)| = |e^x| \leq e^1$. A "crude" but "easy to use" bound for e^1 is $e^1 < (3)^1 = 3 = M$. (A more precise bound is $e^1 < (2.72)^1 < 2.72$.)

Then $|R_3(x)| < M \cdot \frac{|x-c|^{n+1}}{(n+1)!} < 3 \cdot \frac{1}{24} = 0.125$.

For all $-1 < x < 1$, $P_3(x) = 1 + x + \frac{x^2}{2} + \frac{x^3}{6}$ is within 0.125 of e^x.

(b) For x in the interval $[-3, 2]$: $|(x-c)^{n+1}| = |x^4| \leq |(-3)^4| = 81$ and $(n+1)! = 4! = 24$.

For x in $[-3, 2]$, $|f^{(n+1)}(x)| = |e^x| \leq e^2$. A "crude" but "easy to use" bound for e^2 is $e^2 < (3)^2 = 9 = M$. (A more precise bound is $e^2 < (2.72)^2 < 7.4$.)

Then $|R_3(x)| < M \cdot \frac{|x-c|^{n+1}}{(n+1)!} < 9 \cdot \frac{81}{24} = 30.375$. Obviously we cannot have much confidence in our use of $P_3(x)$ to approximate e^x on the interval $[-3, 2]$.

(c) For x in the interval $[-0.2, 0.3]$: $|(x-c)^{n+1}| = |x^4| \leq |0.3^4| = 0.0081$.

$(n+1)! = 4! = 24$.

For x in $[-0.2, 0.3]$, $|f^{(n+1)}(x)| = |e^x| \leq e^{0.3}$. A bound for $e^{0.3}$ is

$e^{0.3} < (2.72)^{0.3} < 1.4 = M$ — obtained using a calculator.

Then $|R_3(x)| < M \cdot \frac{|x-c|^{n+1}}{(n+1)!} < 1.4 \cdot \frac{0.0081}{24} = 0.0004725$.

For all $-0.2 < x < 0.3$, $P_3(x) = 1 + x + \frac{x^2}{2} + \frac{x^3}{6}$ is within 0.0004725 of e^x.

When the interval is small, we can be confident that $P_3(x)$ provides a good approximation of e^x, but as the interval grows, so does our bound on the remainder. To guarantee a good approximation on a larger interval, we typically need $(n+1)!$ to be larger so we need to use a higher degree Taylor Polynomial $P_n(x)$.

Practice 3: Find a value of n to guarantee that $P_n(x)$ is within 0.001 of e^x for x in the interval $[-3, 2]$.

Example 4: We want to approximate the values of $f(x) = \sin(x)$ on the interval $[-\pi/2, \pi/2]$ with an error less that 10^{-10}. How many terms of the Maclaurin series for $\sin(x)$ are needed?

Solution: For every value of n, $|f^{(n+1)}(x)|$ is $|\sin(x)|$ or $|\cos(x)|$ so $M = 1$ in the Bound for the Remainder. Then "error" = $|R_n(x)| < 1 \cdot \frac{|x-0|^{n+1}}{(n+1)!} \leq \frac{(\pi/2)^{n+1}}{(n+1)!}$, and we need to find a value of n so that $\frac{(\pi/2)^{n+1}}{(n+1)!}$ is less than 10^{-10}. A bit of numerical experimentation on a calculator shows that

$\frac{(\pi/2)^{14}}{14!} \approx 6.39 \times 10^{-9}$, $\frac{(\pi/2)^{15}}{15!} \approx 6.69 \times 10^{-10}$, and $\frac{(\pi/2)^{16}}{16!} \approx 6.57 \times 10^{-11}$

so we can take $n = 15$: $P_{15}(x) = x - \frac{x^3}{3!} + \frac{x^5}{5!} - \frac{x^7}{7!} + \frac{x^9}{9!} - \frac{x^{11}}{11!} + \frac{x^{13}}{13!} - \frac{x^{15}}{15!}$.

If $-\pi/2 \leq x \leq \pi/2$, then $|P_{15}(x) - \sin(x)| < 10^{-10}$.

Practice 4: How many terms of the Maclaurin series for e^x are needed to approximate e^x to within 10^{-10} for $0 \leq x \leq 1$?

Calculator Notes

Imagine that you are in charge of designing or selecting an algorithm for a calculator to use when the SIN button is pushed. (Smartest move: find a mathematician who knows about "numerical analysis" and the design and implementation of algorithms.) You know that if the value of x is relatively close to 0, then SIN(x) can be quickly approximated to 10 digits (the size of the display of the calculator) by using a "few" terms of the Taylor series for sin(x): if $-1.57 \leq x \leq 1.57$, then

$$x - \frac{x^3}{3!} + \frac{x^5}{5!} - \frac{x^7}{7!} + \frac{x^9}{9!} - \frac{x^{11}}{11!} + \frac{x^{13}}{13!} - \frac{x^{15}}{15!}$$

$$= x\left(1 - \frac{x^2}{2\cdot 3}\left(1 - \frac{x^2}{4\cdot 5}\left(1 - \frac{x^2}{6\cdot 7}\left(1 - \frac{x^2}{8\cdot 9}\left(1 - \frac{x^2}{10\cdot 11}\left(1 - \frac{x^2}{12\cdot 13}\left(1 - \frac{x^2}{14\cdot 15}\right)\right)\right)\right)\right)\right)\right)$$

gives the value of sin(x) with 10 digits of accuracy.

(The second pattern looks more complicated, but is usually preferred because it uses fewer multiplications and avoids very large values such as x^{15} and 15!) But you also want your algorithm to give 10 digits of accuracy even when x is larger, say 10 or 101.7. Rather than computing <u>many</u> more terms of the Maclaurin series for sine, some algorithms simply shift the problem closer to 0. First they use the fact that sin(x) = sin(x – 2π) to keep shifting the problem until the argument is in the interval [0, 2π]:

sin(10) = sin(10 – 2π) = sin(3.71681469)

sin(101.7) = sin(101.7 – 2π) = sin(101.7 – 4π) = ... = sin(101.7 – 32π) = sin(1.169035085).

Once the argument is between 0 and 2π, additional trigonometric facts are used:

if the new value of x is larger than π, use sin(x) = –sin(x – π) to replace "x" with "x – π" (and keep track of the change in sign of the answer). The new x value is in the interval [0, π].

Finally, we can shift the problem into the interval [0, π/2]:

if the new value of x is larger than π/2, use sin(x) = sin(π – x) to replace "x" with "π – x."

This new x value is in the interval [0, π/2] ≈ [0, 1.57] and the 7 terms of the sine series shown above are sufficient to approximate sin(x) with 10 digits of accuracy.

There are, however, major problems when calculators encounter the sine or exponential of a very large number. Since calculators only store the leading finite number of digits of a number (usually 10 or 12 digits), the calculator can not distinguish large numbers that differ past that leading number of stored digits: one calculator correctly says that (10^12+1) – 10^12 = 1, but it incorrectly reports that (10^13+1) – 10^13 = 0. Since it calculates "10^13+1 = 10^13", it also would falsely report the same values for sin(10^13+1) and sin(10^13). In fact, the people who programmed this particular type of calculator recognized the problem, and the calculator gives an error message if it is asked to calculate sin(10^11). This particular calculator reports e^230 ≈ 7.7 x 10^{99}. It reports an error for e^231 since the largest number it can display is 9.9 x 10^{99} and e^231 exceeds that value. What happens on your calculator?

PROBLEMS

In problems 1 – 10, calculate the Taylor polynomials P_0, P_1, P_2, P_3, and P_4 for the given function centered at the given value of c. Then graph the function and the Taylor polynomials on the given interval.

1. $f(x) = \sin(x)$, $c = 0$, $[-2, 4]$
2. $f(x) = \cos(x)$, $c = 0$, $[-2, 4]$
3. $f(x) = \ln(x)$, $c = 1$, $[0.1, 3]$
4. $f(x) = \arctan(x)$, $c = 0$, $[-3, 3]$
5. $f(x) = \sqrt{x}$, $c = 1$, $[0, 3]$
6. $f(x) = \sqrt{x}$, $c = 9$, $[0, 20]$
7. $f(x) = (1 + x)^{-1/2}$, $c = 0$, $[-2, 3]$
8. $f(x) = e^{2x}$, $c = 0$, $[-2, 4]$
9. $f(x) = \sin(x)$, $c = \pi/2$, $[-1, 5]$
10. $f(x) = \sin(x)$, $c = \pi$, $[-1, 5]$

In problems 11 – 18, a function $f(x)$ and a value of n are given. Determine a formula for $R_n(x)$ and find a bound for $|R_n(x)|$ on the given interval. This bound for $|R_n(x)|$ is our "guaranteed accuracy" for P_n to approximate $f(x)$ on the given interval. (Use $c = 0$.)

11. $f(x) = \sin(x)$, $n = 5$, $[-\pi/2, \pi/2]$
12. $f(x) = \sin(x)$, $n = 9$, $[-\pi/2, \pi/2]$
13. $f(x) = \sin(x)$, $n = 5$, $[-\pi, \pi]$
14. $f(x) = \sin(x)$, $n = 9$, $[-\pi, \pi]$
15. $f(x) = \cos(x)$, $n = 10$, $[-1, 2]$
16. $f(x) = \cos(x)$, $n = 10$, $[-1, 5]$
17. $f(x) = e^x$, $n = 6$, $[-1, 2]$
18. $f(x) = e^x$, $n = 10$, $[-1, 3]$

In problems 19 – 24, determine how many terms of the Taylor series for $f(x)$ are needed to approximate f to within the specified error on the given interval. (For each function use the center $c = 0$.)

19. $f(x) = \sin(x)$ within 0.001 on $[-1, 1]$
20. $f(x) = \sin(x)$ within 0.001 on $[-3, 3]$
21. $f(x) = \sin(x)$ within 0.00001 on $[-1.6, 1.6]$
22. $f(x) = \cos(x)$ within 0.001 on $[-2, 2]$
23. $f(x) = e^x$ within 0.001 on $[0, 2]$
24. $f(x) = e^x$ within 0.001 on $[-1, 4]$

Series Approximations of π

The following problems illustrate some of the ways series have been used to obtain very precise approximations of π. Several of these methods use the series for $\arctan(x)$,

$$\arctan(x) = x - \frac{x^3}{3} + \frac{x^5}{5} - \frac{x^7}{7} + \frac{x^9}{9} - \ldots = \sum_{n=0}^{\infty} (-1)^n \frac{x^{2n+1}}{2n+1},$$

which converges rapidly if $|x|$ is close to zero.

Method I: $\tan(\frac{\pi}{4}) = 1$ so $\frac{\pi}{4} = \arctan(1) = 1 - \frac{1}{3} + \frac{1}{5} - \frac{1}{7} + \frac{1}{9} - \ldots = \sum_{n=0}^{\infty} (-1)^n \frac{1}{2n+1}$ and

$$\pi = 4\left\{ 1 - \frac{1}{3} + \frac{1}{5} - \frac{1}{7} + \frac{1}{9} - \ldots \right\}.$$

25. (a) Approximate π as $4\left\{ 1 - \frac{1}{3} + \frac{1}{5} - \frac{1}{7} + \frac{1}{9} \right\}$ and compare this value with the value your calculator gives for π.

   (b) The series for arctan(1) is an alternating series so we have an "easy" error bound. Use the error bound for an alternating series to find a bound for the error if 50 terms of the arctan(1) series are used.

   (c) Using the error bound for an alternating series, how many terms of the arctan(1) series are needed to guarantee that the series approximation of π is within 0.0001 of the exact value of π?
   (The arctan(1) series converges so slowly that it is not used to approximate π.)

Method II: $\tan(a + b) = \frac{\tan(a) + \tan(b)}{1 - \tan(a)\tan(b)}$ so $\tan(\arctan(\frac{1}{2}) + \arctan(\frac{1}{3})) = \frac{\frac{1}{2} + \frac{1}{3}}{1 - \frac{1}{2} \cdot \frac{1}{3}} = 1$. Then

$\frac{\pi}{4} = \arctan(1) = \arctan(\frac{1}{2}) + \arctan(\frac{1}{3})$, and the series for $\arctan(\frac{1}{2})$ and $\arctan(\frac{1}{3})$ converge much more rapidly than the series for arctan(1).

26. (a) Approximate π as

   $4\left\{ \text{(sum of the first 4 terms of the } \arctan(\frac{1}{2}) \text{ series)} + \text{(sum of the first 4 terms of the } \arctan(\frac{1}{3}) \text{ series)} \right\}$.

   Then compare this value with the value your calculator gives for π.

   (b) The series for $\arctan(\frac{1}{2})$ and $\arctan(\frac{1}{3})$ are each alternating series. Use the error bound for an alternating series to find a bound for the error if 10 terms of each series are used.

   (c) How many terms of each series are needed to guarantee that the series approximation of π is within 0.0001 of the exact value of π?

Other Methods: We will not justify these methods, but they converge to π more rapidly than the first two methods.

   A: $\frac{\pi}{4} = 4 \cdot \arctan(\frac{1}{5}) - \arctan(\frac{1}{239})$ (due to Machin in 1706)

   B: $\pi = 48 \cdot \arctan(\frac{1}{18}) + 32 \cdot \arctan(\frac{1}{57}) - 20 \cdot \arctan(\frac{1}{239})$

27. (a) Use the first 3 terms of each series in formula A to approximate π. How much does it differ from the value your calculator gives you?

   (b) Why does formula A converge more rapidly (using fewer terms) than methods I and II?

28. (a) Use the first 3 terms of each series in formula B to approximate π. How much does it differ from the value your calculator gives you?

   (b) Why does formula B converge more rapidly (using fewer terms) than Methods I and II and formula A?

Practice Answers

Practice 1: $\cos(x) = 1 - \frac{x^2}{2!} + \frac{x^4}{4!} - \frac{x^6}{6!} + \frac{x^8}{8!} - \frac{x^{10}}{10!} + \ldots = \sum_{n=0}^{\infty} (-1)^n \frac{x^{2n}}{(2n)!}$ so

$P_0(x) = 1$, $P_2(x) = 1 - \frac{x^2}{2}$, and $P_4(x) = 1 - \frac{x^2}{2} + \frac{x^4}{24}$. Their graphs are shown in Fig. 3.

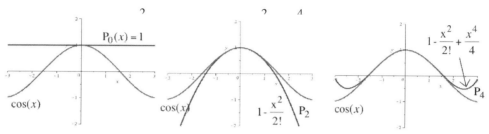

Fig. 3

Practice 2:

$\cos(x) = -(x - \pi/2) + \frac{1}{3!}(x - \pi/2)^3 - \frac{1}{5!}(x - \pi/2)^5 + \frac{1}{7!}(x - \pi/2)^7 - \ldots$

$= \sum_{n=0}^{\infty} (-1)^{n+1} \frac{1}{(2n+1)!}(x - \pi/2)^{2n+1}$.

Then $P_0(x) = 0$, $P_1(x) = -(x - \pi/2)$, and $P_3(x) = -(x - \pi/2) + \frac{1}{6}(x - \pi/2)^3$. The graphs of $\cos(x)$, $P_1(x)$, and $P_3(x)$ are shown in Fig. 4.

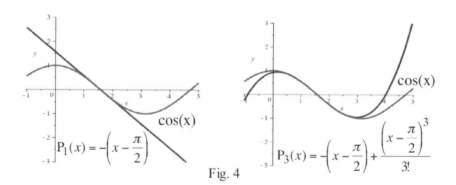

Fig. 4

Practice 3: For x in the interval $[-3, 2]$, $|(x-c)^{n+1}| = |x^3| \leq |(-3)^3| = 27$.

For x in $[-3, 2]$, $|f^{(n+1)}(x)| = |e^x| \leq e^2$. A "crude" bound for e^2 is $e^2 < (3)^2 = 9 = M$.

Then $|R_n(x)| < M \cdot \frac{(x-c)^{n+1}}{(n+1)!} < 9 \cdot \frac{27}{n!}$, and we want a value of n so $9 \cdot \frac{27}{n!} \leq 0.001$:

we want $n! \geq \frac{(9)(27)}{0.001} = 243{,}000$. Using a calculator, we see that $8! = 40{,}320$ is not large enough, but $9! = 362{,}880 > 243{,}000$ so we can use $n = 9$.

For x in the interval $[-3, 2]$, $P_9(x) = 1 + x + \frac{x^2}{2!} + \frac{x^3}{3!} + \frac{x^4}{4!} + \frac{x^5}{5!} + \frac{x^6}{6!} + \frac{x^7}{7!} + \frac{x^8}{8!} + \frac{x^9}{9!}$ is within 0.001 of e^x.

Practice 4: $R_n(x) \leq 10^{-10}$. $f(x) = e^x$, $c = 0$, and for every n, $f^{(n+1)}(x) = e^x$. For $0 \leq x \leq 1$,

$|f^{(n+1)}(x)| = |e^x| \leq e < 2.72 = M$. We want to find a value for n so

$M \cdot \frac{(x-c)^{n+1}}{(n+1)!} = 2.72 \cdot \frac{(1-0)^{n+1}}{(n+1)!} < 10^{-10}$. Some numerical experimentation on a calculator

shows that $2.72 \cdot \frac{1}{15!} \approx 1.58 \times 10^{-9}$ and $2.72 \cdot \frac{1}{16!} \approx 9.9 \times 10^{-11}$ so we can take $n = 15$.

For $0 \leq x \leq 1$, $P_{15}(x) = 1 + x + \frac{x^2}{2!} + \frac{x^3}{3!} + \frac{x^4}{4!} + \ldots + \frac{x^{15}}{15!}$ is within 10^{-10} of e^x.

Appendix: Idea and details of a Proof of Taylor's Formula with Remainder

Main idea of the proof:

We define a new differentiable function $g(t)$ and show that $g(x) = 0$ and $g(c) = 0$.

Then, by Rolle's Theorem, we can conclude that there is a number z, between x and c, so that $g'(z) = 0$. Finally, we set $g'(z) = 0$ and algebraically obtain the given formula for $R_n(x)$.

Let $R_n(x) = f(x) - P_n(x)$ be the difference between $f(x)$ and the nth Taylor polynomial for $P_n(x)$.

Define a differentiable function $g(t)$ to be

$$g(t) = f(x) - \left\{ f(t) + f'(t)(x-t) + \frac{f''(t)}{2!}(x-t)^2 + \frac{f'''(t)}{3!}(x-t)^3 + \ldots + \frac{f^{(n)}(t)}{n!}(x-t)^n \right\} - R_n(x) \frac{(x-t)^{n+1}}{(x-c)^{n+1}}.$$

This may seem to be a strange way to define a function, but it turns out to have the properties we need:

$g(x) = f(x) - \{f(x) + 0 + 0 + 0 + \ldots + 0\} - R_n(x) \frac{0}{(x-c)^{n+1}} = f(x) - f(x) = 0$, and

$g(c) = f(x) - \left\{ f(c) + f'(c)(x-c) + \frac{f''(c)}{2!}(x-c)^2 + \frac{f'''(c)}{3!}(x-c)^3 + \frac{f^{(4)}(c)}{4!}(x-c)^4 + \frac{f^{(n)}(c)}{n!}(x-c)^n \right\}$

$\quad - R_n(x) \frac{(x-c)^{n+1}}{(x-c)^{n+1}}$

$= f(x) - P_n(x) - R_n(x) = 0$ since $R_n(x) = f(x) - P_n(x)$.

Then, by Rolle's Theorem, there is a number z, strictly between x and c, so $g'(z) = 0$.

Notice that g is defined to be a function of t so we treat x and c as constants and differentiate with respect to t. The key pattern is that when we differentiate a term such as $\frac{f'''(t)}{3!} \cdot (x-t)^3$ with respect to t, we need to use the product rule. The resulting derivative has two terms:

$$\frac{d}{dt}\left\{\frac{f'''(t)}{3!}(x-t)^3\right\} = \frac{f'''(t)}{3!} \frac{d}{dt}(x-t)^3 + (x-t)^3 \cdot \frac{d}{dt}\frac{f'''(t)}{3!}$$

$$= \frac{f'''(t)}{3!}(3)(x-t)^2(-1) + \frac{f^{(4)}(t)}{3!}(x-t)^3 = -\frac{f'''(t)}{2!}(x-t)^2 + \frac{f^{(4)}(t)}{3!}(x-t)^3.$$

When we differentiate $g(t)$ with respect to t, we get a complicated pattern, but most of the terms cancel:

$$g'(t) = \frac{d}{dt} g(t) = 0 - \Big\{ \ f'(t)$$

$$- f'(t) + f''(t)(x-t)$$

$$- f''(t)(x-t) + \frac{f'''(t)}{2!}(x-t)^2$$

$$- \frac{f'''(t)}{2!}(x-t)^2 + \frac{f^{(4)}(t)}{3!}(x-t)^3$$

$$- \ldots - \frac{f^{(n)}(t)}{(n-1)!}(x-t)^{n-1} + \frac{f^{(n+1)}(t)}{n!}(x-t)^n \Big\} - R_n(x)\cdot(n+1)\cdot\frac{(x-t)^n(-1)}{(x-c)^{n+1}}$$

$$= -\frac{f^{(n+1)}(t)}{n!}(x-t)^n + R_n(x)\cdot(n+1)\cdot\frac{(x-t)^n}{(x-c)^{n+1}}$$

$$= (x-t)^n \left\{ R_n(x)\cdot(n+1)\cdot\frac{1}{(x-c)^{n+1}} - \frac{f^{(n+1)}(t)}{n!} \right\}.$$

By Rolle's Theorem, there is a value z, between x and c, for the variable t so $g'(z) = 0$. Then

$$(x-z)^n \left\{ R_n(x)\cdot(n+1)\cdot\frac{1}{(x-c)^{n+1}} - \frac{f^{(n+1)}(z)}{n!} \right\} = 0.$$

z is strictly between x and c so $z \neq x$ and we can divide each side by $(x-z)^n$ to get

$$R_n(x)\cdot(n+1)\cdot\frac{1}{(x-c)^{n+1}} - \frac{f^{(n+1)}(z)}{n!} = 0.$$

Finally, $R_n(x)\cdot(n+1)\cdot\frac{1}{(x-c)^{n+1}} = \frac{f^{(n+1)}(z)}{n!}$ so $R_n(x) = \frac{f^{(n+1)}(z)}{n!(n+1)}\cdot(x-c)^{n+1} = \frac{f^{(n+1)}(z)}{(n+1)!}(x-c)^{n+1}$,

the result we wanted to prove.

MAPLE command to plot $\cos(x)$ and $P_3(x)$.

plot({1-x^2/2+x^4/24,cos(x)},x=-3..3,y=-2..2,color=[blue,red],thickness=3);

10.12 INTRODUCTION TO FOURIER SERIES

When we discussed Maclaurin series in earlier sections we used information about the derivatives of a function f(x) to create an infinite series of the form

$$\sum_{n=0}^{\infty} a_n \cdot x^n = a_0 + a_1 \cdot x^1 + a_2 \cdot x^2 + a_3 \cdot x^3 + a_4 \cdot x^4 + \ldots$$

In this case the "building blocks" of the series were powers of x ($x^0, x^1, x^2, x^3, \ldots$). and this is why the series is called a "power series." The coefficients $a_n = \dfrac{f^{(n)}(0)}{n!}$ of the Maclaurin series told us how much of each power to include in our series, and we found the formula involving derivatives that enabled us to calculate the values of the coefficients. For general Taylor series the "building blocks" were powers of (x-c) for some fixed center c, and we had a similar formula to calculate the values of the coefficients for those series. We then used the Taylor and Maclaurin series to approximate functions "near c" and sometimes even for "every value of x."

Fourier series have a number of similarities with power series:
  we will approximate functions using "building blocks"
  we will need to calculate the values of the coefficients of the building blocks, and
  we will need to be concerned about where the approximations are "good."
But our building blocks for the Fourier series will be the trigonometric functions sin(x), sin(2x), sin(3x), ... and cos(x), cos(2x), cos(3x), ... and the result will be a "trigonometric series." Our formula for calculating the values of the coefficients will involve integrals rather than derivatives. Finally, since each building block repeats its values every 2π units, we will only (at first) use Fourier series to approximate 2π-periodic functions. (This is not as serious a restriction as it might seem.)

First goal: Finding the coefficients

Our first goal is to find a relationship between the function f(x) and the coefficients a_n and b_n in the series

$$\frac{a_0}{2} + \sum_{n=1}^{\infty} \{a_n \cos(nx) + b_n \cdot \sin(nx)\}$$

$$= \frac{a_0}{2} + \{a_1 \cdot \cos(x) + b_1 \cdot \sin(x)\} + \{a_2 \cdot \cos(2x) + b_2 \cdot \sin(2x)\} + \{a_3 \cdot \cos(3x) + b_3 \cdot \sin(3x)\} + \ldots$$

The key to finding the values of the coefficients depends on the following results about the integrals of the products of any two of our building blocks, sin(mx) and cos(nx), for positive integer values of m and n.

10.12 Fourier Series   Contemporary Calculus

$$\int_{x=0}^{2\pi} \sin(nx) \cdot \sin(mx)\, dx = \begin{cases} \pi & \text{if } m = n \quad \text{(using integral formula \#13)} \\ 0 & \text{if } m \neq n \quad \text{(using integral formula \#25)} \end{cases}$$

$$\int_{x=0}^{2\pi} \cos(nx) \cdot \cos(mx)\, dx = \begin{cases} \pi & \text{if } m = n \quad \text{(using integral formula \#14)} \\ 0 & \text{if } m \neq n \quad \text{(using integral formula \#26)} \end{cases}$$

$$\int_{x=0}^{2\pi} \sin(nx) \cdot \cos(mx)\, dx = 0 \quad \text{for all integer values of m and n} \quad \text{(using integral formula \#27)}$$

The technical term is that the set of functions { sin(x), cos(x), sin(2x), cos(2x), sin(3x), cos(3x), ...} forms an "orthogonal" family on the interval $[0, 2\pi]$: for any two functions from this set

$$\int_{x=0}^{2\pi} f(x) \cdot g(x)\, dx = \begin{cases} \pi & \text{if f and g are the same member of the set} \\ 0 & \text{if f and g are different members of the set} \end{cases}$$

Example 1: Suppose we have the function $f(x) = 3 + 2\cos(3x) - 7\sin(4x)$ that is already a finite trigonometric series with $a_0 = 6$, $a_3 = 2$, $b_4 = -7$, and all of the other coefficients are 0. Evaluate the integral of the product of f(x) with a general building block, sin(nx) (for every value of n).

Solution:
$$\int_{x=0}^{2\pi} \sin(nx) \cdot f(x)\, dx = \int_{x=0}^{2\pi} \sin(nx) \cdot (3 + 2\cos(3x) - 7\sin(4x))\, dx$$

$$= \int_{x=0}^{2\pi} \sin(nx) \cdot 3\, dx + \int_{x=0}^{2\pi} \sin(nx) \cdot 2\cos(3x) + \int_{x=0}^{2\pi} \sin(nx) \cdot (-7\sin(4x))\, dx$$

$$= 0 + 0 + \begin{cases} -7\pi & \text{if } n = 4 \\ 0 & \text{if } n \neq 4 \end{cases}.$$

It appears that we can find the value of b_n by dividing this last value by π:

$$b_n = \frac{1}{\pi} \int_{x=0}^{2\pi} \sin(nx) \cdot f(x)\, dx$$

Practice 1: Using the same function $f(x) = 3 + 2\cos(3x) - 7\sin(4x)$, evaluate the integral of the product with the cosine blocks, cos(nx), and show that

$$a_3 = \frac{1}{\pi} \int_{x=0}^{2p} \cos(3x) \cdot f(x)\, dx \quad \text{and} \quad a_n = \frac{1}{\pi} \int_{x=0}^{2p} \cos(nx) \cdot f(x)\, dx \quad \text{for } n = 1, 2, 3, \ldots$$

10.12 Fourier Series

The general cases for the Example and Practice problems say that if f(x) is already a trigonometric polynomial with

$$f(x) = \frac{a_0}{2} + \sum_{n=1}^{\infty} \{a_n \cos(nx) + b_n \cdot \sin(nx)\}$$

$$= \frac{a_0}{2} + \{a_1 \cdot \cos(x) + b_1 \cdot \sin(x)\} + \{a_2 \cdot \cos(2x) + b_2 \cdot \sin(2x)\} + \{a_3 \cdot \cos(3x) + b_3 \cdot \sin(3x)\} + \ldots$$

then we can find the coefficients of the terms of f(x) using

$$a_n = \frac{1}{\pi} \int_{x=0}^{2\pi} \cos(nx) \cdot f(x)\, dx \quad \text{and} \quad b_n = \frac{1}{\pi} \int_{x=0}^{2\pi} \sin(nx) \cdot f(x)\, dx\ .$$

What we have just done is similar to our beginning work with Maclaurin series in Section 10.10. There we started with a polynomial P(x) and found that the coefficients were given by the formula

$$a_n = \frac{f^{(n)}(0)}{n!}$$ using derivatives of P(x). Here we started with a trigonometric polynomial f(x) and found that the coefficients a_n and b_n were given by formulas using integrals of products with f(x). In Section 10.10 we then extended the Maclaurin series to functions that were not polynomials. Here we will make a similar extension of trigonometric series to functions that are not trigonometric polynomials.

Definition: If f(x) is integrable on the interval $[0, 2\pi]$, then the

Fourier Series of f(x) is $\dfrac{a_0}{2} + \sum_{n=1}^{\infty} \{a_n \cos(nx) + b_n \cdot \sin(nx)\}$ with

Fourier coefficients $\quad a_n = \dfrac{1}{\pi} \int_{x=0}^{2\pi} \cos(nx) \cdot f(x)\, dx \quad$ and

$$b_n = \frac{1}{\pi} \int_{x=0}^{2\pi} \sin(nx) \cdot f(x)\, dx \quad \text{for n=0, 1, 2, 3, \ldots}$$

Before going further with the development and discussion, let's take a look at how this actually works with some different types of functions – a differentiable function, a function that is continuous but not differentiable, and a function that is not even continuous. At the end of this section is a MAPLE program that automatically calculates the Fourier coefficients and builds the Fourier series.

Fourier Series approximation of a differentiable function $f(x) = x^2 \cdot (2\pi - x)$

The series of graphs below show how the higher degree Fourier series of f become better and better approximations of the graph of f on the interval $[0, 2\pi]$.

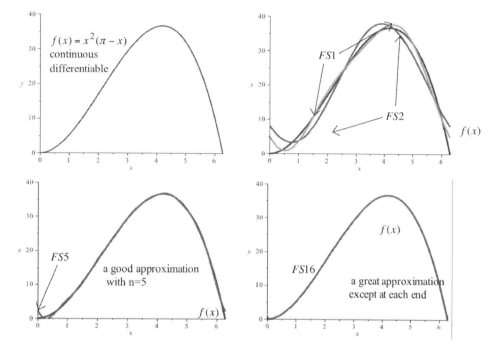

FS5(x) = 20.67085112-12.56637062*cos(x)-12.*sin(x)-3.141592654*cos(2*x)-1.5*sin(2*x)
-1.396263402*cos(3*x)-.4444444444*sin(3*x)-.7853981635*cos(4*x)-.1875*sin(4*x)
-.5026548246*cos(5*x)-0.096*sin(5*x)

Taylor and Maclaurin series require that we have a differentiable function of f or we could not calculate the coefficients for the series. For Fourier series, however, we only need that the function of f be integrable, a much less demanding condition. The next example illustrates the convergence of the Fourier series of a function that has a "corner" so it is not differentiable (at that point).

Note: We only need f to be an integrable function in order to be able to calculate the coefficients of the Fourier series, but that is not enough to guarantee that the Fourier series we get converges to f(x) for every value of x. In fact, there are continuous functions for which the Fourier series does not converge to f(x) for an infinite number of vales of x. The whole study of conditions that do and do not guarantee the convergence of Fourier series led to some very interesting, very beautiful and very deep results in mathematics.

Fourier Series approximation of a continuous function with a corner: $f(x) = 4 - |\pi - x|$

These figures illustrate that the Fourier series can converge to a function that is not differentiable at a point.

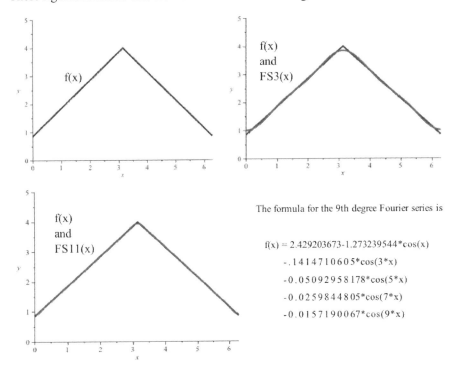

The formula for the 9th degree Fourier series is

$f(x) = 2.429203673 - 1.273239544 \cdot \cos(x)$
  $- .1414710605 \cdot \cos(3 \cdot x)$
  $- 0.05092958178 \cdot \cos(5 \cdot x)$
  $- 0.0259844805 \cdot \cos(7 \cdot x)$
  $- 0.0157190067 \cdot \cos(9 \cdot x)$

Because of the symmetry of f(x) around π, all of the sine terms, the b_n, are 0.

The next example illustrates that the Fourier coefficients can even be found for functions which have some discontinuities (but they can still be integrated), and that the resulting Fourier series can still do a good job of approximating the function between the breaks.

Fourier series approximation of a discontinuous function:

$f(x) = |1.3 - \text{Int}(x/2.3)|$

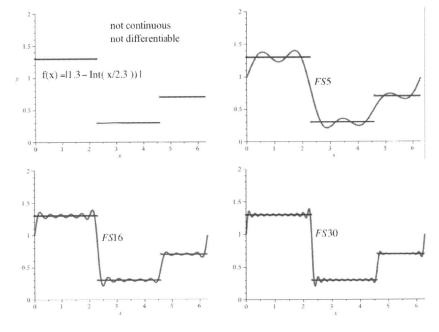

What good are Fourier Series?

Historically Fourier series were important because they provided a method of analyzing data that arose from the interaction of various periodic influences such as the orbits of the planets (influenced by the gravitational attractions of the sun and other planets whose orbits are almost periodic) or the tides (influenced by the attraction of the moon and also by local conditions). More recently, Fourier series provide a way to efficiently store and regenerate the musical tones of various instruments. Each tone for an instrument can be efficiently stored by saving only the coefficients, and we only need to save the first "several" terms of the series since the higher order coefficients correspond to high frequency sounds that are beyond the human hearing range.

Fourier series, and variations on that idea, have also been used extensively in signal processing and to clean up noisy signals. Since "random noise" is usually high frequency, the Fourier series of each piece of the signal can be calculated (automatically) and only the lower degree terms kept in order to reproduce a "clean" result. The following figures illustrate a noisy signal that has been cleaned up in this way. This might represent an attempt to reclaim voice information from a recording that has been damaged or originally contained a lot of background noise.

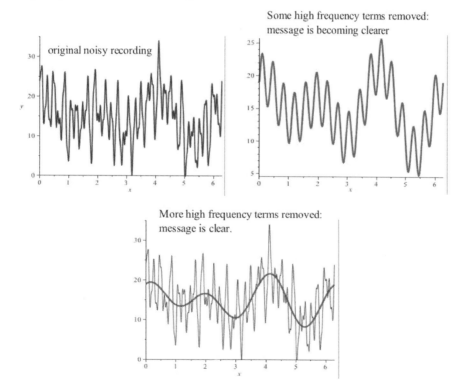

And Fourier series still enable us to solve problems of the type that led Fourier to develop them in the first place: if we know the temperature at each point on the boundary of a region (such as a solid circular disk) what will the temperature become at each point on the interior of the disk? And they are very useful for solving certain types of differential equations.

What if the function is not 2π-periodic?

If the function f is periodic with period P, we can "squeeze" f to create a new function g that does repeat every 2π units. Then after calculating the Fourier series for g, we can "unsqueeze" g to get a series for f.

Problems

The point of this section was only to use the Taylor series approach to create a new kind of infinite series, one whose building blocks were trigonometric functions, and the to show that this approach actually worked to approximate functions that were not even differentiable or continuous. There are still other infinite series that use different building blocks, but the point here was to illustrate that not all useful infinite series are power series.

Main Reference

Fourier Analysis, T.W. Korner, Cambridge University Press, 1988 (available in paperback)

This is a beautifully written book that is willing to forego some of the completeness and generalities of the results in order to present an understandable development of the main ideas and their applications and the personalities involved in this development. The level is aimed at a university student in their 3rd or 4th year with a strong background in mathematics (physics helps too), but the stories and descriptions of the applications and the flow of ideas are mostly accessible to very good and motivated students with a year of calculus.

MAPLE

You don't need a computer to calculate and graph the Fourier series of a function (your calculator can do the work), but Maple does a really nice job. The following program will automatically calculate the coefficients, build the Fourier series and graph both the original function and its Fourier series. One variation will even animate the approximation degree by degree. To use the program you need to enter the formula for the function (f:=x-> ...) and the degree of the approximation you want (N:= ...). The comments in italics are not part of the program.

with(plots):
f:=x->abs(1.3-floor(x/2.3)); *the original function – pick your own*
a0:=evalf((1/Pi)*int(f(x),x=0..2*Pi)):
PF:=plot(f(x), x=0..2*Pi, y=0..2, color=blue, thickness=3, discont=true):
a0:=evalf((1/Pi)*int(f(x),x=0..2*Pi)):

The next commands automatically construct the Fourier series
N:=15: *degree of the approximating Fourier series – pick your own*
fs[0]:=a0/2;
FSplot[0]:=plot(fs[0], x=0..2*Pi, color=red, thickness=2, title="degree = "||i):
R[0]:=display(PF,FSplot[0], title="degree = 0"):
for i from 1 to N do
a:=evalf((1/Pi)*int(cos(i*x)*f(x), x=0..2*Pi)):
b:=evalf((1/Pi)*int(sin(i*x)*f(x), x=0..2*Pi)):
fs[i]:=fs[i-1]+a*cos(i*x)+b*sin(i*x);od:
fs[N]; *prints the approximating formula of the Fourier series of degree N*
plot({f(x),fs[N]}, x=0..2*Pi, color=[red,blue], thickness=2, title="degree = "||N); *plots f and Nth degree Fourier series*

This set of commands will animate the approximation degree by degree
for i from 1 to N do
FSplot[i]:=plot(fs[i], x=0..2*Pi, color=red, thickness=2, title="degree = "||i):
R[i]:=display(PF, FSplot[i]):
od:
M:=[seq(R[i], i=0..N)]:
display(M, axes=normal, insequence=true);

Chapter 10: Odd Answers

Section 10.0

1. (a) 32, 64 (b) 2^5 (c) 2^n

3. (a) $-1, +1$ (b) $-1 = (-1)^5$ (c) $(-1)^n$

5. (a) 120, 720 (b) 5! or 5·24 (c) n! or $n \cdot a_{n-1}$

7. 1, 3/2, 11/6, 25/12, 137/60, 147/60

9. 1, 1/2, 3/4, 5/8, 11/16, 21/32

11. 1, 0, 1, 0, 1, 0

13. (a) $g(5) = -1, g(6) = +1$ (b) see the figure for $g(x)$.

15. (a) $t(5) = 1 - 1/2 + 1/4 - 1/8 + 1/16 - 1/32 = 21/32$,
    $t(6) = 43/64$. The graph of $t(x)$ is shown.

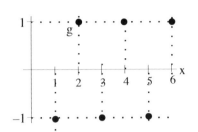

17. (a) $P(x) = 1 - \dfrac{x^2}{2}$ (b) Graphs

x	P(x)	cos(x)	| P(x) − cos(x) |
0	1.0	1.0	0
0.1	0.995	0.99500	0
0.2	0.98	0.98006	0.00006
0.3	0.955	0.95533	0.00033
1.0	0.5	0.54030	0.04030
2.0	−1.0	−0.41615	0.58385

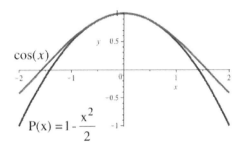

(c) $P(x) = 1 - \dfrac{x^2}{2} + \dfrac{x^4}{24}$

Graphs

x	P(x)	cos(x)	| P(x) − cos(x) |
0	1.0	1.0	0
0.1	0.99500	0.99500	0
0.2	0.98006	0.98006	0
0.3	0.95533	0.95533	0
1.0	0.54167	0.54030	0.00137
2.0	−0.33333	−0.41615	0.08282

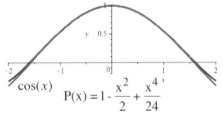

19. $P(x) = Ax + B$. $5 = P(0) = A \cdot 0 + B = B$ so $B = 5$. $3 = P'(0) = A$. $P(x) = 3x + 5$.

21. $P(x) = Ax + B$. $4 = P(0) = A \cdot 0 + B = B$ so $B = 4$. $-1 = P'(0) = A$. $P(x) = -1x + 4$.

23. $P(x) = 0x + 4 = 4$.

25. $P(x) = Ax + B$. $P(0) = B$. $P'(0) = A$.

27. $P(x) = Ax^2 + Bx + C$. $-2 = P(0) = A \cdot 0 + B \cdot 0 + C = C$. $7 = P'(0) = 2A \cdot 0 + B = B$.
    $6 = P''(0) = 2A$ so $A = 6/2 = 3$. $P(x) = 3x^2 + 7x - 2$.

29. $P(x) = Ax^2 + Bx + C$. $8 = P(0) = A \cdot 0 + B \cdot 0 + C = C$. $5 = P'(0) = 2A \cdot 0 + B = B$.
    $10 = P''(0) = 2A$ so $A = 10/2 = 5$. $P(x) = 5x^2 + 5x + 8$.

31. $P(x) = Ax^2 + Bx + C$. $-3 = P(0) = A \cdot 0 + B \cdot 0 + C = C$. $-2 = P'(0) = 2A \cdot 0 + B = B$.
   $4 = P''(0) = 2A$ so $A = 4/2 = 2$. $P(x) = 2x^2 - 2x - 3$.

33. $P(x) = Ax^3 + Bx^2 + Cx + D$. $5 = P(0) = A \cdot 0 + B \cdot 0 + C \cdot 0 + D = D$. $3 = P'(0) = 3A \cdot 0 + 2B \cdot 0 + C = C$.
   $4 = 6A0 + 2B = 2B$ so $B = 4/2 = 2$. $6 = P'''(0) = 6A$ so $A = 6/6 = 1$.
   $P(x) = 1x^3 + 2x^2 + 3x + 5$.

35. $P(x) = Ax^3 + Bx^2 + Cx + D$. $4 = P(0) = A \cdot 0 + B \cdot 0 + C \cdot 0 + D = D$. $-1 = P'(0) = 3A \cdot 0 + 2B \cdot 0 + C = C$.
   $-2 = 6A0 + 2B = 2B$ so $B = -2/2 = -1$. $-12 = P'''(0) = 6A$ so $A = -12/6 = -2$.
   $P(x) = -2x^3 - 1x^2 - 1x + 4$.

37. $P(x) = Ax^3 + Bx^2 + Cx + D$. $4 = P(0) = A \cdot 0 + B \cdot 0 + C \cdot 0 + D = D$. $0 = P'(0) = 3A \cdot 0 + 2B \cdot 0 + C = C$.
   $-4 = 6A0 + 2B = 2B$ so $B = -4/2 = -2$. $36 = P''(0) = 6A$ so $A = 36/6 = 6$.
   $P(x) = 6x^3 - 2x^2 + 0x + 4 = 6x^3 - 2x^2 + 4$.

39. $A = P'''(0)/6$, $B = P''(0)/2$, $C = P'(0)$, and $D = P(0)$.

Section 10.1

1. $\dfrac{1}{n^2}$ 3. $\dfrac{n-1}{n} = 1 - \dfrac{1}{n}$ 5. $\dfrac{n}{2^n}$

7. $\{-1, 0, 1/3, 1/2, 3/5, 2/3, ...\}$ Graph is shown. 9. $\{1, 2/3, 3/5, 4/7, 5/9, 6/11, ...\}$ Graph is shown.

11. $\{2, 3\tfrac{1}{2}, 2\tfrac{2}{3}, 3\tfrac{1}{4}, 2\tfrac{4}{5}, 3\tfrac{1}{6}, ...\}$ Graph is shown. 13. $\{0, \tfrac{1}{2}, -\tfrac{2}{3}, \tfrac{3}{4}, -\tfrac{4}{5}, \tfrac{5}{6}, ...\}$ Graph is shown.

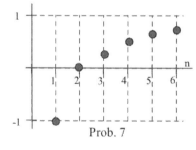
Prob. 7

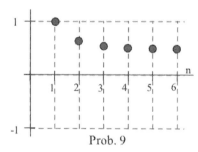
Prob. 9

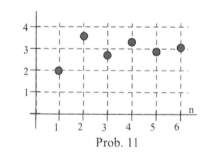
Prob. 11

15. $\{1, \tfrac{1}{2}, \tfrac{1}{3!}, \tfrac{1}{4!}, \tfrac{1}{5!}, \tfrac{1}{6!}, ...\} = \{1, \tfrac{1}{2}, \tfrac{1}{6}, \tfrac{1}{24}, \tfrac{1}{120}, \tfrac{1}{720}, ...\}$ The graph is shown below.

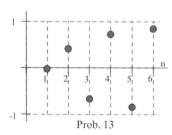
Prob. 13

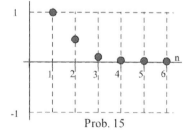
Prob. 15

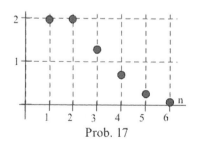
Prob. 17

17. $\{ \frac{2^1}{1!}, \frac{2^2}{2!}, \frac{2^3}{3!}, \frac{2^4}{4!}, \frac{2^5}{5!}, \frac{2^6}{6!}, ... \}$ The graph is shown.

19. $a_1 = 2, a_2 = -2, a_3 = 2, a_4 = -2, a_5 = 2, a_6 = -2, a_7 = 2, a_8 = -2, a_9 = 2, a_{10} = -2$

21. $\{ \sin(2\pi/3), \sin(4\pi/3), \sin(6\pi/3), \sin(8\pi/3), \sin(10\pi/3), \sin(12\pi/3), \sin(14\pi/3), \sin(16\pi/3),$

    $\sin(18\pi/3), \sin(20\pi/3), ... \}$

23. $c_1 = 1, c_2 = 3, c_3 = 6, c_4 = 10, c_5 = 15, c_6 = 21, c_7 = 28, c_8 = 36, c_9 = 45, c_{10} = 55$

25. $\{ a_n \}$ appears to converge. $\{ b_n \}$ does not appear to converge.

27. $\{ e_n \}$ does not appear to converge. $\{ f_n \}$ appears to converge.

29. $\{ 1 - \frac{2}{n} \}$ converges to 1.   31. $\{ \frac{n^2}{n+1} \}$ grows arbitrarily large and diverges.

33. $\{ \frac{n}{2n-1} \}$ converges to $\frac{1}{2}$ 35. $\{ \ln(3 + \frac{7}{n}) \}$ converges to $\ln(3) \approx 1.099$.

37. $\{ 4 + (-1)^n \}$ alternates in value between 3 and 5 and does not approach a single number.

    The sequence diverges.

39. $\{ \frac{1}{n!} \}$ converges to 0.

41. $\{ (1 - \frac{1}{n})^n \}$ converges to $e^{-1} = \frac{1}{e}$. (See Section 3.7, Example 7.)

43. $\{ \frac{(n+2)(n-5)}{n^2} \} = \{ \frac{n^2 - 3n - 10}{n^2} \} = \{ 1 - \frac{3}{n} - \frac{10}{n^2} \}$ converges to 1.

45. Take $N = \sqrt{\frac{3}{\varepsilon}}$. If $n > N = \sqrt{\frac{3}{\varepsilon}}$ then $n^2 > \frac{3}{\varepsilon}$ and $\varepsilon > \frac{3}{n^2} = \left| \frac{3}{n^2} - 0 \right|$.

47. Take $N = \frac{1}{\varepsilon}$. If $n > N = \frac{1}{\varepsilon}$ then $\varepsilon > \frac{1}{n} = |(3 - \frac{1}{n}) - 3| = |\frac{3n-1}{n} - 3|$.

49. $\{ \frac{1}{n^{th} \text{ prime}} \}$ is a subsequence of $\{ \frac{1}{n^{th} \text{ integer}} \} = \{ \frac{1}{n} \}$ which converges to 0, so we can

    conclude that $\{ \frac{1}{n^{th} \text{ prime}} \}$ converges to 0.

51. $\{ (-2)^n (\frac{1}{3})^n \} = \{ (-1)^n (\frac{2}{3})^n \}$.

    If n is even, $\{ (-1)^n (\frac{2}{3})^n \} = \{ (\frac{2}{3})^n \}$ which converges to 0. If n is odd,

    $\{ (-1)^n (\frac{2}{3})^n \} = \{ -(\frac{2}{3})^n \}$ which also converges to 0. Since "n even" and "n odd" account

    for all of the positive integers, we can conclude that $\{ (-2)^n (\frac{1}{3})^n \}$ converges to 0.

53. $\left\{ (1+\frac{5}{n^2})^{(n^2)} \right\}$ is a subsequence of $\left\{ (1+\frac{5}{n})^n \right\}$ which converges to e^5 (Section 3.7, Example 7) so $\left\{ (1+\frac{5}{n^2})^{(n^2)} \right\}$ also converges to e^5.

55. $a_n = 7 - \frac{2}{n}$ so $a_{n+1} = 7 - \frac{2}{n+1}$. $a_{n+1} - a_n = (7 - \frac{2}{n+1}) - (7 - \frac{2}{n}) = \frac{2}{n} - \frac{2}{n+1} = \frac{2}{n(n+1)} > 0$ for all $n \geq 1$.
Therefore, $a_{n+1} > a_n$ and $\{a_n\}$ is monotonically increasing.

57. $a_n = 2^n$ so $a_{n+1} = 2^{n+1}$. $a_{n+1} - a_n = 2^{n+1} - 2^n = 2^n \cdot 2 - 2^n = 2^n(2-1) = 2^n > 0$ for all $n \geq 1$.
Therefore, $a_{n+1} > a_n$ and $\{a_n\}$ is monotonically increasing.

59. $a_n = \frac{n+1}{n!}$ so $a_{n+1} = \frac{(n+1)+1}{(n+1)!} = \frac{n+2}{(n+1)!}$. Then

$$\frac{a_{n+1}}{a_n} = \frac{\frac{n+2}{(n+1)!}}{\frac{n+1}{n!}} = \frac{n+2}{n+1} \cdot \frac{n!}{(n+1)!} = \frac{n+2}{n+1} \cdot \frac{1 \cdot 2 \cdot 3 \cdots n}{1 \cdot 2 \cdot 3 \cdots n \cdot (n+1)} = \frac{n+2}{n+1} \cdot \frac{1}{n+1} < 1 \text{ for all } n \geq 1$$

so $a_{n+1} < a_n$ for all $n \geq 1$ and $\{a_n\}$ is monotonically decreasing.

61. $a_n = (\frac{5}{4})^n$ so $a_{n+1} = (\frac{5}{4})^{n+1}$. Then

$$\frac{a_{n+1}}{a_n} = \frac{(\frac{5}{4})^{n+1}}{(\frac{5}{4})^n} = \frac{5}{4} > 1 \text{ for all } n \text{ so } a_{n+1} > a_n \text{ for all } n > 0 \text{ and } \{a_n\} \text{ is montonically increasing.}$$

63. $a_n = \frac{n}{e^n}$ so $a_{n+1} = \frac{n+1}{e^{n+1}}$. Then

$$\frac{a_{n+1}}{a_n} = \frac{\frac{n+1}{e^{n+1}}}{\frac{n}{e^n}} = \frac{n+1}{n} \cdot \frac{e^n}{e^{n+1}} = \frac{n+1}{n} \cdot \frac{1}{e} < 1 \text{ for } n > 1 \text{ (reason: } e > 2 \text{ so } n \cdot e > 2n > n+1 \text{ so } \frac{n+1}{n \cdot e} < 1).$$

So $a_{n+1} < a_n$ for all $n > 0$ and $\{a_n\}$ is monotonically decreasing.

65. Let $f(x) = 5 - \frac{3}{x}$. Then $f'(x) = \frac{3}{x^2} > 0$ for all x so $f(x)$ is increasing. From that we can conclude that $a_n = f(n)$ is monotonically increasing.

67. Let $f(x) = \cos(\frac{1}{x})$. Then $f'(x) = -\sin(\frac{1}{x}) \cdot (\frac{-1}{x^2}) = \frac{1}{x^2} \cdot \sin(\frac{1}{x}) > 0$ for all $x \geq 1$. From that we can conclude that $a_n = f(n)$ is monotonically increasing.

69. This is similar to problem 59. The ratio method works nicely.

71. One method is to examine $a_{n+1} - a_n = (1 - \frac{1}{2^{n+1}}) - (1 - \frac{1}{2^n}) = \frac{1}{2^n} - \frac{1}{2^{n+1}} = \frac{1}{2^n} - \frac{1}{2 \cdot 2^n} = \frac{1}{2 \cdot 2^n} > 0$

for all n so $a_{n+1} > a_n$ for all n and $\{a_n\}$ is monotonically increasing.

The ratio and derivative methods also work.

$(\ D(1 - \frac{1}{2^x}) = D(1 - 2^{-x}) = 0 - 2^{-x} \cdot \ln(2) \cdot D(-x) = 2^{-x} \ln(2) = \frac{\ln(2)}{2^x} > 0$ for $x > 0$.)

73. This is similar to problem 63. The ratio method works nicely.

75. N = 4: $a_1 = 4$, $a_2 = \frac{1}{2}(4 + \frac{4}{4}) = \frac{5}{2} = 2.5$, $a_3 = \frac{1}{2}(2.5 + \frac{4}{2.5}) = 2.05$, $a_n = \frac{1}{2}(2.05 + \frac{4}{2.05}) \approx 2.00061$.

N = 9: $a_1 = 9$, $a_2 = \frac{1}{2}(9 + \frac{9}{9}) = \frac{10}{2} = 5$, $a_3 = \frac{1}{2}(5 + \frac{9}{5}) = 3.2$, $a_n = \frac{1}{2}(3.2 + \frac{9}{3.2}) = 3.00625$.

N = 5: $a_1 = 5$, $a_2 = \frac{1}{2}(5 + \frac{5}{5}) = \frac{6}{2} = 3$. $a_3 = \frac{1}{2}(3 + \frac{5}{3}) \approx 2.333$, $a_n = \frac{1}{2}(2.333 + \frac{5}{2.333}) \approx 2.238$.

77. (a) $p = 0.02$, and we want to solve $0.01 = \frac{0.02}{0.02k + 1}$ for k. Then $0.02k + 1 = \frac{0.02}{0.01} = 2$ so

$0.02k = 1$ and $k = \frac{1}{0.02} = 50$ generations.

(b) We want to solve $\frac{1}{2} p = \frac{p}{kp + 1}$ for k in terms of p. $kp + 1 = \frac{p}{0.5p} = 2$ so $kp = 1$ and

$k = \frac{1}{p}$ generations.

79. (a) The first "few" grains can be anywhere on the x–axis.

(b) After a "lot of grains" have been placed, there will be a large pile of sand close to 3 on the x–axis.

81. $-1 \leq \sin(n) \leq 1$ for all integers n.

(a) The first few grains will be scattered between -1 and $+1$ on the x–axis.

(b) After a "lot of grains" have been placed, the sand will be scattered "uniformly" along the interval from -1 to $+1$. (See part (c).)

(c) This argument is rather sophisticated, but the result is interesting: no two grains ever end up on the same point.

We assume that two grains do end up on the same point, and then derive a contradiction. From this we conclude that our original assumption (two grains on one point) was false.

Assume that two grains do end up on the same point so $a_m = a_n$ for distinct integers m and n.

Then $\sin(m) = \sin(n)$ so $0 = \sin(m) - \sin(n) = 2 \cdot \sin(\frac{m-n}{2}) \cdot \cos(\frac{m+n}{2})$ and either $\sin(\frac{m-n}{2}) = 0$ or $\cos(\frac{m+n}{2}) = 0$. If $\sin(\frac{m-n}{2}) = 0$, then $\frac{m-n}{2} = \pi K$ for some integer K and $\pi = \frac{m-n}{2K}$ where m, n, and K are integers. Then π is a rational number, a contradiction of the fact that π is irrational.

If $\cos(\frac{m+n}{2}) = 0$, then $\frac{m+n}{2} = \frac{\pi}{2} + K\pi = \pi(\frac{1}{2} + K)$ for some integer K so $\pi = \frac{m+n}{1 + 2K}$, a rational number. This again contradicts the irrationality of π, so out original assumption (two grains on the same point) was false.

Section 10.2

1. $\sum_{k=1}^{\infty} \frac{1}{k}$

3. $\sum_{k=1}^{\infty} \frac{2}{3k}$

5. $\sum_{k=1}^{\infty} \left(-\frac{1}{2}\right)^k$ or $\sum_{k=1}^{\infty} (-1)^k \cdot \frac{1}{2^k}$

7. The graph is given.

9. The graph is given.

11. The graph is given.

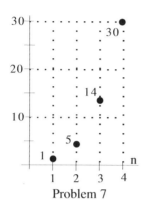
Problem 7

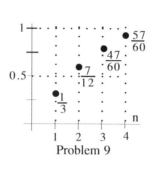
Problem 9

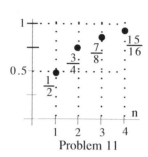
Problem 11

13. $a_1 = 3$, $a_2 = -1$, $a_3 = 2$, $a_4 = 1$

15. $a_1 = 4$, $a_2 = 0.5$, $a_3 = -0.2$, $a_4 = 0.5$

17. $a_1 = 1$, $a_2 = 0.1$, $a_3 = 0.01$, $a_4 = 0.001$

19. $\sum_{k=1}^{\infty} \frac{8}{10^k}$

21. $\sum_{k=1}^{\infty} \frac{5}{10^k}$

23. $\sum_{k=1}^{\infty} \frac{a}{10^k}$

25. $\sum_{k=1}^{\infty} \frac{17}{100^k}$

27. $\sum_{k=1}^{\infty} \frac{7}{100^k}$

29. $\sum_{k=1}^{\infty} \frac{abc}{1000^k}$

31. $\sum_{k=0}^{\infty} 30 \cdot (0.8)^k$

33. $80, 64, 51.2, 100 \cdot (0.8)^n$

35. $(1/4)^n \to 0$ so the series may converge or may diverge. (Later we will see that it converges.)

37. $(4/3)^n \to \infty \neq 0$ so the series diverges.

39. $\frac{\sin(n)}{n} \to 0$ so the series may converge or may diverge.

41. $\cos(1/n) \to \cos(0) = 1 \neq 0$ so the series diverges.

43. $\frac{n^2 - 20}{n^5 + 4} \to 0$ so the series may converge or may diverge. (Later we will see that it converges.)

Section 10.3

1. $\sum_{k=0}^{\infty} \left(\frac{1}{3}\right)^k = \frac{1}{1-\left(\frac{1}{3}\right)} = \frac{3}{2}$

2. $\sum_{k=0}^{\infty} \left(\frac{2}{3}\right)^k = \frac{1}{1-\left(\frac{2}{3}\right)} = 3$

3. $\frac{1}{8} \cdot \sum_{k=0}^{\infty} \left(\frac{1}{2}\right)^k = \frac{1}{8} \cdot \frac{1}{1-\left(\frac{1}{2}\right)} = \frac{1}{4}$

5. $-\frac{2}{3} \cdot \sum_{k=0}^{\infty} \left(-\frac{2}{3}\right)^k = -\frac{2}{3} \cdot \frac{1}{1-\left(-\frac{2}{3}\right)} = -\frac{2}{5}$

7. (a) $\frac{1}{2} \cdot \sum_{k=0}^{\infty} \left(\frac{1}{2}\right)^k = \frac{1}{2} \cdot \frac{1}{1-\left(\frac{1}{2}\right)} = \frac{1}{2} \cdot \frac{2}{1} = 1$, $\quad \frac{1}{3} \cdot \sum_{k=0}^{\infty} \left(\frac{1}{3}\right)^k = \frac{1}{3} \cdot \frac{1}{1-\left(\frac{1}{3}\right)} = \frac{1}{3} \cdot \frac{3}{2} = \frac{1}{2}$

   (b) $\frac{1}{a} \cdot \sum_{k=0}^{\infty} \left(\frac{1}{a}\right)^k = \frac{1}{a} \cdot \frac{1}{1-\left(\frac{1}{a}\right)} = \frac{1}{a} \cdot \frac{a}{a-1} = \frac{1}{a-1}$

9. (a) $40 \cdot (0.4)^{n-1}$ (b) $40 \cdot \sum_{k=0}^{\infty} (0.4)^k$ (c) $40 \cdot \frac{1}{1-0.4} = \frac{40}{0.6} = 66\frac{2}{3}$ ft.

11. (a) $\sum_{k=1}^{\infty} \left(\frac{1}{2}\right)^k$ (b) $\frac{1}{2}, \frac{1}{4}, \left(\frac{1}{2}\right)^n$ (c) All of the cake.

13. $1 + \frac{1}{4} + \left(\frac{1}{4}\right)^2 + \left(\frac{1}{4}\right)^3 + \ldots = \sum_{k=0}^{\infty} \left(\frac{1}{4}\right)^k = \frac{4}{3}$.

15. (a) Area $= 1 + \frac{3}{9} + \frac{3}{9} \cdot \frac{4}{9} + \frac{3}{9}\left(\frac{4}{9}\right)^2 + \frac{3}{9}\left(\frac{4}{9}\right)^3 + \ldots = 1 + \frac{3}{9}\left\{1 + \frac{4}{9} + \left(\frac{4}{9}\right)^2 + \left(\frac{4}{9}\right)^3 + \ldots\right\}$

    $= 1 + \frac{3}{9}\left\{\frac{1}{1-(4/9)}\right\} = 1 + \frac{3}{5} = \frac{8}{5} = 1.6$.

   (b) Let L be the length of the original triangle ($L = 3\sqrt{\frac{4}{\sqrt{3}}}$) and P_n be the perimeter at the nth step. Then $P_0 = 3L$. $P_1 = 3 \cdot 4 \cdot \left(\frac{L}{3}\right) = 4L$,

   $P_2 = 3 \cdot 4^2 \cdot \left(\frac{L}{3^2}\right) = 3L\left(\frac{4}{3}\right)^2$

   $P_3 = 3 \cdot 4^3 \cdot \left(\frac{L}{3^3}\right) = 3L\left(\frac{4}{3}\right)^3$

   $P_4 = 3 \cdot 4^4 \cdot \left(\frac{L}{3^4}\right) = 3L\left(\frac{4}{3}\right)^4$, and, in general,

   $P_n = 3 \cdot 4^n \cdot \left(\frac{L}{3^n}\right) = 3L\left(\frac{4}{3}\right)^n$.

   Since $\frac{4}{3} > 1$, the sequence of terms $3L\left(\frac{4}{3}\right)^n$ grows without bound, and the perimeter "approaches infinity."

17. (a) Height $= 2 + 2(\frac{1}{2}) + 2(\frac{1}{4}) + 2(\frac{1}{8}) + ... = 2\{ 1 + \frac{1}{2} + (\frac{1}{2})^2 + (\frac{1}{2})^3 + ... \} = 2 \cdot \frac{1}{1 - \frac{1}{2}} = 4$.

(b) Surface area $= 4\pi(1)^2 + 4\pi(\frac{1}{2})^2 + 4\pi(\frac{1}{4})^2 + 4\pi(\frac{1}{8})^2 + ...$

$= 4\pi\{ 1 + \frac{1}{4} + (\frac{1}{4})^2 + (\frac{1}{4})^3 + ... \} = 4\pi \cdot \frac{1}{1 - \frac{1}{4}} = \frac{16\pi}{3} \approx 16.755$.

(c) Volume $= \frac{4\pi}{3}(1)^3 + \frac{4\pi}{3}(\frac{1}{2})^3 + \frac{4\pi}{3}(\frac{1}{4})^3 + \frac{4\pi}{3}(\frac{1}{8})^3 + ...$

$= \frac{4\pi}{3} \{ 1 + \frac{1}{8} + (\frac{1}{8})^2 + (\frac{1}{8})^3 + ... \} = \frac{4\pi}{3} \cdot \frac{1}{1 - \frac{1}{8}} = \frac{32\pi}{21} \approx 4.787$.

19. $0.8888... = \frac{8}{10} + \frac{8}{10^2} + \frac{8}{10^3} + ... = \frac{8}{10} \{ 1 + \frac{1}{10} + (\frac{1}{10})^2 + (\frac{1}{10})^3 + ...\} = \frac{8}{10} \{ \frac{10}{9} \} = \frac{8}{9}$.

$0.9999... = \frac{9}{10} + \frac{9}{10^2} + \frac{9}{10^3} + ... = \frac{9}{10} \{ 1 + \frac{1}{10} + (\frac{1}{10})^2 + (\frac{1}{10})^3 + ...\} = \frac{9}{10} \{ \frac{10}{9} \} = 1$.

$0.285714... = \frac{285714}{1000000} + \frac{285714}{1000000^2} + \frac{285714}{1000000^3} + ...$

$= \frac{285714}{1000000} \{ 1 + \frac{1}{1000000} + (\frac{1}{1000000})^2 + (\frac{1}{1000000})^3 + ...\}$

$= \frac{285714}{1000000} \{ \frac{1000000}{999999} \} = \frac{285714}{999999}$.

21. Series converges for $|2x + 1| < 1$: $-1 < x < 0$.

23. Series converges for $|1 - 2x| < 1$: $0 < x < 1$.

25. Series converges for $|7x| < 1$: $-\frac{1}{7} < x < \frac{1}{7}$.

27. Series converges for $|\frac{x}{2}| < 1$: $-2 < x < 2$.

29. Series converges for $|2x| < 1$: $-\frac{1}{2} < x < \frac{1}{2}$.

31. Series converges for $|\sin(x)| < 1$: for all $x \ne \frac{\pi}{2} \pm N\pi$ for integer values of N.

33. The formula is correct if $|x| < 1$. The value $x = 2$ does not satisfy the condition $|x| < 1$, so the formula does not apply.

35. $s_4 = (\frac{1}{3} - \frac{1}{4}) + (\frac{1}{4} - \frac{1}{5}) = \frac{1}{3} - \frac{1}{5}$, $s_5 = \frac{1}{3} - \frac{1}{6}$, $s_n = \frac{1}{3} - \frac{1}{n+1} \to \frac{1}{3}$

37. $s_4 = (1^3 - 2^3) + (2^3 - 3^3) + (3^3 - 4^3) = 1 - 4^3$, $s_5 = 1 - 5^3$, $s_n = 1 - (n+1)^3 \to -\infty$.

39. $s_4 = (f(3) - f(4)) + (f(4) - f(5)) = f(3) - f(5)$, $s_5 = f(3) - f(6)$, $s_n = f(3) - f(n+1)$

41. $s_4 = \sin(1) - \sin(\frac{1}{5}) \approx 0.643$, $s_5 = \sin(1) - \sin(\frac{1}{6}) \approx 0.676$, $s_n = \sin(1) - \sin(\frac{1}{n+1}) \to \sin(1) \approx 0.841$.

43. $s_4 = (\frac{1}{2^2} - \frac{1}{3^2}) + (\frac{1}{3^2} - \frac{1}{4^2}) + (\frac{1}{4^2} - \frac{1}{5^2}) = \frac{1}{4} - \frac{1}{25}$, $s_5 = \frac{1}{4} - \frac{1}{36}$, $s_n = \frac{1}{4} - \frac{1}{(n+1)^2} \to \frac{1}{4}$

45. & 47. On your own.

Section 10.3.5

Integrals and sums

1. The sum.
2. The integral.
3. $\sum_{k=1}^{\infty} f(k)$
4. $\sum_{k=1}^{\infty} f(k)$
5. $\sum_{k=2}^{\infty} f(k)$
6. (b) $f(1) + f(2)$
7. (b) $f(1) + f(2)$
8. (d) $f(3) + f(4)$
9. (c) $f(2) + f(3)$
10. (b), (d), (a), (c) : $\int_2^4 f(x)\,dx < f(2) + f(3) < \int_1^3 f(x)\,dx < f(1) + f(2)$
11. (c), (a), (d), (b) : $f(1) + f(2) + f(3) < \int_1^4 f(x)\,dx < f(2) + f(3) + f(4) < \int_2^5 f(x)\,dx$

Comparisons

12. (a) No. You are definitely too tall. (b) Apply. You **may** meet the requirements.
    (c) No. Definitely too short. (d) Apply. You **may** meet the requirements.
    (e) You do not meet the requirements if you are {shorter than Sam} or {taller than Tom}.
    (f) You do not have enough information if you are {shorter than Tom} or {taller than Sam}.

13. (a) You did well (better than Wendy). (b) You may have done well or poorly.
    (c) You may have done well or poorly. (d) You did poorly (worse than Paula).

14. (a) Baker is too easy (easier than Index). (b) Baker may be right for you.
    (c) Baker may be right for you. (d) Baker is too hard for you (harder than Liberty Bell).
    (e) Baker may be a good climb for you if it is harder than Index and easier than Liberty Bell.
    (f) Baker is too easy if Baker is easier than Index. Baker is too hard if Baker is harder than Liberty Bell.

15. (a) Expect Unknown to be (very) good. (b) Unknown is still unknown.
    (c) Unknown is still unknown. (d) Unknown is bad.

16. $\sum_{k=1}^{\infty} \frac{1}{k^2+1} < \sum_{k=1}^{\infty} \frac{1}{k^2}$

17. $\sum_{k=2}^{\infty} \frac{1}{k^3-5} > \sum_{k=2}^{\infty} \frac{1}{k^3}$

18. $\sum_{k=1}^{\infty} \frac{1}{k^2+3k-1} < \sum_{k=1}^{\infty} \frac{1}{k^2}$

19. $\sum_{k=3}^{\infty} \frac{1}{k^2+5k} > \sum_{k=3}^{\infty} \frac{1}{k^3+k-1}$

Ratios of successive terms

20. $a_k = 3k$, $a_{k+1} = 3(k+1)$, $\frac{a_{k+1}}{a_k} = \frac{k+1}{k}$.

21. $a_k = k+3$, $a_{k+1} = (k+1)+3$, $\frac{a_{k+1}}{a_k} = \frac{k+4}{k+3}$.

22. $a_k = 2k + 5$, $a_{k+1} = 2(k+1) + 5$, $\dfrac{a_{k+1}}{a_k} = \dfrac{2k+7}{2k+5}$.

23. $a_k = 3/k$, $a_{k+1} = \dfrac{3}{k+1}$, $\dfrac{a_{k+1}}{a_k} = \dfrac{\frac{3}{k+1}}{\frac{3}{k}} = \dfrac{k}{k+1}$.

24. $a_k = k^2$, $a_{k+1} = (k+1)^2$, $\dfrac{a_{k+1}}{a_k} = \dfrac{(k+1)^2}{k^2}$.   25. $a_k = 2^k$, $a_{k+1} = 2^{k+1}$, $\dfrac{a_{k+1}}{a_k} = \dfrac{2^{k+1}}{2^k} = 2$.

26. $a_k = (1/2)^k$, $a_{k+1} = (1/2)^{k+1}$, $\dfrac{a_{k+1}}{a_k} = \dfrac{(1/2)^{k+1}}{(1/2)^k} = \dfrac{1}{2}$.

27. $a_k = x^k$, $a_{k+1} = x^{k+1}$, $\dfrac{a_{k+1}}{a_k} = \dfrac{x^{k+1}}{x^k} = x$.

28. $a_k = (x-1)^k$, $a_{k+1} = (x-1)^{k+1}$, $\dfrac{a_{k+1}}{a_k} = \dfrac{(x-1)^{k+1}}{(x-1)^k} = x - 1$.

29. $\sum\limits_{k=1}^{\infty} \left(\dfrac{1}{2}\right)^k$ is a geometric series with $r = 1/2$ so the series converges. $\dfrac{a_{k+1}}{a_k} = \dfrac{(1/2)^{k+1}}{(1/2)^k} = \dfrac{1}{2}$.

30. $\sum\limits_{k=1}^{\infty} \left(\dfrac{1}{5}\right)^k$ is a geometric series with $r = 1/5$ so the series converges. $\dfrac{a_{k+1}}{a_k} = \dfrac{(1/5)^{k+1}}{(1/5)^k} = \dfrac{1}{5}$.

31. $\sum\limits_{k=1}^{\infty} 2^k$ is a geometric series with $r = 2$ so the series diverges. $\dfrac{a_{k+1}}{a_k} = \dfrac{2^{k+1}}{2^k} = 2$.

32. $\sum\limits_{k=1}^{\infty} (-3)^k$ is a geometric series with $r = -3$ so the series diverges. $\dfrac{a_{k+1}}{a_k} = \dfrac{(-3)^{k+1}}{(-3)^k} = -3$.

33. $\sum\limits_{k=1}^{\infty} 4 = 4 + 4 + 4 + \ldots$ diverges by the Nth Term Test for Divergence since $a_n = 4$ for all n, and

$a_n = 4$ does not approach 0. $\dfrac{a_{k+1}}{a_k} = \dfrac{4}{4} = 1$.

34. $\sum\limits_{k=1}^{\infty} (-1)^k$ diverges by the Nth Term Test for Divergence since a_n does not approach 0. (It also is a

geometric series with $r = -1$ and $|r| = 1$.) $\dfrac{a_{k+1}}{a_k} = \dfrac{(-1)^{k+1}}{(-1)^k} = -1$.

35. $\sum\limits_{k=1}^{\infty} \dfrac{1}{k}$ is the harmonic series which diverges. $\dfrac{a_{k+1}}{a_k} = \dfrac{\frac{1}{k+1}}{\frac{1}{k}} = \dfrac{k}{k+1}$.

36. $\sum\limits_{k=1}^{\infty} \dfrac{7}{k} = 7 \cdot \sum\limits_{k=1}^{\infty} \dfrac{1}{k}$ is the divergent harmonic series. $\dfrac{a_{k+1}}{a_k} = \dfrac{\frac{7}{k+1}}{\frac{7}{k}} = \dfrac{k}{k+1}$.

Alternating terms

37. If $a_5 > 0$, then $s_4 < s_5$.

38. If $a_5 = 0$, then $s_4 = s_5$.

39. If $a_5 < 0$, then $s_4 > s_5$.

40. If $a_{n+1} > 0$ for all n, then $s_n < s_{n+1}$ for all n.

41. If $a_{n+1} < 0$ for all n, then $s_n > s_{n+1}$ for all n.

42. If $a_4 > 0$ and $a_5 < 0$, then $s_3 < s_4$ and $s_4 > s_5$.

43. If $a_4 = 0.2$ and $a_5 = -0.1$ and $a_6 = 0.2$, then $s_3 < s_5 < s_4$.

44. If $a_4 = -0.3$ and $a_5 = 0.2$ and $a_6 = -0.1$, then $s_3 > s_5 > s_4$.

45. If $a_4 = -0.3$ and $a_5 = -0.2$ and $a_6 = 0.1$, then $s_3 > s_4 > s_5$.

46. $s_1 = 2, s_2 = 1, s_3 = 3, s_4 = 2, s_5 = 4, s_6 = 3, s_7 = 5, s_8 = 4$.

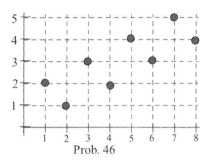
Prob. 46

47. $s_1 = 2, s_2 = 1, s_3 = 1.9, s_4 = 1.1, s_5 = 1.8, s_6 = 1.2$,
    $s_7 = 1.7, s_8 = 1.3$. The graph is given.

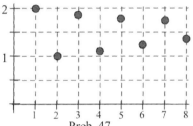
Prob. 47

48. $s_1 = 2, s_2 = 1, s_3 = 2, s_4 = 1, s_5 = 2, s_6 = 1, s_7 = 2, s_8 = 1$.
    The graph is given.

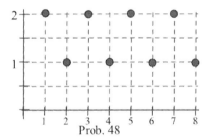
Prob. 48

49. $s_1 = -2, s_2 = -0.5, s_3 = -1.3, s_4 = -0.7, s_5 = -1.1, s_6 = -0.9$,
    $s_7 = 1.1, s_8 = 1.0$. The graph is given.

50. $s_1 = 5, s_2 = 6, s_3 = 5.4, s_4 = 5.0, s_5 = 5.2, s_6 = 5.3$,
    $s_7 = 5.4, s_8 = 5.2$. The graph is given.

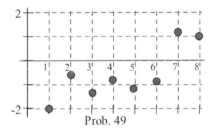
Prob. 49

51. If the a_k alternate in sign, then the graph of the partial sums s_n follows an "up–down–up–down" pattern?

52. If the a_k alternate in sign and decrease in magnitude (the $|a_k|$ is decreasing), then the graph of the partial sums s_n forms a "narrowing funnel" pattern.

53. (a) The terms a_k alternate in sign for the graphs D, E, and F.
    (b) The $|a_k|$ decrease for the graphs A and D.
    (c) The terms a_k alternate in sign and decrease in absolute value for the graph D.

54. (a) The terms a_k alternate in sign for the graphs C, D, E, and F in Fig. 14.
    (b) The $|a_k|$ decrease for the graphs B and C.
    (c) The terms a_k alternate in sign and decrease in absolute value for the graph C.

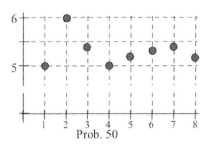
Prob. 50

55. The graph of the first eight partial sums of

$$\sum_{k=0}^{\infty} (-\tfrac{1}{2})^k = 1 - \tfrac{1}{2} + \tfrac{1}{4} - \ldots = \tfrac{2}{3}\ \text{is given.}$$

Notice the "narrowing funnel" shape of the graph.

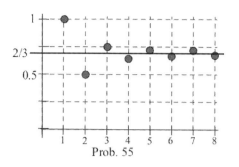
Prob. 55

56. The graph of the first eight partial sums of

$$\sum_{k=0}^{\infty} (-0.6)^k = 1 - 0.6 + 0.36 - \ldots = 0.625$$

is given.

Notice the "narrowing funnel" shape of the graph.

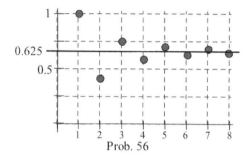
Prob. 56

57. The graph is given of the first eight partial sums of the divergent series

$$\sum_{k=0}^{\infty} (-2)^k = 1 - 2 + 4 - \ldots$$

This is a divergent series (N^{th} Term Test for Divergence).

Notice the "widening funnel" shape of the graph.

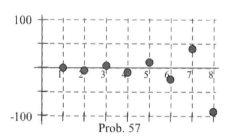
Prob. 57

Section 10.4 (Odd numbered problem solutions followed by even numbered problem answers.)

1. $\int_{1}^{\infty} \dfrac{1}{2x+5}\,dx = \tfrac{1}{2}\ln|2x+5|\Big|_{1}^{A} = \tfrac{1}{2}\ln|2A+5| - \tfrac{1}{2}\ln|7| \to \infty$ (as $A \to \infty$) so $\sum_{k=1}^{\infty} \dfrac{1}{2k+5}$ diverges.

3. $\int_{1}^{\infty} (2x+5)^{-3/2}\,dx = -(2x+5)^{-1/2}\Big|_{1}^{A} = \dfrac{-1}{\sqrt{2A+5}} - \dfrac{-1}{\sqrt{7}} \to \dfrac{1}{\sqrt{7}}$ (as $A \to \infty$) so $\sum_{k=1}^{\infty} \dfrac{1}{(2k+5)^{3/2}}$ converges.

5. $\int \dfrac{1}{x \cdot (\ln(x))^2}\,dx = \dfrac{-1}{\ln(x)} + C$ (using a u–substitution with $u = \ln(x)$ and $du = \tfrac{1}{x}\,dx$) so

$\int_{2}^{\infty} \dfrac{1}{x \cdot (\ln(x))^2}\,dx = \dfrac{-1}{\ln(x)}\Big|_{2}^{A} = \dfrac{-1}{\ln(A)} - \dfrac{-1}{\ln(2)} \to \dfrac{1}{\ln(2)}$ (as $A \to \infty$) so $\sum_{k=2}^{\infty} \dfrac{1}{k \cdot (\ln(k))^2}$ converges.

7. $\int_{1}^{\infty} \dfrac{1}{x^2+1}\,dx = \arctan(x)\Big|_{1}^{A} = \arctan(A) - \arctan(1) \to \dfrac{\pi}{2} - \dfrac{\pi}{4}$ (as $A \to \infty$) so $\sum_{k=1}^{\infty} \dfrac{1}{k^2+1}$ converges.

9. This is a telescoping series:

$$\sum_{k=1}^{\infty} \left\{ \frac{1}{k} - \frac{1}{k+3} \right\} = \left\{ 1 - \frac{1}{4} \right\} + \left\{ \frac{1}{2} - \frac{1}{5} \right\} + \left\{ \frac{1}{3} - \frac{1}{6} \right\} + \left\{ \frac{1}{4} - \frac{1}{7} \right\} + \left\{ \frac{1}{5} - \frac{1}{8} \right\} + \ldots \to 1 + \frac{1}{2} + \frac{1}{3}.$$

The Integral Test also works:

$$\int_{1}^{\infty} \frac{1}{x} - \frac{1}{x+3} \, dx = \ln|x| - \ln|x+3| \Big|_{1}^{A} = \{ \ln|A| - \ln|A+3| \} - \{ \ln(1) - \ln(4) \}$$

$$= \ln\left|\frac{4A}{A+3}\right| \to \ln(4) \text{ (as } A \to \infty\text{) so } \sum_{k=1}^{\infty} \left\{ \frac{1}{k} - \frac{1}{k+3} \right\} \text{ converges.}$$

(Notice that the "telescoping series" method gives the value of the series, but the Integral Test only tells us that the series converges. For this series, the "telescoping series" method is both easier and more precise.)

11. $\int_{1}^{\infty} \frac{1}{x(x+5)} dx = \int_{1}^{\infty} \frac{1}{5} \left\{ \frac{1}{x} - \frac{1}{x+5} \right\} dx$ (using the method of Partial Fraction Decomposition)

$$= \frac{1}{5} \{ \ln|x| - \ln|x+5| \} = \frac{1}{5} \left\{ \ln\left|\frac{x}{x+5}\right| \right\} \Big|_{1}^{A} = \frac{1}{5} \left\{ \ln\left|\frac{A}{A+5}\right| - \ln\left|\frac{1}{6}\right| \right\} \to \frac{1}{5} \left\{ \ln(1) - \ln\left(\frac{1}{6}\right) \right\}$$

(as $A \to \infty$) so $\sum_{k=1}^{\infty} \frac{1}{k \cdot (k+5)}$ converges.

13. $\int_{1}^{\infty} x \cdot e^{-(x^2)} dx = \frac{-1}{2} e^{-(x^2)} \Big|_{1}^{A} = \left(\frac{-1}{2} e^{-(A^2)} \right) - \left(\frac{-1}{2} e^{-1} \right) = \frac{1}{2e} - \frac{1}{2e^{(A^2)}} \to \frac{1}{2e}$ (as $A \to \infty$) so

$\sum_{k=1}^{\infty} k \cdot e^{-(k^2)}$ converges.

15. $\int \frac{1}{\sqrt{6x+10}} dx = \frac{1}{3} \sqrt{6x+10} + C$ (using a u–substitution with $u = 6x+10$ and $du = 6 \, dx$). Then

$$\int_{1}^{\infty} \frac{1}{\sqrt{6x+10}} dx = \frac{1}{3} \sqrt{6x+10} \Big|_{1}^{A} = \frac{1}{3}\sqrt{6A+10} - \frac{1}{3}\sqrt{16} \to \infty \text{ (as } A \to \infty\text{)}$$

so $\sum_{k=1}^{\infty} \frac{1}{\sqrt{6k+10}}$ diverges.

17. $p = 3 > 1$ so $\sum_{k=1}^{\infty} \frac{1}{k^3}$ converges.   19. $p = 1/2 < 1$ so $\sum_{k=2}^{\infty} \frac{1}{\sqrt{k}}$ diverges.

21. $p = 3/2 > 1$ so $\sum_{k=3}^{\infty} \frac{1}{k^{3/2}}$ converges.

23. $\sum_{k=1}^{\infty} \frac{1}{k^3}$: $\int_{1}^{11} \frac{1}{x^3} dx = 0.4958677$, $1 + \int_{1}^{10} \frac{1}{x^3} dx = 1.495$ so $0.4958677 < s_{10} < 1.495$.

$\int_{1}^{101} \frac{1}{x^3} dx = 0.499951$, $1 + \int_{1}^{100} \frac{1}{x^3} dx = 1.49995$ so $0.499951 < s_{100} < 1.49995$.

$\int_{1}^{1000001} \frac{1}{x^3} dx = 0.5000000$, $1 + \int_{1}^{1000000} \frac{1}{x^3} dx = 1.5000000$ so $0.5000000 < s_{1000000} < 1.5000000$.

25. $\sum_{k=1}^{\infty} \frac{1}{k+1000}$: $\int_{1}^{11} \frac{1}{x+1000} dx = 0.0099404$, $\frac{1}{1001} + \int_{1}^{10} \frac{1}{x+1000} dx = 0.0099498$

so $\mathbf{0.00994}04 < s_{10} < \mathbf{0.0099 4}98$. (This is a very precise estimate of s_{10}.)

$\int_{1}^{101} \frac{1}{x+1000} dx = 0.09522$, $\frac{1}{1001} + \int_{1}^{100} \frac{1}{x+1000} dx = 0.0953$ so $\mathbf{0.09522} < s_{100} < \mathbf{0.0953}$.

$\int_{1}^{1000001} \frac{1}{x+1000} dx = 6.90776$, $\frac{1}{1001} + \int_{1}^{1000000} \frac{1}{x+1000} dx = 6.90875$ so $\mathbf{6.90}776 < s_{1000000} < \mathbf{6.90}875$.

27. $\sum_{k=1}^{\infty} \frac{1}{k^2+100}$: $\int_{1}^{11} \frac{1}{x^2+100} dx = \frac{1}{10} \arctan(\frac{x}{10}) \Big|_{1}^{11} = 0.0733$, $\frac{1}{101} + \int_{1}^{10} \frac{1}{x^2+100} dx = 0.0783$

so $\mathbf{0.073} < s_{10} < \mathbf{0.078}$. Also, $\mathbf{0.137} < s_{100} < \mathbf{0.147}$ and $\mathbf{0.1471} < s_{1000000} < \mathbf{0.157}$.

29. For $q \neq 1$, let $u = \ln(x)$ and $du = \frac{1}{x} dx$.

Then $\int \frac{1}{x \cdot (\ln(x))^q} dx = \int \frac{1}{(u)^q} du = \frac{1}{1-q} u^{-q+1} = \frac{1}{1-q} (\ln(x))^{1-q} + C$.

Then $\int_{2}^{\infty} \frac{1}{x \cdot (\ln(x))^q} dx = \frac{1}{1-q} (\ln(x))^{1-q} \Big|_{2}^{A} = \frac{1}{1-q} (\ln(A))^{1-q} - \frac{1}{1-q} (\ln(2))^{1-q}$.

If $q < 1$, then $\frac{1}{1-q} (\ln(A))^{1-q} - \frac{1}{1-q} (\ln(2))^{1-q} \to \infty$ (as $A \to \infty$) so $\sum_{k=2}^{\infty} \frac{1}{k \cdot (\ln k)^q}$ diverges.

If $q > 1$, then $\frac{1}{1-q} (\ln(A))^{1-q} - \frac{1}{1-q} (\ln(2))^{1-q} \to -\frac{1}{1-q} (\ln(2))^{1-q}$ (as $A \to \infty$)

so $\sum_{k=2}^{\infty} \frac{1}{k \cdot (\ln k)^q}$ converges.

If $q = 1$, then

$$\int_{2}^{\infty} \frac{1}{x \cdot (\ln(x))^q} \, dx = \int_{2}^{\infty} \frac{1}{x \cdot \ln(x)} \, dx = \ln|\ln(x)| \Big|_{2}^{A} = |\ln|\ln(A)| - |\ln|\ln(2)|| \to \infty \text{ (as } A \to \infty)$$

so $\sum_{k=2}^{\infty} \frac{1}{k \cdot (\ln k)}$ diverges.

"Q–Test:" $\sum_{k=2}^{\infty} \frac{1}{k \cdot (\ln k)^q}$ $\begin{cases} \text{diverges} & \text{if } q \leq 1 \\ \text{converges} & \text{if } q > 1 \end{cases}$

31. $q = 3 > 1$ so $\sum_{k=2}^{\infty} \frac{1}{k \cdot (\ln k)^3}$ converges.

33. $\sum_{k=2}^{\infty} \frac{1}{k \cdot \ln(k^3)} = \sum_{k=2}^{\infty} \frac{1}{3k \cdot \ln(k)} = \frac{1}{3} \sum_{k=2}^{\infty} \frac{1}{k \cdot \ln(k)}$ which diverges ($q = 1$).

Section 10.4 Some Even Answers

2. Converge   4. Diverge   6. Converge   8. Converge   10. Converge

12. Converge   14. Converge   16. Converge   18. Diverge   20. Diverge

30. Diverge   32. Diverge

Section 10.5 (Odd numbered problem solutions followed by even numbered problem answers.)

1. $\sum_{k=1}^{\infty} \frac{\cos^2(k)}{k^2} \leq \sum_{k=1}^{\infty} \frac{1}{k^2}$ which converges by the P–Test (p=2) so $\sum_{k=1}^{\infty} \frac{\cos^2(k)}{k^2}$ converges.

3. $\sum_{n=3}^{\infty} \frac{5}{n-1} > 5 \sum_{n=3}^{\infty} \frac{1}{n}$ which is the harmonic series and is divergent, so $\sum_{n=3}^{\infty} \frac{5}{n-1}$ diverges.

5. $-1 \leq \cos(x) \leq 1$ so $2 \leq 3 + \cos(x) \leq 4$. Then $\sum_{j=1}^{\infty} \frac{3 + \cos(j)}{j} > 2 \sum_{j=1}^{\infty} \frac{1}{j}$ which is the harmonic series and is divergent, so $\sum_{j=1}^{\infty} \frac{3 + \cos(j)}{j}$ diverges.

7. $\sum_{k=1}^{\infty} \frac{\ln(k)}{k} > \sum_{k=1}^{\infty} \frac{1}{k}$ which is the harmonic series and is divergent, so $\sum_{k=1}^{\infty} \frac{\ln(k)}{k}$ diverges.

9. $\sum_{k=1}^{\infty} \frac{k+9}{k \cdot 2^k} = \sum_{k=1}^{\infty} \frac{k+9}{k} \cdot \frac{1}{2^k} < 10 \sum_{k=1}^{\infty} \frac{1}{2^k}$ which is a convergent geometric series ($r = \frac{1}{2}$)

so $\sum_{k=1}^{\infty} \frac{k+9}{k \cdot 2^k}$ converges.

11. $\sum_{n=1}^{\infty} \frac{1}{1+2+3+\ldots+(n-1)+n} = \sum_{n=1}^{\infty} \frac{1}{\frac{n(n+1)}{2}} = \sum_{n=1}^{\infty} \frac{2}{n(n+1)} < 2\sum_{n=1}^{\infty} \frac{1}{n^2}$ which converges by

    the P–Test (p = 2) so $\sum_{n=1}^{\infty} \frac{1}{1+2+3+\ldots+(n-1)+n}$ converges.

13. Let $a_k = \frac{k+1}{k^2+4}$ and $b_k = \frac{1}{k}$. Then $\frac{a_k}{b_k} = \frac{k^2+k}{k^2+4} \to 1$ and $\sum_{k=3}^{\infty} \frac{1}{k}$ diverges so $\sum_{k=3}^{\infty} \frac{k+1}{k^2+4}$ diverges.

15. Let $a_w = \frac{5}{w+1}$ and $b_w = \frac{5}{w}$. Then $\frac{a_w}{b_w} = \frac{w}{w+1} \to 1$ and $\sum_{w=1}^{\infty} \frac{1}{w}$ diverges so $\sum_{w=1}^{\infty} \frac{5}{w+1}$ diverges.

17. $\sum_{k=1}^{\infty} \frac{k^3}{(1+k^2)^3} = \sum_{k=1}^{\infty} \left(\frac{k}{1+k^2}\right)^3$. Let $a_k = \left(\frac{k}{1+k^2}\right)^3$ and $b_k = \frac{1}{k^3}$.

    Then $\frac{a_k}{b_k} = \left(\frac{k}{1+k^2}\right)^3 \cdot \frac{k^3}{1} = \left(\frac{k^2}{1+k^2}\right)^3 \to (1)^3 = 1$ and $\sum_{k=1}^{\infty} \frac{1}{k^3}$ converges by the

    P–Test so $\sum_{k=1}^{\infty} \frac{k^3}{(1+k^2)^3}$ converges.

19. $\sum_{n=1}^{\infty} \left(\frac{5-\frac{1}{n}}{n}\right)^3 = \sum_{n=1}^{\infty} \left(\frac{5}{n} - \frac{1}{n^2}\right)^3 = \sum_{n=1}^{\infty} \left(\frac{5n-1}{n^2}\right)^3$. Let $a_n = \left(\frac{5n-1}{n^2}\right)^3$ and $b_n = \frac{1}{n^3}$.

    Then $\frac{a_n}{b_n} = \left(\frac{5n-1}{n^2}\right)^3 \frac{n^3}{1} = \left(\frac{5n-1}{n^2} \cdot \frac{n}{1}\right)^3 = \left(\frac{5n-1}{n}\right)^3 \to 5^3 = 125$ (positive and finite).

    $\sum_{k=1}^{\infty} \frac{1}{k^3}$ converges by the P–Test so $\sum_{n=1}^{\infty} \left(\frac{5-\frac{1}{n}}{n}\right)^3$ converges.

21. $\sum_{j=1}^{\infty} \left(1 - \frac{1}{j}\right)^j$ diverges by the Nth Term Test for Divergence since $a_j = \left(1 - \frac{1}{j}\right)^j \to e^{-1} \neq 0$ as $j \to \infty$.

    We could use the Limit Comparison Test by taking $b_j = e^{-1}$ and showing that $\frac{a_j}{b_j} \to 1$, but the Nth Term Test for Divergence is more direct for this series.

23. $\sum_{k=1}^{\infty} \frac{7k}{\sqrt{k^3+5}}$: The dominant term series is $\sum_{k=1}^{\infty} \frac{k}{k^{3/2}} = \sum_{k=1}^{\infty} \frac{1}{k^{1/2}}$ which diverges by the P–Test (p=1/2).

25. $\sum_{j=1}^{\infty} \dfrac{j^3 - 4j + 3}{2j^4 + 7j^6 + 9}$: The dominant term series is $\sum_{j=1}^{\infty} \dfrac{j^3}{j^6} = \sum_{j=1}^{\infty} \dfrac{1}{j^3}$ which converges by the P–Test (p=3).

27. $\sum_{n=1}^{\infty} \left(\dfrac{\arctan(3n)}{2n}\right)^2$: The dominant term series is $\sum_{n=1}^{\infty}\left(\dfrac{\pi/2}{2n}\right)^2 = \dfrac{\pi^2}{16}\sum_{n=1}^{\infty}\dfrac{1}{n^2}$ which converges by the P–Test (p=2).

29. $\sum_{j=1}^{\infty} \dfrac{\sqrt{j^3 + 4j^2}}{j^2 + 3j - 2}$: The dominant term series is $\sum_{j=1}^{\infty} \dfrac{j^{3/2}}{j^2} = \sum_{j=1}^{\infty} \dfrac{1}{j^{1/2}}$ which diverges by the P–Test.

31. $\sum_{n=2}^{\infty} \dfrac{n^2 + 10}{n^3 - 2}$ diverges using dominant terms and the P–Test (p=1).

33. $\sum_{k=1}^{\infty} \dfrac{3}{2k + 1}$ diverges using dominant terms and the P–Test (p=1).

35. $\sum_{n=1}^{\infty} \dfrac{2n^3 + n^2 + 5}{(3 + n^2)^2}$ diverges using dominant terms and the P–Test (p=1).

37. $\sum_{k=1}^{\infty}\left(\dfrac{1 - \frac{2}{k}}{k}\right)^3 = \sum_{k=1}^{\infty}\left(\dfrac{k - 2}{k^2}\right)^3$ converges using dominant terms $\left(\dfrac{k}{k^2}\right)^3 = \left(\dfrac{1}{k}\right)^3$ and the P–Test (p=3).

39. $\sum_{k=1}^{\infty} \dfrac{k + 5}{k \cdot 3^k}$ converges by using dominant terms $\dfrac{k}{k \cdot 3^k} = \dfrac{1}{3^k} = \left(\dfrac{1}{3}\right)^k$ and the Geometric Series Test (r = 1/3).

41. $\sum_{k=1}^{\infty} \dfrac{k + 2}{\sqrt{k^2 + 1}}$ diverges using dominant terms and the Nth Term Test for Divergence.

43. $\sum_{j=1}^{\infty} \dfrac{3}{e^j + j}$ converges using dominant terms $\dfrac{1}{e^j} = \left(\dfrac{1}{e}\right)^j$ and the Geometric Series Test (r = 1/e).

45. $\sum_{n=1}^{\infty}\left(\dfrac{\tan(3)}{2 + n}\right)^2$ converges by using dominant terms $\dfrac{1}{n^2}$ and the P–Test (p=2).

47. $\sum_{k=1}^{\infty} \sin^3\left(\dfrac{1}{n}\right)$ converges by comparison with the convergent series $\sum_{n=1}^{\infty} \dfrac{1}{n^3}$

49. $\sum_{j=1}^{\infty} \cos^3(\frac{1}{n})$ diverges using the Nth Term Test for Divergence: the terms approach $1 \neq 0$.

51. $\sum_{n=1}^{\infty} (1 - \frac{2}{n})^n$ diverges using the Nth Term Test for Divergence: the terms approach $\frac{1}{e^2} \neq 0$.

Section 10.5 Some Even Answers

2. Converges	4. Converges	6. Converges	8. Converges	10. Diverges
12. Converges	14. Converges	16. Diverges	18. Converges	20. Diverges
22. Diverges	24. Diverges	26. Converges	28. Converges	30. Converges
32. Converges	34. Converges	36. Converges	38. Converges	40. Converges
42. Converges	44. Converges	46. Converges	48. Diverges	50. Converges

You still need to supply reasons for each answer given below.

R1. Converge	R2. Converge	R3. Diverge	R4. Diverge	R5. Converge
R6. Converge	R7. Converge	R8. Diverge	R9. Converge	R10. Diverge
R11. Converge	R12. Diverge	R13. Diverge	R14. Converge	R15. Converge
R16. Converge	R17. Diverge	R18. Diverge	R19. Diverge	R20. Converge
R21. Diverge				

Section 10.6 (Odd numbered problem solutions followed by some even numbered problem answers.)

1. $1, 0.2, 0.8, 0.4$ Alternating (so far). The graph of s_n is shown.

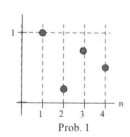
Prob. 1

3. $-1, 1, -2, 2$ Alternating (so far). The graph of s_n is shown.

5. $-1, -1.6, -1.2, -1$ Not alternating. The graph of s_n is shown.

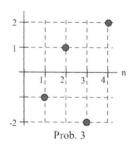
Prob. 3

7. Alternating: $a_1 = 2, a_2 = -1, a_3 = 2, a_4 = -1, a_5 = 2$

9. Not alternating: $a_1 = 2, a_2 = 1, a_3 = -0.9, a_4 = 0.8, a_5 = -0.1$

11. Not alternating: $a_1 = -1, a_2 = 2, a_3 = -1.2, a_4 = 0.2, a_5 = 0.2$

13. Graphs A and C are not the graphs of partial sums of alternating series.

15. Graph B is not the graph of an alternating series.

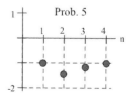
Prob. 5

17. Converges.   19. Converges.   21. Diverges.   23. Converges.

25. Converges.   27. Diverges.   29. Converges.   31. Converges.

33. $s_4 = -\frac{1}{\ln(2)} + \frac{1}{\ln(3)} - \frac{1}{\ln(4)} + \frac{1}{\ln(5)} \approx -0.63247$.

   $|s_4 - S| < \frac{1}{\ln(6)} \approx 0.55811$. $-1.19058 < S < -0.07436$

35. $s_4 = (-0.8)^2 + (-0.8)^3 + (-0.8)^4 + (-0.8)^5 \approx 0.20992$. $|s_4 - S| < (0.8)^6 \approx 0.26214$.

   $-0.05222 < S < 0.47206$. (For your information: $s_{10} \approx 0.317378$, $s_{50} \approx 0.355550$, $s_{100} \approx 0.355556$)

37. $s_4 = \sin(1) - \sin(\frac{1}{2}) + \sin(\frac{1}{3}) - \sin(\frac{1}{4}) \approx 0.441836$. $|s_4 - S| < \sin(\frac{1}{5}) \approx 0.198669$.

   $0.243167 < S < 0.640505$. ($s_{10} \approx 0.503356, s_{50} \approx 0.0.540897, s_{100} \approx 0.5458219$)

39. $s_4 = -1 + \frac{1}{8} - \frac{1}{27} + \frac{1}{64} \approx -0.896412$. $|s_4 - S| < \frac{1}{125} = 0.008$. $-0.904412 < S < -0.888412$.

   ($s_{10} \approx -0.901116, s_{50} \approx -0.901539, s_{100} \approx -0.901542$)

41. $|\frac{1}{(n+1)+6}| \leq 0.01$ so $n = 93$ works. Use s_{93}.

43. $|\frac{1}{\sqrt{n+1}}| \leq \frac{1}{100}$ so $\sqrt{n+1} \geq 100$ and $n = 10{,}000 - 1$ works. Use s_{9999}.

45. $|\frac{1}{3^{n+1}}| \leq \frac{2}{1000}$ so $3^{n+1} \geq 500$ and $n = 5$ works. Use s_5.

47. $|\frac{1}{(n+1)^2}| \leq \frac{1}{1000}$ so $(n+1)^2 \geq 1000$ and $n = 31$ works. Use s_{31}.

49. $|\frac{1}{(n+1) + \ln(n+1)}| \leq \frac{4}{100}$ so $(n+1) + \ln(n+1) \geq 25$. Some "calculator experimentation" shows that $n = 21$ works. $(20+1) + \ln(20+1) \approx 24.04$ so $n = 20$ is too small. $(21+1) + \ln(21+1) \approx 25.09$ so $n = 21$ works. Use s_{21}.

51. (a) $S(0.3) = x - \frac{(0.3)^3}{2 \cdot 3} + \frac{(0.3)^5}{2 \cdot 3 \cdot 4 \cdot 5} - \frac{(0.3)^7}{2 \cdot 3 \cdot 4 \cdot 5 \cdot 6 \cdot 7} + \ldots + (-1)^n \frac{(0.3)^{2n+1}}{(2n+1)!} + \ldots$

   (b) $s_3 = (0.3) - \frac{(0.3)^3}{2 \cdot 3} + \frac{(0.3)^5}{2 \cdot 3 \cdot 4 \cdot 5} \approx 0.29552025$.

   (c) $|S - s_3| \leq \frac{(0.3)^7}{2 \cdot 3 \cdot 4 \cdot 5 \cdot 6 \cdot 7} \approx 0.000000043$. ($\sin(0.3) \approx 0.295520206661$)

53. (a) $S(0.1) = x - \frac{(0.1)^3}{2 \cdot 3} + \frac{(0.1)^5}{2 \cdot 3 \cdot 4 \cdot 5} - \frac{(0.1)^7}{2 \cdot 3 \cdot 4 \cdot 5 \cdot 6 \cdot 7} + \ldots + (-1)^n \frac{(0.1)^{2n+1}}{(2n+1)!} + \ldots$

   (b) $s_3 = (0.1) - \frac{(0.1)^3}{2 \cdot 3} + \frac{(0.1)^5}{2 \cdot 3 \cdot 4 \cdot 5} \approx 0.099833416667$.

   (c) $|S - s_3| \leq \frac{(0.1)^7}{2 \cdot 3 \cdot 4 \cdot 5 \cdot 6 \cdot 7} \approx 1.98 \cdot 10^{-11}$. ($\sin(0.1) \approx 0.099833416647$)

55. (a) $C(1) = 1 - \frac{1^2}{2} + \frac{1^4}{2 \cdot 3 \cdot 4} - \frac{1^6}{2 \cdot 3 \cdot 4 \cdot 5 \cdot 6} + \ldots + (-1)^n \frac{1^{2n}}{(2n)!} + \ldots = 1 - \frac{1}{2} + \frac{1}{24} - \frac{1}{720} + \ldots + (-1)^n \frac{1^{2n}}{(2n)!} + \ldots$

(b) $s_3 = 1 - \frac{1^2}{2} + \frac{1^4}{2 \cdot 3 \cdot 4} \approx 0.5416667$.

(c) $|S - s_3| \leq \frac{(1)^7}{2 \cdot 3 \cdot 4 \cdot 5 \cdot 6 \cdot 7} \approx 0.0013889\ldots$ ($\cos(1) \approx 0.5403023$)

57. (a) $C(-0.2) = 1 - \frac{(-0.2)^2}{2} + \frac{(-0.2)^4}{2 \cdot 3 \cdot 4} - \frac{(-0.2)^6}{2 \cdot 3 \cdot 4 \cdot 5 \cdot 6} + \ldots + (-1)^n \frac{1^{2n}}{(2n)!} + \ldots$

(b) $s_3 = 1 - \frac{(-0.2)^2}{2} + \frac{(-0.2)^4}{2 \cdot 3 \cdot 4} \approx 0.98006666667$.

(c) $|S - s_3| \leq \frac{(-0.2)^6}{2 \cdot 3 \cdot 4 \cdot 5 \cdot 6} \approx 8.9 \cdot 10^{-8} \ldots$ ($\cos(-0.2) \approx 0.980066577841$)

59. (a) $E(-1) = 1 + (-1) + \frac{(-1)^2}{2} + \frac{(-1)^3}{2 \cdot 3} + \frac{(-1)^4}{2 \cdot 3 \cdot 4} + \ldots + \frac{(-1)^n}{n!} + \ldots = 1 - 1 + \frac{1}{2} - \frac{1}{6} + \frac{1}{24} - \frac{1}{120} + \ldots$

(b) $s_3 = 1 - 1 + \frac{1}{2} = 0.5$.

(c) $|S - s_3| \leq \frac{1}{6} \approx 0.16667\ldots$ ($e^{-1} \approx 0.36787944$)

61. (a) $E(-0.2) = 1 + (-0.2) + \frac{(-0.2)^2}{2} + \frac{(-0.2)^3}{2 \cdot 3} + \frac{(-0.2)^4}{2 \cdot 3 \cdot 4} + \ldots + \frac{(-1)^n}{n!} + \ldots$

(b) $s_3 = 1 + (-0.2) + \frac{(-0.2)^2}{2} = 0.82$ (c) $|S - s_3| \leq \frac{(0.2)^3}{6} \approx 0.0013333\ldots$ ($e^{-0.2} \approx 0.8187307$)

Section 10.6 Some Even Answers

2. Alternating (so far)   4. Not alternating   6. Alternating (so far)

8. Alternating (so far)   10. Alternating (so far)   12. Not alternating

14. B and C.   16. A.   18. Converges.   20. Converges.

22. Converges.   24. Converges (to 0).   26. Diverges.   28. Converges.

30. Converges.   42. $n \approx 2.7 \cdot 10^{43}$ 44. $n = 26$ works.   46. $n = 16$ works.

48. $n = 21$ works.

Section 10.7 (Odd numbered problem solutions followed by some even numbered problem answers.)

1. Conditionally convergent   3. Absolutely convergent   5. Absolutely convergent

7. Absolutely convergent   9. Conditionally convergent   11. Conditionally convergent

13. Conditionally convergent   15. Absolutely convergent   17. Divergent

19. Conditionally convergent   21. Divergent   23. Absolutely convergent

25. Conditionally convergent   27. Divergent   29. Absolutely convergent

31. $\dfrac{n!}{(n+1)!} = \dfrac{1\cdot 2\cdot 3 \ldots \cdot n}{1\cdot 2\cdot 3 \ldots \cdot n\cdot (n+1)} = \dfrac{1}{n+1}$ 33. $\dfrac{n!}{(n+3)!} = \dfrac{1\cdot 2\cdot 3 \ldots \cdot n}{1\cdot 2\cdot 3 \ldots \cdot n\cdot (n+1)\cdot (n+2)\cdot (n+3)} = \dfrac{1}{(n+1)(n+2)(n+3)}$

35. $\dfrac{(n-1)!}{(n+1)!} = \dfrac{1\cdot 2\cdot 3 \ldots \cdot (n-1)}{1\cdot 2\cdot 3 \ldots \cdot (n-1)\cdot (n)\cdot (n+1)} = \dfrac{1}{(n)(n+1)}$

37. $\dfrac{(2n!)}{(2n+1)!} = \dfrac{1\cdot 2\cdot 3 \ldots \cdot n\cdot (n+1)\cdot \ldots \cdot (2n)}{1\cdot 2\cdot 3 \ldots \cdot n\cdot (n+1)\cdot \ldots \cdot (2n)\cdot (2n+1)} = \dfrac{1}{2n+1}$ 39. $\dfrac{n^n}{n!} = \dfrac{n\cdot n\cdot n\cdot n\cdot \ldots \cdot n\cdot n}{1\cdot 2\cdot 3\cdot 4\cdot \ldots \cdot (n-1)\cdot n}$

41. Ratio $= \left| \dfrac{n}{n+1} \right| \to 1 = L$. Diverges (Harmonic series)

43. Ratio $= \left| \left(\dfrac{n}{n+1}\right)^3 \right| \to 1 = L$. Converges by the P–Test (p = 3).

45. Ratio $= \left| \dfrac{1}{2} \right| \to \dfrac{1}{2} = L$. Converges (Geometric series with r = 1/2).

47. Ratio $= |\,1\,| \to 1 = L$. Diverges by the Nth Term Test for Divergence.

49. Ratio $= \left| \dfrac{n!}{(n+1)!} \right| = \left| \dfrac{1}{n+1} \right| \to 0 = L$. Converges by the Ratio Test.

51. Ratio $= \left| \dfrac{2^{n+1}}{(n+1)!} \dfrac{n!}{2^n} \right| = \left| \dfrac{2}{n+1} \right| \to 0 = L$. Converges by the Ratio Test.

53. Ratio $= \left| \dfrac{(1/2)^{3(n+1)}}{(1/2)^{3n}} \right| = \left| (\dfrac{1}{2})^3 \right| = \dfrac{1}{8} \to \dfrac{1}{8} = L$. Converges by the Ratio Test.

55. Ratio $= \left| \dfrac{(0.9)^{2(n+1)+1}}{(0.9)^{2n+1}} \right| = |(0.9)^2| = 0.81 \to 0.81 = L$. Converges by the Ratio Test.

57. Ratio $= \left| \dfrac{(x-5)^{n+1}}{(x-5)^n} \right| = |x-5| \to |x-5| = L$. Series converges absolutely if and only if $|x-5| < 1$: $4 < x < 6$.

59. Ratio $= \left| \dfrac{(x-5)^{n+1}}{(n+1)^2} \dfrac{n^2}{(x-5)^n} \right| = \left| (x-5)\cdot (\dfrac{n}{n+1})^2 \right| \to |x-5| = L$. Series converges absolutely if and only if $|x-5| < 1$: $4 < x < 6$.

61. Ratio $= \left| \dfrac{(x-2)^{n+1}}{(n+1)!} \dfrac{n!}{(x-2)^n} \right| = \left| (x-2)\cdot \dfrac{1}{n+1} \right| \to 0 = L$ for all values of x so the series converges absolutely for all values of x.

63. Ratio $= \left| \dfrac{(2x-12)^{n+1}}{(n+1)^2} \dfrac{n^2}{(2x-12)^n} \right| = \left| (2x-12)\cdot (\dfrac{n}{n+1})^2 \right| \to |2x-12| = L$. Series converges absolutely

if and only if $|2x - 12| < 1$: $\frac{11}{2} < x < \frac{13}{2}$.

65. Ratio = $\left| \frac{(6x-12)^{n+1}}{(n+1)!} \frac{n!}{(6x-12)^n} \right| = \left| (6x-12) \cdot \frac{1}{n+1} \right| \to 0 = L$ for all values of x so the series converges absolutely for all values of x.

67. Ratio = $\left| \frac{(x+1)^{2(n+1)}}{n+1} \frac{n}{(x+1)^{2n}} \right| = \left| (x+1)^2 \cdot \frac{n}{n+1} \right| \to (x+1)^2 = L$ for all values of x. Series converges absolutely if and only if $(x+1)^2 < 1$: $-2 < x < 0$.

69. Ratio = $\left| \frac{(x-5)^{3(n+1)+1}}{(n+1)^2} \frac{n^2}{(x-5)^{3n+1}} \right| = \left| (x-5)^3 \cdot \left(\frac{n}{n+1}\right)^2 \right| \to \left| (x-5)^3 \right| = L$ for all values of x. Series converges absolutely if and only if $\left| (x-5)^3 \right| < 1$: $4 < x < 6$.

71. Ratio = $\left| \frac{(x+3)^{2(n+1)-1}}{((n+1)+1)!} \frac{(n+1)!}{(x+3)^{2n-1}} \right| = \left| (x+3)^2 \cdot \frac{1}{n+2} \right| \to 0 = L$ for all values of x so the series converges absolutely for all values of x.

73. Ratio = $\left| \frac{x^{2(n+1)}}{(2(n+1))!} \frac{(2n)!}{x^{2n}} \right| = \left| x^2 \cdot \frac{1}{(2n+1)(2n+2)} \right| \to 0 = L$ for all values of x so the series converges absolutely for all values of x.

Section 10.7 Some Even Answers

2. Conditionally convergent
4. Conditionally convergent
6. Divergent
8. Absolutely convergent
10. Absolutely convergent
12. Conditionally convergent
14. Absolutely convergent
16. Conditionally convergent
18. Divergent
20. Absolutely convergent
22. Absolutely convergent
24. Absolutely convergent
26. Divergent
28. Divergent
30. Divergent

42. $L = 1$. Convergent by P–Test.
44. $L = 1$. Divergent by P–Test.
46. $L = 1/3$. Convergent Geo. series.
48. Divergent by N^{th} Term Test.
50. $L = 0$. Convergent by Ratio Test.
52. $L = 0$. Convergent by Ratio Test.
54. $L = (1/3)^2$. Convergent by Ratio Test.
56. $L = 0.16$. Convergent by Ratio Test.
58. Absolutely convergent for $4 < x < 6$.
60. Absolutely convergent for $1 < x < 3$.
62. Absolutely convergent for all x.
64. Absolutely convergent for $11/4 < x < 13/4$.
66. Absolutely convergent for $2 < x < 4$.
68. Absolutely convergent for $-3 < x < -1$.
70. Absolutely convergent for all x.
72. Absolutely convergent for all x.
74. Absolutely convergent for all x.

Section 10.8 (Odd numbered problem solutions)

1. Ratio test: $\left| \dfrac{a_{n+1}}{a_n} \right| = \left| \dfrac{x^{n+1}}{x^n} \right| = |x| \to |x| = L$. $|x| < 1$ if and only if $-1 < x < 1$.

   Endpoints: if $x = -1$ or $x = 1$, then the terms do not approach 0 so the series diverges.

   Interval of convergence: $-1 < x < 1$. (You can provide the graph of this interval.)

3. Ratio test: $\left| \dfrac{a_{n+1}}{a_n} \right| = \left| \dfrac{(x+2)^{n+1}}{(x+2)^n} \right| = |x+2| \to |x+2| = L$. $|x+2| < 1$ if and only if

   $-1 < x+2 < 1$ so $-3 < x < -1$.

   Endpoints: if $x = -3$ or $x = -1$, then the terms do not approach 0 so the series diverges.

   Interval of convergence: $-3 < x < -1$. (You can provide the graph of this interval.)

5. Ratio test: $\left| \dfrac{a_{n+1}}{a_n} \right| = \left| \dfrac{x^{n+1}/(n+1)}{x^n/n} \right| = \left| x \cdot \dfrac{n}{n+1} \right| \to |x| = L$. $|x| < 1$ if and only if $-1 < x < 1$.

   Endpoints: if $x = -1$, then $\sum_{n=1}^{\infty} \dfrac{x^n}{n} = \sum_{n=1}^{\infty} \dfrac{(-1)^n}{n}$ which converges by the Alternating Series Test

   if $x = 1$, then $\sum_{n=1}^{\infty} \dfrac{x^n}{n} = \sum_{n=1}^{\infty} \dfrac{(1)^n}{n}$ which diverges (harmonic series)

   Interval of convergence: $-1 \leq x < 1$. (You can provide the graph of this interval.)

7. Ratio test: $\left| \dfrac{a_{n+1}}{a_n} \right| = \left| \dfrac{(x+3)^{n+1}/(n+1)}{(x+3)^n/n} \right| = \left| (x+3) \cdot \dfrac{n}{n+1} \right| \to |x+3| = L$. $|x+3| < 1$ if and only if

   $-1 < x+3 < 1$ or $-4 < x < -2$.

   Endpoints: if $x = -4$, then $\sum_{n=1}^{\infty} \dfrac{(x+3)^n}{n} = \sum_{n=1}^{\infty} \dfrac{(-1)^n}{n}$ which converges by the Alternating Series Test

   if $x = -2$, then $\sum_{n=1}^{\infty} \dfrac{(x+3)^n}{n} = \sum_{n=1}^{\infty} \dfrac{(1)^n}{n}$ which diverges (harmonic series)

   Interval of convergence: $-4 \leq x < -2$. (You can provide the graph of this interval.)

9. Ratio test: $\left| \dfrac{a_{n+1}}{a_n} \right| = \left| \dfrac{(x-7)^{2(n+1)+1}/(n+1)^2}{(x-7)^{2n+1}/n^2} \right| = \left| (x-7)^2 \cdot \left(\dfrac{n}{n+1}\right)^2 \right| \to (x-7)^2 = L$. $(x-7)^2 < 1$ if

   and only if $-1 < x-7 < 1$ or $6 < x < 8$.

   Endpoints: if $x = 6$, then $\sum_{n=1}^{\infty} \dfrac{(x-7)^{2n+1}}{n^2} = \sum_{n=1}^{\infty} \dfrac{(-1)^{2n+1}}{n^2} = \sum_{n=1}^{\infty} \dfrac{-1}{n^2}$ which converges by the P–Test

   if $x = 8$, then $\sum_{n=1}^{\infty} \dfrac{(x-7)^{2n+1}}{n^2} = \sum_{n=1}^{\infty} \dfrac{(1)^{2n+1}}{n^2}$ which converges by the P–Test (p=2).

   Interval of convergence: $6 \leq x \leq 8$. (You can provide the graph of this interval.)

Chapter 10: Odd Answers   Contemporary Calculus   24

11. Ratio = $|2x| \to |2x| = L$. $|2x| < 1$ if and only if $-1 < 2x < 1$ or $-1/2 < x < 1/2$.

   Endpoints: if $x = -1/2$, then the series = $\sum_{n=1}^{\infty} (2 \cdot (\frac{-1}{2}))^n = \sum_{n=1}^{\infty} (-1)^n$ which diverges

   if $x = 1/2$, then the series = $\sum_{n=1}^{\infty} (2 \cdot (\frac{1}{2}))^n = \sum_{n=1}^{\infty} (1)^n$ which diverges

   Interval of convergence: $-1/2 < x < 1/2$. (You can provide the graph of this interval.)

13. Ratio = $|(\frac{x}{3})^2| \to |\frac{x^2}{9}| = L$. $|\frac{x^2}{9}| < 1$ if and only if $-1 < \frac{x^2}{9} < 1$ or $-3 < x < 3$.

   The series diverges at both endpoints, $x = -3$ and $x = 3$, so the interval of convergence is $-3 < x < 3$.

15. Ratio = $|2x - 6| \to |2x - 6| = L$. $|2x - 6| < 1$ if and only if $-1 < 2x - 6 < 1$ or $5/2 < x < 7/2$.

   The series diverges at both endpoints, $x = 5/2$ and $x = 7/2$, so the interval of convergence is $5/2 < x < 7/2$.

17. Ratio = $|\frac{n!}{(n+1)!} \frac{x^{n+1}}{x^n}| = |\frac{1}{n+1} \cdot x| \to 0 = L$, and $L < 1$ for all x so the interval of convergence is

   the entire real number line.

19. Ratio = $|\frac{(n+1)^2}{n^2} \frac{3^n}{3^{n+1}} \frac{x^{n+1}}{x^n}| = |(\frac{n+1}{n})^2 \cdot \frac{x}{3}| \to |\frac{x}{3}| = L$. $|\frac{x}{3}| < 1$ if and only if $-1 < \frac{x}{3} < 1$.

   The series diverges at both endpoints, $x = -3$ and $x = 3$, so the interval of convergence is $-3 < x < 3$.

21. Ratio = $|\frac{(n+1)!}{n!} \frac{x^{n+1}}{x^n}| = |(n+1) \cdot x| \to \infty > 1$ if $x \neq 0$ and $|(n+1) \cdot x| \to 0 < 1$ if $x = 0$.

   The series diverges for $x \neq 0$, and the series converges (and is boring) when $x = 0$. The "interval" of
   convergence is a single point: $\{0\}$.

23. Ratio = $|\frac{(n+1)!}{n!} \frac{(x-7)^{n+1}}{(x-7)^n}| = |(n+1) \cdot (x-7)| \to \infty > 1$ if $x \neq 7$ and $|(n+1) \cdot (x-7)| \to 0 < 1$ if $x = 7$.

   The series diverges for $x \neq 7$, and the series converges (and is boring) when $x = 7$. The "interval" of
   convergence is a single point: $\{7\}$.

25. Ratio = $|\frac{(x-a)^{n+1}}{(x-a)^n}| = |x - a| \to |x - a| = L$. $|x - a| < 1$ if and only if $a - 1 < x < a + 1$.

   The series diverges at both endpoints, $x = a-1$ and $x = a+1$, so the interval of convergence is $a-1 < x < a+1$.

27. Ratio = $|\frac{(x-a)^{n+1}/(n+1)}{(x-a)^n/n}| = |\frac{n}{n+1} (x-a)| \to |x - a| = L$. $|x - a| < 1$ if and only if $a - 1 < x < a + 1$.

   Endpoints: if $x = a-1$, then $\sum_{n=1}^{\infty} \frac{(x-a)^n}{n} = \sum_{n=1}^{\infty} \frac{(-1)^n}{n}$ which converges by the Alternating Series Test

   if $x = a+1$, then $\sum_{n=1}^{\infty} \frac{(x-a)^n}{n} = \sum_{n=1}^{\infty} \frac{(1)^n}{n}$ which diverges (harmonic series)

   Interval of convergence: $a-1 \leq x < a+1$.

29. Ratio = $\left|\dfrac{(ax)^{n+1}}{(ax)^n}\right| = |ax| \to |ax| = L$. $|ax| < 1$ if and only if $\dfrac{-1}{a} < x < \dfrac{1}{a}$.

The series diverges at both endpoints, $x = -1/a$ and $x = 1/a$, so the interval of convergence is $\dfrac{-1}{a} < x < \dfrac{1}{a}$.

31. Ratio = $\left|\dfrac{(ax-b)^{n+1}}{(ax-b)^n}\right| = |ax - b| \to |ax - b| = L$. $|ax - b| < 1$ if and only if $-1 < ax - b \leq 1$ or $\dfrac{b-1}{a} < x < \dfrac{b+1}{a}$.

The series diverges at both endpoints so the interval of convergence is $\dfrac{b-1}{a} < x < \dfrac{b+1}{a}$.

33. We can be certain the friend is wrong because the interval of convergence must be symmetric about the point $x = 4$, and the friend's interval, $1 < x < 9$, is not symmetric about $x = 4$.

35. $(5, 9), [1, 13], (-1, 15], [3, 11), [0, 14), x = 7$ are all possible intervals of convergence for the series.

37. $(0, \mathbf{2}), (-5, 7), [1, \mathbf{1}], (-3, 5], [-9, 11], [0, \mathbf{2}), x = \mathbf{1}$.

Note: There are many possible correct answers for problems 38 – 45.

39. $\displaystyle\sum_{n=1}^{\infty} \dfrac{1}{n}\left(\dfrac{x}{3}\right)^n$

41. $\displaystyle\sum_{n=1}^{\infty} \dfrac{(x-3)^n}{3^n} = \sum_{n=1}^{\infty} \left(\dfrac{x-3}{3}\right)^n$

43. $\displaystyle\sum_{n=1}^{\infty} \dfrac{1}{n}\left(\dfrac{x-5}{3}\right)^n$

45. $\displaystyle\sum_{n=1}^{\infty} n!\cdot(x-3)^n$

47. Interval of convergence: $2 < x < 4$. $\sum = \dfrac{1}{1 - (x-3)} = \dfrac{1}{4 - x}$.

49. Interval of convergence: $\dfrac{-1}{3} < x < \dfrac{1}{3}$. $\sum = \dfrac{1}{1 - 3x}$.

51. Interval of convergence: $-1 < x < 1$. $\sum = \dfrac{1}{1 - x^3}$.

53. Interval of convergence: $1 < x < 11$. $\sum = \dfrac{1}{1 - \dfrac{x-6}{5}} = \dfrac{5}{11 - x}$.

55. Interval of convergence: $-5 < x < 5$. $\sum = \dfrac{1}{1 - \dfrac{x}{5}} = \dfrac{5}{5 - x}$.

57. Interval of convergence: $-2 < x < 2$. $\sum = \dfrac{1}{1 - (x/2)^3} = \dfrac{8}{8 - x^3}$.

59. $\left|\dfrac{1}{3}\cos(x)\right| < 1$ **for all x**, so for all x the sum $\sum = \dfrac{1}{1 - \dfrac{1}{3}\cos(x)} = \dfrac{3}{3 - \cos(x)}$.

Section 10.9 (Odd numbered problem solutions)

1. $\dfrac{1}{1-x^4} = 1 + x^4 + x^8 + x^{12} + x^{16} + \ldots = \sum_{n=0}^{\infty} x^{4n}$

3. $\dfrac{1}{1+x^4} = 1 - x^4 + x^8 - x^{12} + x^{16} + \ldots = \sum_{n=0}^{\infty} (-1)^n x^{4n}$

5. $\dfrac{1}{5+x} = \dfrac{1}{5} \cdot \dfrac{1}{1+(x/5)} = \dfrac{1}{5}\left\{ 1 - \dfrac{x}{5} + \left(\dfrac{x}{5}\right)^2 - \left(\dfrac{x}{5}\right)^3 + \left(\dfrac{x}{5}\right)^4 + \ldots \right\} = \dfrac{1}{5} \sum_{n=0}^{\infty} (-1)^n \left(\dfrac{x}{5}\right)^n$

7. $\dfrac{x^2}{1+x^3} = x^2\left\{\dfrac{1}{1+x^3}\right\} = x^2\left\{ 1 - x^3 + x^6 - x^9 + x^{12} + \ldots \right\} = x^2 \sum_{n=0}^{\infty} (-1)^n x^{3n} = \sum_{n=0}^{\infty} (-1)^n x^{3n+2}$

9. $\ln(1+x^2) = x^2 - \dfrac{x^4}{2} + \dfrac{x^6}{3} - \dfrac{x^8}{4} + \ldots = \sum_{n=1}^{\infty} (-1)^{n+1} \dfrac{x^{2n}}{n}$

11. $\arctan(x^2) = x^2 - \dfrac{x^6}{3} + \dfrac{x^{10}}{5} - \dfrac{x^{14}}{7} + \dfrac{x^{18}}{9} - \ldots = \sum_{n=0}^{\infty} (-1)^n \dfrac{x^{2(2n+1)}}{2n+1} = \sum_{n=0}^{\infty} (-1)^n \dfrac{x^{4n+2}}{2n+1}$

13. $\dfrac{1}{(1-x^2)^2} = 1 + 2x^2 + 3x^4 + 4x^6 + 5x^8 + \ldots = \sum_{n=1}^{\infty} n \cdot x^{2(n-1)} = \sum_{n=1}^{\infty} n \cdot x^{2n-2}$

15. $\displaystyle\int_0^{0.5} \dfrac{1}{1-x^3}\, dx \approx \int_0^{0.5} 1 + x^3 + x^6 + x^9 + \ldots\, dx = x + \dfrac{x^4}{4} + \dfrac{x^7}{7} + \ldots \Big|_0^{0.5} \approx 0.516741$

17. $\displaystyle\int_0^{0.6} \ln(1+x)\, dx \approx \int_0^{0.6} x - \dfrac{x^2}{2} + \dfrac{x^3}{3} - \ldots\, dx = \dfrac{x^2}{2} - \dfrac{x^3}{2\cdot 3} + \dfrac{x^4}{3\cdot 4} - \ldots \Big|_0^{0.6} = 0.1548$

19. $\displaystyle\int_0^{0.3} \dfrac{1}{(1-x)^2}\, dx \approx \int_0^{0.3} 1 + 2x + 3x^2 + \ldots\, dx = x + x^2 + x^3 + \ldots \Big|_0^{0.3} = 0.417$

21. $\displaystyle\lim_{x\to 0} \dfrac{\arctan(x)}{x} = \lim_{x\to 0} \dfrac{1}{x}\left\{ x - \dfrac{x^3}{3} + \dfrac{x^5}{5} - \dfrac{x^7}{7} + \dfrac{x^9}{9} - \ldots \right\} = \lim_{x\to 0} 1 - \dfrac{x^2}{3} + \dfrac{x^4}{5} - \ldots = \mathbf{1}$

23. $\displaystyle\lim_{x\to 0} \dfrac{\ln(1+x)}{2x} = \lim_{x\to 0} \dfrac{1}{2x}\left\{ x - \dfrac{x^2}{2} + \dfrac{x^3}{3} - \dfrac{x^4}{4} + \ldots \right\} = \lim_{x\to 0} \dfrac{1}{2} - \dfrac{x}{4} + \dfrac{x^2}{6} - \dfrac{x^3}{8} + \ldots = \mathbf{\dfrac{1}{2}}$

25. $\displaystyle\lim_{x\to 0} \dfrac{\ln(1-x^2)}{3x} = \lim_{x\to 0} \dfrac{1}{3x}\left\{ -x^2 - \dfrac{x^4}{2} - \dfrac{x^6}{3} - \dfrac{x^8}{4} - \ldots \right\} = \lim_{x\to 0} -\dfrac{x}{3} - \dfrac{x^3}{6} - \dfrac{x^5}{9} - \ldots = \mathbf{0}$

27. $\frac{1}{1+x} = \sum_{n=0}^{\infty} (-1)^n x^n$. Using the Ratio Test, the ratio $= \left| \frac{x^{n+1}}{x^n} \right| = |x| \to |x| = L$. $|x| < 1$ if and only if $-1 < x < 1$.

The series diverges at the endpoints $x = -1$ and $x = 1$ so the interval of convergence is $-1 < x < 1$.

29. $\ln(1-x) = -\sum_{n=1}^{\infty} \frac{x^n}{n}$. Using the Ratio Test, the ratio $= \left| \frac{x^{n+1}}{x^n} \frac{n}{n+1} \right| = \left| x \cdot \frac{n}{n+1} \right| \to |x| = L$.

$|x| < 1$ if and only if $-1 < x < 1$. The series converges when $x = -1$ and diverges when $x = 1$ so the interval of convergence is $-1 \leq x < 1$.

31. $\arctan(x) = \sum_{n=0}^{\infty} (-1)^n \frac{x^{2n+1}}{2n+1}$.

Using the Ratio Test, the ratio $= \left| \frac{x^{2(n+1)+1}}{x^{2n+1}} \frac{2n+1}{2(n+1)+1} \right| = \left| x^2 \cdot \frac{2n+1}{2n+3} \right| \to |x^2| = L$. $|x^2| < 1$

if and only if $-1 < x^2 < 1$ so $-1 < x < 1$. The series converges when $x = -1$ and when $x = 1$ so the interval of convergence is $-1 \leq x \leq 1$.

33. $\sin(x^2) = (x^2) - \frac{(x^2)^3}{3!} + \frac{(x^2)^5}{5!} - \frac{(x^2)^7}{7!} + \ldots = x^2 - \frac{x^6}{3!} + \frac{x^{10}}{5!} - \frac{x^{14}}{7!} + \ldots$

$= \sum_{k=0}^{\infty} (-1)^k \frac{(x^2)^{2k+1}}{(2k+1)!} = \sum_{k=0}^{\infty} (-1)^k \frac{x^{4k+2}}{(2k+1)!}$

35. $e^{(-x^2)} = 1 + (-x^2) + \frac{(-x^2)^2}{2!} + \frac{(-x^2)^3}{3!} + \frac{(-x^2)^4}{4!} + \ldots = 1 - x^2 + \frac{x^4}{2!} - \frac{x^6}{3!} + \frac{x^8}{4!} + \ldots = \sum_{k=0}^{\infty} (-1)^k \frac{x^{2k}}{k!}$

37. $\cos(x) = D(\sin(x)) = D\left(x - \frac{x^3}{3!} + \frac{x^5}{5!} - \frac{x^7}{7!} + \ldots \right) = 1 - \frac{x^2}{2!} + \frac{x^4}{4!} - \frac{x^6}{6!} + \ldots = \sum_{k=0}^{\infty} (-1)^k \frac{x^{2k}}{(2k)!}$

39. Using the result of Problem 33,

$\int_0^1 \sin(x^2)\, dx = \int_0^1 x^2 - \frac{x^6}{3!} + \frac{x^{10}}{5!} - \frac{x^{14}}{7!} + \ldots\, dx = \left. \frac{x^3}{3} - \frac{x^7}{3! \cdot 7} + \frac{x^{11}}{5! \cdot 11} - \frac{x^{15}}{7! \cdot 15} + \ldots \right|_0^1$

$= \left\{ \frac{1}{3} - \frac{1}{3! \cdot 7} + \frac{1}{5! \cdot 11} - \frac{1}{7! \cdot 15} + \ldots \right\} - \{ 0 \}$.

41. $\lim_{x \to 0} \frac{\sin(x)}{x} = \lim_{x \to 0} \frac{x - \frac{x^3}{3!} + \frac{x^5}{5!} - \frac{x^7}{7!} + \ldots}{x} = \lim_{x \to 0} 1 - \frac{x^2}{3!} + \frac{x^4}{5!} - \frac{x^6}{7!} + \ldots = 1 + (\text{all 0s}) = 1$.

Section 10.10 (Odd numbered problem solutions)

1. $P(x) = 4x^2 - 5x + 7$
2. $P(x) = 2x^3 - \frac{5}{2}x^2 + 2x - 1$
3. $P(x) = 2(x-3)^2 + 5(x-3) - 2$

5. $\ln(1 + x) = x - \frac{x^2}{2} + \frac{x^3}{3} - \frac{x^4}{4} + \ldots$

7. $\arctan(x) = x - \frac{x^3}{3} + \ldots$

9. $\cos(x) = 1 - \frac{x^2}{2!} + \frac{x^4}{4!} - \frac{x^6}{6!} + \ldots$

11. $\sec(x) = 1 + \frac{x^2}{2!} + \frac{5x^4}{4!} + \ldots$ (The higher derivatives of $\sec(x)$ get "messy.")

13. Around $c = 1$, $\ln(x) = (x-1) - \frac{(x-1)^2}{2} + \frac{(x-1)^3}{3} - \frac{(x-1)^4}{4} + \ldots$

15. Around $c = \pi/2$, $\sin(x) = 1 - \frac{(x - \pi/2)^2}{2!} + \frac{(x - \pi/2)^4}{4!} - \frac{(x - \pi/2)^6}{6!} + \ldots$

17. Around $c = 9$, $\sqrt{x} = 3 + \frac{1}{2 \cdot 3}(x - 9) - \frac{1}{4 \cdot 3^3}\frac{(x-9)^2}{2!} + \frac{3}{8 \cdot 3^5}\frac{(x-9)^3}{3!} - \frac{15}{16 \cdot 3^7}\frac{(x-9)^4}{4!} + \ldots$

19. Using the first three nonzero terms for $\cos(x)$,

    $P(x) = 1 - \frac{x^2}{2!} + \frac{x^4}{4!}$. See the Cosine Table.

21. Using the first three nonzero terms for $\arctan(x)$,

    $P(x) = x - \frac{x^3}{3} + \frac{x^5}{5}$. See the Arctan Table.

x	cos(x)	P(x)
0.1	0.995004165	0.995004167
0.2	0.98006657	0.98006666
0.5	0.87758	0.87604
1	0.54030	0.54167
2	–0.4161	–0.3333

x	arctan(x)	P(x)
0.1	0.09966865	0.09966867
0.2	0.197396	0.197397
0.5	0.4636	0.4646
1	0.7854	0.8667
2	1.1071	5.7333

23. $\int \sin(x^2)\, dx = \int x^2 - \frac{x^6}{6} + \frac{x^{10}}{120}\, dx$

    $= \frac{x^3}{3} - \frac{x^7}{42} + \frac{x^{11}}{1320} + C$

    $\int \sin(x^3)\, dx = \int x^3 - \frac{x^9}{6} + \frac{x^{15}}{120}\, dx$

    $= \frac{x^4}{4} - \frac{x^{10}}{60} + \frac{x^{16}}{1920} + C$

25. $\int e^{(-x^2)}\, dx = \int 1 + (-x^2) + \frac{(-x^2)^2}{2}\, dx = \int 1 - x^2 + \frac{x^4}{2}\, dx = x - \frac{x^3}{3} + \frac{x^5}{10} + C$

    $\int e^{(-x^3)}\, dx = \int 1 + (-x^3) + \frac{(-x^3)^2}{2}\, dx = \int 1 - x^3 + \frac{x^6}{2}\, dx = x - \frac{x^4}{4} + \frac{x^7}{14} + C$

27. $\int x\sin(x)\,dx = \int x\{x - \frac{x^3}{6} + \frac{x^5}{120}\}\,dx = \int x^2 - \frac{x^4}{6} + \frac{x^6}{120}\,dx = \frac{x^3}{3} - \frac{x^5}{30} + \frac{x^7}{840} + C$

$\int x^2\sin(x)\,dx = \int x^2\{x - \frac{x^3}{6} + \frac{x^5}{120}\}\,dx = \int x^3 - \frac{x^5}{6} + \frac{x^7}{120}\,dx = \frac{x^4}{4} - \frac{x^6}{36} + \frac{x^8}{960} + C$

29. $\lim_{x\to 0} \frac{1 - \cos(x)}{x^2} = \lim_{x\to 0} \frac{1 - \{1 - \frac{x^2}{2!} + \frac{x^4}{4!} - \frac{x^6}{6!} + \ldots\}}{x^2}$

$= \lim_{x\to 0} \frac{\frac{x^2}{2!} - \frac{x^4}{4!} + \frac{x^6}{6!} - \ldots}{x^2} = \lim_{x\to 0} \frac{1}{2!} - \frac{x^2}{4!} + \frac{x^4}{6!} - \ldots = \mathbf{\frac{1}{2}}$.

31. $\lim_{x\to 0} \frac{1 - e^x}{x} = \lim_{x\to 0} \frac{1 - \{1 + x + \frac{x^2}{2!} + \frac{x^3}{3!} + \ldots\}}{x}$

$= \lim_{x\to 0} \frac{-x - \frac{x^2}{2!} - \frac{x^3}{3!} - \ldots}{x} = \lim_{x\to 0} -1 - \frac{x}{2!} - \frac{x^2}{3!} - \ldots = \mathbf{-1}$

33. $\lim_{x\to 0} \frac{\sin(x)}{x} = \lim_{x\to 0} \frac{x - \frac{x^3}{3!} + \frac{x^5}{5!} - \ldots}{x} = \lim_{x\to 0} 1 - \frac{x^2}{3!} + \frac{x^4}{5!} - \ldots = 1$

35. $\lim_{x\to 0} \frac{x - \frac{x^3}{6} - \sin(x)}{x^5} = \lim_{x\to 0} \frac{x - \frac{x^3}{6} - \{x - \frac{x^3}{3!} + \frac{x^5}{5!} - \frac{x^7}{7!} \ldots\}}{x^5}$

$= \lim_{x\to 0} \frac{-\frac{x^5}{5!} + \frac{x^7}{7!} \ldots}{x^5} = \lim_{x\to 0} -\frac{1}{5!} + \frac{x^2}{7!} \ldots = \mathbf{-\frac{1}{120}}$

37. $\sinh(x) = \frac{1}{2}(e^x - e^{-x}) = \frac{1}{2}(\{1 + x + \frac{x^2}{2!} + \frac{x^3}{3!} + \frac{x^4}{4!} + \ldots\} - \{1 - x + \frac{x^2}{2!} - \frac{x^3}{3!} + \frac{x^4}{4!} + \ldots\})$

$= \frac{1}{2}(2x + 2\frac{x^3}{3!} + 2\frac{x^5}{5!} + 2\frac{x^7}{7!} + \ldots) = x + \frac{x^3}{3!} + \frac{x^5}{5!} + \frac{x^7}{7!} + \ldots$

39. $\mathbf{D}(x + \frac{x^3}{3!} + \frac{x^5}{5!} + \frac{x^7}{7!} + \ldots) = 1 + 3\frac{x^2}{3!} + 5\frac{x^4}{5!} + 7\frac{x^6}{7!} + \ldots = 1 + \frac{x^2}{2!} + \frac{x^4}{4!} + \frac{x^6}{6!} + \ldots$

41. $e^{ix} = \cos(x) + i\cdot\sin(x)$. When $x = \pi/2$, then $e^{ix} = e^{i(\pi/2)} = \cos(\pi/2) + i\cdot\sin(\pi/2) = i$.

If $x = \pi$, then $e^{ix} = e^{\pi i} = \cos(\pi) + i\cdot\sin(\pi) = -1$. This result is often restated in the form

$e^{\pi i} + 1 = 0$, a formula that relates the 5 most common constants in all of mathematics, $e, \pi, i, 1$, and 0.

43. $\binom{3}{0} = 1$ (by the definition), $\binom{3}{1} = \dfrac{3!}{1! \cdot 2!} = \dfrac{3 \cdot 2 \cdot 1}{(1) \cdot (2 \cdot 1)} = 3$,

$\binom{3}{2} = \dfrac{3!}{2! \cdot 1!} = \dfrac{3 \cdot 2 \cdot 1}{(2 \cdot 1) \cdot (1)} = 3$, and $\binom{3}{3} = \dfrac{3!}{0! \cdot 3!} = \dfrac{3 \cdot 2 \cdot 1}{(1) \cdot (3 \cdot 2 \cdot 1)} = 1$.

45. The Maclaurin series for $(1+x)^{5/2}$ is

$$1 + \frac{5}{2}x + \left(\frac{5}{2}\right)\left(\frac{3}{2}\right)\frac{x^2}{2!} + \left(\frac{5}{2}\right)\left(\frac{3}{2}\right)\left(\frac{1}{2}\right)\frac{x^3}{3!} + \left(\frac{5}{2}\right)\left(\frac{3}{2}\right)\left(\frac{1}{2}\right)\left(\frac{-1}{2}\right)\frac{x^4}{4!} + \left(\frac{5}{2}\right)\left(\frac{3}{2}\right)\left(\frac{1}{2}\right)\left(\frac{-1}{2}\right)\left(\frac{-3}{2}\right)\frac{x^5}{5!} + $$

47. If $f(x) = (1+x)^m$, then $f(0) = 1$. $f'(x) = m(1+x)^{m-1}$ so $f'(0) = m$.

$f''(x) = m(m-1)(1+x)^{m-2}$ so $f''(0) = m(m-1)$. $f'''(x) = m(m-1)(m-2)(1+x)^{m-3}$

so $f'''(0) = m(m-1)(m-2)$. $f^{(4)}(x) = m(m-1)(m-2)(m-3)(1+x)^{m-4}$ so $f'''(0) = m(m-1)(m-2)(m-3)$.

And so on. Then the Maclaurin series for $(1+x)^m$ is

$$1 + mx + m(m-1)\frac{x^2}{2!} + m(m-1)(m-2)\frac{x^3}{3!} + m(m-1)(m-2)(m-3)\frac{x^4}{4!} + \ldots$$

Section 10.11 (Odd numbered problem solutions)

1. $f(x) = \sin(x)$, $c = 0$, $[-2, 4]$: $f'(x) = \cos(x)$, $f''(x) = -\sin(x)$, $f'''(x) = -\cos(x)$, $f^{(iv)}(x) = \sin(x)$ so
$a_0 = 0, a_1 = 1, a_2 = 0, a_3 = -1/6, a_4 = 0$. Then $P_0(x) = 0$, $P_1(x) = 0 + x = x$,
$P_2(x) = 0 + x + 0 = x$, $P_3(x) = 0 + x + 0 - x^3/6 = x - x^3/6$,
and $P_4(x) = 0 + x + 0 - x^3/6 + 0 = x - x^3/6$.

3. $f(x) = \ln(x)$, $c = 1$, $[0.1, 3]$: $f'(x) = 1/x$, $f''(x) = -1/x^2$, $f'''(x) = 2/x^3$, $f^{(iv)}(x) = -6/x^4$ so
(using $c = 1$) $a_0 = 0, a_1 = 1, a_2 = -1/2, a_3 = 1/3, a_4 = -1/4$. Then $P_0(x) = 0$, $P_1(x) = (x-1)$,
$P_2(x) = (x-1) - \frac{1}{2}(x-1)^2$, $P_3(x) = (x-1) - \frac{1}{2}(x-1)^2 + \frac{1}{3}(x-1)^3$, and
$P_4(x) = (x-1) - \frac{1}{2}(x-1)^2 + \frac{1}{3}(x-1)^3 - \frac{1}{4}(x-1)^4$.

5. $f(x) = \sqrt{x} = x^{1/2}$, $c = 1$, $[1, 3]$: $f'(x) = \frac{1}{2}x^{-1/2}$, $f''(x) = \frac{-1}{4}x^{-3/2}$, $f'''(x) = \frac{3}{8}x^{-5/2}$, and
$f^{(iv)}(x) = \frac{-15}{16}x^{-7/2}$ so $a_0 = 1, a_1 = \frac{1}{2}, a_2 = -\frac{1}{8}, a_3 = \frac{1}{16}, a_4 = \frac{-5}{128}$. Then $P_0(x) = 1$,
$P_1(x) = 1 + \frac{1}{2}(x-1)$, $P_2(x) = 1 + \frac{1}{2}(x-1) - \frac{1}{8}(x-1)^2$,
$P_3(x) = 1 + \frac{1}{2}(x-1) - \frac{1}{8}(x-1)^2 + \frac{1}{16}(x-1)^3$, and
$P_4(x) = 1 + \frac{1}{2}(x-1) - \frac{1}{8}(x-1)^2 + \frac{1}{16}(x-1)^3 - \frac{5}{128}(x-1)^4$.

7. $f(x) = (1+x)^{-1/2}$, $c = 0$, $[-2, 3]$: $f'(x) = \frac{-1}{2}(1+x)^{-3/2}$, $f''(x) = \frac{3}{4}(1+x)^{-5/2}$, $f'''(x) = \frac{-15}{8}(1+x)^{-7/2}$, $f^{(iv)}(x) = \frac{105}{16}(1+x)^{-9/2}$ so $a_0 = 1$, $a_1 = \frac{-1}{2}$, $a_2 = \frac{3}{8}$, $a_3 = \frac{-5}{16}$, and $a_4 = \frac{35}{128}$. Then $P_0(x) = 1$, $P_1(x) = 1 - \frac{1}{2}x$, $P_2(x) = 1 - \frac{1}{2}x + \frac{3}{8}x^2$, $P_3(x) = 1 - \frac{1}{2}x + \frac{3}{8}x^2 - \frac{5}{16}x^3$, and $P_4(x) = 1 - \frac{1}{2}x + \frac{3}{8}x^2 - \frac{5}{16}x^3 + \frac{35}{128}x^4$.

9. $f(x) = \sin(x)$, $c = \pi/2$, $[-1, 5]$: $f'(x) = \cos(x)$, $f''(x) = -\sin(x)$, $f'''(x) = -\cos(x)$, $f^{(iv)}(x) = \sin(x)$ so (using $c = \pi/2$) $a_0 = 1$, $a_1 = 0$, $a_2 = -1/2$, $a_3 = 0$, $a_4 = 1/4!$. Then $P_0(x) = 1$, $P_1(x) = 1$, $P_2(x) = 1 - \frac{1}{2}(x - \frac{\pi}{2})^2$, $P_3(x) = 1 - \frac{1}{2}(x - \frac{\pi}{2})^2$, and $P_4(x) = 1 - \frac{1}{2}(x - \frac{\pi}{2})^2 + \frac{1}{4!}(x - \frac{\pi}{2})^4$.

11. $f(x) = \sin(x)$, $c = 0$, $n = 5$, $-\pi/2 \leq x \leq \pi/2$: $R_5(x) = \frac{f^{(6)}(z)}{6!}(x-c)^6 = \frac{f^{(6)}(z)}{720}x^6$ for some z between x and 0.

$|R_5(x)| \leq \frac{|-\sin(z)|}{720}|x|^6 \leq \frac{1}{720}(\frac{\pi}{2})^6 \approx 0.02086$.

13. same as problem 11 except $-\pi \leq x \leq \pi$: $R_5(x) = \frac{f^{(6)}(z)}{6!}(x-c)^6 = \frac{f^{(6)}(z)}{720}x^6$ for some z between x and 0. $|R_5(x)| \leq \frac{|-\sin(z)|}{720}|x|^6 \leq \frac{1}{720}(\pi)^6 \approx 1.335$.

15. $f(x) = \cos(x)$, $c = 0$, $n = 10$, $-1 \leq x \leq 2$: $R_{10}(x) = \frac{f^{(11)}(z)}{11!}(x-c)^{11} = \frac{\sin(z)}{11!}x^{11}$ for some z between x and 0. $|R_{10}(x)| = \frac{|\sin(z)|}{11!}x^{11} \leq \frac{1}{11!}(2)^{11} = \frac{2048}{39916800} \approx 0.000051$.

Since the only nonzero terms of the power series for $f(x) = \cos(x)$ (about $c = 0$) are the even powers of x, P_{10} is the same as P_{11} so we could use the error term for P_{11} instead of the one for P_{10}. The advantage of using the P_{11} error term is that we have a larger factorial in the denominator of R_{11} and get a smaller bound on the error of the approximation.

Taking $n = 11$, $-1 \leq x \leq 2$: $R_{11}(x) = \frac{f^{(12)}(z)}{12!}(x-c)^{12} = \frac{|\cos(z)|}{12!}x^{12}$ for some z between x and 0. $|R_{11}(x)| = \frac{|\cos(z)|}{12!}x^{12} \leq \frac{1}{12!}(2)^{12} = \frac{4096}{479001600} \approx 0.00000855$.

17. $f(x) = e^x$, $c = 0$, $n = 6$, $-1 \leq x \leq 2$: $R_6(x) = \frac{f^{(7)}(z)}{7!}(x-c)^7 = \frac{f^{(7)}(z)}{5040}x^7$ for some z between x and 0. Using the estimate $e < 3$, $|R_6(x)| = \frac{|e^z|}{7!}|x|^7 < \frac{3^2}{7!}(2)^7 = \frac{1152}{5040} \approx 0.2286$.

19. $f(x) = \sin(x), c = 0, [-1, 1], E = $ "error" $= 0.001$. The derivatives of $\sin(x)$ are $\pm\sin(x)$ or $\pm\cos(x)$ and all of them have maximum absolute value less than or equal to 1.

$|R_n(x)| = \frac{|f^{(n+1)}(z)|}{(n+1)!}|x-c|^{n+1} \leq \frac{1}{(n+1)!}|1|^{n+1} = \frac{1}{(n+1)!}$ so we want to find a value of n that makes $\frac{1}{(n+1)!} \leq 0.001$. It is difficult to solve factorial equations algebraically, but a "calculator investigation" shows that $\frac{1}{(5+1)!} = \frac{1}{720} \approx 0.00139 > 0.001$, and $\frac{1}{(6+1)!} = \frac{1}{5040} \approx 0.000198 < 0.001$ so we should **use n = 6**. That means we need to use the 3 terms involving x, x^3 and x^5.

21. $f(x) = \sin(x), c = 0, [-1.6, 1.6], E = $ "error" $= 0.00001$. The derivatives of $\sin(x)$ are $\pm\sin(x)$ or $\pm\cos(x)$ and all of them have maximum absolute value less than or equal to 1.

$|R_n(x)| = \frac{|f^{(n+1)}(z)|}{(n+1)!}|x-c|^{n+1} \leq \frac{1}{(n+1)!}|1.6|^{n+1} = \frac{1}{(n+1)!}(1.6)^{n+1}$ so we want to find a value of n that makes $\frac{1}{(n+1)!}(1.6)^{n+1} \leq 0.00001$. Some calculator investigation shows that $\frac{1}{(9+1)!}(1.6)^{9+1} \approx 0.00003 > 0.00001$ and $\frac{1}{(10+1)!}(1.6)^{10+1} \approx 0.0000044 < 0.00001$ so we should **use n = 10**. That means we need to use the 5 terms involving x, x^3, x^5, x^7 and x^9.

23. $f(x) = e^x, c = 0, [0, 2], E = $ "error" $= 0.001$. The derivatives of e^x are all e^x.

$|R_n(x)| = \frac{|f^{(n+1)}(z)|}{(n+1)!}|x-c|^{n+1} \leq \frac{|e^z|}{(n+1)!}|x|^{n+1} \leq \frac{3^2}{(n+1)!}(2)^{n+1}$ (using $e < 3$).

Some calulator investigation shows $\frac{3^2}{(9+1)!}(2)^{9+1} \approx 0.0025 > 0.001$ and $\frac{3^2}{(10+1)!}(2)^{10+1} \approx 0.00046 < 0.001$ so we should **use n = 10**. (Using a better upper bound for the value of e such as 2.75 or 2.72 does not change the conclusion: use n = 10.)

25. (a) $4\{1 - \frac{1}{3} + \frac{1}{5} - \frac{1}{7} + \frac{1}{9}\} \approx 4\{0.83492063\} = 3.33968252$ compared with $\pi \approx 3.14159265$ from a calculator.

(b) Let $A = 4\{1 - \frac{1}{3} + \frac{1}{5} - \frac{1}{7} + ... - \frac{1}{99}\}$. Then $|A - \pi| < |51^{st}$ term $| = \frac{4}{101} \approx 0.0396$.

(c) Let $B = 4\{1 - \frac{1}{3} + \frac{1}{5} - \frac{1}{7} + ... \pm \frac{1}{2n-1}\}$.

Then $|B - \pi| < |$ next term $| = \frac{4}{2(n+1)-1} = \frac{4}{2n+1}$. We want $\frac{4}{2n+1} \leq 0.0001$ so $n \geq 1999.5$. Take **n = 2000 terms** to get the precision we want.

27. (a) A: Put $C = 16 \cdot \arctan(\frac{1}{5}) - 4 \cdot \arctan(\frac{1}{239})$

$= 16\{ \frac{1}{5} - (\frac{1}{3})(\frac{1}{5})^3 + (\frac{1}{5})(\frac{1}{5})^5 \} - 4\{ \frac{1}{239} - (\frac{1}{3})(\frac{1}{239})^3 + (\frac{1}{5})(\frac{1}{239})^5 \}$

$\approx 16\{ 0.197397333333 \} - 4\{ 0.004184076002 \} = 3.14162102879$.

($C - \pi \approx 0.0000284$.)

(b) Formula A converges more rapidly than Methods I and II because we are using smaller values of x, x = 1/5 and x = 1/239, and the powers of these smaller values of x approach 0 more quickly than the values of x, x = 1 and x = 1/2 and x = 1/3, used in Methods I and II.

TRIGONOMETRY FACTS

Appendix A

Right Angle Trigonometry

$$\sin(\theta) = \frac{opp}{hyp} \quad \cos(\theta) = \frac{adj}{hyp}$$

$$\tan(\theta) = \frac{opp}{adj} \quad \cot(\theta) = \frac{adj}{opp} \quad \sec(\theta) = \frac{hyp}{adj} \quad \csc(\theta) = \frac{hyp}{opp}$$

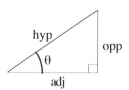

Trigonometric Functions

$$\sin(\theta) = \frac{y}{r} \quad \cos(\theta) = \frac{x}{r}$$

$$\tan(\theta) = \frac{y}{x} \quad \cot(\theta) = \frac{x}{y} \quad \sec(\theta) = \frac{r}{x} \quad \csc(\theta) = \frac{x}{y}$$

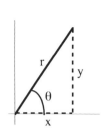

Fundamental Identities

$$\tan(\theta) = \frac{\sin(\theta)}{\cos(\theta)} \quad \cot(\theta) = \frac{\cos(\theta)}{\sin(\theta)} \quad \sec(\theta) = \frac{1}{\cos(\theta)}$$

$$\csc(\theta) = \frac{1}{\sin(\theta)}$$

$$\sin^2(\theta) + \cos^2(\theta) = 1 \qquad \tan^2(\theta) + 1 = \sec^2(\theta) \qquad 1 + \cot^2(\theta) = \csc^2(\theta)$$

$$\sin(-\theta) = -\sin(\theta) \qquad \cos(-\theta) = \cos(\theta) \qquad \tan(-\theta) = -\tan(\theta)$$

$$\sin(\tfrac{\pi}{2} - \theta) = \cos(\theta) \qquad \cos(\tfrac{\pi}{2} - \theta) = \sin(\theta) \qquad \tan(\tfrac{\pi}{2} - \theta) = \cot(\theta)$$

Law of Sines: $\dfrac{\sin(A)}{a} = \dfrac{\sin(B)}{b} = \dfrac{\sin(C)}{c}$

Law of Cosines: $a^2 = b^2 + c^2 - 2bc \cdot \cos(A)$

$$b^2 = a^2 + c^2 - 2ac \cdot \cos(B)$$

$$c^2 = a^2 + b^2 - 2ab \cdot \cos(C)$$

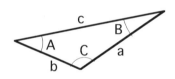

Angle Addition and Subtraction Formulas

$$\sin(x+y) = \sin(x)\cos(y) + \cos(x)\sin(y) \qquad \sin(x-y) = \sin(x)\cos(y) - \cos(x)\sin(y)$$

$$\cos(x+y) = \cos(x)\cos(y) - \sin(x)\sin(y) \qquad \cos(x-y) = \cos(x)\cos(y) + \sin(x)\sin(y)$$

$$\tan(x+y) = \frac{\tan(x) + \tan(y)}{1 - \tan(x)\tan(y)} \qquad \tan(x-y) = \frac{\tan(x) - \tan(y)}{1 + \tan(x)\tan(y)}$$

Function Product Formulas

$$\sin(x)\sin(y) = \tfrac{1}{2}\cos(x-y) - \tfrac{1}{2}\cos(x+y)$$

$$\cos(x)\cos(y) = \tfrac{1}{2}\cos(x-y) + \tfrac{1}{2}\cos(x+y)$$

$$\sin(x)\cos(y) = \tfrac{1}{2}\sin(x+y) + \tfrac{1}{2}\sin(x-y)$$

Function Sum Formulas

$$\sin(x) + \text{ain}(y) = 2\sin(\tfrac{x+y}{2})\cos(\tfrac{x-y}{2})$$

$$\cos(x) + \cos(y) = 2\cos(\tfrac{x+y}{2})\cos(\tfrac{x-y}{2})$$

$$\tan(x) + \tan(y) = \frac{\sin(x+y)}{\cos(x)\cos(y)}$$

Double Angle Formulas

$$\sin(2x) = 2\sin(x)\cos(x)$$

$$\cos(2x) = \cos^2(x) - \sin^2(x) = 2\cos^2(x) - 1$$

$$\tan(2x) = \frac{2\tan(x)}{1 - \tan^2(x)}$$

Half Angle Formulas

$$\sin(\tfrac{x}{2}) = \pm\sqrt{\tfrac{1}{2}(1 - \cos(x))}$$

$$\cos(\tfrac{x}{2}) = \pm\sqrt{\tfrac{1}{2}(1 + \cos(x))}$$

$$\tan(\tfrac{x}{2}) = \frac{1 - \cos(x)}{\sin(x)}$$

± depends on quadrant of x/2

Calculus Reference Facts

Derivatives

Notation: $D(f) = Df$ represents the derivative with respect to x of the function $f(x)$

$$D(f) = Df = \frac{d}{dx} f(x) = f'(x)$$

$D(k) = 0$ k a constant $D(f + g) = Df + Dg$ $D(f \cdot g) = f \cdot Dg + g \cdot Df$

$D(k \cdot f) = k \cdot Df$ k a constant $D(f - g) = Df - Dg$ $D(f/g) = \dfrac{g \cdot Df - f \cdot Dg}{g^2}$

$D(f^n) = n\, f^{n-1} \cdot Df$

$D(f(g(x))) = f'(g(x)) \cdot Dg$ (Chain Rule)

$D(\sin(f)) = \cos(f) \cdot Df$ $D(\tan(f)) = \sec^2(f) \cdot Df$ $D(\sec(f)) = \sec(f) \tan(f) \cdot Df$

$D(\cos(f)) = -\sin(f) \cdot Df$ $D(\cot(f)) = -\csc^2(f) \cdot Df$ $D(\csc(f)) = -\csc(f) \cot(f) \cdot Df$

$D(e^f) = e^f \cdot Df$ $D(a^f) = a^f \ln(a) \cdot Df$

$D(\ln|f|) = \dfrac{1}{f} \cdot Df$ Logarithmic Differentiation: $Df = f \cdot D(\ln|f|)$

$D(\log_a |f|) = \dfrac{1}{f \cdot \ln(a)} \cdot Df$

$D(\arcsin(f)) = \dfrac{1}{\sqrt{1-f^2}} \cdot Df$ $D(\arctan(f)) = \dfrac{1}{1+f^2} \cdot Df$ $D(\text{arcsec}(f)) = \dfrac{1}{|f|\sqrt{f^2-1}} \cdot Df$

$D(\arccos(f)) = \dfrac{-1}{\sqrt{1-f^2}} \cdot Df$ $D(\text{arccot}(f)) = \dfrac{-1}{1+f^2} \cdot Df$ $D(\text{arccsc}(f)) = \dfrac{-1}{|f|\sqrt{f^2-1}} \cdot Df$

Integrals — General Properties

$\displaystyle\int k \cdot f\, dx = k \cdot \int f\, dx$ $\displaystyle\int (f - g)\, dx = \int f\, dx - \int g\, dx$

$\displaystyle\int (f + g)\, dx = \int f\, dx + \int g\, dx$ $\displaystyle\int f \cdot (g')\, dx = f \cdot g - \int g \cdot (f')\, dx$

Calculus Reference Facts

Integrals — Particular Functions

1. $\int k\,dx = k\cdot x + C$

2. $\int x^n\,dx = \frac{1}{n+1}\cdot x^{n+1} + C \quad \text{if } n \ne -1$ $\qquad \int x^{-1}\,dx = \int \frac{1}{x}\,dx = \ln|x| + C$

3. $\int (ax+b)^n\,dx = \frac{1}{n+1}\cdot\frac{1}{a}\cdot(ax+b)^{n+1} + C \quad n \ne -1$

   (a) $\int (ax+b)^{-1}\,dx = \int \frac{1}{ax+b}\,dx = \frac{1}{a}\cdot\ln|ax+b| + C$

   (b) $\int \sqrt{ax+b}\,dx = \frac{2}{3a}\cdot(ax+b)^{3/2} + C$

   (c) $\int \frac{1}{\sqrt{ax+b}}\,dx = \frac{2}{a}\cdot(ax+b)^{1/2} + C$

4. $\int \frac{1}{x(ax+b)}\,dx = \frac{1}{b}\cdot\ln\left|\frac{x}{ax+b}\right| + C$

5. $\int \frac{1}{(x+a)(x+b)}\,dx = \frac{1}{b-a}\{\ln|x+a| - \ln|x+b|\} + C = \frac{1}{b-a}\ln\left|\frac{x+a}{x+b}\right| + C \quad a \ne b$

   (a) $\int \frac{1}{(x+a)(x+a)}\,dx = \int \frac{1}{(x+a)^2}\,dx = -\frac{1}{x+a} + C$

6. $\int x(ax+b)^n\,dx = \frac{(ax+b)^{n+1}}{a}\cdot\left\{\frac{ax+b}{n+2} - \frac{b}{n+1}\right\} + C \quad n \ne -1, -2$

   (a) $\int x(ax+b)^{-1}\,dx = \int \frac{x}{ax+b}\,dx = \frac{x}{a} - \frac{b}{a^2}\cdot\ln|ax+b| + C$

   (b) $\int x(ax+b)^{-2}\,dx = \int \frac{x}{(ax+b)^2}\,dx = \frac{1}{a^2}\cdot\left\{\ln|ax+b| + \frac{b}{ax+b}\right\} + C$

7. $\int \sin(ax)\,dx = -\frac{1}{a}\cos(ax) + C$ $\qquad$ 8. $\int \cos(ax)\,dx = \frac{1}{a}\sin(ax) + C$

9. $\int \tan(ax)\,dx = \int \frac{\sin(ax)}{\cos(ax)}\,dx = -\frac{1}{a}\ln|\cos(ax)| + C = \ln|\sec(ax)| + C$

10. $\int \cot(ax)\,dx = \int \frac{\cos(ax)}{\sin(ax)}\,dx = \frac{1}{a}\ln|\sin(x)| + C$

11. $\int \sec(ax)\,dx = \frac{1}{a}\ln|\sec(ax) + \tan(ax)| + C$ $\qquad$ 12. $\int \csc(ax)\,dx = \frac{1}{a}\ln|\csc(ax) - \cot(ax)| + C$

13. $\int \sin^2(ax)\,dx = \frac{x}{2} - \frac{\sin(2ax)}{4a} + C = \frac{x}{2} - \frac{\sin(ax)\cos(ax)}{2a} + C$

14. $\int \cos^2(ax)\,dx = \dfrac{x}{2} + \dfrac{\sin(2ax)}{4a} + C = \dfrac{x}{2} + \dfrac{\sin(ax)\cos(ax)}{2a} + C$

15. $\int \tan^2(ax)\,dx = \dfrac{1}{a}\tan(ax) - x + C$
16. $\int \cot^2(ax)\,dx = -\dfrac{1}{a}\cot(ax) - x + C$

17. $\int \sec^2(ax)\,dx = \dfrac{1}{a}\tan(ax) + C$
18. $\int \csc^2(ax)\,dx = -\dfrac{1}{a}\cot(ax) + C$

19. $\int \sin^n(ax)\,dx = \dfrac{-\sin^{n-1}(ax)\cos(ax)}{na} + \dfrac{n-1}{n}\int \sin^{n-2}(ax)\,dx$

   (a) $\int \sin^3(ax)\,dx = \dfrac{-\sin^2(ax)\cos(ax)}{3a} - \dfrac{2}{3a}\cos(ax) + C$

   (b) $\int \sin^4(ax)\,dx = \dfrac{-\sin^3(ax)\cos(ax)}{4a} + \dfrac{3x}{8} - \dfrac{3\sin(ax)\cos(ax)}{8a} + C$

20. $\int \cos^n(ax)\,dx = \dfrac{\cos^{n-1}(ax)\sin(ax)}{na} + \dfrac{n-1}{n}\int \cos^{n-2}(ax)\,dx$

21. $\int \tan^n(ax)\,dx = \dfrac{\tan^{n-1}(ax)}{(n-1)a} - \int \tan^{n-2}(ax)\,dx \qquad n \ne 1$

   (a) $\int \tan^3(ax)\,dx = \dfrac{1}{2a}\tan^2(ax) + \dfrac{1}{a}\ln|\cos(ax)| + C$

   (b) $\int \tan^4(ax)\,dx = \dfrac{1}{3a}\tan^3(ax) - \dfrac{1}{a}\tan(ax) + C$

22. $\int \cot^n(ax)\,dx = -\dfrac{\cot^{n-1}(ax)}{n-1} - \int \cot^{n-2}(ax)\,dx \qquad n \ne 1$

23. $\int \sec^n(ax)\,dx = \dfrac{\sec^{n-2}(ax)\tan(ax)}{(n-1)a} + \dfrac{n-2}{n-1}\int \sec^{n-2}(ax)\,dx \quad n \ne 1$

   (a) $\int \sec^3(ax)\,dx = \dfrac{1}{2a}\sec(ax)\tan(ax) + \dfrac{1}{2a}\ln|\sec(ax) + \tan(ax)| + C$

   (b) $\int \sec^4(ax)\,dx = \dfrac{1}{3a}\sec^2(ax)\tan(ax) + \dfrac{2}{3a}\tan(ax) + C$

24. $\int \csc^n(ax)\,dx = -\dfrac{\csc^{n-2}(ax)\cot(ax)}{(n-1)a} + \dfrac{n-2}{n-1}\int \csc^{n-2}(ax)\,dx \quad n \ne 1$

25. $\int \sin(ax)\sin(bx)\,dx = \dfrac{\sin((a-b)x)}{2(a-b)} - \dfrac{\sin((a+b)x)}{2(a+b)} + C \qquad a^2 \ne b^2$

26. $\int \cos(ax)\cos(bx)\,dx = \dfrac{\sin((a-b)x)}{2(a-b)} + \dfrac{\sin((a+b)x)}{2(a+b)} + C \qquad a^2 \ne b^2$

27. $\int \sin(ax)\cos(bx)\,dx = \dfrac{-\cos((a+b)x)}{2(a+b)} - \dfrac{\cos((a-b)x)}{2(a-b)} + C \qquad a^2 \ne b^2$

28. $\int x^n \sin(ax)\, dx = -\frac{1}{a} x^n \cos(ax) + \frac{n}{a^2} \int x^{n-1} \cos(ax)\, dx$

29. $\int x^n \cos(ax)\, dx = \frac{1}{a} x^n \sin(ax) - \frac{n}{a^2} \int x^{n-1} \sin(ax)\, dx$

30. $\int e^{ax}\, dx = \frac{1}{a} e^{ax} + C$
31. $\int b^{ax}\, dx = \frac{b^{ax}}{a \cdot \ln(b)} + C$

32. $\int x^n e^{ax}\, dx = \frac{1}{a} x^n e^{ax} - \frac{n}{a} \int x^{n-1} e^{ax}\, dx + C$

(a) $\int x\, e^{ax}\, dx = \frac{1}{a} x\, e^{ax} - \frac{1}{a^2} e^{ax} = \frac{e^{ax}}{a^2}(ax - 1) + C$

(b) $\int x^2 e^{ax}\, dx = \frac{1}{a} x^2 e^{ax} - \frac{2}{a^2} x\, e^{ax} + \frac{2}{a^3} e^{ax} = \frac{e^{ax}}{a^3}(a^2 x^2 - 2ax + 2) + C$

33. $\int x^n b^{ax}\, dx = \frac{1}{\ln(b)} \left\{ \frac{1}{a} x^n b^{ax} - \frac{n}{a} \int x^{n-1} b^{ax}\, dx \right\} + C$

34. $\int \frac{1}{\sqrt{a^2 - x^2}}\, dx = \arcsin\left(\frac{x}{a}\right) + C$
35. $\int \frac{1}{a^2 + x^2}\, dx = \frac{1}{a} \arctan\left(\frac{x}{a}\right) + C$

36. $\int \frac{1}{|x|\sqrt{x^2 - a^2}}\, dx = \frac{1}{a} \operatorname{arcsec}\left(\frac{x}{a}\right) + C$
37. $\int \frac{1}{a^2 - x^2}\, dx = \frac{1}{2a} \ln\left|\frac{x+a}{x-a}\right| + C$

38. $\int \ln(x)\, dx = x \cdot \ln(x) - x + C$
39. $\int x^n \cdot \ln(x)\, dx = \frac{x^{n+1}}{(n+1)^2} \left\{ (n+1)\cdot \ln(x) - 1 \right\} + C$

40. $\int \frac{1}{x \cdot \ln(x)}\, dx = \ln|\ln(x)| + C$

41. $\int e^{ax} \sin(nx)\, dx = \frac{e^{ax}}{a^2 + n^2} \left\{ a \cdot \sin(nx) - n \cdot \cos(nx) \right\} + C$

42. $\int e^{ax} \cos(nx)\, dx = \frac{e^{ax}}{a^2 + n^2} \left\{ a \cdot \cos(nx) + n \cdot \sin(nx) \right\} + C$

43. $\int \frac{1}{\sqrt{x^2 \pm a^2}}\, dx = \ln|x + \sqrt{x^2 \pm a^2}| + C$

44. $\int \sqrt{x^2 \pm a^2}\, dx = \frac{x}{2}\sqrt{x^2 \pm a^2} + \frac{1}{2} a^2 \ln|x + \sqrt{x^2 \pm a^2}| + C$

45. $\int x^2 \sqrt{x^2 \pm a^2}\, dx = \frac{1}{8} x \cdot (2x^2 \pm a^2)\sqrt{x^2 \pm a^2} - \frac{1}{8} a^4 \ln|x + \sqrt{x^2 \pm a^2}| + C$

Made in the USA
San Bernardino, CA
09 June 2015